VOLUME ONE HUNDRED AND TWENTY FIVE

CURRENT TOPICS IN DEVELOPMENTAL BIOLOGY

Nuclear Receptors in Development and Disease

CURRENT TOPICS IN DEVELOPMENTAL BIOLOGY

"A meeting-ground for critical review and discussion of developmental processes"
A.A. Moscona and Alberto Monroy (Volume 1, 1966)

VOLUME ONE HUNDRED AND TWENTY FIVE

CURRENT TOPICS IN DEVELOPMENTAL BIOLOGY

Nuclear Receptors in Development and Disease

Edited by

DOUGLAS FORREST

Laboratory of Endocrinology and Receptor Biology, National Institute of Diabetes and Digestive and Kidney Diseases (NIDDK), National Institutes of Health (NIH), MD, United States

SOPHIA TSAI

Baylor College of Medicine, Department of Molecular and Cellular Biology, Program in Developmental Biology, Houston, TX, United States

ELSEVIER

ACADEMIC PRESS
An imprint of Elsevier

Academic Press is an imprint of Elsevier
50 Hampshire Street, 5th Floor, Cambridge, MA 02139, United States
525 B Street, Suite 1800, San Diego, CA 92101-4495, United States
The Boulevard, Langford Lane, Kidlington, Oxford OX5 1GB, United Kingdom
125 London Wall, London, EC2Y 5AS, United Kingdom

First edition 2017

ISBN: 978-0-12-802172-9
ISSN: 0070-2153

For information on all Academic Press publications
visit our website at https://www.elsevier.com/books-and-journals

Publisher: Zoe Kruze
Acquisition Editor: Zoe Kruze
Editorial Project Manager: Shellie Bryant
Production Project Manager: Vignesh Tamil
Cover Designer: Greg Harris

Typeset by SPi Global, India

CONTENTS

CONTRIBUTORS

Michihiko Aramaki
Laboratory of Endocrinology and Receptor Biology, NIDDK, National Institutes of Health, Bethesda, MD, United States

Yukitomo Arao
Reproductive and Developmental Biology Laboratory, National Institute of Environmental Health Sciences/NIH, Research Triangle Park, NC, United States

Jonathan T. Busada
Molecular Endocrinology Group, Signal Transduction Laboratory, National Institute of Environmental Health Sciences, National Institutes of Health, Research Triangle Park, NC, United States

Chawnshang Chang
George Whipple Lab for Cancer Research, The Wilmot Cancer Center, University of Rochester Medical Center, Rochester, NY, United States; Sex Hormone Research Center, China Medical University/Hospital, Taichung, Taiwan

Hong-Chiang Chang
George Whipple Lab for Cancer Research, The Wilmot Cancer Center, University of Rochester Medical Center, Rochester, NY, United States

Krishna Chatterjee
Wellcome-MRC Institute of Metabolic Science, University of Cambridge, Cambridge, United Kingdom

Alys M. Cheatle Jarvela
University of Maryland, College Park, MD, United States

John A. Cidlowski
Molecular Endocrinology Group, Signal Transduction Laboratory, National Institute of Environmental Health Sciences, National Institutes of Health, Research Triangle Park, NC, United States

Qi Cui
Beckman Research Institute of City of Hope; Irell & Manella Graduate School of Biological Sciences, Beckman Research Institute of City of Hope, Duarte, CA, United States

Francesco J. DeMayo
Reproductive and Developmental Biology Laboratory, National Institute of Environmental Health Sciences, National Institute of Health, Research Triangle Park, NC, United States

Su Feng
Institute of Life Science, Nanchang University, Nanchang, Jiangxi, China

Frédéric Flamant
Institut de Génomique Fonctionnelle de Lyon, Université de Lyon, Université Lyon 1, CNRS UMR 5242, INRA USC 1370, Ecole Normale Supérieure de Lyon, Lyon cedex, France

Douglas Forrest
Laboratory of Endocrinology and Receptor Biology, NIDDK, National Institutes of Health, Bethesda, MD, United States

Yulong Fu
Laboratory of Endocrinology and Receptor Biology, NIDDK, National Institutes of Health, Bethesda, MD, United States

Karine Gauthier
Institut de Génomique Fonctionnelle de Lyon, Université de Lyon, Université Lyon 1, CNRS UMR 5242, INRA USC 1370, Ecole Normale Supérieure de Lyon, Lyon cedex, France

Norbert B. Ghyselinck
Institut de Génétique et de Biologie Moléculaire et Cellulaire (IGBMC), Illkirch; Centre National de la Recherche Scientifique (CNRS); Institut National de la Santé et de la Recherche Médicale (INSERM), Paris; Université de Strasbourg (UNISTRA), Strasbourg, France

Katherine J. Hamilton
Reproductive and Developmental Biology Laboratory, National Institute of Environmental Health Sciences/NIH, Research Triangle Park, NC, United States

Sylvia C. Hewitt
Reproductive and Developmental Biology Laboratory, National Institute of Environmental Health Sciences/NIH, Research Triangle Park, NC, United States

Guillaume Holzer
Institut de Génomique Fonctionnelle de Lyon, Université de Lyon, Université Lyon 1, CNRS, Ecole Normale Supérieure de Lyon, Lyon Cedex 07, France

Kenneth S. Korach
Reproductive and Developmental Biology Laboratory, National Institute of Environmental Health Sciences/NIH, Research Triangle Park, NC, United States

Vincent Laudet
Observatoire Océanologique de Banyuls-sur-Mer, UMR7232, Université Pierre et Marie Curie, Paris, Banyuls-sur-Mer, France

Gonghui Li
George Whipple Lab for Cancer Research, The Wilmot Cancer Center, University of Rochester Medical Center, Rochester, NY, United States; Sir Run Run Shaw Hospital, Zhejiang University, Hangzhou, China

Chang-Yi Lin
George Whipple Lab for Cancer Research, The Wilmot Cancer Center, University of Rochester Medical Center, Rochester, NY, United States

Shin-Jen Lin
George Whipple Lab for Cancer Research, The Wilmot Cancer Center, University of Rochester Medical Center, Rochester, NY, United States

Hong Liu
Laboratory of Endocrinology and Receptor Biology, NIDDK, National Institutes of Health, Bethesda, MD, United States

Manuel Mark
Institut de Génétique et de Biologie Moléculaire et Cellulaire (IGBMC), Illkirch; Centre National de la Recherche Scientifique (CNRS); Institut National de la Santé et de la Recherche Médicale (INSERM), Paris; Université de Strasbourg (UNISTRA); Hôpitaux Universitaires de Strasbourg (HUS), Strasbourg, France

Gabriel V. Markov
Sorbonne Universités, UPMC Université Paris 06, CNRS, UMR 8227, Integrative Biology of Marine Models, Station Biologique de Roscoff, Place Georges Teissier, Roscoff Cedex, France

Marcel E. Meima
Erasmus University Medical Center, Rotterdam, The Netherlands

Carla Moran
Wellcome-MRC Institute of Metabolic Science, University of Cambridge, Cambridge, United Kingdom

Robin P. Peeters
Erasmus University Medical Center, Rotterdam, The Netherlands

Leslie Pick
University of Maryland, College Park, MD, United States

Sabine Richard
Institut de Génomique Fonctionnelle de Lyon, Université de Lyon, Université Lyon 1, CNRS UMR 5242, INRA USC 1370, Ecole Normale Supérieure de Lyon, Lyon cedex, France

Yanhong Shi
Beckman Research Institute of City of Hope; Irell & Manella Graduate School of Biological Sciences, Beckman Research Institute of City of Hope, Duarte, CA, United States

Guoqiang Sun
Beckman Research Institute of City of Hope, Duarte, CA, United States

Ke Tang
Institute of Life Science, Nanchang University, Nanchang, Jiangxi, China

Marius Teletin
Institut de Génétique et de Biologie Moléculaire et Cellulaire (IGBMC), Illkirch; Centre National de la Recherche Scientifique (CNRS); Institut National de la Santé et de la Recherche Médicale (INSERM), Paris; Université de Strasbourg (UNISTRA); Hôpitaux Universitaires de Strasbourg (HUS), Strasbourg, France

Ming-Jer Tsai
Baylor College of Medicine; Program in Developmental Biology, Baylor College of Medicine, Houston, TX, United States

Sophia Tsai
Baylor College of Medicine; Department of Molecular and Cellular Biology, Program in Developmental Biology, Baylor College of Medicine, Houston, TX, United States

Anja L.M. van Gucht
Erasmus University Medical Center, Rotterdam, The Netherlands

Nadège Vernet
Institut de Génétique et de Biologie Moléculaire et Cellulaire (IGBMC), Illkirch; Centre National de la Recherche Scientifique (CNRS); Institut National de la Santé et de la Recherche Médicale (INSERM), Paris; Université de Strasbourg (UNISTRA), Strasbourg, France

Theo J. Visser
Erasmus University Medical Center, Rotterdam, The Netherlands

W. Edward Visser
Erasmus University Medical Center, Rotterdam, The Netherlands

Yihong Wan
The University of Texas Southwestern Medical Center, Dallas, TX, United States

San-Pin Wu
Reproductive and Developmental Biology Laboratory, National Institute of Environmental Health Sciences, National Institute of Health, Research Triangle Park, NC, United States

Xin Xie
Baylor College of Medicine, Houston, TX, United States

Dong-Rong Yang
George Whipple Lab for Cancer Research, The Wilmot Cancer Center, University of Rochester Medical Center, Rochester, NY, United States; The Second Affiliated Hospital of Suzhou University, Suzhou, China

Guosheng Yang
George Whipple Lab for Cancer Research, The Wilmot Cancer Center, University of Rochester Medical Center, Rochester, NY, United States; Guangdong 2nd Provincial People's Hospital, Guangzhou, China

Xiong Yang
Institute of Life Science, Nanchang University, Nanchang, Jiangxi, China

Hao Zuo
The University of Texas Southwestern Medical Center, Dallas, TX, United States

PREFACE

Nuclear receptors form a large family of transcription factors whose functions and molecular mechanisms have been extensively studied over the past 30 years. A remarkable range of functions has been described for these receptors in many physiological systems, but somewhat less attention has focused on their roles in development. The articles in this volume attempt to cover various functions of nuclear receptors in development and their potential impact on diseases. The articles highlight not only differentiation and disease but also seek to give an evolutionary context for this superfamily of receptors. The article by Holzer, Markov, and Laudet presents an overview of the evolution of nuclear receptors and ligand signaling across animal species, whereas that by Jarvela and Pick discusses the function and evolution of nuclear receptors in insect species.

The next articles shift the focus onto mammalian systems and include the article by Zuo and Wan on the participation of a number of nuclear receptors in bone formation and remodeling. Several articles consider the roles of classical steroid hormone receptors and other receptors with defined ligands in the reproductive, nervous, endocrine, and other systems. Hamilton, Hewitt, Arao, and Korach discuss functions of the estrogen receptor, Busada and Cidlowski discuss the glucocorticoid receptor, and Wu and DeMayo discuss the progesterone receptor. Teletin, Vernet, Ghyselinck, and Mark describe the role of retinoic acid receptors in germ cell differentiation, whereas Flamant, Gauthier, and Richard focus on thyroid hormone receptors in brain development. The article by van Gucht, Moran, Meima, Visser, Chatterjee, Visser, and Peeters reviews recent findings on mutations in the *THRA* thyroid hormone receptor gene in human disease.

Several articles address the actions of orphan nuclear receptors and focus on differentiation in neuronal and other systems. Liu, Aramaki, Fu, and Forrest review the functions of the *RORB* retinoid-related orphan receptor gene in neuronal cell fate decisions and neurological disease, Sun, Cui, and Shi discuss the *TLX* orphan receptor in neurogenesis and neurodegeneration, and Yang, Feng, and Tang discuss *COUP-TF* genes in neurodevelopment and disease. The concluding articles discuss the roles of orphan receptors in other systems. Lin, Yang, Yang, Lin, Chang, Li, and Chang review the involvement of TR2/TR4 receptors in diverse tissues, and Xie, Wu, Tsai, and Tsai review the role of COUP-TF2 in muscle development.

We hope this volume will draw the attention of readers to the critical roles of nuclear receptors in development and will stimulate interest in the potential of these receptors as therapeutic targets for treatment of various diseases.

Douglas Forrest
Laboratory of Endocrinology and Receptor Biology,
National Institute of Diabetes and Digestive
and Kidney Diseases (NIDDK),
National Institutes of Health (NIH), MD, United States

Sophia Tsai
Baylor College of Medicine,
Department of Molecular and Cellular Biology,
Program in Developmental Biology,
Houston, TX, United States

CHAPTER ONE

Evolution of Nuclear Receptors and Ligand Signaling: Toward a Soft Key–Lock Model?

Guillaume Holzer*, Gabriel V. Markov†, Vincent Laudet‡,1
*Institut de Génomique Fonctionnelle de Lyon, Université de Lyon, Université Lyon 1, CNRS, Ecole Normale Supérieure de Lyon, Lyon Cedex 07, France
†Sorbonne Universités, UPMC Université Paris 06, CNRS, UMR 8227, Integrative Biology of Marine Models, Station Biologique de Roscoff, Place Georges Teissier, Roscoff Cedex, France
‡Observatoire Océanologique de Banyuls-sur-Mer, UMR7232, Université Pierre et Marie Curie, Paris, Banyuls-sur-Mer, France
[1]Corresponding author: e-mail address: vincent.laudet@obs-banyuls.fr

Contents

Abstract

Nuclear receptors (NRs) are a family of ligand-regulated transcription factors that modulate a wide variety of physiological functions in a ligand-dependent manner. The first NRs were discovered as receptors of well-known hormones such as 17β-estradiol, corticosteroids, or thyroid hormones. In these cases a direct activation of the receptor transcriptional activity by a very specific ligand, with nanomolar affinity, was demonstrated, providing a strong conceptual framework to understand the mechanism of action of these hormones. However, the discovery that some NRs are able to bind different ligands with micromolar affinity was a first sign that the univocal relationship between a specific receptor (e.g., TR) and a specific ligand (e.g., thyroid hormone) should not be generalized to the whole family. These discussions about the nature of NR ligands have been reinforced by the study of the hormone/receptor couple evolution. Indeed when the ligand is not a protein but a small molecule derived from a biochemical pathway, a simple coevolution mechanism between the ligand and the receptor cannot operate. We and others have recently shown that the ligands acting for a given NR early on during evolution were often different from the classical mammalian ligands. This suggests that the NR/ligand evolutionary relationship is more dynamic than anticipated and that the univocal relationship between a receptor and a specific molecule may be an

Current Topics in Developmental Biology, Volume 125
ISSN 0070-2153
http://dx.doi.org/10.1016/bs.ctdb.2017.02.003

oversimplification. Moreover, classical NRs can have different ligands acting in a tissue-specific fashion with significant impact on their function. This also suggests that we may have to reevaluate the pharmacology of the ligand/receptor couple.

1. INTRODUCTION

The developmental and physiological effects of hydrophobic hormones within organisms are mediated through a superfamily of conserved ligand-activated transcription factors, the nuclear receptors (NRs) which include 48 genes in the human genome (Laudet & Gronemeyer, 2001). This superfamily includes receptors for steroid hormones and steroid derivatives (e.g., estrogens, glucocorticoids, progesterone, mineralocorticoids, androgens, but also vitamin D, oxysterols, bile acids, and ecdysteroids in insects, as well as dafachronic acids in nematodes), retinoic acids, thyroid hormones, fatty acids, and some derivatives such as leukotrienes and prostaglandins. Because of the essential role played by NRs in virtually all aspects of animal development, metabolism, and physiology, dysfunctions of signaling pathways controlled by these receptors are associated with many diseases in particular cancer and metabolic syndrome (Stump, Mukohda, Hu, & Sigmund, 2015; Tenbaum & Baniahmad, 1997; Wu, Cheung, Wang, Yu, & Chan, 2016). NRs are, therefore, major drug targets.

All NRs share a common modular structure composed of a highly conserved DNA-binding domain (DBD) and a less conserved ligand-binding domain (LBD). These two domains are connected by a flexible hinge playing an important role in the selection of the repertoire of DNA-binding sites (Billas & Moras, 2013). In addition, most NRs also contain an N-terminal A/B domain and a C-terminal F domain, which are highly variable (Germain, Staels, Dacquet, Spedding, & Laudet, 2006; Gronemeyer, Gustafsson, & Laudet, 2004). NRs bind to the regulatory regions of target genes as homodimers, heterodimers, and more rarely as monomers (Germain et al., 2006; Gronemeyer et al., 2004). When NRs are forming heterodimers, they do so with RXR, which is a common partner of many closely related receptors, including TR, RAR, or VDR (see Table 1 for abbreviation) as well as metabolic receptors such as PPARs, LXRs, or FXR (Fig. 1). Although it is difficult to generalize at the level of the entire, very diverse, superfamily, the most classical mode of action of NRs suggests that in the absence of their ligands they behave as transcriptional repressors

Table 1 List of Nuclear Receptors

Nomenclature	Name	Abbreviation	Canonical Ligand	Alternative
NR1A1	Thyroid hormone receptor α	TRα	Triiodothyronine (T3)	Thyroxine (T4) 3,5-diiodothyronine (3,5-T2) Triac
NR1A2	Thyroid hormone receptor β	TRβ	Triiodothyronine (T3)	Thyroxine (T4) 3,5-diiodothyronine (3,5-T2) Triac
NR1B1	Retinoic acid receptor α	RARα	All-*trans* retinoic acid, 9-*cis* RA	13-*cis* RA
NR1B2	Retinoic acid receptor β	RARβ	All-*trans* retinoic acid, 9-*cis* RA	13-*cis* RA
NR1B3	Retinoic acid receptor γ	RARγ	All-*trans* retinoic acid, 9-*cis* RA	13-*cis* RA
NR1C1	Peroxisome proliferator-activated receptor α	PPARα	Leukotriène, palmitic acid	
NR1C2	Peroxisome proliferator-activated receptor β	PPARβ	Fatty acids	
NR1C3	Peroxisome proliferator-activated receptor γ	PPARγ	Fatty acids, prostaglandin	
NR1D1	Rev-Erb α	Rev-Erb α	Former orphan	Heme
NR1D2	Rev-Erb β	Rev-Erb β	Former orphan	Heme
NR1F1	RAR-related orphan receptor α	RORα	Cholesterol derivatives, oxysterols	Melatonin
NR1F2	RAR-related orphan receptor β	RORβ		Retinoic acid

Continued

Table 1 List of Nuclear Receptors—cont'd

Nomenclature	Name	Abbreviation	Canonical Ligand	Alternative
NR1F3	RAR-related orphan receptor γ	RORγ	Oxysterols	Retinoic acid
NR1H1	Ecdysone receptor	EcR	Ecdysone, 20H-ecdysone	
NR1H2	Liver X receptor α	LXRα	Oxysterols	
NR1H3	Liver X receptor β	LXRβ	Oxysterols	
NR1H4	Farnesoid X receptor	FXR	Lithocholic acid, fexaramine, lanosterol	
NR1I1	Vitamin D receptor	VDR	25-dihydroxyvitamin D3	Lithocholic acid
NR1I2	Pregnane X receptor	PXR	Oxysterols, pregnenolone 16α-carbonitrile.	
NR1I3	Constitutive androstane receptor	CAR	Androstenols	
NR2A1	Hepatocyte nuclear factor 4 receptor α	HNF4α	Orphan	
NR2A2	Hepatocyte nuclear factor 4 receptor γ	HNF4γ	Orphan	
NR2B1	Retinoid X receptor α	RXRα	9-*cis* RA	
NR2B2	Retinoid X receptor β	RXRβ	9-*cis* RA	
NR2B3	Retinoid X receptor γ	RXRγ	9-*cis* RA	
NR2C1	Testicular receptor 2	TR2	Orphan	
NR2C2	Testicular receptor 4	TR4	Orphan	

NR2E1	Tailless homolog	TLX	Orphan	
NR2E2	Tailless	TLL	Orphan	
NR2E3	Photoreceptor-specific nuclear receptor	PNR	Orphan	
NR2F1	Chicken ovalbumin upstream promoter transcription factor I	COUP-TF I	Orphan	
NR2F2	Chicken ovalbumin upstream promoter transcription factor II	COUP-TF II	Former orphan	ATRA, 9-*cis* RA
NR2F6	V-erbA-related protein	EAR-2	Orphan	
NR3A1	Estrogen receptor α	ER α	17β-Estradiol	27-Hydroxycholesterol, 3βAdiol
NR3A2	Estrogen receptor γ	ER γ	17β-Estradiol	27-Hydroxycholesterol, 3βAdiol
NR3B1	Estrogen-related receptor α	ERR	Orphan	
NR3B2	Estrogen-related receptor β	ERR	Orphan	
NR3B3	Estrogen-related receptor γ	ERR	Orphan	
NR3C1	Glucocorticoid receptor	GR	Cortisol, hydrocortisone	
NR3C2	Mineralocorticoid receptor	MR	Aldosterone, spirolactone	
NR3C3	Progesterone receptor	PR	Progesterone	
NR3C4	Androgen receptor	AR	Dihydrotestosterone (DHT)	Diindolylmethane

Continued

Table 1 List of Nuclear Receptors—cont'd

Nomenclature	Name	Abbreviation	Canonical Ligand	Alternative
NR4A1	NGF-induced factor B/nuclear hormone receptor NUR77	NGFIB/ UR77	Orphan	
NR4A2	Nuclear receptor-related protein 1	NURR1	Orphan	
NR4A3	Neuron-derived orphan receptor 1	NOR1	Orphan	
NR5A1	Steroidogenic factor 1	SF-1	Former orphan	Phosphatidyl inositol, 25-hydroxycholesterol
NR5A2	Liver receptor homolog-1	LRH-1	Former orphan	Phosphatidyl inositol
NR6A1	Germ cell nuclear factor 1	GCNF	Orphan	
NR0B1	Dosage-sensitive sex reversal-adrenal hypoplasia congenital critical region on the X chromosome, gene 1	DAX1	Orphan	
NR0B2	Small heterodimeric partner	SHP	Orphan	

The nomenclature, name, abbreviation, canonical ligand(s), and alternative ligand(s) are indicated. Here, we consider the canonical ligand as the ones with established biological activity. Alternatively, we consider the alternative ligand as the ones that can bind the receptor but with a more elusive biological role.

Adapted from Gronemeyer, H., Gustafsson, J. -A. A., & Laudet, V. (2004). Principles for modulation of the nuclear receptor superfamily. *Nature Reviews. Drug Discovery*, *3*, 950–964 and Becnel, L., Darlington, Y., Ochsner, S., Easton-Marks, J., Watkins, C., McOwiti, A., et al. (2015). Nuclear receptor signaling Atlas: Opening access to the biology of nuclear receptor signaling pathways. *Plos One*, *10*, e0135615.

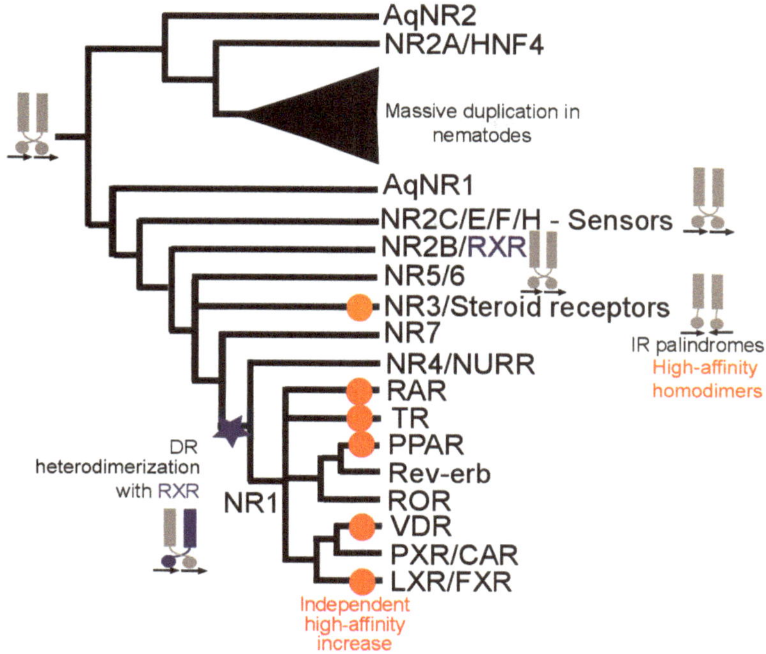

Fig. 1 Dendrogram of NRs and evolutive scenario of the superfamily. The main evolutionary events are plotted on the tree branch. Increase in ligand affinity is observed in two separate branches, the NR3 and the NR1. Structurally, the main event is the acquisition of heterodimerization with RXR in the common ancestor of NR4 and NR1 receptors, thus long time before the increase in ligand-binding affinity. AqNR1 and AqNR stand for the nuclear receptor of *Amphimedon queenslandica* 1 and 2. DR stand for direct repeat and IR for inverted repeat, two types on DNA-response element that NRs bind. The *blue star* at the base of the NR1/NR4 subtree indicates the heterodimerization with RXR (in *blue*).

through the recruitment of specific corepressors. Ligand binding inside the specific ligand-binding pocket induces a conformational change of the receptor, allowing the release of corepressors and the recruitment of coactivators and the transactivation of target genes (Germain et al., 2006; Gronemeyer et al., 2004). The availability of the ligand, therefore, controls NR activity in space and in time, and this is why recent analysis has focused on the study of the mechanisms that allow the formation of ligands in vivo (Grygiel-Górniak, 2014; Mahanti et al., 2014). It has to be noted that the mechanism of action of steroid receptors (ER, GR, PR, MR, and AR) is quite different from the classical model (Ritter & Mueller, 2014). In these cases, the unliganded steroid receptor appears to be in the cytoplasm. In this context, ligand binding promotes a conformational change of the receptor that unmasks the DBD and allows the receptor to bind to DNA. Several recent genome-wide characterizations of the DNA-binding activity of RXR heterodimers tend to suggest that for some cases (RXR–RAR, for example) the ligand binding also increases the number of DNA-binding sites on which the receptor interacts, suggesting that the classical mode of action described earlier is at least an oversimplified view of a much more complex, and dynamic, reality (Chatagnon et al., 2015; Lalevée et al., 2011; Fig. 2).

Indeed, we still have a static and quite schematic view of NRs action. Until recently, our observation of the effect of ligand binding on the conformation of the receptor only came from structural analysis of isolated domains, in particular the LBD. In recent years, we have now structures of full-size NRs heterodimers (Chandra et al., 2008; Orlov, Rochel, Moras, & Klaholz, 2012). This has strongly modified our view of the receptors, the role of ligand binding, the recruitment of cofactors, etc. (Nwachukwu & Nettles, 2012). Thanks to the new flow of data coming

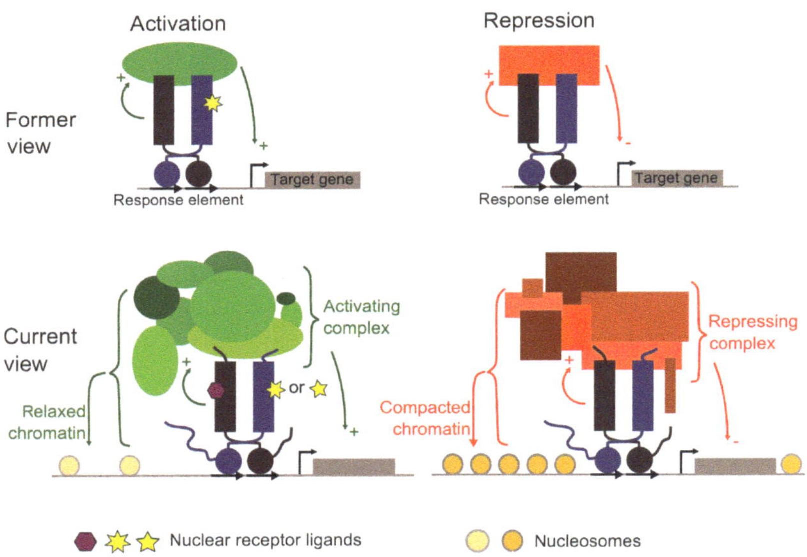

Fig. 2 Representation of NR transactivation. The former and current views of NR functioning are, respectively, indicated on the *upper* and *lower panel* for both activating (*left panel*) and repressing (*right panel*) function. Nuclear receptors are schematized in *blue* and *gray*, coactivators in *green*, and corepressors in *red*. Nucleosomes are represented by the *gray circle* with *light* (*left panel*) or *dark* (*right panel*) shade of *yellow*. Nuclear receptor ligands are represented by the *star* or *hexagon shape*. Promotion or inhibition activities are indicated by *arrows* with (+) or (–) symbols.

from integrated structure and function analysis (Moras, Billas, Rochel, & Klaholz, 2015); the heterodimer now appears as an allosteric rheostat that reacts to several inputs: ligand binding of course but also interaction with DNA, heterodimerization partner, cofactors (either repressors or activators) as well as protein-modifying enzymes (kinases, phosphatases, acetylases, methylases, etc.). The full conceptual impact of this integrated view is still difficult to fully grasp, notably at the evolutionary level, but it is likely that it will strongly impact our vision on NR diversification. In particular, we can expect that, as for other developmental mechanisms investigated in a comparative perspective, there is a significant amount of developmental system drift (True & Haag, 2001). Developmental system drift is a concept coined to generalize the observation that various phenotypic traits are conserved even if details of the molecular pathways implicated are different even between closely related species (True & Haag, 2001). Although this was initially described at the level of organismal phenotypes, similar cases are also known at the level of molecular processes. For example, the transcription from a given gene can be maintained even if the transcription factors that are activating it are changing, and there is also variation in the domain structure of NRs involved in transcriptional repression (reviewed in Markov & Sommer, 2012). Similarly, the emerging picture of allosteric activation of NRs is that there are plenty of mechanical solutions to trigger the allosteric switch to a transcriptionally active receptor, using either ligand binding (in or out the canonical pocket), interactions with other proteins or DNA, and even posttranslational modifications of some amino acids in the receptor. As we will discuss further, even ligand binding in the canonical pocket can be subject to more variation than initially thought.

NRs form an ancient and conserved family that arose early in metazoan lineage (Bertrand, Belgacem, & Escriva, 2011; Bertrand et al., 2004; Bridgham et al., 2010). For example, 48 NRs have been identified in the human genome (Robinson-Rechavi, Carpentier, Duffraisse, & Laudet, 2001), 21 in the fruit fly (Adams et al., 2000), 33 in the genome of the amphioxus (Bertrand et al., 2011), and 2 in the sponge *Amphimedon queenslandica* (Bridgham et al., 2010), whereas they are totally absent in the genomes of choanoflagellates, the closest metazoan relatives. The overall history of this gene family is now relatively well understood in terms of molecular evolution with major events of gene duplication and gene losses well located in the evolutionary tree (Bertrand et al., 2004). One of the most spectacular events is the lineage-specific amplification of HNF4 in nematodes, leading to about 250 genes in *Caenorhabditis elegans*

(Robinson-Rechavi, Maina, Gissendanner, Laudet, & Sluder, 2005). Phylogenetic analyses have proposed that one or two NRs were present at the base of metazoans, 25 at the base of bilaterians, and 23 at the base of chordates (Bertrand et al., 2004; Bridgham et al., 2010). According to the phylogeny, the NR superfamily has been organized into seven distinct subfamilies (Fig. 1). It has to be mentioned that one of these subfamilies, subfamily II, is in fact nonmonophyletic, that is, it does not form a clade that would unite all its members deriving from a unique ancestral receptor. This is because, based on the existence of only two NR genes in sponges, it has been proposed that this family contains the root of the tree (Bridgham et al., 2010). This tree clearly implies that the origin of the superfamily was within subfamily II and separates the family into HNF4 on one hand and all the rest on the other (Fig. 1). We can, therefore, infer that the primordial NR was probably a receptor similar to HNF4 and/or the group comprising COUP-TF and its close relatives. Secondarily, the whole diversity of mode of action of modern NRs diversified from this situation. But before we can propose a scenario linking the NR phylogeny with their main functional features (DNA binding, ligand binding, etc.), we need to refine the concept of NR ligand as it is much more complex, and fascinating, than anticipated.

Any attempt to trace back the main steps of the evolution of the superfamily, notably in terms of ligand binding must take into account the peculiar nature of NR ligands. If we consider classical membrane receptors whose ligands are of peptidic nature, that is ligand/receptor couples in which both are encoded by genes, we have a relatively good understanding of their evolution (Fig. 3). In those cases, there is a precise structural interface between the ligand and the receptor, and any mutation in one partner can be compensated by a mutation on the other, leading to the concept of ligand/receptor coevolution (Goh, Bogan, Joachimiak, Walther, & Cohen, 2000). In case of gene duplication, the two pairs will slowly diverge, giving rise to specific pairs of receptors after some times as this has been beautifully shown for FSH and LH systems (Moyle et al., 1994). More recently, other examples (ghrelin/growth hormone and prolactin/NPY) have shown how such a system allows the diversification of divergent receptors (Lagerström et al., 2005; Tine, Kuhl, Teske, Tschöp, & Jastroch, 2016). For NRs, as well as for membrane receptors whose ligands are small molecules not encoded by genes, the situation is radically different. In that case, the ligand is not the product of a single gene but the product of a metabolic pathway, often complex, often intricate with other pathways. Thus, evolution does not occur on a single gene by slow accumulation of mutations but on a network of genes.

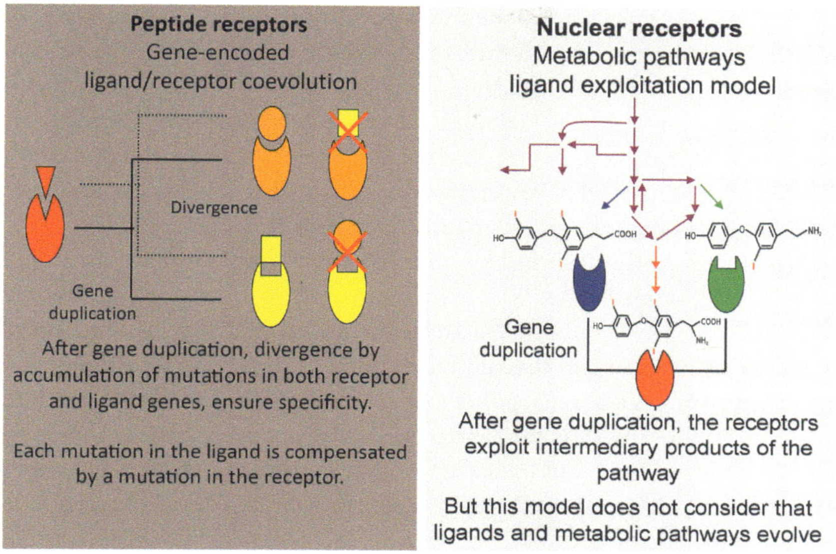

Fig. 3 Differences between coevolution of peptide receptors, peptide ligands, and nuclear receptors with small-molecule ligands. Direct coevolution through simultaneous gene duplications is possible for peptides and their receptors. On the contrary, small-molecule ligands for nuclear receptors come from complex pathways encoded by many enzymes, which makes the evolutionary scenario more complicated.

Therefore, it is not simple to predict how the specificity of a receptor for its ligand may evolve (Markov & Laudet, 2011; Markov, Lecointre, Demeneix, & Laudet, 2008). Several models were proposed to account for this specificity but before we come back to this question, we need to present in more detail the variety of NR ligands.

2. THE LIGAND–RECEPTOR COUPLE FOR NRs

If we want to fully understand the nature of NR ligands and the impact it has on our view about the origin and diversification of this superfamily, it is useful to take an historical point of view. This will illustrate the very strong historical bias that has skewed our view of the superfamily and our understanding of its nature and its evolution.

NRs have been discovered by endocrinologists. In the 1970s and 1980s, these scientists were actively searching for the receptors of major human hormones such as steroids and thyroid hormones (Chambon, 2005; Evans, 2005; Gustafsson, 2005; O'Malley, 2005). Using several methods, but most often through biochemical purification, the genes coding the receptors for glucocorticoids and estrogens, and soon after, thyroid hormones were cloned (Evans & Hollenberg, 1988; Green & Chambon, 1988; Koehler, Helguero, Haldosén, Warner, & Gustafsson, 2005). It was a surprise at that time to realize that the receptors for hormones as different as estrogens and thyroid hormones were in fact members of the same family. However, in all the cases, it was rapidly clear that the receptors were ligand-activated transcription factors able to interact very specifically with a given molecule (17β-estradiol for ER, cortisol for GR, and triiodothyronine for TR; see Table 1). The retinoic acid receptors (RARs) that were identified soon after the steroid and thyroid receptors represent a first enlargement of the primary concept of nuclear hormone receptors because their ligands are locally diffusing growth factors rather than hormones involved in long distance communication. But basically, these receptors are not that different in terms of interaction with the ligands: again, they bind with nanomolar affinity to very specific molecules. Therefore, the concept of the NR superfamily is based on the detailed characterization of a relatively small number of highly specific high-affinity receptors, steroid receptors in particular. Our current understanding of NR and their ligands is derived from these investigations (Sladek, 2011) and as we will see later this has induced a strong bias in our knowledge because primarily the NR superfamily is not a family of hormone receptors.

From this start, and because most members of the NR superfamily share the well conserved DBD, a flurry of discovery allowed the identification of the 48 NR genes known to be present in the human genome (Gronemeyer et al., 2004) as well as some NRs in other species, in particular in *Drosophila*. Among these, there were other high-affinity receptors for important hormones (PR, AR, and MR but also VDR and the insect EcR) as well as some more unexpected cases such as the RARs. However, it was rapidly clear with the discovery of ERRs in human that for some members of the superfamily it was not obvious to find ligands (Giguère, Yang, Segui, & Evans, 1988). These receptors were called orphan receptors (Giguère, 1999). For some of them, it was rapidly demonstrated that they had a ligand, but others remained for years without known ligands (Parks et al., 1999). In the case of the ERRs, no bona fide endogenous ligand has been identified yet. Of course, it is impossible to know if an orphan receptor is a true orphan or if there is indeed an unknown molecule that could bind to the receptor and modulate its activity. Even structural studies showing that the ligand pocket of a given receptor is filled with specific amino acid are not a proof that the orphan cannot become able to bind a ligand in specific conditions (Benoit, Malewicz, & Perlmann, 2004; Raghuram et al., 2007; Renaud, Harris, Downes, Burke, & Muscat, 2000).

For most pharmacologists, orphan receptors were a unique opportunity to discover new active molecules and rapidly high-throughput screens were devoted to searching high-affinity ligands specific for these NRs. In most cases, however, and to the great disappointment of the pharmaceutical industry, these screens were not successful. Gradually the field had to recognize that the situation was not that simple and the discovery of metabolic promiscuous receptors helped this recognition. In the 1990s, it became clear with the discovery of the PPARs (Issemann & Green, 1990) that beside the high-affinity receptors for specific ligands and the orphan receptor with no known ligands, other types of relationship between NRs and small molecules exist. The PPARs, for example, were demonstrated to be able to bind to several different compounds (nafenopin, clofibric acid, Wy 14643, MEHP, and methyl clofenapate) but only when these molecules were present at a high concentration (micromolar range). The later discovery of LXR, FXR, and even more of PXR and CAR demonstrated that the case of PPARs was not an exception. In all those cases, the receptors bind a quite a diverse set of molecules with a micromolar affinity. It has to be noted, however, that we cannot formally exclude the possibility that these receptors might have high-affinity-specific ligands, which are yet to be discovered (Higgins & Depaoli, 2010). However, physiologically speaking, these NRs effectively act as metabolic sensors that finely tune the amount of key metabolic products (cholesterol and fatty acids) by exerting feedback and feedforward regulation on the enzymes responsible of their production and/or degradation. Therefore, their metabolic function, which is well known from a number of genetic and metabolic studies, is clearly in accordance with this sensor function. Their ligands are very different from classical hormones and in fact some of them, such as fatty acids and their derivatives, including leukotrienes and prostaglandins, derive directly from food (Raman, Kaplan, Thompson, Heuvel, & Kaminski, 2011). PXR and CAR are even specialized receptors for interacting with xenobiotics since their main physiological role is to promote the degradation of toxic compounds by activating cytochrome P450s enzymes that metabolize the xenobiotics (reviewed in Amacher, 2010).

The different properties between the high-affinity receptors and the metabolic sensors can clearly be seen in the structure of the ligand-binding pocket. Indeed in high-affinity receptors, the ligand-binding pocket is small (ca. 400–600 Å^3), and largely filled (ca. 60%) by the ligand, which allows its very precise positioning and, therefore, the high affinity and high degree of binding specificity. In contrast, in metabolic sensors, the ligand-binding pocket is huge (1000–1600 Å^3) and filled only at ca. 20% by a variety of ligands. There are even cases with two molecules of ligands present inside the pockets. These molecules can be identical (Rocchi et al., 2001) or even different (Delfosse et al., 2015). In other cases, three different positions of the same ligand within the ligand-binding pocket of the same receptor have been observed (Watkins et al., 2001). Another level of complexity has been reached with the discovery that some NR can bind their canonical ligand not in the LBD but in secondary hormone-binding site (Souza et al., 2014).

The discovery of metabolic sensors was very useful in illustrating that the type of interaction between NR and their ligands was more diverse than anticipated and helps to realize that screening programs should probably not focus solely on high-affinity and very specific ligands. Data coming from structural biology helped to further blur the frontier between liganded receptors and orphan receptors. In three cases (HNF4, RORγ, and USP, the *Drosophila* ortholog of RXR) when the 3D structure of the LBD of the receptor was determined in the absence of exogenous ligand (as those receptors were orphans), a lipidic ligand was found inside the ligand-binding pocket to the general surprise (Billas, Moulinier, Rochel, & Moras, 2000; Clayton, Peak-Chew, Evans, & Schwabe, 2001; Dhe-Paganon, Duda, Iwamoto, Chi, & Shoelson, 2002; Kallen et al., 2002; Wisely et al., 2002). Strikingly, these ligands were interacting permanently (that is, they were not exchangeable) with the ligand-binding pocket and were necessary for the maintenance of the structure of the pocket, suggesting that they are example of prosthetic groups. It is still unclear in these cases (HNF4, ROR, and USP) if the ligands are effectively present inside the pocket when the receptor is present in its native environment, in the animal cell. For HNF4, it has been suggested that linolenic acid (a fatty acid) interacting with the LBP may be exchangeable (Yuan et al., 2009). It has been shown in this case that receptor occupancy is dramatically reduced in the fasted mouse as well as in a receptor carrying a mutation derived from diabetic patients (Yuan et al., 2009). However, this case is also puzzling since, paradoxically, ligand occupancy does not appear to have a significant effect on the transcriptional activity of the receptor. However in another case, a clear physiological function has been assigned to the prosthetic group bound to the receptor. This case is the *Drosophila* E75 orphan receptor, which is the ortholog of Rev-erbs in mammals. E75 LBD is able to bind heme, a large molecule that appears to be permanently bound in the structure (Reinking et al., 2005). However, the redox status of heme, and therefore, indirectly the conformation of the LBD and thus its transcriptional activity is regulated by the interaction with a gas, either nitric oxide (NO) or carbon monoxide (CO). Therefore E75 in *Drosophila*, but also Rev-erbs in vertebrates are biological sensors for diatomic gasses (Raghuram et al., 2007; Reinking et al., 2005). In that case, it is difficult to determine what the real ligand is? Is it the heme that interacts permanently inside the LBP or is the gas that controls the activity of the receptor by interacting only with reduced form of E75 (Thummel, 2005)?

Another aspect should be taken into account in this discussion about NR ligands: the origin of the ligands themselves. Again our vision has considerably evolved in the recent years. As mentioned earlier, primarily NR ligands were considered as classical hormones that are secreted by a gland, transported by the blood flow, and delivered to target organs when their cognate receptor is expressed. The RARs that were identified soon after the steroid and thyroid receptors were a first indication that NR ligands were more diverse as retinoic acid is fundamentally a growth factor which is not secreted by a specialized organ, not released in the blood but acts at a short range from the cells that produce them. In other words they are not hormones per se. However, we have to stress that even classical hormones are sometimes involved in local paracrine signaling. For example, the testosterone produced by the Leydig cells has an effect on the spermatocytes maturation in the neighboring Sertoli cells (Lejeune, Skalli, Chatelain, Avallet, & Saez, 1992).

The discovery of metabolic receptors makes the issue even more complex. Their ligands are in fact intermediary products of fatty acid and/or cholesterol metabolism and are, therefore, directly related to food intake. These, in fact, are quite far from the classical endogenous ligand produced by a dedicated biochemical pathway. Things became fuzzier when we consider the case of other compounds deriving from food, which regulate NRs, such as the phytoestrogens, genistein, or coumestrol that are ligands of ERs. In mammals, these compounds are not endogenous since they are not produced in their bodies. Nevertheless, they are natural products that many animals regularly ingest with their food that regulate the activity of a receptor, carrying specific information and having a precise transcriptional and hence physiological effect (Mueller, Simon, Chae, Metzler, & Korach, 2004). Therefore, genistein and other phytoestrogens are not fundamentally different from estrogens as NR ligands and should be taken into account as estrogen receptor ligands. It appears that other cases of natural products coming with our food are real phyto modulators that regulate other members of the NR superfamily. This is the case, for example, of amorfrutins that activate the PPARγ receptor (Weidner et al., 2012) or the gugulipid, extract from the guggul tree (*Commiphora mukul*) that acts as an antagonist of FXR (Urizar et al., 2002). Another example is the 3,3′-diindolylmethane, which is a product of the digestion of cruciferous vegetables (cole crops) that acts as an AR antagonist. It inhibits the transactivation of target gene by the AR ligand dihydrotestosterone (DHT) and prevents AR translocation in the nucleus (Le, Schaldach, Firestone, & Bjeldanes, 2003). More generally, a full range of plant products is increasingly recognized as NR ligands (reviewed in Li, Bonneton, Chen, & Laudet, 2015).

Another category of ligands is the endocrine disruptors (EDCs; reviewed in Le Maire, Bourguet, & Balaguer, 2010). These are synthetic compounds that regulate inappropriately a wide variety of NRs and produce harmful effects. These compounds are not natural products but are artificially made compounds not different from drugs (that are themselves becoming EDCs; Albert et al., 2013) and are accumulating in the environment as industrial byproducts. EDCs exploit the natural propensity of many, if not all, NRs to finely tune the physiological balance with both endogenous information (bona fide NR ligands) and exogenous signal (environmental molecule found in food). Therefore, they are inherent to the physiological role of most NRs and should be taken into account when analyzing in vivo function of NRs.

To fully understand the importance of this new vision of NR ligands, one should remember that the origin of all metazoans was in the ocean. The complexity of the ocean environment is extreme as it contains many types of organisms from virus to blue whale and these organisms in overall release many types of molecules in the water (reviewed in Maxwell, 2005). The ocean can, therefore, be seen as a very large soup of molecules and even if the dilution factor is enormous, since there is a number of very specific ways of interaction between organisms (symbiosis, commensalism, parasitism, competition, substrate sharing, etc.), there are many occasions for them to exchange chemical information. In particular, the ocean has an overall chemical composition different from freshwater and terrestrial substrates, and its richness in halogens (chlorine, bromide, and iodine) means that organic molecules incorporating those chemical elements are very widespread (Gribble, 2003). Up to date, halogenated compounds are mainly studied as NR ligands because some of them are synthetically made by the chemical industry, like the famous PCBs, that are known to disrupt metamorphosis in amphibians and teleost fishes (Gutleb, Appelman, Bronkhorst, Berg, & Murk, 2000; Lerner, Björnsson, & McCormick, 2007). Other important halogenated endocrine disruptors are the halogenated derivatives of bisphenol A that are ligands for ERs and PPARγ (Riu et al., 2011). Some marine natural compounds have structures that are strikingly similar to those of synthetic EDCs. To date, apart from thyroid hormones, no natural halogenated molecule has been demonstrated to be a ligand for a NR, but this may very well be the result, once again, of a conceptual as well as experimental bias. With more sequence data becoming available for marine animal models, it will be possible in the near future to experimentally check if, among the halogenated compounds that modulate metamorphosis in various marine animals, some of them turn to be ligands for TRs and even for other NRs.

Then, where are we now? When we consider the whole superfamily and the very diverse type of interactions that are observed we see clearly that the high-affinity receptors (that is the classical NRs presented in textbooks and present in the mind of most endocrinologists) are the exception and not the rule inside the superfamily (Fig. 4). Several orphan receptors are believed to

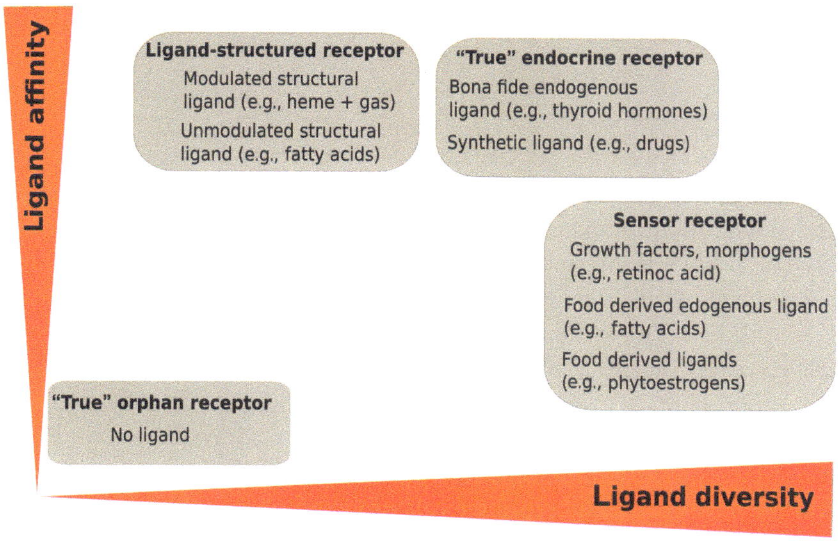

Fig. 4 Illustration of the different pharmacological sensitivities of nuclear receptors. Representation of the different NR pharmacology discussed in this review, regarding their ligand diversity (*X*-axis) and ligand affinity (*Y*-axis). Each category is grouped under a *gray frame* and examples are listed below.

be "adopted orphans" that is orphan receptors interacting with small synthetic molecules. This is the case of ERRs as well as NR4 and NR5 subfamily members (Hammond, Safe, & Tjalkens, 2015; Patch et al., 2011; Whitby et al., 2011). In all these cases, the nature of the bona fide endogenous ligand, if any, is still unknown but all seems to indicate that these adopted orphans would have an atypical ligand (Gallastegui, Mackinnon, Fletterick, & Estébanez-Perpiñá, 2015). And as we will see later even those high-affinity receptors are also behaving strangely when another dimension is taken into account, evolution.

3. NR DIVERSIFICATION DURING ANIMAL EVOLUTION

Section 1 has shown that the basal type of NR is not a high-affinity receptor but rather a sensor binding with relatively low affinity and selectivity. Given the known phylogeny of the superfamily depicted in Fig. 1, it is now possible to propose a scenario about the diversification of the whole superfamily and the emergence of the basic functions of the receptors, in particular in terms of ligand binding.

Thanks to the multiplicity of genome projects, we have now a good idea of the repertoire of NR present at various key steps of metazoan evolution. It is clear since a long time that NRs are specific to metazoans since they are absent outside metazoan and present in all metazoans (Bridgham et al., 2010; Escriva et al., 1997). As there are only two NRs in sponges, apparently the most basal group of metazoan, it is widely accepted that these two receptors may be use to root the NR tree (Fig. 1). As these two NRs, AqNR1 and AqNR2, are clear subfamily II members and are sensors of fatty acids (and not high-affinity receptors), it has been proposed that the ancestral NR was a low-affinity sensor to food-derived metabolic compounds such as fatty acids (Bridgham et al., 2010; Eick & Thornton, 2011; Markov & Laudet, 2011). This would be in line with the absence of an internal circulatory system in sponges that makes by definition, hormonal signaling impossible in such an organism. This view must, however, be tempered by the fact that there was no in vivo characterization of these receptors whose biological function remains unknown. We, therefore, have to wait until we have such data before we can safely infer the status of the ancestral NR. In addition, sponges are only one of the phyla at the base of metazoan and two other important phyla should be taken into account, the Placozoans (with *Trichoplax* as a developmental model) and the Ctenophores, a phylum that was originally believed to be close to Cnidarians. Unfortunately the precise phylogenetic position of these two phyla, and therefore, their contribution to our knowledge of NR status at the base of metazoan is still a matter of controversy. The functional characterization of their NR would be of great interest in addition to the resolution of their phylogeny. The *Trichoplax* genome contains four NRs: one HNF4, one RXR, one at the basis of the subfamily III (ERR and steroid receptors), and a last one that is a member of subfamily II, basal to COUP, and the NR2E sensors (Baker, 2008; Srivastava et al., 2008). This confirms that subfamily II is the ancestral one. Ctenophores are even stranger. One still hotly debated view suggests this phylum is the earliest lineage of metazoan tree (Moroz et al., 2014), opposite to the more classical view that they would diverge after sponges or as Cnidarian relatives (Pisani et al., 2015). Strikingly their NRs are very unusual as they contain only an LBD with no DBD (Reitzel et al., 2011). So far, the three sequenced ctenophore NRs (from two different species) are related to HNF4, again reinforcing the notion that this receptor corresponds to the origin of the superfamily. The striking absence of a zinc-finger DBD in the ctenophore NRs may suggest two hypotheses: (i) there was a secondary loss of this domain within the ctenophore lineage or (ii) if ctenophores are really an early offshoot of metazoans, the original NR may have lacked the canonical DBD and the junction between DDB and LBD would have taken place between the split of ctenophore and the split of sponges (Reitzel et al., 2011). It is too soon to date to know which hypothesis is correct.

Despite the uncertainties at the basis of animal phylogeny, high-affinity hormone-binding receptors are unambiguously present only in some members of family III and family I that are specific to bilaterian animals. This is consistent with the anatomical data indicating that only those animals have internally circulating body fluids that can carry on hormones. Increase in affinity to ligands occurred independently in families III and I, where additionally there are differences in the way the receptors bind to DNA. Indeed, steroid receptors from subfamily III bind on inverted repeat (IR) palindromes as homodimers, whereas most receptors from family I still bind on directed repeats, but forming heterodimers with RXR (Fig. 1). Even in family I, there are many vertebrate receptors that are binding their ligands with micromolar affinity (PPARs and PXR/CAR) and even orphan receptors/adopted orphans (RORs and Rev-erbs). Therefore, even inside family I, there must have been several independent increases in ligand-binding affinity. This is exemplified by the amphioxus RAR that binds RA with less affinity than in mammals, although it is already considered as high-affinity receptor (Escriva et al., 2006). What may have pushed for this increase in affinity? At the molecular level, this is a matter of mutations gradually increasing the affinity of the receptor for the ligand (Gutierrez-Mazariegos et al., 2016). At a physiological level, one of the factors contributing to the fixation of those affinity-increasing mutations may have been the structural constraints linked with distant communication between organs, through body fluid transport. Because NR ligands are hydrophobic, they have to be transported bound to transporter proteins. Therefore, to dissociate from the transporter at the target organ, higher affinity to the receptor than to the transporter is necessary to enable the transfer of the signaling molecule (Baker, 2002). Another factor could be linked to the biological role of RA in development. As a morphogen, it exhibits an anterior–posterior gradient (Godsave et al., 1998; Shimozono, Iimura, Kitaguchi, Higashijima, & Miyawaki, 2013) that appears to be overall conserved throughout evolution (Escriva, Holland, Gronemeyer, Laudet, & Holland, 2002). Thus, higher affinity receptors might be able to differentiate more precisely between two close doses of RA, and thus fine-tune gene responses more precisely than a lower-affinity receptor.

4. EVOLUTION OF LIGAND BINDING

As we discussed earlier, NRs cannot evolve with their ligands by a simple mechanism of coevolution as those ligands are not encoded by genes but are the product of, often complex, metabolic pathways. Therefore, the question of how the specificity of NRs can change remained virtually unexplored until it has been tackled by three different lines of studies.

The first line tackles the problem at a global level. The Thornton lab has proposed the so-called ligand exploitation model to explain how the specificity of the various steroid receptors was modified during evolution and in particular how novel hormone/receptor couples appeared (Eick & Thornton, 2011; Thornton, 2001; Thornton, Need, & Crews, 2003). His model proposes that within the steroid receptors (SR that group the NR3A: ER and NR3C: GR, PR, MR, and AR receptors), the estrogen receptor was the most ancient one. When we observe the structure of the steroidogenic pathway, it can be considered that the estrogen (17β-estradiol) is the terminal molecule of the pathway (Fig. 5). Therefore, the model suggests that the pathway was originally used to produce 17β-estradiol, which was the ligand of ERs, and this was sufficient to maintain a selective pressure for the existence of the pathway. When by gene/genome duplication the most recent steroid receptors (NR3C) emerged, they exploited

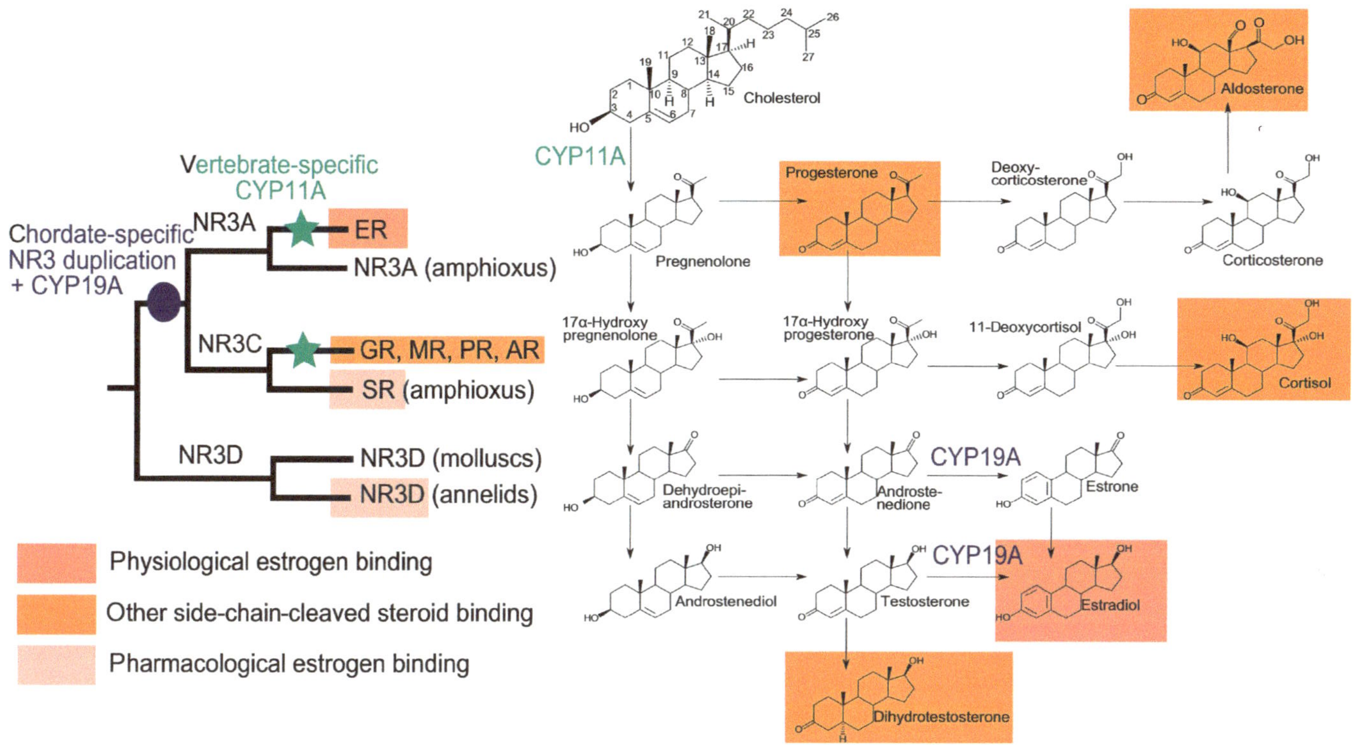

Fig. 5 The ligand exploitation model in front of genomic data. The dendrogram on the *left* shows the relationships between steroid receptors, and their ability to bind various steroids. The apparition date of two key enzymes (CYP11A and CYP19A) is also indicated. On the *right*, steroidogenic pathway showing the relationship between the different metabolites of the pathway. Aldosterone, cortisol, dihydrotestosterone, and 17β-estradiol are highlighted in *red* as "terminal" compounds that cannot be processed into another active compound, whereas progesterone is a synthetic intermediate. The reactions catalyzed by CYP11A and CYP19A are also indicated.

intermediary products along this pathway. These products must have been present because estrogens need to be produced because of their ancestral function. Of course, the model does not exclude that once the first intermediary product, supposed to be progesterone, was locked as NR3C ligand (Thornton, 2001), further refinement and complexification allowed for the divergence of four distinct, although partially overlapping, ligand/receptor couples. This model is obviously very elegant but has two implications that must be verified: (i) ER must be the most ancient steroid receptor and (ii) estrogens should effectively be the most ancient steroids, at least as ancient as their own receptor. Unfortunately none of these predictions are verified. After a period of uncertainty and debate regarding alternative possible topologies (Eick & Thornton, 2011; Markov & Laudet, 2011), most recent data clearly indicate that the protostome NR3Ds, formally called "ERs" are actually orthologous to the common ancestor of chordate ERs and SRs (i.e., the ancestor of AR, GR, MR, and PR; Bridgham, Keay, Ortlund, & Thornton, 2014), which means that ERs are not older than SRs. There is only one unduplicated NR3C (SR) characterized to date in amphioxus and it has been shown to bind estrogen, whereas the so-called estrogen receptor in the same species (NR3A) is unable to bind this compound (Bridgham, Brown, Rodríguez-Marí, Catchen, & Thornton, 2008; Paris, Pettersson, et al., 2008). The second prediction of the model has also been clearly shown to be incorrect. The model indeed suggests that estrogens were ancestral steroids that must have been present in invertebrate deuterostomes as well as in protostomes. However, there is no unequivocal detection of endogenous estrogens outside vertebrates and the claim for the presence of this molecule outside this taxonomic group is experimentally dubious (Scott, 2012, 2013). When biochemists use unbiased methods to detect and characterize steroids in invertebrates they do not find estrogens but rather steroids with long lateral chain such as dafachronic acid (Motola et al., 2006). In addition, careful genomic analyses of steroidogenic enzymes done by us and others have revealed that the key steps of estrogen production are clearly vertebrate specific (see, for example, CYP11A in Fig. 5), suggesting that vertebrate-type steroids are specific from vertebrates (Markov et al., 2009). The reports showing binding of estrogens to NR3D (former protostome "ER") receptors in annelids (Keay & Thornton, 2008) are, therefore, only pharmacological evidences with no biological significance. The ligand exploitation model fails because it considers only the evolution of one member of the couple: the receptor. But the ligands are also evolving and to resolve this issue the evolution of the biochemical pathways controlling ligand synthesis must be performed. This is currently what we are doing in our laboratory.

The second line of study is exploring how the ligand-binding pocket can change by the slow accumulation of mutations. By doing that, the binding pocket could accommodate different ligands, and therefore, exhibit a slow change in selectivity. This type of evidence was gathered combining structural analysis, mutagenesis, pharmacology as well as reconstruction of ancestral sequences. This was done by our group on the RARs (Escriva et al., 2006; Gutierrez-Mazariegos et al., 2016) as well as by Thornton group in an elegant set of papers (Eick, Colucci, Harms, Ortlund, & Thornton, 2012; Harms & Thornton, 2014; Harms et al., 2013) on the steroid receptors (NR3A, C, and D) and by the Krasowski lab on LXR and FXR (Krasowski, Ni, Hagey, & Ekins, 2011; Reschly et al., 2008). We cannot describe in detail these various data but they all point to the same direction: the slow accumulation of mutations can change the ligand-binding pockets, and therefore its selectivity. But this process is not random and key changes (large effect mutations) open new possibilities of evolution in terms of ligand selectivity. Concerning RARs, there is some flexibility in the substitution pattern, enabling some amino acids substitutions to revert to ancestral state (Gutierrez-Mazariegos et al., 2016). On the contrary, for the glucocorticoid receptor, reversion was made impossible by epistatic interactions: some intermediate mutations were necessary to maintain the overall stability of the protein during the structural rearrangement that enabled the switch in ligand-binding specificity (Harms & Thornton, 2014).

The third line of study is probably the most obvious one: this is the characterization of orthologs of liganded NRs in various metazoans and in particular in those from protostomes and invertebrate chordates that were virtually unexplored at that level. One of the first observations on this aspect came from the characterization of the TR in the cephalochordate amphioxus in which TH controls the metamorphosis of the pelagic larvae. This study has shown that it is not triiodothyronine (T3) but Triac, a deaminated version of T3 that is the active compound (Fig. 6). Triac is not present in mammals because of its extremely short half-life (Gosling, Schwart, Dillmann, Surks, & Oppenheimer, 1976). Thus, it is not considered as a true endogenous ligand. In amphioxus Triac (not T3) has been found to be the TR ligand controlling the metamorphosis (Bridgham et al., 2008; Paris, Escriva, et al., 2008). In addition, amphioxus has a deiodinase that is able to use Triac, and not T3, as a substrate reinforcing the notion that in this species Triac is the complete equivalent of T3

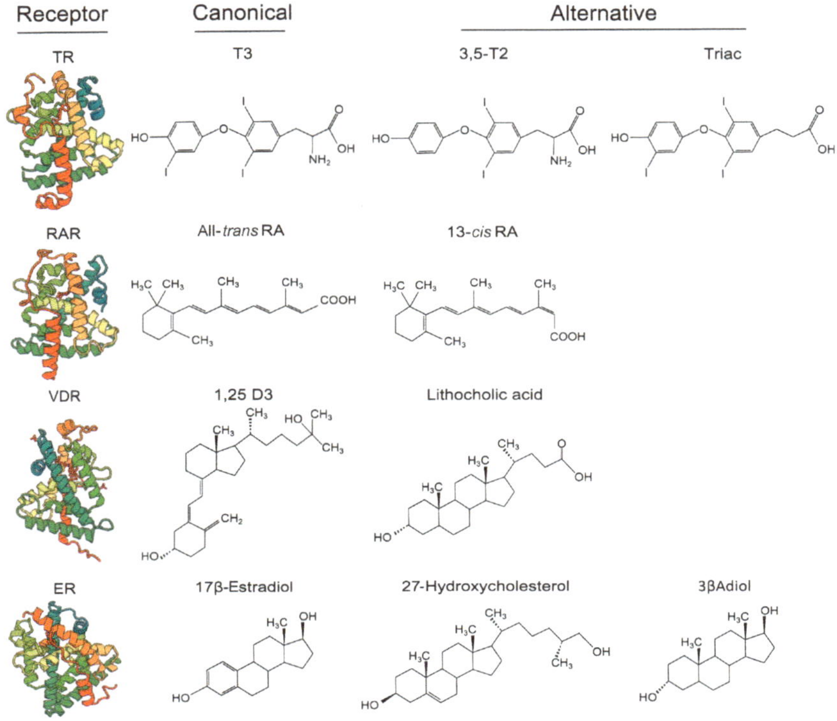

Fig. 6 Example of ligand derivatives for TR, RAR, VDR, and ER. The *left panel* shows the crystal structure of the receptor LBD (TR, PDB number: 3GWS; RAR, PDB number: 2LBD; VDR, PDB number: 3A78; and ER, PDB number 1ERE). The *middle panel* shows the bona fide ligand and the *right panel* shows the example of alternative ligands.

(Klootwijk, Friesema, & Visser, 2011). In addition when labeled T4 is added to amphioxus cultures Triac can be detected in animals extracts suggesting that it is effectively produced in vivo, even if the biochemistry behind this production is totally unknown (Paris et al., 2010). Given the basal position of amphioxus in chordates, these observations lead to the notion that Triac and not T3 could be the basal and original TR ligand in chordates (Paris, Escriva, et al., 2008; Paris, Pettersson, et al., 2008). Nevertheless, no new study investigates the putative role of Triac in thyroid disorder. Together, these examples highlight that evolutionary perspective can improve our understanding of endocrinological pathway and may be allow the discovery of new endogenous ligands and new mechanisms of action.

It has to be mentioned also that the evolution of NRs is full of secondary losses and lineage-specific events, as for any gene family, and peculiar situations can be found. For example, in the subfamily V, grouping the LRH-1 and SF-1 genes, it was originally thought that those were orphan receptors since murine LRH-1 was shown to enhance target gene activity without any treatment (Zhang & Mellon, 1996). Deeper investigation on SF-1 activity shed light on the 25-hydroxycholesterol, which was proposed (Lala et al., 1997) or denied (Mellon & Bair, 1998) as a bona fide ligand. The crystal structure of the mouse LRH-1 receptor favors the hypothesis of an orphan receptor since the ligand-biding pocket of the receptor was found empty with the receptor in an active conformation (Sablin, Krylova, Fletterick, & Ingraham, 2003). A final turnabout came with the analysis of the crystal structure of mouse SF-1 and human SF-1 and LRH-1 (Krylova et al., 2005). These structures show that phosphatidyl inositol binds in the pocket of mouse and human SF-1 and human LRH-1, and this binding is required for their maximal activity. This indicates that the murine LRH-1 has lost the ability to bind a ligand and became an exception in this gene family (Krylova et al., 2005; Sablin et al., 2008).

5. ALTERNATIVE LIGANDS

Recent discoveries in the NR field spark the idea that several NRs do have more than one endogenous ligand. This is striking as it concerns some of the most well-known and characterized receptors with strong nanomolar affinity for their bona fide "historical" ligand such as TR, ER, or VDR.

Strikingly, one of the very first cloned NR, ER is probably the one illustrating the best case of alternative ligands. ERs are at the crossroads of many pathologies, such as cancer or cardiovascular diseases, because many groups searched for ligands able to selectively modulate ER activity. In this context, the 27-hydroxycholesterol (27HC) (Fig. 6) was described as an endogenous ER ligand (DuSell & McDonnell, 2008; Umetani et al., 2007). 27HC is a primary metabolite of cholesterol (Duane & Javitt, 1999). Interestingly, it has effect opposite to those of 17β-estradiol, the bona fide ER ligand, on ER activity according to the cellular and promoter context. For example, it has been described as a competitive ER antagonist in HEK cells (Umetani et al., 2007), but surprisingly as a partial agonist of ER in MCF-7 cells. Another alternative ligand of ER is the DHT derivative 5α-androstan-3β,17β-diol (3βAdiol; Fig. 6; Kuiper et al., 1998). Intriguingly, this compound has a greater affinity for ERs (particularly ERβ) than for AR, the DHT receptor (Handa, Pak, Kudwa, Lund, & Hinds, 2008), meaning that the "male" derivative compound (i.e., 3βAdiol from DHT) has a greater affinity for the "female" endocrine pathway (i.e., ER) than the "male" endocrine pathway. This illustrates the complexity of the sex hormone pathway and regulation in which NRs have a central role. 3βAdiol is involved in many biological functions in a tissue-specific manner, such as regulation of prostate growth by inhibiting cell proliferation (Weihua, Lathe, Warner, & Gustafsson, 2002) and promoting breast cancer cell proliferation (Hackenberg, Turgetto, Filmer, & Schulz, 1993). It also acts in the central nervous system by partially inhibiting the stress-induced release of corticosteroids (Lund, Hinds, & Handa, 2006).

VDR is the receptor of the vitamin D, 1,25-dihydroxycholecalciferol (1,25 D3), which plays a critical role in calcium metabolism and bone development (Holick, 1996). It has been shown that VDR can also bind the bile acid derivative lithocholic acid (LCA; Fig. 6; Makishima et al., 2002). LCA is found in the liver and the intestine, which corresponds to important VDR expression territories (Uhlén et al., 2015). Interestingly, in this organ VDR appears to have a role in the detoxification of LCA (which is a carcinogenic compound) by controlling the expression of LCA catalytic genes (Nehring, Zierold, & DeLuca, 2007), suggesting that it can play a role in cancer protection. The adopted orphan receptor FXR also binds LCA and plays a role in regulating bile acid homeostasis since LCA-bound FXR represses the LCA synthesis process (Makishima et al., 1999). Thus, these two receptors together regulate the LCA metabolism. VDR acts as a tissue-specific LCA receptor in the intestine instead of a vitamin D receptor.

Two molecules are collectively referred as thyroid hormones, the thyroxine precursor (T4) and its derivative T3. As T3 induces a greater response than T4 on both TRα and TRβ, it is considered to be the active ligand (Chopra, 1996). The transformation of T4 into T3 is done by specific deiodinases and in particular deiodinase 2 (Darras, Hume, & Visser, 1999). Deiodination mechanisms also transform T3 into T2 and in particular 3,5-T2 (Fig. 6). Until recently, this compound was considered as an inactive product of T3 degradation because of its low affinity for TR (reviewed in Goglia, 2005). However, several studies in fish have observed that TR can transactivate target genes in presence of 3,5-T2 (García-G, Jeziorski, Valverde-R, & Orozco, 2004; Mendoza et al., 2013). In addition, the teleost specific TRβ long isoform binds 3,5-T2 with better affinity than its shorter counterpart (García-G et al., 2004; Mendoza et al., 2013). In mammals, 3,5-T2 appears to have a TH-like activity and mimic T3 effect on the hypothalamus–pituitary–thyroid axis: the central rheostat of TH level control

(Moreno, Lombardi, Lombardi, Goglia, & Lanni, 1998; Padron et al., 2014). Nevertheless, in mammals, the formal demonstration of 3,5-T2 binding to TR is still lacking and many studies focus on the nongenomic effect of this compound, which is not mediated by TR-induced gene transcription (reviewed in Goglia, 2005). However, the finding that 3,5-T2 can modulate RNA level of TSHR, a gene involved in the central regulation of TH level, could indicate that 3,5-T2 might have a classical genomic action (Padron et al., 2014). Overall, if the biological effects of 3,5-T2 are accepted, the mechanisms underlying these effects, and their difference with T3, remain to be fully understood (Orozco, Navarrete-Ramírez, Olvera, & García-G, 2014). As discussed earlier, evolutionary studies shed light on another TH derivative, Triac, which can be considered as the ancestral ligand of the proto-TR in invertebrates. Therefore, the TR–T3 relationship, which of course is fully relevant in human must, however, be refined by these observations about 3,5-T2 and Triac.

Retinoic acid is an important signaling molecule, often discussed as a morphogen, involved in embryonic development (Gutierrez-Mazariegos, Theodosiou, Campo-Paysaa, & Schubert, 2011). Retinoic acid corresponds in fact to several molecules among which all-*trans* retinoic acid is considered as the endogenous ligand for RAR. Among the derivatives of retinoic acid, 13-*cis* retinoic acid (13-*cis* RA; Fig. 6) is considered as inactive because of its low affinity for the RARs. Nevertheless 13-*cis* RA is an agonist of all three RARs and effectively induce the transactivation of target genes by the RARs (Idres, Marill, Flexor, & Chabot, 2002). 13-*cis* RA is found in blood and may inhibit steroid metabolism (Biswas & Russell, 1997; Blaner, 2001). It has been shown that excess of 13-*cis* RA can have deleterious effects in brain cell division and learning ability (Crandall et al., 2004) as well as in sebocytes proliferation (Tsukada et al., 2000). Nevertheless, 13-*cis* RA can isomerize into all-*trans* retinoic acid in vivo in a cell-dependent manner (Tsukada et al., 2000) and conversely (IARC, 1999). Thus, it remains a disputed question whether the biological effect observed upon 13-*cis* RA treatment is a direct consequence of the binding of the 13-*cis* RA upon RARs, or it is the consequence of the all-*trans* isomer that binds RAR. Nevertheless, 13-*cis* RA has some unique properties that are not shared by other retinoic acid (Bushue & Wan, 2010), such as antiinflammatory effect or better outcome in some thyroid cancer. This may suggest that 13-*cis* RA mediates its own biological effect aside from the other retinoic acids. Interestingly 13-*cis* RA is found in invertebrates (Gutierrez-Mazariegos et al., 2014) and is strongly active on some invertebrates RAR (unpublished observation,

V. Laudet), reinforcing the notion that it may have been, and is perhaps still, more important than originally thought.

As a conclusion, alternative ligands bring another level of modulation in our understanding of NR functioning. When they bind their receptor, they induce a slightly different conformation than the canonical ligand, by interacting differently with the binding pocket residues (Jurutka et al., 2005). As a consequence, alternative ligands induce promoter-specific (Choi, Yamada, & Makishima, 2011) and coactivator-specific (Jurutka et al., 2005) effects, which results in a gene regulation different from the canonical one (DuSell, Umetani, Shaul, Mangelsdorf, & McDonnell, 2007; Jurutka et al., 2005; Sharma, Handa, & Uht, 2012). Moreover, alternative ligands often entail tissue-specific activities (Pak et al., 2005), certainly because they are often synthesized in a tissue-specific manner (Pelletier, Luu-The, Li, & Labrie, 2005). This means that the NR activity in a given tissue does not necessarily go through its canonical ligand but through the ligand actually present within this tissue.

6. GENERALIZATION

Then, where does this lead us? As often, the more we study NRs the more we understand that there is still a lot to be understood! With the notion of alternative ligand (e.g., 3βAdiol), tissue specificity action (e.g., VDR in liver and intestine), differential evolution of NRs (e.g., LRH-1 in mouse), the notion of environmental sensor (e.g., xenobiotics), and the chemical diversity of ligand (e.g., steroid and heme) the ligand/receptor relationship appears to be much more complex than what was proposed in the early models.

The possibility that a single NR can bind several ligands, with different biological activities, led to the conceptualization of the selective NR modulators; that is, to say molecules that can selectively activate a NR in a tissue- or target-specific manner (Flamant, Gauthier, & Samarut, 2006; Gronemeyer et al., 2004; Negro-Vilar, 1999). This concept was originally proposed from a pharmacological perspective, with exogenous artificial molecules. However, with the many examples of endogenous alternative ligands (and the one that remains to be discovered), it is possible to propose the similar idea with endogenous selective NR modulator (endogenous SNuRMs): as a corpus of endogenous ligands able to activate a given NR in a tissue- and/or target-specific manner. These modulators can be closely related (e.g., T3/T2) or completely different (e.g., 1,25 D3/LCA) but must bind an NR and modulate an activity.

Such a diversity of ligands is not surprising from an evolutionary perspective. Indeed, our current understanding of NR evolution favors the hypothesis that early NRs were sensors with low affinity, low selectivity but possibly with a high diversity of ligands. Thus, finding in some receptors the ability to bind different ligands with different selectivities is not that odd. Moreover, ligand can change during evolution as it is exemplified with Triac that was the basal bona fide TR ligand in chordate (Paris, Escriva, et al., 2008; Paris, Pettersson, et al., 2008), whereas T3 is the vertebrate ligand. Given that the evolutionary perspective is also interesting to investigate new ligands as all species may not share the same ligand for a given NR. Thus, a comparative endocrinology approach might be interesting to discover new ligands. But, to fully grasp the implications of this new vision, comparative endocrinology should be transformed in a real evolutionary science, that is to say fully abandon its anthropomorphism and its propensity to compare animal situations with the human one. Indeed, we should now do evolutionary endocrinology. Instead of searching only for "intermediate steps" in the evolution of human signaling in other animals, we should gather detailed functional information in various representative models (as done in insects and nematodes) in order to put the collected data in appropriate physiological and ecological context. NRs are important players in regulating life-history transitions, integrating information about the nutritional status of the organism. Deregulation in that system leads to metabolic diseases including during the aging process (Della Torre, Benedusi, Fontana, & Maggi, 2014). We can hypothesize that some of the plant-produced natural compounds that bind to NRs may also contribute in fine-tuning that system, either in a favorable way or in an antagonist one, depending of the type of ecological interactions between those organisms (Miller & Heyland, 2010). This chemical arms race could explain the tremendous diversity of observed and yet to be discovered NR ligands, as well as the difficulty to find strict correlations between pairs of receptors and ligands.

As a conclusion, we propose a soft key/soft lock model to understand NRs, with a continuum between orphan receptors which could bind some endogenous molecule (e.g., heme on Rev-erb; Reinking et al., 2005) to higher affinity receptor that can bind different ligands for different biological activities (e.g., 17β-estradiol and 27HC for ER; Umetani et al., 2007). Ligands also represent a wide diversity of molecules from food acquired ones (e.g., phytoestrogen) to specifically synthesized ligands (e.g., thyroid hormone). Within the ligand, close molecules deriving from the same metabolic pathway can have completely different biological activities (e.g., DHT and 3βAdiol; Sharma et al., 2012).

REFERENCES

Adams, M., Celniker, S., Holt, R., Evans, C., Gocayne, J., Amanatides, P., et al. (2000). The genome sequence of Drosophila melanogaster. *Science, 287*, 2185–2195.

Albert, O., Desdoits-Lethimonier, C., Lesne, L., Legrand, A., Guille, F., Bensalah, K., et al. (2013). Paracetamol, aspirin and indomethacin display endocrine disrupting properties in the adult human testis in vitro. *Human Reproduction, 28*, 1890–1898.

Amacher, D. (2010). The effects of cytochrome P450 induction by xenobiotics on endobiotic metabolism in pre-clinical safety studies. *Toxicology Mechanisms and Methods, 20*, 159–166.

Baker, M. E. (2002). Albumin, steroid hormones and the origin of vertebrates. *The Journal of Endocrinology, 175*, 121–127.

Baker, M. (2008). Trichoplax, the simplest known animal, contains an estrogen-related receptor but no estrogen receptor: Implications for estrogen receptor evolution. *Biochemical and Biophysical Research Communications, 375*, 623–627.

Benoit, G., Malewicz, M., & Perlmann, T. (2004). Digging deep into the pockets of orphan nuclear receptors: Insights from structural studies. *Trends in Cell Biology, 14*, 369–376.

Bertrand, S., Belgacem, M., & Escriva, H. (2011). Nuclear hormone receptors in chordates. *Molecular and Cellular Endocrinology, 334*, 67–75.

Bertrand, S., Brunet, F., Escriva, H., Parmentier, G., Laudet, V., & Robinson-Rechavi, M. (2004). Evolutionary genomics of nuclear receptors: From twenty-five ancestral genes to derived endocrine systems. *Molecular Biology and Evolution, 21*, 1923–1937.

Billas, I., & Moras, D. (2013). Allosteric controls of nuclear receptor function in the regulation of transcription. *Journal of Molecular Biology, 425*, 2317–2329.

Billas, I., Moulinier, L., Rochel, N., & Moras, D. (2000). Crystal structure of the ligand binding domain of the ultraspiracle protein USP, the ortholog of RXRs in insects. *Journal of Biological Chemistry, 276*, 7465–7474.

Biswas, M. G., & Russell, D. W. (1997). Expression cloning and characterization of oxidative 17beta- and 3alpha-hydroxysteroid dehydrogenases from rat and human prostate. *The Journal of Biological Chemistry, 272*, 15959–15966.

Blaner, W. (2001). Cellular metabolism and actions of 13-cis-retinoic acid. *Journal of the American Academy of Dermatology, 45*, S129–S135.

Bridgham, J., Brown, J., Rodríguez-Marí, A., Catchen, J., & Thornton, J. (2008). Evolution of a new function by degenerative mutation in cephalochordate steroid receptors. *PLoS Genetics, 4,* e1000191.

Bridgham, J., Eick, G., Larroux, C., Deshpande, K., Harms, M., Gauthier, M., et al. (2010). Protein evolution by molecular tinkering: Diversification of the nuclear receptor superfamily from a ligand-dependent ancestor. *PLoS Biology, 8,* e1000497.

Bridgham, J. T., Keay, J., Ortlund, E. A., & Thornton, J. W. (2014). Vestigialization of an allosteric switch: Genetic and structural mechanisms for the evolution of constitutive activity in a steroid hormone receptor. *PLoS Genetics, 10,* e1004058.

Bushue, N., & Wan, Y.-J. Y. J. (2010). Retinoid pathway and cancer therapeutics. *Advanced Drug Delivery Reviews, 62*, 1285–1298.

Chambon, P. (2005). The nuclear receptor superfamily: A personal retrospect on the first two decades. *Molecular Endocrinology (Baltimore, Md.), 19*, 1418–1428.

Chandra, V., Huang, P., Hamuro, Y., Raghuram, S., Wang, Y., Burris, T., et al. (2008). Structure of the intact PPAR-γ–RXR-α nuclear receptor complex on DNA. *Nature, 456*, 350–356.

Chatagnon, A., Veber, P., Morin, V., Bedo, J., Triqueneaux, G., Sémon, M., et al. (2015). RAR/RXR binding dynamics distinguish pluripotency from differentiation associated cis-regulatory elements. *Nucleic Acids Research, 43*, 4833–4854. highwire.

Choi, M., Yamada, S., & Makishima, M. (2011). Dynamic and ligand-selective interactions of vitamin D receptor with retinoid X receptor and cofactors in living cells. *Molecular Pharmacology, 80*, 1147–1155.

Chopra, I. J. (1996). *Nature, source, and relative significance of circulating thyroid hormones.* Philadelphia: Lippincott Raven.

Clayton, G. M., Peak-Chew, S. Y., Evans, R. M., & Schwabe, J. W. (2001). The structure of the ultraspiracle ligand-binding domain reveals a nuclear receptor locked in an inactive conformation. *Proceedings of the National Academy of Sciences of the United States of America, 98*, 1549–1554.

Crandall, J., Sakai, Y., Zhang, J., Koul, O., Mineur, Y., Crusio, W., et al. (2004). 13-cis-retinoic acid suppresses hippocampal cell division and hippocampal-dependent learning in mice. *Proceedings of the National Academy of Sciences of the United States of America, 101*, 5111–5116.

Darras, V., Hume, R., & Visser, T. (1999). Regulation of thyroid hormone metabolism during fetal development. *Molecular and Cellular Endocrinology, 151*, 37–47.

Delfosse, V., Dendele, B., Huet, T., Grimaldi, M., Boulahtouf, A., Gerbal-Chaloin, S., et al. (2015). Synergistic activation of human pregnane X receptor by binary cocktails of pharmaceutical and environmental compounds. *Nature Communications, 6*, 8089.

Della Torre, S., Benedusi, V., Fontana, R., & Maggi, A. (2014). Energy metabolism and fertility: A balance preserved for female health. *Nature Reviews. Endocrinology, 10*, 13–23.

Dhe-Paganon, S., Duda, K., Iwamoto, M., Chi, Y.-I., & Shoelson, S. (2002). Crystal structure of the HNF4α ligand binding domain in complex with endogenous fatty acid ligand. *Journal of Biological Chemistry, 277*, 37973–37976.

Duane, W. C., & Javitt, N. B. (1999). 27-Hydroxycholesterol: Production rates in normal human subjects. *Journal of Lipid Research, 40*, 1194–1199.

DuSell, C., & McDonnell, D. (2008). 27-Hydroxycholesterol: A potential endogenous regulator of estrogen receptor signaling. *Trends in Pharmacological Sciences, 29*, 510–514.

DuSell, C., Umetani, M., Shaul, P., Mangelsdorf, D., & McDonnell, D. (2007). 27-Hydroxycholesterol is an endogenous selective estrogen receptor modulator. *Molecular Endocrinology (Baltimore, Md.), 22*, 65–77.

Eick, G. N., Colucci, J. K., Harms, M. J., Ortlund, E. A., & Thornton, J. W. (2012). Evolution of minimal specificity and promiscuity in steroid hormone receptors. *PLoS Genetics, 8*, e1003072.

Eick, G., & Thornton, J. (2011). Evolution of steroid receptors from an estrogen-sensitive ancestral receptor. *Molecular and Cellular Endocrinology, 334*, 31–38.

Escriva, H., Bertrand, S., Germain, P., Robinson-Rechavi, M., Umbhauer, M., Cartry, J., et al. (2006). Neofunctionalization in vertebrates: The example of retinoic acid receptors. *PLoS Genetics, 2*, e102.

Escriva, H., Holland, N., Gronemeyer, H., Laudet, V., & Holland, L. (2002). The retinoic acid signaling pathway regulates anterior/posterior patterning in the nerve cord and pharynx of amphioxus, a chordate lacking neural crest. *Development (Cambridge, England), 129*, 2905–2916.

Escriva, H., Safi, R., Hänni, C., Langlois, M. C., Saumitou-Laprade, P., Stehelin, D., et al. (1997). Ligand binding was acquired during evolution of nuclear receptors. *Proceedings of the National Academy of Sciences of the United States of America, 94*, 6803–6808.

Evans, R. (2005). The nuclear receptor superfamily: A Rosetta stone for physiology. *Molecular Endocrinology, 19*, 1429–1438.
Evans, R. M., & Hollenberg, S. M. (1988). Cooperative and positional independent transactivation domains of the human glucocorticoid receptor. *Cold Spring Harbor Symposia on Quantitative Biology, 53*, 813–818.
Flamant, F., Gauthier, K., & Samarut, J. (2006). Thyroid hormones signaling is getting more complex: STORMs are coming. *Molecular Endocrinology, 21*, 321–333.
Gallastegui, N., Mackinnon, J., Fletterick, R., & Estébanez-Perpiñá, E. (2015). Advances in our structural understanding of orphan nuclear receptors. *Trends in Biochemical Sciences, 40*, 25–35.
García-G, C., Jeziorski, M., Valverde-R, C., & Orozco, A. (2004). Effects of iodothyronines on the hepatic outer-ring deiodinating pathway in killifish. *General and Comparative Endocrinology, 135*, 201–209.
Germain, P., Staels, B., Dacquet, C., Spedding, M., & Laudet, V. (2006). Overview of nomenclature of nuclear receptors. *Pharmacological Reviews, 58*, 685–704.
Giguère, V. (1999). Orphan nuclear receptors: From gene to function. *Endocrine Reviews, 20*, 689–725.
Giguère, V., Yang, N., Segui, P., & Evans, R. M. (1988). Identification of a new class of steroid hormone receptors. *Nature, 331*, 91–94.
Godsave, S. F., Koster, C. H., Getahun, A., Mathu, M., Hooiveld, M., Wees, J., et al. (1998). Graded retinoid responses in the developing hindbrain. *Developmental Dynamics, 213*, 39–49.
Goglia, F. (2005). Biological effects of 3,5-diiodothyronine (T2). *Biochemistry (Moscow), 70*, 164–172.
Goh, C.-S., Bogan, A., Joachimiak, M., Walther, D., & Cohen, F. (2000). Co-evolution of proteins with their interaction partners. 11 Edited by B. Honig *Journal of Molecular Biology, 299*, 283–293.
Gosling, B., Schwart, H., Dillmann, W., Surks, M., & Oppenheimer, J. (1976). Comparison of the metabolism and distribution of L-triiodothyronine and triiodothyroacetic acid in the rat: A possible explanation of differential hormonal potency. *Endocrinology, 98*, 666–675.
Green, S., & Chambon, P. (1988). Nuclear receptors enhance our understanding of transcription regulation. *Trends in Genetics: TIG, 4*, 309–314.
Gribble, G. (2003). The diversity of naturally produced organohalogens. *Chemosphere, 52*, 289–297.
Gronemeyer, H., Gustafsson, J.-A. A., & Laudet, V. (2004). Principles for modulation of the nuclear receptor superfamily. *Nature Reviews. Drug Discovery, 3*, 950–964.
Grygiel-Górniak, B. (2014). Peroxisome proliferator-activated receptors and their ligands: Nutritional and clinical implications—A review. *Nutrition Journal, 13*, 1–10.
Gustafsson, J.-A. (2005). Steroids and the scientist. *Molecular Endocrinology, 19*, 1412–1417.
Gutierrez-Mazariegos, J., Nadendla, E., Lima, D., Pierzchalski, K., Jones, J., Kane, M., et al. (2014). A mollusk retinoic acid receptor (RAR) ortholog sheds light on the evolution of ligand binding. *Endocrinology, 155*, 4275–4286.
Gutierrez-Mazariegos, J., Nadendla, E. K., Studer, R. A., Alvarez, S., de Lera, A. R., Kuraku, S., et al. (2016). Evolutionary diversification of retinoic acid receptor ligand-binding pocket structure by molecular tinkering. *Royal Society Open Science, 3*, 150484.
Gutierrez-Mazariegos, J., Theodosiou, M., Campo-Paysaa, F., & Schubert, M. (2011). Vitamin A: A multifunctional tool for development. *Seminars in Cell & Developmental Biology, 22*, 603–610.
Gutleb, A., Appelman, J., Bronkhorst, M., Berg, J., & Murk, A. (2000). Effects of oral exposure to polychlorinated biphenyls (PCBs) on the development and metamorphosis of two

amphibian species (Xenopus laevis and Rana temporaria). *Science of the Total Environment, 262*, 147–157.
Hackenberg, R., Turgetto, I., Filmer, A., & Schulz, K.-D. (1993). Estrogen and androgen receptor mediated stimulation and inhibition of proliferation by androst-5-ene-3β, 17β-diol in human mammary cancer cells. *The Journal of Steroid Biochemistry and Molecular Biology, 46*, 597–603.
Hammond, S., Safe, S., & Tjalkens, R. (2015). A novel synthetic activator of Nurr1 induces dopaminergic gene expression and protects against 6-hydroxydopamine neurotoxicity in vitro. *Neuroscience Letters, 607*, 83–89.
Handa, R., Pak, T., Kudwa, A., Lund, T., & Hinds, L. (2008). An alternate pathway for androgen regulation of brain function: Activation of estrogen receptor beta by the metabolite of dihydrotestosterone, 5α-androstane-3β,17β-diol. *Hormones and Behavior, 53*, 741–752.
Harms, M., Eick, G., Goswami, D., Colucci, J., Griffin, P., Ortlund, E., et al. (2013). Biophysical mechanisms for large-effect mutations in the evolution of steroid hormone receptors. *Proceedings of the National Academy of Sciences of the United States of America, 110*, 11475–11480.
Harms, M., & Thornton, J. (2014). Historical contingency and its biophysical basis in glucocorticoid receptor evolution. *Nature, 512*, 203–207.
Higgins, L. S., & Depaoli, A. M. (2010). Selective peroxisome proliferator-activated receptor gamma (PPARgamma) modulation as a strategy for safer therapeutic PPARgamma activation. *The American Journal of Clinical Nutrition, 91*, 267S–272S.
Holick, M. F. (1996). Vitamin D and bone health. *The Journal of Nutrition, 126*, 1159–1164.
IARC. (1999). *Handbook 2, 13-cis-Retinoic acid*. Lyon: International Agency for Research on Cancer, 145–185.
Idres, N., Marill, J., Flexor, M., & Chabot, G. (2002). Activation of retinoic acid receptor-dependent transcription by all-trans-retinoic acid metabolites and isomers. *Journal of Biological Chemistry, 277*, 31491–31498.
Issemann, I., & Green, S. (1990). Activation of a member of the steroid hormone receptor superfamily by peroxisome proliferators. *Nature, 347*, 645–650.
Jurutka, P., Thompson, P., Whitfield, K., Eichhorst, K., Hall, N., Dominguez, C., et al. (2005). Molecular and functional comparison of 1,25-dihydroxyvitamin D3 and the novel vitamin D receptor ligand, lithocholic acid, in activating transcription of cytochrome P450 3A4. *Journal of Cellular Biochemistry, 94*, 917–943.
Kallen, J. A., Schlaeppi, J.-M. M., Bitsch, F., Geisse, S., Geiser, M., Delhon, I., et al. (2002). X-ray structure of the hRORalpha LBD at 1.63 A: Structural and functional data that cholesterol or a cholesterol derivative is the natural ligand of RORalpha. *Structure (London England: 1993), 10*, 1697–1707.
Keay, J., & Thornton, J. (2008). Hormone-activated estrogen receptors in annelid invertebrates: Implications for evolution and endocrine disruption. *Endocrinology, 150*, 1731–1738.
Klootwijk, W., Friesema, E. C., & Visser, T. J. (2011). A nonselenoprotein from amphioxus deiodinates triac but Not T3: Is triac the primordial bioactive thyroid hormone? *Endocrinology, 152*, 3259–3267.
Koehler, K., Helguero, L., Haldosén, L.-A., Warner, M., & Gustafsson, J.-A. (2005). Reflections on the discovery and significance of estrogen receptor. *Endocrine Reviews, 26*, 465–478.
Krasowski, M., Ni, A., Hagey, L., & Ekins, S. (2011). Evolution of promiscuous nuclear hormone receptors: LXR, FXR, VDR, PXR, and CAR. *Molecular and Cellular Endocrinology, 334*, 39–48.

Krylova, I., Sablin, E., Moore, J., Xu, R., Waitt, G., MacKay, J., et al. (2005). Structural analyses reveal phosphatidyl inositols as ligands for the NR5 orphan receptors SF-1 and LRH-1. *Cell, 120*, 343–355.

Kuiper, G. G., Lemmen, J. G., Carlsson, B., Corton, J. C., Safe, S. H., van der Saag, P. T., et al. (1998). Interaction of estrogenic chemicals and phytoestrogens with estrogen receptor β. *Endocrinology, 139*, 4252–4263.

Lagerström, M., Fredriksson, R., Bjarnadóttir, T., Fridmanis, D., Holmquist, T., Andersson, J., et al. (2005). Origin of the prolactin-releasing hormone (PRLH) receptors: Evidence of coevolution between PRLH and a redundant neuropeptide Y receptor during vertebrate evolution. *Genomics, 85*, 688–703.

Lala, D. S., Syka, P. M., Lazarchik, S. B., Mangelsdorf, D. J., Parker, K. L., & Heyman, R. A. (1997). Activation of the orphan nuclear receptor steroidogenic factor 1 by oxysterols. *Proceedings of the National Academy of Sciences of the United States of America, 94*, 4895–4900.

Lalevée, S., Anno, Y., Chatagnon, A., Samarut, E., Poch, O., Laudet, V., et al. (2011). Genome-wide in silico identification of new conserved and functional retinoic acid receptor response elements (direct repeats separated by 5 bp). *Journal of Biological Chemistry, 286*, 33322–33334. highwire.

Laudet, V., & Gronemeyer, H. (2001). *The nuclear receptor FactsBook*. Cambridge, MA, Academic Press, 462.

Le Maire, A., Bourguet, W., & Balaguer, P. (2010). A structural view of nuclear hormone receptor: Endocrine disruptor interactions. *Cellular and Molecular Life Sciences: CMLS, 67*, 1219–1237.

Le, H., Schaldach, C., Firestone, G., & Bjeldanes, L. (2003). Plant-derived 3,3′-diindolylmethane is a strong androgen antagonist in human prostate cancer cells. *Journal of Biological Chemistry, 278*, 21136–21145.

Lejeune, H., Skalli, M., Chatelain, P., Avallet, O., & Saez, J. (1992). The paracrine role of sertoli cells on Leydig cell function. *Cell Biology and Toxicology, 8*, 73–83.

Lerner, D., Björnsson, B., & McCormick, S. (2007). Effects of aqueous exposure to polychlorinated biphenyls (Aroclor 1254) on physiology and behavior of smolt development of Atlantic salmon. *Aquatic Toxicology, 81*, 329–336.

Li, L., Bonneton, F., Chen, X. Y., & Laudet, V. (2015). Botanical compounds and their regulation of nuclear receptor action: The case of traditional Chinese medicine. *Molecular and Cell Endocrinology, 401*, 221–237.

Lund, T. D., Hinds, L. R., & Handa, R. J. (2006). The androgen 5alpha-dihydrotestosterone and its metabolite 5alpha-androstan-3beta, 17beta-diol inhibit the hypothalamo-pituitary-adrenal response to stress by acting through estrogen receptor beta-expressing neurons in the hypothalamus. *The Journal of Neuroscience: The Official Journal of the Society for Neuroscience, 26*, 1448–1456.

Mahanti, P., Bose, N., Bethke, A., Judkins, J. C., Wollam, J., Dumas, K. J., et al. (2014). Comparative metabolomics reveals endogenous ligands of DAF-12, a nuclear hormone receptor, regulating C. elegans development and lifespan. *Cell Metabolism, 19*, 73–83.

Makishima, M., Lu, T., Xie, W., Whitfield, G., Domoto, H., Evans, R., et al. (2002). Vitamin D receptor as an intestinal bile acid sensor. *Science, 296*, 1313–1316.

Makishima, M., Okamoto, A. Y., Repa, J. J., Tu, H., Marc Learned, R., Luk, A., et al. (1999). Identification of a nuclear receptor for bile acids. *Science, 284*, 1362–1365.

Markov, G. V., & Laudet, V. (2011). Origin and evolution of the ligand-binding ability of nuclear receptors. *Molecular and Cellular Endocrinology, 334*, 21–30.

Markov, G., Lecointre, G., Demeneix, B., & Laudet, V. (2008). The 'street light syndrome', or how protein taxonomy can bias experimental manipulations. *BioEssays, 30*, 349–357.

Markov, G. V., & Sommer, R. J. (2012). The evolution of novelty in conserved gene families. *International Journal of Evolutionary Biology, 2012*, 490894.
Markov, G. V., Tavares, R., Dauphin-Villemant, C., Demeneix, B. A., Baker, M. E., & Laudet, V. (2009). Independent elaboration of steroid hormone signaling pathways in metazoans. *Proceedings of the National Academy of Sciences of the United States of America, 106*, 11913–11918.
Maxwell, S. (2005). An aquatic pharmacy: The biomedical potential of the deep sea. *The Journal of Marine Education, 21*, 31–32.
Mellon, S. H., & Bair, S. R. (1998). 25-Hydroxycholesterol is not a ligand for the orphan nuclear receptor steroidogenic factor-1 (SF-1). *Endocrinology, 139*, 3026–3029.
Mendoza, A., Navarrete-Ramírez, P., Hernández-Puga, G., Villalobos, P., Holzer, G., Renaud, J. P., et al. (2013). 3,5-T2 is an alternative ligand for the thyroid hormone receptor β1. *Endocrinology, 154*, 2948–2958.
Miller, A. E. M., & Heyland, A. (2010). Endocrine interactions between plants and animals: Implications of exogenous hormone sources for the evolution of hormone signaling. *General and Comparative Endocrinology, 166*, 455–461.
Moras, D., Billas, I. M. L., Rochel, N., & Klaholz, B. P. (2015). Structure-function relationships in nuclear receptors: The facts. *Trends in Biochemical Sciences, 40*, 287–290.
Moreno, M., Lombardi, A., Lombardi, P., Goglia, F., & Lanni, A. (1998). Effect of 3,5-diiodo-L-thyronine on thyroid stimulating hormone and growth hormone serum levels in hypothyroid rats. *Life Sciences, 62*, 2369–2377.
Moroz, L., Kocot, K., Citarella, M., Dosung, S., Norekian, T., Povolotskaya, I., et al. (2014). The ctenophore genome and the evolutionary origins of neural systems. *Nature, 510*, 109–114.
Motola, D., Cummins, C., Rottiers, V., Sharma, K., Li, T., Li, Y., et al. (2006). Identification of ligands for DAF-12 that govern dauer formation and reproduction in C. elegans. *Cell, 124*, 1209–1223.
Moyle, W. R., Campbell, R. K., Myers, R. V., Bernard, M. P., Han, Y., & Wang, X. (1994). Co-evolution of ligand-receptor pairs. *Nature, 368*, 251–255.
Mueller, S., Simon, S., Chae, K., Metzler, M., & Korach, K. (2004). Phytoestrogens and their human metabolites show distinct agonistic and antagonistic properties on estrogen receptor α (ERα) and ERβ in human cells. *Toxicological Sciences, 80*, 14–25.
Negro-Vilar, A. (1999). Selective androgen receptor modulators (SARMs): A novel approach to androgen therapy for the new millennium. *The Journal of Clinical Endocrinology and Metabolism, 84*, 3459–3462.
Nehring, J. A., Zierold, C., & DeLuca, H. F. (2007). Lithocholic acid can carry out in vivo functions of vitamin D. *Proceedings of the National Academy of Sciences of the United States of America, 104*, 10006–10009.
Nwachukwu, J., & Nettles, K. (2012). The nuclear receptor signalling scaffold: Insights from full-length structures. *The EMBO Journal, 31*, 251–253.
O'Malley, B. (2005). A life-long search for the molecular pathways of steroid hormone action. *Molecular Endocrinology, 19*, 1402–1411.
Orlov, I., Rochel, N., Moras, D., & Klaholz, B. (2012). Structure of the full human RXR/VDR nuclear receptor heterodimer complex with its DR3 target DNA. *The EMBO Journal, 31*, 291–300.
Orozco, A., Navarrete-Ramírez, P., Olvera, A., & García-G, C. (2014). 3,5-Diiodothyronine (T2) is on a role. A new hormone in search of recognition. *General and Comparative Endocrinology, 203*, 174–180.
Padron, A., Neto, R., Pantaleão, T., Santos, M., Araujo, R., Andrade, B., et al. (2014). Administration of 3,5-diiodothyronine (3,5-T2) causes central hypothyroidism and stimulates thyroid-sensitive tissues. *Journal of Endocrinology, 221*, 415–427.

Pak, T., Chung, W., Lund, T., Hinds, L., Clay, C., & Handa, R. (2005). The androgen metabolite, 5-androstane-3,17-diol, is a potent modulator of estrogen receptor-1-mediated gene transcription in neuronal cells. *Endocrinology, 146*, 147–155.

Paris, M., Escriva, H., Schubert, M., Brunet, F., Brtko, J., Ciesielski, F., et al. (2008). Amphioxus postembryonic development reveals the homology of chordate metamorphosis. *Current Biology, 18*, 825–830.

Paris, M., Hillenweck, A., Bertrand, S., Delous, G., Escriva, H., Zalko, D., et al. (2010). Active metabolism of thyroid hormone during metamorphosis of amphioxus. *Integrative and Comparative Biology, 50*, 63–74.

Paris, M., Pettersson, K., Schubert, M., Bertrand, S., Pongratz, I., Escriva, H., et al. (2008). An amphioxus orthologue of the estrogen receptor that does not bind estradiol: Insights into estrogen receptor evolution. *BMC Evolutionary Biology, 8*, 219.

Parks, D. J., Blanchard, S. G., Bledsoe, R. K., Chandra, G., Consler, T. G., Kliewer, S. A., et al. (1999). Bile acids: Natural ligands for an orphan nuclear receptor. *Science, 284*, 1365–1368.

Patch, R., Searle, L., Kim, A., De, D., Zhu, X., Askari, H. B., et al. (2011). Identification of diaryl ether-based ligands for estrogen-related receptor α as potential antidiabetic agents. *Journal of Medical Chemistry, 54*, 788–808.

Pelletier, G., Luu-The, V., Li, S., & Labrie, F. (2005). Localization of type 7 17β-hydroxysteroid dehydrogenase in mouse tissues in situ hybridization studies. *The Journal of Steroid Biochemistry and Molecular Biology, 93*, 49–57.

Pisani, D., Pett, W., Dohrmann, M., Feuda, R., Rota-Stabelli, O., Philippe, H., et al. (2015). Genomic data do not support comb jellies as the sister group to all other animals. *Proceedings of the National Academy of Sciences of the United States of America, 112*, 15402–15407.

Raghuram, S., Stayrook, K., Huang, P., Rogers, P., Nosie, A., McClure, D., et al. (2007). Identification of heme as the ligand for the orphan nuclear receptors REV-ERBα and REV-ERBβ. *Nature Structural & Molecular Biology, 14*, 1207–1213.

Raman, P., Kaplan, B., Thompson, J., Heuvel, J., & Kaminski, N. (2011). 15-Deoxy-Δ12,14-prostaglandin J2-glycerol ester, a putative metabolite of 2-arachidonyl glycerol, activates peroxisome proliferator activated receptor γ. *Molecular Pharmacology, 80*, 201–209.

Reinking, J., Lam, M., Pardee, K., Sampson, H., Liu, S., Yang, P., et al. (2005). The drosophila nuclear receptor E75 contains heme and is gas responsive. *Cell, 122*, 195–207.

Reitzel, A., Pang, K., Ryan, J., Mullikin, J., Martindale, M., Baxevanis, A., et al. (2011). Nuclear receptors from the ctenophore Mnemiopsis leidyi lack a zinc-finger DNA-binding domain: Lineage-specific loss or ancestral condition in the emergence of the nuclear receptor superfamily? *EvoDevo, 2*, 3.

Renaud, J.-P., Harris, J., Downes, M., Burke, L., & Muscat, G. (2000). Structure-function analysis of the Rev-erbA and RVR ligand-binding domains reveals a large hydrophobic surface that mediates corepressor binding and a ligand cavity occupied by side chains. *Molecular Endocrinology, 14*, 700–717.

Reschly, E., Ai, N., Ekins, S., Welsh, W., Hagey, L., Hofmann, A., et al. (2008). Evolution of the bile salt nuclear receptor FXR in vertebrates. *Journal of Lipid Research, 49*, 1577–1587.

Ritter, H., & Mueller, C. (2014). Expression microarray identifies the unliganded glucocorticoid receptor as a regulator of gene expression in mammary epithelial cells. *BMC Cancer, 14*, 275.

Riu, A., Maire, A., Grimaldi, M., Audebert, M., Hillenweck, A., Bourguet, W., et al. (2011). Characterization of novel ligands of ERα, Erβ, and PPARγ: The case of halogenated bisphenol A and their conjugated metabolites. *Toxicological Sciences, 122*, 372–382.

Robinson-Rechavi, M., Carpentier, A. S., Duffraisse, M., & Laudet, V. (2001). How many nuclear hormone receptors are there in the human genome? *Trends in Genetics: TIG, 17*, 554–556.
Robinson-Rechavi, M., Maina, C. V., Gissendanner, C. R., Laudet, V., & Sluder, A. (2005). Explosive lineage-specific expansion of the orphan nuclear receptor HNF4 in nematodes. *Journal of Molecular Evolution, 60*, 577–586.
Rocchi, S., Picard, F., Vamecq, J., Gelman, L., Potier, N., Zeyer, D., et al. (2001). A unique PPARγ ligand with potent insulin-sensitizing yet weak adipogenic activity. *Molecular Cell, 8*, 737–747.
Sablin, E. P., Blind, R. D., Krylova, I. N., Ingraham, J. G., Cai, F., Williams, J. D., et al. (2008). Structure of SF-1 bound by different phospholipids: Evidence for regulatory ligands. *Molecular Endocrinology, 23*, 25–34. endocrine.
Sablin, E. P., Krylova, I. N., Fletterick, R. J., & Ingraham, H. A. (2003). Structural basis for ligand-independent activation of the orphan nuclear receptor LRH-1. *Molecular Cell, 11*, 1575–1585.
Scott, A. P. (2012). Do mollusks use vertebrate sex steroids as reproductive hormones? Part I: Critical appraisal of the evidence for the presence, biosynthesis and uptake of steroids. *Steroids, 77*, 1450–1468.
Scott, A. P. (2013). Do mollusks use vertebrate sex steroids as reproductive hormones? II. Critical review of the evidence that steroids have biological effects. *Steroids, 78*, 268–281.
Sharma, D., Handa, R. J., & Uht, R. M. (2012). The ERβ ligand 5α-androstane, 3β,17β-diol (3β-diol) regulates hypothalamic oxytocin (Oxt) gene expression. *Endocrinology, 153*, 2353–2361.
Shimozono, S., Iimura, T., Kitaguchi, T., Higashijima, S., & Miyawaki, A. (2013). Visualization of an endogenous retinoic acid gradient across embryonic development. *Nature, 496*, 363–366.
Sladek, F. (2011). What are nuclear receptor ligands? *Molecular and Cellular Endocrinology, 334*, 3–13.
Souza, P. C. T., Puhl, A. C., Martínez, L., Aparício, R., Nascimento, A. S., Figueira, A. C., et al. (2014). Identification of a new hormone-binding site on the surface of thyroid hormone receptor. *Molecular Endocrinology, 28*, 534–545.
Srivastava, M., Begovic, E., Chapman, J., Putnam, N., Hellsten, U., Kawashima, T., et al. (2008). The trichoplax genome and the nature of placozoans. *Nature, 454*, 955–960.
Stump, M., Mukohda, M., Hu, C., & Sigmund, C. (2015). PPARγ regulation in hypertension and metabolic syndrome. *Current Hypertension Reports, 17*.
Tenbaum, S., & Baniahmad, A. (1997). Nuclear receptors: Structure, function and involvement in disease. *The International Journal of Biochemistry & Cell Biology, 29*, 13251341.
Thornton, J. (2001). Evolution of vertebrate steroid receptors from an ancestral estrogen receptor by ligand exploitation and serial genome expansions. *Proceedings of the National Academy of Sciences of the United States of America, 98*, 5671–5676.
Thornton, J., Need, E., & Crews, D. (2003). Resurrecting the ancestral steroid receptor: Ancient origin of estrogen signaling. *Science (New York, N.Y.), 301*, 1714–1717.
Thummel, C. (2005). Powered by gas—A ligand for a fruit fly nuclear receptor. *Cell, 122*, 151–153.
Tine, M., Kuhl, H., Teske, P., Tschöp, M., & Jastroch, M. (2016). Diversification and coevolution of the ghrelin/growth hormone secretagogue receptor system in vertebrates. *Ecology and Evolution, 6*, 2516–2535.
True, J. R., & Haag, E. S. (2001). Developmental system drift and flexibility in evolutionary trajectories. *Evolution & Development, 3*, 109–119.
Tsukada, M., Schröder, M., Roos, T., Chandraratna, R., Reichert, U., Merk, H., et al. (2000). 13-cis retinoic acid exerts its specific activity on human sebocytes through selective intracellular isomerization to all-trans retinoic acid and binding to retinoid acid receptors. *Journal of Investigative Dermatology, 115*, 321–327.

Uhlén, M., Fagerberg, L., Hallström, B., Lindskog, C., Oksvold, P., Mardinoglu, A., et al. (2015). Tissue-based map of the human proteome. *Science, 347*, 1260419.

Umetani, M., Domoto, H., Gormley, A., Yuhanna, I., Cummins, C., Javitt, N., et al. (2007). 27-Hydroxycholesterol is an endogenous SERM that inhibits the cardiovascular effects of estrogen. *Nature Medicine, 13*, 1185–1192.

Urizar, N. L., Liverman, A. B., Dodds, D. T., Silva, F. V., Ordentlich, P., Yan, Y., et al. (2002). A natural product that lowers cholesterol as an antagonist ligand for FXR. *Science (New York, N.Y.), 296*, 1703–1706.

Watkins, R. E., Wisely, G. B., Moore, L. B., Collins, J. L., Lambert, M. H., Williams, S. P., et al. (2001). The human nuclear xenobiotic receptor PXR: Structural determinants of directed promiscuity. *Science (New York, N.Y.), 292*, 2329–2333.

Weidner, C., de Groot, J. C., Prasad, A., Freiwald, A., Quedenau, C., Kliem, M., et al. (2012). Amorfrutins are potent antidiabetic dietary natural products. *Proceedings of the National Academy of Sciences of the United States of America, 109*, 7257–7262.

Weihua, Z., Lathe, R., Warner, M., & Gustafsson, J.-Å. (2002). An endocrine pathway in the prostate, ERβ, AR, 5α-androstane-3β,17β-diol, and CYP7B1, regulates prostate growth. *Proceedings of the National Academy of Sciences of the United States of America, 99*, 13589–13594.

Whitby, R. J., Stec, J., Blind, R. D., Dixon, S., Leesnitzer, L. M., Orband-Miller, L. A., et al. (2011). Small molecule agonists of the orphan nuclear receptors steroidogenic factor-1 (SF-1, NR5A1) and liver receptor homologue-1 (LRH-1, NR5A2). *Journal of Medicinal Chemistry, 54*, 2266–2281.

Wisely, G., Miller, A., Davis, R., Thornquest, A., Johnson, R., Spitzer, T., et al. (2002). Hepatocyte nuclear factor 4 is a transcription factor that constitutively binds fatty acids. *Structure (London England: 1993), 10*, 1225–1234.

Wu, D., Cheung, A., Wang, Y., Yu, S., & Chan, F. (2016). The emerging roles of orphan nuclear receptors in prostate cancer. *Biochimica et Biophysica Acta, 1866*, 23–36.

Yuan, X., Ta, T., Lin, M., Evans, J., Dong, Y., Bolotin, E., et al. (2009). Identification of an endogenous ligand bound to a native orphan nuclear receptor. *PloS One, 4*, e5609.

Zhang, P., & Mellon, S. H. (1996). The orphan nuclear receptor steroidogenic factor-1 regulates the cyclic adenosine 3',5'-monophosphate-mediated transcriptional activation of rat cytochrome P450c17 (17 alpha-hydroxylase/c17-20 lyase). *Molecular Endocrinology (Baltimore, Md.), 10*, 147–158.

CHAPTER TWO

The Function and Evolution of Nuclear Receptors in Insect Embryonic Development

Alys M. Cheatle Jarvela, Leslie Pick[1]
University of Maryland, College Park, MD, United States
[1]Corresponding author: e-mail address: lpick@umd.edu

Contents

Abstract

Nuclear receptors are a family of transcription factors that are often responsive to small ligands, allowing for efficient gene expression-level responses to a stimulus. The average insect has 21 genes encoding nuclear receptors, whose functions are especially well studied in developmental transitions during the insect life cycle, such as metamorphosis and molting. However, their utility as well-controlled transcriptional regulators also lends them to important roles in embryogenesis, neurogenesis, metabolism, and organogenesis. Such developmental functions have been explored in depth in the model organism *Drosophila melanogaster*. More recently, advances in genomic resources and functional genomic methodologies have allowed for comparison of nuclear receptor function among a wider range of insect species. As has been the trend throughout the field of Evo-Devo, these new data sets reveal that many genes are shared, but the

Current Topics in Developmental Biology, Volume 125
ISSN 0070-2153
http://dx.doi.org/10.1016/bs.ctdb.2017.01.003

ways in which they are utilized in different lineages are more variable. In this chapter, we describe the suite of nuclear receptor genes found in *Drosophila* and their developmental functions. We then compare and contrast these genes and their functions in diverse insects.

1. INTRODUCTION: NUCLEAR RECEPTOR STRUCTURE AND FUNCTION

Nuclear receptors (NRs) are a superfamily of transcription factors that regulate gene expression and are often controlled by the binding of lipophilic ligands, such as steroid hormones or other small molecules. Because they play roles in many aspects of homeostasis, and their ligand-binding properties offer a mechanism to target them through small molecules, there has been much interest in designing therapeutics to control NR activity. However, their ability to exert precisely controlled transcriptional responses also makes them ideal for executing developmental genetic programs. In insects, they are used extensively in oogenesis, embryonic development, molting, and metamorphosis. These functions are well characterized in the model organism *Drosophila melanogaster*, and this knowledge is being applied to understand the development of other insect species. The NRs are divided into seven subfamilies based on the similarity of their functional domains, as determined by rigorous phylogenetic studies (for review, see Bertrand, Belgacem, & Escriva, 2011; Laudet, 1997; Laudet, Hänni, Coll, Catzeflis, & Stéhelin, 1992; Nuclear Receptors Nomenclature Committee, 1999).

These functional domains are referred to by the letters A–F: transactivation (A,B), DNA-binding domain (C), hinge (D), ligand-binding domain (LBD) and transactivation 2 (E), variable (F). In general, the N-termini are variable. Some have a transactivation domain at their N-terminus called the activation function-1 (AF-1) domain. The central portion of the protein is the DNA-binding domain (DBD), composed of two C_4-zinc fingers. This domain allows NRs to recognize and bind specific DNA sequences in the genome, critical to their transcription factor function. This is followed by a hinge region. Hinges are variable in sequence but may include several important features, such as nuclear localization signals or sites of posttranslational modification. Additionally, this region promotes synergy between AF-1 and activation function-2 (AF-2) (see below) domains and can influence DNA-binding function by altering affinity, specificity, or polarity of heterodimer binding (reviewed in Pawlak, Lefebvre, & Staels, 2012). Finally, at the C-terminus lies the LBD, allowing for the regulation of the NR activity through the binding of ligand. Upon ligand binding, this domain undergoes a conformational change, which in general displaces corepressors and exposes interfaces for interaction with coactivators through the AF-2 domain. Both the DBD and LBD confer dimerization abilities upon NRs, but different NRs operate as monomers, homodimers, or heterodimers. Furthermore, while monomers bind to RGGTCA-DNA sequences, also called hormone response elements, the orientation and spacing of double sites for dimers are variable. Dimerization and binding site spacing preferences do seem to track roughly with NR subfamily (Bonneton & Laudet, 2012). The modular structure and flexible binding properties of NRs allow them to have broad utility as regulators of gene expression.

Many NRs are controlled by the binding of small molecules, especially steroids, vitamins, and lipids, to their LBDs (Gronemeyer, Gustafsson, & Laudet, 2004; Sladek, 2011). Ligand binding allows NRs to translate cues from the environment or the organism's diet into a biological response, through gene-expression changes. As mentioned earlier, ligand binding can change the conformation of the receptor from inactive to active (true or activating ligands), often by altering the ability of the protein to interact with corepressors vs coactivators. Surprisingly, some NRs, such as E75 in *Drosophila*, can be regulated by gases such as NO, as the gas changes the redox state of heme, which in turn influences binding to a protein partner (Reinking et al., 2005). In this case, the heme is a structural ligand, which stabilizes the active conformation, but its binding per se does not switch the receptor between "on" and "off" states. Other examples of stabilizing ligands include fatty acids, phospholipids, and other small molecules (reviewed in Evans & Mangelsdorf, 2014).

Some NRs do not seem to have any ligand and are referred to as "orphan" receptors. Over the years, many former orphan receptors have been adopted, that is, matched to ligand cognates, but others retain their orphan status in spite of intensive study (see for review Gronemeyer et al., 2004; Mullican, DiSpirito, & Lazar, 2013; Sladek, 2011). It has been argued that a true orphan receptor lacks a pocket to bind a ligand (Bonneton & Laudet, 2012). For example, for the red flour beetle, *Tribolium castaneum,* the crystal structure of Ultraspiracle (Usp) revealed that loops not present in ligand-dependent Usp homologs fill the region that would have been the ligand-binding pocket (Iwema et al., 2007). Similarly, *Drosophila*

Ftz-F1's ligand-binding pocket is filled by its own LBD helix 6, presumably occluding binding to any small molecule (Yoo et al., 2011). It is hypothesized that unliganded NRs are controlled through other means, such as protein–protein interactions (e.g., Yussa, Löhr, Su, & Pick, 2001), phosphorylation (reviewed in Rochette-Egly, 2003), sumolyation (e.g., Poukka, Karvonen, Jänne, & Palvimo, 2000), ubiquitination (reviewed in Dennis, Haq, & Nawaz, 2001), and acetylation (e.g., Jacob, Lund, Martinez, & Hedin, 2001). There are also atypical NRs that do not have an LBD at all and therefore *must* operate independently of any ligand.

In this chapter, we summarize the current understanding of NR function in *D. melanogaster*, the most extensively studied insect model system. We focus on the multiple roles of NRs in *Drosophila* embryogenesis and compare their functions in other insect species. So far, experimental data from non-*Drosophila* insects are relatively sparse. Much study of other insect species has revolved around understanding NR function during ecdysone signaling (MartÍn, 2010), which controls molting and metamorphosis of insects. However, studies investigating other aspects of NR biology are increasing rapidly due to large-scale projects to sequence insect genomes and transcriptomes, as well as the application of RNAi and gene editing approaches to additional organisms. A broader understanding of how NR developmental roles are conserved or flexible over evolutionary time may provide insights into how insect diversity is generated and opportunities to design pest control strategies against specific insect species or lineages.

2. ROLES OF NRs IN *DROSOPHILA* EMBRYONIC DEVELOPMENT

D. melanogaster has 18 NR genes plus another 3 atypical subfamily members (listed in Table 1), and many have been the subject of intensive functional analysis (for review, see King-Jones & Thummel, 2005). The sophistication of genetic tools in this model system, coupled with the lack of whole-genome duplication found in vertebrates, makes *Drosophila* an ideal system for investigating NR function during development. NRs originated in metazoans, as they are not present in sister taxa, the choanoflagellates, but are present in all extant metazoans (King et al., 2008). Bertrand and colleagues mapped the history of the NR superfamily through metazoan evolution and found that 25 is the ancestral number, but insects have 19–22 depending on lineage, representing six ancestral subfamilies, including

Table 1 Nomenclature of Insect Nuclear Receptors and Their Vertebrate Counterparts

Insect NRNC Nomenclature	Insect Nuclear Receptor	Abbreviations	Vertebrate NRNC Nomenclature	Abbreviations
NR0A1	Knirps	Kni		
NR0A2	Knirps-like	Knrl		
NR0A3	Eagle	Eg		
NR1D3	Ecdysone-induced protein 75	E75	NR1D1–2	Rev-erbα, Rev-erbβ
NR1E1	Ecdysone-induced protein 78	E78		
NR1F4	Hormone receptor 3	DHR3	NR1F1–3	ROR, RZR
NR1H1	Ecdysone receptor	EcR	NR1H3–4	LXR, FXR
NR1J1	Hormone receptor-like in 96	DHR96	NR111–4	VDR, PXR, CAR
NR2A4	Hepatocyte nuclear factor 4	Hnf4	NR2A1–3	HNF4
NR2B4	Ultraspiracle	Usp	NR2B1–3	RXR
NR2D1	Hormone-receptor-like in 78	DHR78	NR2C1–2	TR2, TR4
NR2E2	Tailless	Tll	NR2E1	TLX
NR2E3	Hormone receptor 51	DHR51	NR2E3	PNR
NR2E4	Dissatisfaction	Dsf		
NR2E5	Hormone receptor 83	DHR83		

Continued

Table 1 Nomenclature of Insect Nuclear Receptors and Their Vertebrate Counterparts—cont'd

Insect NRNC Nomenclature	Insect Nuclear Receptor	Abbreviations	Vertebrate NRNC Nomenclature	Abbreviations
NR2E6				
NR2F3	Seven-up	Svp	NR2F1–2	COUP-TFA, COUP-TFB
NR3B4	Estrogen-related receptor	ERR	NR3B1–2	ERR1, ERR2
NR4A4	Hormone receptor-like in 38	DHR38	NR4A1–3	NURR1, NGF1B, NOR1
NR5A3	Ftz transcription factor 1	Ftz-F1	NR5A1–2	SF-1, LRH
NR5B1	Hormone receptor-like in 39	DHR39		
NR6A1	Hormone receptor 4	DHR4	NR6A1	GCNF1

Insect nuclear receptors are listed according to their official name, with their common *Drosophila* name and abbreviation listed as well. NR2E6 is an exception, as it is absent from the *Drosophila* genome but is present in other insect genomes and does not have a common name. Vertebrate orthologs are listed next to each insect nuclear receptor when one exists. Orthology assignment and nomenclature based on information from Bertrand et al. (2004), Bonneton and Laudet (2012), King-Jones and Thummel (2005), and Committee (1999).

2–3 "atypical" NRs (subfamily 0) (Bertrand et al., 2004). This is illustrated in Fig. 1. A seventh subfamily of NRs (NR7) was later detected in some metazoan lineages, but it is not present in insects (Bertrand et al., 2011). Insect NRs do not always have a vertebrate counterpart, and vice versa, due to independent gene duplication and divergence, and gene loss events. However, studies of NRs in *Drosophila* often reveal striking similarity to vertebrate NR function, thus advancing our understanding of NRs in less tractable systems (King-Jones & Thummel, 2005). Initial investigations suggested that only a subset of *Drosophila* NRs are expressed during embryonic development (Sullivan & Thummel, 2003). A recent detailed analysis of NR localization by fluorescent in situ hybridization revealed that genes encoding all *Drosophila* NRs are expressed at some point between oviposition and hatching, especially in tissues and organs important to metabolic processes,

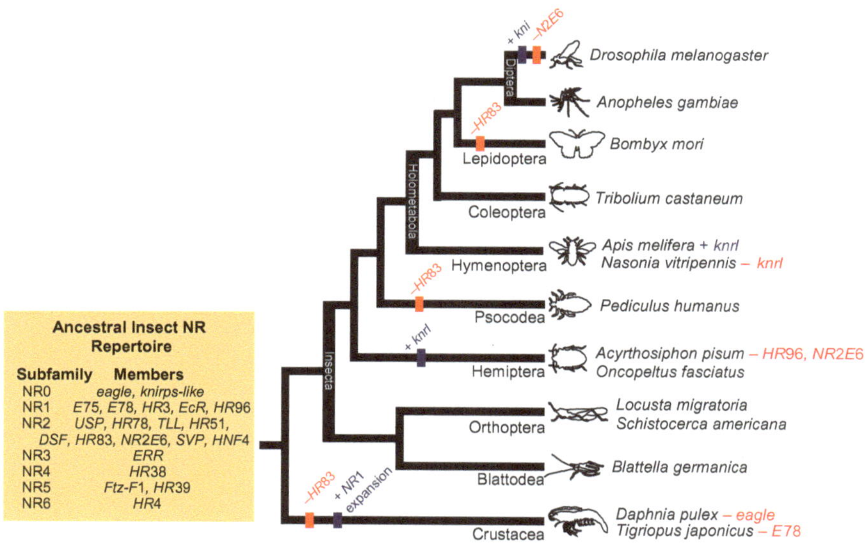

Fig. 1 *A comparison of nuclear receptor repertoires among insect lineages.* Insects share a conserved set of nuclear receptor genes (listed in the *yellow box*) across diverse lineages and with their closest sister taxa, the crustaceans. The phylogenetic tree depicts the relationships between some of the major insect lineages. Species discussed in this chapter are listed and diagramed next to the order they belong to. Gains and losses of particular nuclear receptor genes are indicated by *blue* and *red* font, respectively. These are indicated on the phylogenetic tree branches where enough data are available to suggest a change is common to that entire lineage. Otherwise, changes are listed next to the specific species in which they are known to occur. Tree topology based on Misof et al. (2014) and Regier et al. (2010); insect nuclear receptor gene sets previously described in Bonneton and Laudet (2012) and Fahrbach, Smagghe, and Velarde (2012).

but the functional significance of many of these expression patterns is yet to be investigated (Wilk, Hu, & Krause, 2013). In this section, we describe the well-characterized roles of 12 *Drosophila* NRs in a variety of embryonic process, including early patterning, morphogenesis, organogenesis, and neurogenesis (summarized in Fig. 2).

2.1 Nuclear Receptors and Early Patterning Mechanisms

Typical of metazoan embryonic development, the *Drosophila* egg is preloaded with many localized maternal proteins and RNAs. These maternal products determine the basic axes of the embryo (e.g., anterior–posterior, dorsal–ventral), which is then divided into repeated segmental units along the anterior–posterior axis by the sequential activity of gap, pair rule, and segment polarity classes of segmentation genes (Nüsslein-Volhard, Kluding, & Jürgens, 1985). Most of these segmentation genes encode transcription factors, three of which are NRs: *tailless* (*tll*, *NR2E2*), *knirps* (*kni*, *NR0A1*), and *ftz-factor 1* (*ftz-f1*, *NR5A3*).

Acting as a gap gene, *tll* encodes an orphan NR required for specifying the anterior-most and posterior-most ends of the embryo, where it is expressed (Jürgens, Wieschaus, Nüsslein-Volhard, & Kluding, 1984; Pignoni et al., 1990; Strecker, Merriam, & Lengyel, 1988) (see Fig. 2A, cyan). Unlike many NRs, which can act as both activators and repressors of transcription depending on which cofactors they are bound to, Tll is an obligate repressor (Haecker et al., 2007; Wang, Rajan, Pitman, McKeown, & Tsai, 2006). In liganded NRs, the last helix of AF-2 (H12) switches to an active conformation upon ligand binding to optimize binding to coactivator amino-acid motifs (LXXLL) vs corepressor motifs (LXXXIXXX) (Shiau et al., 1998; Xu et al., 2002). However, Tll binds its protein partner Atrophin irrespective of any ligand via an unrelated motif, the Atro box, which fits into a pocket underneath H12, locking it in an autorepressed state (Zhi et al., 2015).

Mutations in the gap gene *kni* result in embryos lacking abdominal segments 1–7 (Nauber et al., 1988; Nüsslein-Volhard & Wieschaus, 1980). At the syncytial and cellular blastoderm stages, *kni* is expressed in a broad posterior domain, which corresponds to its gap phenotype (Fig. 2A, pink), in addition to an anterior domain (Rothe et al., 1989). *kni* and its paralog, *knirps-like* (*knrl*, *NR0A2*), are the result of a recent duplication event (Fig. 1) (Perl, Schmid, Schwirz, & Chipman, 2013). While *Drosophila kni* and *knrl* share expression patterns and functional capabilities, *kni* is the

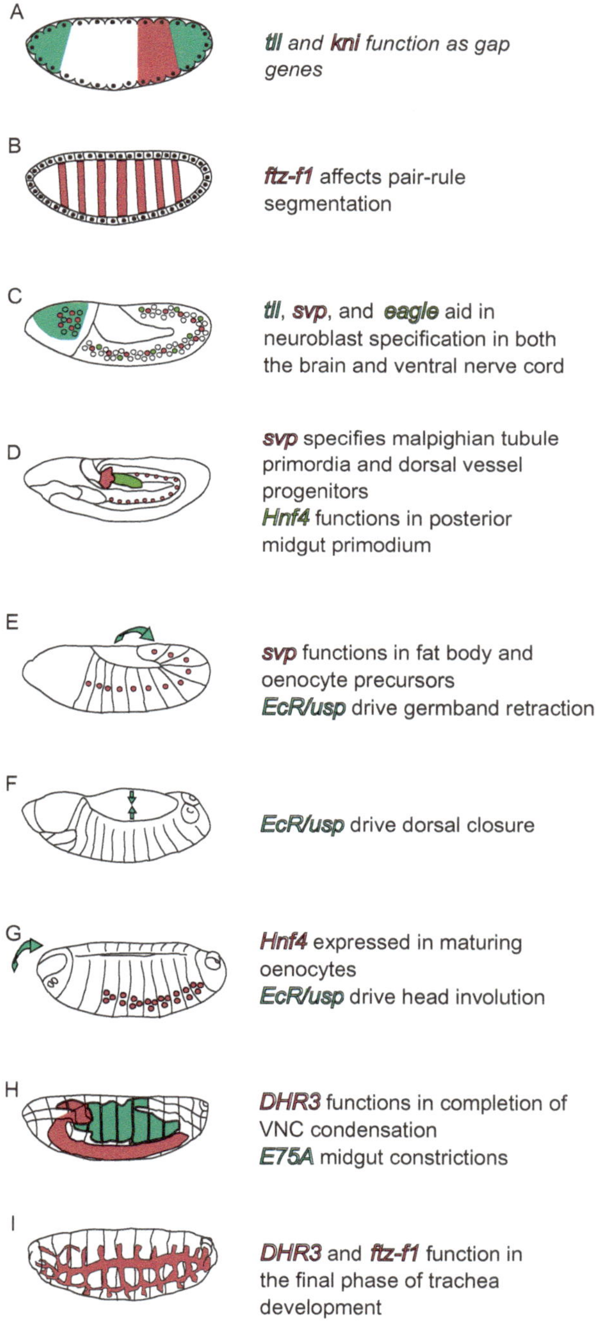

Fig. 2 See legend on next page.

one that actually exerts the gap function (González-Gaitán, Rothe, Wimmer, Taubert, & Jäckle, 1994; Rothe, Pehl, Taubert, & Jäckle, 1992). The difference appears to be a result of intron size; *knrl* has very long introns, which are not conducive to keeping up with the rapid mitotic cycles in early fly development, whereas *kni* has evolved a more streamlined intron/exon structure (Rothe et al., 1992). This mode of gene regulation, a phenomenon known as intron delay, was long hypothesized to be important during early *Drosophila* development, but only recently shown to be a general feature of the early embryonic transcriptome in this organism (Artieri & Fraser, 2014; Gubb, 1986).

Ftz-F1 was first identified biochemically based on its direct binding to a *cis*-regulatory element of the *fushi tarazu* (*ftz*) gene (Ueda, Sonoda, Brown, Scott, & Wu, 1990), hence its name. There are two isoforms of Ftz-F1, α and β, which are used in different aspects of *Drosophila* development: pair-rule segmentation and metamorphosis, respectively (reviewed in

Fig. 2 *Functions of nuclear receptors during Drosophila embryonic development.* A summary of the well-defined activities of nuclear receptor genes in particular regions, tissues, and cell types, where gene function is indicated by *pink*, *green*, and *cyan shading*. Morphogenetic movements, such as germband retraction, are indicated by *arrows*. For simplicity, only the earliest functional expression is shown for each gene, although many of these expression patterns and functions continue for several stages. Note that the *colored regions* denote specific sites of gene activity and not total expression patterns. (A) *tll* (*cyan*) functions at termini and *kni* (*pink*) as an abdominal gap gene in syncytial blastoderm stage embryos. (B) Ftz-F1 functions as a pair-rule gene at the cellular blastoderm stage, where its ubiquitous expression overlaps with Ftz pair-rule stripes, promoting the formation of alternate segments. (C) During early germband extension, *tll* (*cyan*) is used broadly in brain, *svp* (*pink*) in a subset of VNC and anterior CNS neuroblasts, and *eagle* (*green*) in subset of VNC NBs. (D) During later germband extension, *svp* (*pink*) functions in malpighian tubule primordia, which stem from the hindgut, and dorsal vessel progenitors, while *Hnf4* (*green*) functions in posterior midgut primodium. (E) *svp* (*pink*) is used in the development of the fat body and oenocyte precursors, while *EcR/usp* (*cyan*) and its downstream NR cascade drive germband retraction. (F) *EcR/usp* and downstream NR cascade drive dorsal closure. (G) *Hnf4* (*pink*) is expressed by maturing oenocytes. *EcR/usp* and downstream NR cascade (*cyan*) drive head involution. (H) *DHR3* (*pink*) functions in the completion of VNC condensation, while *E75A* (*cyan*) is needed for midgut constrictions. (I) *DHR3* and *ftz-f1* (both *pink*) are required for the final phase of trachea development. Gene functions based upon data described in Bilder and Scott (1995), Chavoshi, Moussian, and Uv (2010), Guichet et al. (1997), Higashijima, Shishido, Matsuzaki, and Saigo (1996), Hoshizaki et al. (1994), Kerber, Fellert, and Hoch (1998), Kozlova and Thummel (2003), Pignoni et al. (1990), Rothe, Nauber, and Jäckle (1989), Ruaud, Lam, and Thummel (2010), Younossi-Hartenstein et al. (1997), Yu et al. (1997), and Zhong, Sladek, and Darnell (1993). *Drosophila* embryonic stages drawn according to Campos-Ortega and Hartenstein (1997).

Pick, Anderson, Shultz, & Woodard, 2006). αFtz-F1 is maternally expressed and functions as a pair-rule protein (Fig. 2B) due to its interaction with protein partner Ftz, a homeodomain-containing transcription factor. Ftz-F1 and Ftz form a stable complex in vivo and bind adjacent sites in *cis*-regulatory regions of target genes, activating their expression (Bowler, Kosman, Licht, & Pick, 2006; Field, Xiang, Anderson, Graham, & Pick, 2016; Florence, Guichet, Ephrussi, & Laughon, 1997; Hou et al., 2009; Yu et al., 1997). While Ftz-F1 is expressed ubiquitously, Ftz is expressed in stripes in early embryos in the primordia of alternate segmental regions missing in either *ftz* or *ftz-f1* mutant embryos (Guichet et al., 1997; Nüsslein-Volhard et al., 1985; Wakimoto, Turner, & Kaufman, 1984; Yu et al., 1997). The interaction between Ftz and Ftz-F1 explains the Ftz-F1 pair-rule phenotype: although it is expressed throughout the embryo, its activity is limited to those cells in which Ftz is expressed. Ftz appears to function as an NR coactivator, binding the Ftz-F1 AF-2 domain through an LXXLL motif in Ftz (Schwartz et al., 2001; Yussa et al., 2001). Although phospholipid ligands have been identified for some vertebrate Ftz-F1 orthologs (Krylova et al., 2005), the occupancy of the ligand-binding pocket in *Drosophila* Ftz-F1 by a portion of its LBD (Yoo et al., 2011) and the fact that putative ligand-dependent and -independent mammalian orthologs rescued *Drosophila ftz-f1* mutants (Lu, Anderson, Zhang, Feng, & Pick, 2013) suggest that *Drosophila* Ftz-F1 is a true orphan NR. The level at which Ftz limits Ftz-F1 transcriptional activity is still under investigation.

An Ftz-F1-related NR, initially named Ftz-F1beta (not to be confused with the late, βFtz-F1 isoform mentioned earlier) but now referred to as DHR39 (NR5B1), has similar DNA-binding properties to Ftz-F1 (Ayer et al., 1993; Ohno & Petkovich, 1993; Ohno, Ueda, & Petkovich, 1994). It is expressed during embryogenesis, but its best-studied roles to date are in the ecdysone pathway (Horner, Chen, & Thummel, 1995; see below).

In sum, cascades of transcription factors regulate early embryonic patterning in *Drosophila* and some of these are NRs. In keeping with the mosaic nature of *Drosophila* development and the rapid pace at which cell division and determination take place, these NRs appear to act in ligand-independent fashions to regulate gene expression.

2.2 Functions of NRs During Neurogenesis

The early patterning domains established in the first few stages of embryogenesis set the stage for neurogenesis to occur, both within the newly defined trunk segments and in the region set aside to become the animal's head. In the ventral nerve cord (VNC), neuroblasts (NBs), which are neural precursor cells, are formed in each segment under the control of pair-rule patterning. These NBs divide asymmetrically as neural stem cells, to both replenish themselves and allow for progression toward a differentiated neuronal fate. Each newly generated neural precursor takes on a different fate depending upon birth order, which is regulated by sequential expression of different transcription factors in the NB. *Seven-up* (*svp, NR2F3*) is an NR, first identified for its role in *Drosophila* eye development (Mlodzik, Hiromi, Weber, Goodman, & Rubin, 1990), that provides the cue for transition from "early-born" to "late-born" cell types in early neurogenesis (Fig. 2C, pink). *svp* mutants display an excess of early born-type neurons at the expense of later types (Kanai, Okabe, & Hiromi, 2005), and *svp* is expressed in the procephalic NBs that give rise to the central nervous system (Younossi-Hartenstein, Nassif, Green, & Hartenstein, 1996).

In addition to its role as a gap protein, Tll functions in the late stages of embryonic development in the primordia of the larval visual system (Daniel, Dumstrei, Lengyel, & Hartenstein, 1999). After initial expression of *tll*, its anterior expression domain shifts to the region that will become the protocerebrum (anterior brain, Fig. 2C). Correspondingly, *tll* mutants fail to develop protocerebrum NBs (Younossi-Hartenstein et al., 1997). Additionally, in *tll* mutants, the region that is meant to become optic lobe instead develops into Bolwig's organ (Daniel et al., 1999). The larval eye is ultimately made up of two photoreceptor subtypes: those that express *rhodopsin-5*, and another set that express *rhodopsin-6*. *tll* mutants develop an excessive number of *rhodopsin-6*-expressing photoreceptors because Tll is needed to repress *rhodopsin-5* fate, while also promoting *rhodopsin-6* (Mishra et al., 2013; Sprecher, Pichaud, & Desplan, 2007). Tll remains important as optic lobe development continues during larval stages (Guillermin, Perruchoud, Sprecher, & Egger, 2015).

Eagle (*Eg, NR0A3*) was initially isolated as a homolog of *kni* and was originally named "*egon*" because prominent expression was detected in the embryonic gonad (Rothe et al., 1989). However, the function of this transient expression domain remains unclear. *Eagle* is better known for its requirement during the development of serotonergic neurons in the VNC (Fig. 2C, green), where it promotes terminal differentiation (Dittrich, Bossing, Gould, Technau, & Urban, 1997; Higashijima et al., 1996; Lundell & Hirsh, 1998).

In sum, several NRs are known to have important roles in *Drosophila* embryonic neurogenesis. Because abnormalities in particular neural cell types are typically not lethal to the embryo, and gene expression often occurs transiently in a small number of cells, additional NR neurogenic functions likely remain to be discovered.

2.3 A Cascade of NR Activity in Response to Ecdysone Mediates Mid-Embryogenesis Morphogenetic Movements and Recalibrates Developmental Timing

Perhaps the best-studied role of NRs in insects is in metamorphosis and molting (reviewed in MartÍn, 2010). Here, a cascade of NRs mediates the animals' response to pluses of the steroid hormone, 20-hydroxyecdysone (20E). This regulatory cascade, including NRs *EcR* (*NR1H1*)/*usp* (*NR2B4*), *DHR3* (*NR1F4*), *DHR4* (*NR6A*), *βftz-f1*, and *E75A* (*NR1D3*), transduces temporal information to ensure proper timing of the onset of metamorphosis in *Drosophila* (reviewed in Thummel, 2001). This cascade is initiated by a pulse of 20E received and transduced by its receptor, an EcR/Usp heterodimer. Interestingly, it was found that a similar cascade including *DHR3*, *βftz-f1*, and *E75A*, also following a strong 20E pulse, occurs during mid-embryogenesis as germband retraction is underway (Beck, Pecasse, & Richards, 2004; Sullivan & Thummel, 2003). In fact, reuse of a sequence of NRs occurs at several key developmental transitions: mid-embryogenesis, second to third instar molt, and third instar to pupae (Sullivan & Thummel, 2003). Repetition of this cascade at these points suggests that its function is to time developmental transitions.

Further investigation into the role of this NR cascade during embryonic development revealed that the interactions between these genes closely resemble the interconnections known to occur during metamorphosis (Ruaud et al., 2010). As this response occurs before the embryo develops its endocrine organ (the ring gland), the 20E initiating this response is thought to be maternally deposited and stored (Kozlova & Thummel, 2003). This 20E signal comes from the amnioserosa, an extra-embryonic tissue essential for germband retraction and dorsal closure (Kozlova & Thummel, 2003; Lamka & Lipshitz, 1999). Disrupting ecdysone signaling, either by reducing ecdysone levels or by expressing a dominant-negative EcR, blocks germband retraction and other morphogenic movements such as head involution (Kozlova & Thummel, 2003), but does not cause embryogenesis to arrest before this point (Fig. 2E–G). Correspondingly, EcR mutants do not hatch and have defects in head involution and dorsal closure (Chavoshi et al., 2010). Originally, it was thought that EcR operated without its usual ecdysone-signaling partner, Usp, during these morphogenetic movements because *usp* mutants did not exhibit these defects (Kozlova & Thummel, 2003; Oro, McKeown, & Evans, 1992; Perrimon, Engstrom, & Mahowald, 1985). However, it was later determined that the mild *usp* phenotypes observed in these studies are due to the hypomorphic nature of the alleles used (Henrich et al., 1994). Embryos derived from germline clones with a strong *usp* mutant allele exhibited the same set of defects seen in EcR mutants (Chavoshi et al., 2010). This implicates Usp and EcR working together as a heterodimer to mediate 20E-controlled tissue movements and organ morphogenesis during embryogenesis, in addition to their better-known roles in metamorphosis.

There is a significant overlap between the genes involved in the ecdysone cascade and several aspects of organogenesis. Usp has high maternal expression and is distributed uniformly in embryos, until germband retraction, at which point expression is elevated in the ventral nervous system and midgut (Oro et al., 1992). Similarly, EcR mutants have defects in midgut morphogenesis and also exhibit bloated trachea (Chavoshi et al., 2010). 20E-defective mutants fail to express *E75* and *DHR3* during late embryogenesis in the midgut and trachea, while exogenous 20E applied to early embryos causes ectopic activation of these genes (Chavoshi et al., 2010). Thus, EcR/Usp direct 20E-coordinated regulatory cascades in multiple tissues follow the mid-embryonic 20E pulse.

Additionally, DHR3 and βFtz-F1 have embryonic roles not related to their participation in this regulatory cascade (Fig. 2H and I), as evidenced by the presence of nonoverlapping, in addition to shared, target genes identified by microarray analysis, and differences in defects seen in mutant embryos (Ruaud et al., 2010). Namely, *βftz-f1* mutants produce denticles of abnormal size and pigmentation, while *DHR3* mutants fail to complete VNC condensation. Likewise, while both genes are needed for the final stages of tracheal development, their roles in this process are independent; tracheal defects in *DHR3* mutants cannot be rescued by ectopic *βftz-f1* expression. As both *βftz-f1* and *DHR3* mutants are embryonic lethal (Carney, Wade, Sapra, Goldstein, & Bender, 1997; Ruaud et al., 2010; Yamada et al., 2000), additional roles of this regulatory cascade, likely related to regulation of developmental timing, perhaps a prehatching developmental checkpoint, are expected (Ruaud et al., 2010).

2.4 Functions of NRs in Morphogenesis and Maturation of Metabolic Organs

Recent work demonstrated that nearly every NR is expressed during development in the gut, and many others are expressed in organs related to metabolism and energy storage (Wilk et al., 2013). Some of these NRs already had well-studied functions in the development of those organs. For example, *kni* and *knrl* are expressed together at the hindgut borders (Rothe et al., 1989), where they control cell cycle, resulting in polyploid cells, and aid in gut regionalization (Fuss et al., 2001).

Hnf4 (*NR2A1*) is expressed during embryogenesis in *Drosophila* midgut, fat body, maliphagan tubes, and salivary gland (Zhong et al., 1993) (Fig. 2D). A mutant with a large deletion encompassing the *Hnf4* locus disrupts the development of these same tissues, starting at the stage when zygotic *Hnf4* can first be detected. Larval oenocytes are a special embryonically derived cell type used for lipid homeostasis. They initially form as segmentally iterated clusters in abdominal segments (Gutzwiller et al., 2010). After they mature, they express *Hnf4*. This NR uses long-chain fatty acids as a ligand (Wisely et al., 2002), which makes it ideal for sensing energy availability. *Hnf4* mutant larvae have decreased tolerance to starvation, because they fail to generate energy from stored lipids. Additionally, a ligand-sensor assay demonstrated that *Hnf4* can be activated both by starvation and by excess long-chain fatty acid (Palanker, Tennessen, Lam, & Thummel, 2009). *svp* is also expressed in the oenocytes (Fig. 2E), but its function there has not been characterized (Kanai et al., 2005). Additionally, *svp* is expressed in the early fat body where it is needed for the expression of terminal differentiation genes and is important in the development of the malpighian tubules, which function as the insect equivalent of a kidney (Fig. 2D) (Hoshizaki et al., 1994; Kerber et al., 1998).

ERR (NR3B) is critical for carbohydrate metabolism. Although *ERR* transcript is present throughout embryogenesis, protein and ERR activity are not detected until mid-embryogenesis (Palanker et al., 2006; Tennessen, Baker, Lam, Evans, & Thummel, 2011). At that time, ERR directs the upregulation of a variety of genes involved in glycolysis so that they attain peak levels by hatching, when the organism transitions from maternally provided nutrition to external sources of energy (Tennessen et al., 2011). Activity in the fat body, muscle, and midgut is the most critical, in accordance with the roles of these tissues in glucose metabolism. ERR promotes distinct metabolic programs in each of these tissues (Tennessen et al., 2011). In sum, several different *Drosophila* NRs play roles in metabolism, as do NRs in vertebrates.

2.5 Other NR Functions During Organogenesis

During late development, *kni* and *knrl* are expressed in the stomodeum (the embryonic mouth) and function redundantly in head formation (González-Gaitán et al., 1994). An additional late embryonic function of *kni* and *knrl* was originally suggested by expression domains in the trachea (Rothe et al., 1989). *kni* and *knrl* function redundantly in trachea development, by regulating the cell migration needed for branching, and aiding in the establishment of the border between the trunk and branches (Chen et al., 1998).

In the dorsal vessel (heart), *svp* is needed to repress expression of *tinman* in a subset of cardioblasts (the contractile cells that pump hemolymph) (Gajewski, Choi, Kim, & Schulz, 2000; Lo & Frasch, 2001). This is critical to allow for different cardioblast identities: the *svp*+ cells become the ostia, or openings which allow hemolymph to enter the dorsal vessel, while the remainder form the muscular wall of the dorsal vessel (Molina & Cripps, 2001). Only the posterior-most three sets of *svp*+ cells become larval ostia, while the remaining four are "saved" until pupation and give rise to the adult ostia (Molina & Cripps, 2001).

2.6 Future Directions

In spite of the long and productive history of *Drosophila* research, there is still much left to learn from this model system (Bellen, Tong, & Tsuda, 2010; Cheatle Jarvela & Pick, 2016). Several NRs may have developmental functions that remain to be discovered, as suggested by their tissue-specific expression patterns (Wilk et al., 2013). For example, an enhancer trap screen demonstrated *DHR4* expression in the embryonic gut and pharynx (Bourbon et al., 2002) and as part of the repeated ecdysone-triggered NR cascade (Sullivan & Thummel, 2003). Likewise, *DHR96* is present in the hindgut, consistent with its known roles in sensing metabolites, nutrients, and toxins in adult flies, but functional roles in embryonic gut morphogenesis or gut function in first-instar larvae have not been documented (Afschar et al., 2016; King-Jones, Horner, Lam, & Thummel, 2006; Wilk et al.,

2013). *Dissatisfaction* (*dsf, NR3E4*), *DHR38* (*NR4A4*), and *DHR83* (*NR2E5*) are detected late in embryogenesis in trachea, salivary glands, and posterior spiracles by in situ (Wilk et al., 2013), but none of these genes have a known function in embryogenesis of *Drosophila*, or any insect for that matter (Bonneton & Laudet, 2012). The continued development of functional genomics tools in *Drosophila* and interest in using this organism to advance our understanding of NR biology may result in additional studies of these NRs that reveal previously unsuspected developmental functions.

3. FUNCTIONAL ANALYSIS OF NRs IN NONMODEL INSECTS

3.1 Insects Are the Most Diverse Group on the Planet

Insects are a class within the phylum Arthropoda that encompasses approximately 30 different orders, including over a million described species (Gullan & Cranston, 2009). They constitute on the order of half of global species diversity and occupy essentially every terrestrial and freshwater environment. Fig. 1 depicts a subset of this diversity, including the division of holometabolous insects, which undergo metamorphosis (including *Drosophila*) from the hemimetabolous insects, so-called direct developers, that develop to maturity without metamorphosis. While studies of NR function in *Drosophila* have been an informative starting point, more recent data from a greater representation of insect species have demonstrated the ubiquitous importance and diverse functions of NRs during insect development.

3.2 Genomic Inventory of NRs Across Insect Species

There is a general, core set of 19 plus 2 atypical NR genes shared among diverse insect species. The atypical NR0 family is the most flexible among these families. Duplication of *knrl* in brachyceran flies (the lineage that includes *Drosophila*) produced *kni*, which is not present in other insect lineages (Perl et al., 2013). However, other lineages have also had independent duplications and losses of NR0 genes (Fig. 1) (Christiaens, Iga, Velarde, Rougé, & Smagghe, 2010; Perl et al., 2013; Velarde, Robinson, & Fahrbach, 2006). Lineage-specific losses of other individual NRs have occurred, as summarized in Fig. 1, but generally the set is present in insect genomes sequenced to data. Thus, although there are still several basally branching insect lineages that lack sufficient genomic data to know for certain if any major expansions or losses of NR genes have occurred within the insect class, it is suspected that the currently known core set is fairly stable.

The genomes of crustacean species, the water flea, *Daphnia pulex*, and intertidal harpacticoid copepod, *Tigriopus japonicus*, outgroups to the insects, help to resolve ancestral states. The NR repertoire of these crustaceans largely mirrors that of insects with known genomic sequence (Fig. 1). *HR83* is missing from crustacean genomes, but it is present in other animal taxa and so was presumably in the ancestral arthropod. There has been one dramatic change in that *Daphnia* has evolved a new family of NR1s not found in major model systems (Thomson, Baldwin, Wang, Kwon, & LeBlanc, 2009). *Tigriopus* also has members of this new NR1L subfamily, but they are divergent and more numerous than those found in *Daphnia* (Hwang et al., 2014). Further investigation into NR complements in insect lineages might also reveal previously unnoticed NR expansions and gene duplication events. Coupled with the expansion of functional genomic techniques available in a growing number of nonmodel insects, additional developmental functions of NRs that reveal insights into insect biology are sure to be discovered.

3.3 Functional Studies of NRs in Emerging Insect Model Systems

Although the repertoire of NR genes has remained relatively stable during insect evolution, the functions of specific NRs during embryonic development demonstrate both conservation and divergence. In particular, NR function during organogenesis and neurogenesis is very similar between *Drosophila* and distantly related insects. For example, expression of *Hnf4* in mature oenocytes is conserved in *Tribolium*, even though there are differences in oenocyte number and patterning when compared to *Drosophila* (Burns, Gutzwiller, Tomoyasu, & Gebelein, 2012). RNAi knockdown of *ftz-f1* in *Tribolium* revealed function in cuticle formation during late embryogenesis (Heffer, Grubbs, Mahaffey, & Pick, 2013), as was seen in *Drosophila* (Ruaud et al., 2010). In *Drosophila* and *Schistocerca americana* (American grasshopper), *svp* exhibits similar spatiotemporal expression patterns in NBs, even though upstream pair-rule patterning is not conserved (Broadus & Doe, 1995). Additionally, transient expression of *svp* in *Tribolium* early-born NBs suggests a conserved function in NB temporal identity specification (Biffar & Stollewerk, 2014).

By contrast, the expression and function of NRs during early patterning are labile among insects. For example, the role of *tll* in terminal patterning has changed during insect evolution. Analysis of a variety of other insects reveals that as evolutionary distance from *Drosophila* increases, so do differences in Tll's terminal patterning roles. In another dipteran, the malaria vector mosquito, *Anopheles gambiae*, there is a minor difference in *tll* expression compared to *Drosophila*, with a temporary expansion into the abdominal region that might impact the region of the embryo that Tll patterns (Goltsev, Hsiong, Lanzaro, & Levine, 2004). In the more distant Order Coleoptera, *tll* is expressed in the anterior and posterior portions of the *Tribolium* embryo, but its exact role must be different because the posterior expression in *Tribolium* is not maintained long enough to implicate it in the regulation of the same genes as in *Drosophila* (Schroder, Eckert, Wolff, & Tautz, 2000). Additionally, the anterior expression does not occur in the early embryo. Instead, it appears to be reflecting a conserved expression in the brain, which, in *Drosophila*, is needed for optic lobe development. Similarly, in *Nasonia vitripennis* (jewel wasp, Order Hymenoptera), although *tll* is expressed at both poles, knockdown of function by RNAi suggested that the posterior domain contributes to terminal patterning, while the anterior expression domain corresponds to nervous system (Lynch, Olesnicky, & Desplan, 2006). Studies in *Apis mellifera* (honeybee) demonstrated that expression and function of *tll* are well conserved among hymenoptera (Wilson & Dearden, 2009). In the hemipteran, *Oncopeltus fasciatus* (milkweed bug), *tll* is expressed in the anterior of the early embryo, and ultimately in the eye analgen and head lobes (Weisbrod, Cohen, & Chipman, 2013). RNAi results in truncation of anterior structures. No expression or function in posterior patterning was detected. Similarly, *tll* was expressed in cephalic, but not posterior regions of both oviparous and viviparous *Acyrthosiphon pisum* (pea aphid) embryos (Bickel et al., 2013). Together, these results indicate that *tll* was probably used to specify anterior nervous system structures in ancestral insects and evolved different functions in anterior and posterior early terminal patterning over the course of holometabolous insect radiations. This fits well with the observation that *tll* homologs also have anterior nervous system function in other animals, such as vertebrates (Bonneton & Laudet, 2012).

Much like *tll*, *kni*'s gap function is not highly conserved among insects. Phylogenetic analysis suggests that after duplication, *kni* evolved more quickly than *knrl*, and therefore the single ortholog found in other insects is more similar to *knrl* (Cerny, Grossmann, Bucher, & Klingler, 2008; Perl et al., 2013). In spite of this, *knrl* is expressed in a *Drosophila*-like gap pattern in distantly related flies (García-Solache, Jaeger, & Akam, 2010; Goltsev et al., 2004; Lemke et al., 2010). In *Tribolium*, *kni* functions have changed considerably; unlike in *Drosophila*, it is critical for head patterning, with mandibular and antennal segments lost after knockdown with RNAi (Cerny et al., 2008). Additionally, *kni* does not function as an abdominal gap gene in *Tribolium*, even though it does have a phase of posterior expression in the growth zone that initially suggested a possibility of conserved gap function. *Tribolium kni* RNAi sometimes resulted in embryos that have abdominal segment defects, but they are never missing multiple consecutive segments. Even the head phenotype is not gap like; *kni* initiates and maintains the parasegmental boundaries of the affected head segments only (Peel et al., 2013). A gap function of *kni* is not conserved in Hymenoptera either. *Drosophila*-like gap gene expression of *kni* homologs is not seen in *Apis*, and *Nasonia* does not even have a *kni* ortholog in its genome (Dearden et al., 2006; Perl et al., 2013). In *Oncopeltus*, one of its two *knrl* copies is expressed during early embryogenesis, where it appears as an anterior stripe and a posterior domain in what will be the growth zone, but RNAi did not reveal a noticeable phenotype (Ben-David & Chipman, 2010). This is perhaps in-line with the very subtle phenotypes observed in *Tribolium* abdominal segmentation (Cerny et al., 2008). Together, these results suggest *kni* gap function may be a fly-specific innovation.

Eagle's function is also completely different from *Drosophila* in *Tribolium* embryos, where it is maternally loaded into the anterior pole and was originally speculated to be an anterior determinant, analogous to *Drosophila bicoid* (Bucher, Farzana, Brown, & Klingler, 2005). However, neither RNAi nor other knockdown approaches revealed an anterior–posterior patterning role, or any other significant developmental role that could be observed by cuticle phenotype. Expression in neurons, which would suggest a conserved function, was not observed.

Surprisingly, Ftz-F1's pair-rule function is conserved in *Tribolium* (Heffer et al., 2013). As in *Drosophila*, *Tribolium-ftz-f1* expression begins with a ubiquitous maternal phase, but unlike *Drosophila*, this is followed by a phase of pair-rule expression in which stripes occur in every other developing segment. Because Ftz-F1's segmentation function is so tightly coupled to that of Ftz in *Drosophila*, and Ftz evolved its role in segmentation fairly recently (reviewed in Pick, 2016), it was unclear whether Ftz-F1 would demonstrate pair-rule function in other insects upon molecular perturbation. However, knockdown of *Tribolium-ftz-f1* expression by RNAi resulted in the loss of alternating segments. In contrast, functional studies of *ftz* in *Tribolium* showed that it does not have pair-rule function even though it is expressed stripes (Brown, Hilgenfeld, & Denell, 1994; Choe, Miller, & Brown, 2006). *Tribolium* and *Drosophila ftz-f1* differ in their expression patterns, but not their functions, because Ftz limits Ftz-F1 activity to pair-rule stripes in *Drosophila*, whereas *Tribolium* Ftz-F1's activity is confined to stripes by transcriptional regulation. These and other findings suggest that of the partner pair, *ftz* and *ftz-f1*, *ftz-f1* has a more ancestral role in pair-rule patterning. Future studies will test this hypothesis and determine whether *ftz-f1* is required for segmentation in embryos of nonholometabolous insect species.

Interestingly, a different NR, E75A, functions as a pair-rule gene in *Oncopeltus* (Erezyilmaz, Kelstrup, & Riddiford, 2009). *Oncopeltus-E75A* is expressed in a pair-rule pattern and knockdown by parental RNAi resulted in embryos with fewer En stripes and fused segments. Previous studies of E75 during *Drosophila* embryogenesis did not reveal a pair-rule function, but instead, a role in midgut development (Bilder & Scott, 1995). Likewise, E75 has never come up in the extensive screens that isolated the other *Drosophila* pair-rule genes, including screens that specifically looked for maternal effect genes (e.g., Chou, Noll, & Perrimon, 1993; Nüsslein-Volhard & Wieschaus, 1980; Perrimon, Lanjuin, Arnold, & Noll, 1996). Knockdown of E75 by RNAi in *Tribolium* blocked oogenesis, so its effect on embryogenesis remains unknown (Xu, Tan, & Palli, 2010). However, in the German cockroach, *Blattella germanica*, a detailed RT-PCR followed by southern blotting of *E75* isoforms during embryogenesis revealed three peaks of *E75* expression: two occurred in correlation with ecdysone pulses, but the first occurred independently during very early development (Mané-Padrós et al., 2008). Furthermore, the pulse occurred at day 2 and lingered into day 3, which should be an active period of segment addition (Bell, 1981). Until studies of *E75* function are performed in a variety of insects, it will be impossible to know whether *E75* had pair-rule function ancestrally and lost it in the lineage leading to *Drosophila*, or if a gain of pair-rule function occurred in the lineage leading to *Oncopeltus*.

Finally, some NR genes that do not have known embryonic functions in *Drosophila* were found to affect embryogenesis in a parental RNAi screen in *Tribolium*: knockdown of E78 caused embryogenesis to arrest at the beginning of germband growth, and knockdown of HR51 caused embryos to fail to hatch, even though they develop into first-instar larvae. In *Drosophila*, *DHR51* (*unfulfilled*) is expressed throughout embryogenesis, although at low levels, and even has a transcript that is specific to early embryogenesis (Sung et al., 2009). However, in situ hybridization did not show significant expression until late development, when it is found in ectodermal segmental stripes, followed by expression in the salivary glands, posterior spiracles, and pharynx (Wilk et al., 2013). The significance of HR51 expression is not known for any of these tissues. *HR96* RNAi in *Tribolium* caused embryogenesis to arrest in the middle of germband growth (Xu et al., 2010). In *Drosophila*, this gene has no characterized embryonic function. Similarly, *Tribolium dsf*, *HR38*, and *HR39* RNAi results in arrest of embryogenesis near the end of germband growth (Xu et al., 2010). In the case of *HR39*, this might be consistent with its function during mid-embryogenesis in *Drosophila* (Sullivan & Thummel, 2003), but functions for the other two NRs during *Drosophila* embryogenesis have not been described.

In spite of dramatic differences in how adulthood is achieved, both holometabolous and hemimetabolous insects rely on 20E and the associated NR regulatory cascade for developmental transitions and progress toward maturity (reviewed in Martín, 2010). In all insects, Usp functions as a heterodimer with EcR to bind 20E and initiate developmental and molting events (Bonneton & Laudet, 2012; Yao, Segraves, Oro, McKeown, & Evans, 1992). This is likely an ancient interaction as it has been detected in other arthropods, including ticks (Guo et al., 1998). Usp is homologous to vertebrate RXR (Oro, McKeown, & Evans, 1990), yet Usp does not have a ligand in holometabolous insects, while vertebrate RXR and Usp from the orthopteran, *Locusta migratoria* (migratory locust), bind 9-*cis*-retanoic acid and related retinoids (Allenby et al., 1993; Nowickyj et al., 2008; Rühl et al., 2015). This ligand independence correlates with divergent USP LBDs in holometabolous insects, while the LBDs of other insect lineages are more vertebrate-like in sequence (Bonneton, Zelus, Iwema, Robinson-Rechavi, & Laudet, 2003; Hayward et al., 1999). Interestingly, in *Locusta* and *Blattella*, both of which are hemimetabolous insects, embryos express a short *usp* isoform during mid-embryogenesis, during peak 20E levels, and a long form with additional 22 amino acids in the LBD in early development, before EcR is present but coinciding with maternally loaded 9-*cis*-retinoic acid, indicating a potential 20E-independent function (Hayward et al., 2003; Maestro, Cruz, Pascual, Martín, & Bellés, 2005; Nowickyj et al., 2008).

Given the critical role of ecdysone signaling in a wide variety of developmental events and transitions (oogenesis, embryogenesis, molting, metamorphosis) and diversity of these features among insects (ovary organization, germ band type, number of molts, presence/absence of metamorphosis in life cycle), better understanding of the functional consequences of Ecr/Usp evolution among insects could help explain developmental differences. Moreover, a detailed analysis of the evolution of other NRs at both the sequence and functional level, made possible by the expansion of available sequence data and techniques, may reveal new insights into other aspects of insect developmental, morphological, and metabolic diversity that has allowed them to colonize virtually every habitat type on the planet.

ACKNOWLEDGMENTS

Work in the authors' lab on *Drosophila* nuclear receptors was supported by the National Science Foundation (IOS-1457145). Thanks to Katie Reding and Faith Kung for comments on the manuscript.

REFERENCES

Afschar, S., Toivonen, J. M., Hoffmann, J. M., Tain, L. S., Wieser, D., Finlayson, A. J., … Partridge, L. (2016). Nuclear hormone receptor DHR96 mediates the resistance to xenobiotics but not the increased lifespan of insulin-mutant Drosophila. *Proceedings of the National Academy of Sciences of the United States of America*, *113*(5), 1321–1326. https://doi.org/10.1073/pnas.1515137113.

Allenby, G., Bocquel, M. T., Saunders, M., Kazmer, S., Speck, J., Rosenberger, M., … Chambon, P. (1993). Retinoic acid receptors and retinoid X receptors: Interactions with endogenous retinoic acids. *Proceedings of the National Academy of Sciences of the United States of America*, *90*(1), 30–34.

Artieri, C. G., & Fraser, H. B. (2014). Transcript length mediates developmental timing of gene expression across drosophila. *Molecular Biology and Evolution*, *31*(11), 2879–2889. https://doi.org/10.1093/molbev/msu226.

Ayer, S., Walker, N., Mosammaparast, M., Nelson, J. P., Shilo, B. Z., & Benyajati, C. (1993). Activation and repression of Drosophila alcohol dehydrogenase distal transcription by two steroid hormone receptor superfamily members binding to a common response element. *Nucleic Acids Research*, *21*(7), 1619–1627.

Beck, Y., Pecasse, F., & Richards, G. (2004). Krüppel-homolog is essential for the coordination of regulatory gene hierarchies in early Drosophila development. *Developmental Biology*, *268*(1), 64–75. https://doi.org/10.1016/j.ydbio.2003.12.017.

Bell, W. J. (1981). *The laboratory cockroach. Experiments in cockroach anatomy, physiology and behavior.* London; New York: Springer.

Bellen, H. J., Tong, C., & Tsuda, H. (2010). 100 years of Drosophila research and its impact on vertebrate neuroscience: A history lesson for the future. *Nature Reviews. Neuroscience*, *11*(7), 514–522. https://doi.org/10.1038/nrn2839.

Ben-David, J., & Chipman, A. D. (2010). Mutual regulatory interactions of the trunk gap genes during blastoderm patterning in the hemipteran Oncopeltus fasciatus. *Developmental Biology*, *346*(1), 140–149. https://doi.org/10.1016/j.ydbio.2010.07.010.

Bertrand, S., Belgacem, M. R., & Escriva, H. (2011). Nuclear hormone receptors in chordates. *Molecular and Cellular Endocrinology*, *334*(1–2), 67–75. https://doi.org/10.1016/j.mce.2010.06.017.

Bertrand, S., Brunet, F. G., Escriva, H., Parmentier, G., Laudet, V., & Robinson-Rechavi, M. (2004). Evolutionary genomics of nuclear receptors: From twenty-five ancestral genes to derived endocrine systems. *Molecular Biology and Evolution*, *21*(10), 1923–1937. https://doi.org/10.1093/molbev/msh200.

Bickel, R. D., Cleveland, H. C., Barkas, J., Jeschke, C. C., Raz, A. A., Stern, D. L., & Davis, G. K. (2013). The pea aphid uses a version of the terminal system during oviparous, but not viviparous, development. *EvoDevo*, *4*(1), 10. https://doi.org/10.1186/2041-9139-4-10.

Biffar, L., & Stollewerk, A. (2014). Conservation and evolutionary modifications of neuroblast expression patterns in insects. *Developmental Biology*, *388*(1), 103–116. https://doi.org/10.1016/j.ydbio.2014.01.028.

Bilder, D., & Scott, M. P. (1995). Genomic regions required for morphogenesis of the Drosophila embryonic midgut. *Genetics*, *141*(3), 1087–1100.

Bonneton, F., & Laudet, V. (2012). Chapter 6—Evolution of nuclear receptors in insects A2. In L. I. Gilbert (Ed.), *Insect endocrinology* (pp. 219–252). San Diego: Academic Press. Retrieved from http://www.sciencedirect.com/science/article/pii/B9780123847492100068.

Bonneton, F., Zelus, D., Iwema, T., Robinson-Rechavi, M., & Laudet, V. (2003). Rapid divergence of the ecdysone receptor in Diptera and Lepidoptera suggests coevolution between ECR and USP-RXR. *Molecular Biology and Evolution*, *20*(4), 541–553. https://doi.org/10.1093/molbev/msg054.

Bourbon, H.-M., Gonzy-Treboul, G., Peronnet, F., Alin, M.-F., Ardourel, C., Benassayag, C., … Vincent, A. (2002). A P-insertion screen identifying novel X-linked essential genes in Drosophila. *Mechanisms of Development*, *110*(1–2), 71–83. https://doi.org/10.1016/S0925-4773(01)00566-4.

Bowler, T., Kosman, D., Licht, J. D., & Pick, L. (2006). Computational identification of Ftz/Ftz-F1 downstream target genes. *Developmental Biology*, *299*(1), 78–90. https://doi.org/10.1016/j.ydbio.2006.07.007.

Broadus, J., & Doe, C. Q. (1995). Evolution of neuroblast identity: Seven-up and prospero expression reveal homologous and divergent neuroblast fates in Drosophila and Schistocerca. *Development (Cambridge, England)*, *121*(12), 3989–3996.

Brown, S. J., Hilgenfeld, R. B., & Denell, R. E. (1994). The beetle Tribolium castaneum has a fushi tarazu homolog expressed in stripes during segmentation. *Proceedings of the National Academy of Sciences of the United States of America*, *91*(26), 12922–12926.

Bucher, G., Farzana, L., Brown, S. J., & Klingler, M. (2005). Anterior localization of maternal mRNAs in a short germ insect lacking bicoid. *Evolution & Development*, 7(2), 142–149. https://doi.org/10.1111/j.1525-142X.2005.05016.x.

Burns, K. A., Gutzwiller, L. M., Tomoyasu, Y., & Gebelein, B. (2012). Oenocyte development in the red flour beetle Tribolium castaneum. *Development Genes and Evolution*, *222*(2), 77–88. https://doi.org/10.1007/s00427-012-0390-z.

Campos-Ortega, J. A., & Hartenstein, V. (1997). *The embryonic development of Drosophila melanogaster.* Berlin, Heidelberg: Springer.

Carney, G. E., Wade, A. A., Sapra, R., Goldstein, E. S., & Bender, M. (1997). DHR3, an ecdysone-inducible early-late gene encoding a Drosophila nuclear receptor, is required for embryogenesis. *Proceedings of the National Academy of Sciences of the United States of America*, *94*(22), 12024–12029.

Cerny, A. C., Grossmann, D., Bucher, G., & Klingler, M. (2008). The Tribolium ortholog of knirps and knirps-related is crucial for head segmentation but plays a minor role during abdominal patterning. *Developmental Biology*, *321*(1), 284–294. https://doi.org/10.1016/j.ydbio.2008.05.527.

Chavoshi, T. M., Moussian, B., & Uv, A. (2010). Tissue-autonomous EcR functions are required for concurrent organ morphogenesis in the Drosophila embryo. *Mechanisms of Development*, *127*(5–6), 308–319. https://doi.org/10.1016/j.mod.2010.01.003.

Cheatle Jarvela, A. M., & Pick, L. (2016). Evo-Devo: Discovery of diverse mechanisms regulating development. *Current Topics in Developmental Biology*, *117*, 253–274. https://doi.org/10.1016/bs.ctdb.2015.11.014.

Chen, C. K., Kühnlein, R. P., Eulenberg, K. G., Vincent, S., Affolter, M., & Schuh, R. (1998). The transcription factors KNIRPS and KNIRPS RELATED control cell migration and branch morphogenesis during Drosophila tracheal development. *Development (Cambridge, England)*, *125*(24), 4959–4968.

Choe, C. P., Miller, S. C., & Brown, S. J. (2006). A pair-rule gene circuit defines segments sequentially in the short-germ insect Tribolium castaneum. *Proceedings of the National Academy of Sciences of the United States of America*, *103*(17), 6560–6564. https://doi.org/10.1073/pnas.0510440103.

Chou, T. B., Noll, E., & Perrimon, N. (1993). Autosomal P[ovoD1] dominant female-sterile insertions in Drosophila and their use in generating germ-line chimeras. *Development*, *119*(4), 1359–1369.

Christiaens, O., Iga, M., Velarde, R. A., Rougé, P., & Smagghe, G. (2010). Halloween genes and nuclear receptors in ecdysteroid biosynthesis and signalling in the pea aphid. *Insect Molecular Biology*, *19*, 187–200. https://doi.org/10.1111/j.1365-2583.2009.00957.x.

Daniel, A., Dumstrei, K., Lengyel, J. A., & Hartenstein, V. (1999). The control of cell fate in the embryonic visual system by atonal, tailless and EGFR signaling. *Development (Cambridge, England)*, *126*(13), 2945–2954.

Dearden, P. K., Wilson, M. J., Sablan, L., Osborne, P. W., Havler, M., McNaughton, E., … Beye, M. (2006). Patterns of conservation and change in honey bee developmental genes. *Genome Research*, *16*(11), 1376–1384. https:/doi.org/10.1101/gr.5108606.

Dennis, A. P., Haq, R. U., & Nawaz, Z. (2001). Importance of the regulation of nuclear receptor degradation. *Frontiers in Bioscience: A Journal and Virtual Library*, *6*, D954–D959.

Dittrich, R., Bossing, T., Gould, A. P., Technau, G. M., & Urban, J. (1997). The differentiation of the serotonergic neurons in the Drosophila ventral nerve cord depends on the combined function of the zinc finger proteins Eagle and Huckebein. *Development (Cambridge, England)*, *124*(13), 2515–2525.

Erezyilmaz, D. F., Kelstrup, H. C., & Riddiford, L. M. (2009). The nuclear receptor E75A has a novel pair-rule-like function in patterning the milkweed bug, Oncopeltus fasciatus. *Developmental Biology*, *334*(1), 300–310. https://doi.org/10.1016/j.ydbio.2009.06.038.

Evans, R. M., & Mangelsdorf, D. J. (2014). Nuclear receptors, RXR, and the Big Bang. *Cell*, *157*(1), 255–266. https://doi.org/10.1016/j.cell.2014.03.012.

Fahrbach, S. E., Smagghe, G., & Velarde, R. A. (2012). Insect nuclear receptors. *Annual Review of Entomology*, *57*(1), 83–106. https://doi.org/10.1146/annurev-ento-120710-100607.

Field, A., Xiang, J., Anderson, W. R., Graham, P., & Pick, L. (2016). Activation of Ftz-F1-responsive genes through Ftz/Ftz-F1 dependent enhancers. *PLoS One*, *11*(10). e0163128. https://doi.org/10.1371/journal.pone.0163128

Florence, B., Guichet, A., Ephrussi, A., & Laughon, A. (1997). Ftz-F1 is a cofactor in Ftz activation of the Drosophila engrailed gene. *Development (Cambridge, England)*, *124*(4), 839–847.

Fuss, B., Meissner, T., Bauer, R., Lehmann, C., Eckardt, F., & Hoch, M. (2001). Control of endoreduplication domains in the Drosophila gut by the knirps and knirps-related genes. *Mechanisms of Development*, *100*(1), 15–23.

Gajewski, K., Choi, C. Y., Kim, Y., & Schulz, R. A. (2000). Genetically distinct cardial cells within the Drosophila heart. *Genesis*, *28*(1), 36–43. http://dx.doi.org/10.1002/1526-968X(200009)28:1<36::AID-GENE50>3.0.CO;2-4.

García-Solache, M., Jaeger, J., & Akam, M. (2010). A systematic analysis of the gap gene system in the moth midge Clogmia albipunctata. *Developmental Biology*, *344*(1), 306–318. https://doi.org/10.1016/j.ydbio.2010.04.019.

Goltsev, Y., Hsiong, W., Lanzaro, G., & Levine, M. (2004). Different combinations of gap repressors for common stripes in Anopheles and Drosophila embryos. *Developmental Biology*, *275*(2), 435–446. https://doi.org/10.1016/j.ydbio.2004.08.021.

González-Gaitán, M., Rothe, M., Wimmer, E. A., Taubert, H., & Jäckle, H. (1994). Redundant functions of the genes knirps and knirps-related for the establishment of anterior Drosophila head structures. *Proceedings of the National Academy of Sciences of the United States of America*, *91*(18), 8567–8571.

Gronemeyer, H., Gustafsson, J.-A., & Laudet, V. (2004). Principles for modulation of the nuclear receptor superfamily. *Nature Reviews. Drug Discovery*, *3*(11), 950–964. https://doi.org/10.1038/nrd1551.

Gubb, D. (1986). Intron-delay and the precision of expression of homoeotic gene products in Drosophila. *Developmental Genetics*, 7(3), 119–131. https://doi.org/10.1002/dvg.1020070302.

Guichet, A., Copeland, J. W., Erdélyi, M., Hlousek, D., Závorszky, P., Ho, J., … Ephrussi, A. (1997). The nuclear receptor homologue Ftz-F1 and the homeodomain protein Ftz are mutually dependent cofactors. *Nature*, *385*(6616), 548–552. https://doi.org/10.1038/385548a0.

Guillermin, O., Perruchoud, B., Sprecher, S. G., & Egger, B. (2015). Characterization of tailless functions during Drosophila optic lobe formation. *Developmental Biology*, *405*(2), 202–213. https://doi.org/10.1016/j.ydbio.2015.06.011.

Gullan, P. J., & Cranston, P. S. (2009). *The insects: An outline of entomology*. Chichester: John Wiley & Sons.

Guo, X., Xu, Q., Harmon, M. A., Jin, X., Laudet, V., Mangelsdorf, D. J., & Palmer, M. J. (1998). Isolation of two functional retinoid X receptor subtypes from the Ixodid tick, Amblyomma americanum (L.). *Molecular and Cellular Endocrinology*, *139*(1–2), 45–60. https://doi.org/10.1016/S0303-7207(98)00073-2.

Gutzwiller, L. M., Witt, L. M., Gresser, A. L., Burns, K. A., Cook, T. A., & Gebelein, B. (2010). Proneural and abdominal Hox inputs synergize to promote sensory organ formation in the Drosophila abdomen. *Developmental Biology*, *348*(2), 231–243. https://doi.org/10.1016/j.ydbio.2010.09.014.

Haecker, A., Qi, D., Lilja, T., Moussian, B., Andrioli, L. P., Luschnig, S., & Mannervik, M. (2007). Drosophila brakeless interacts with atrophin and is required for tailless-mediated transcriptional repression in early embryos. *PLoS Biology*, *5*(6). e145. https://doi.org/10.1371/journal.pbio.0050145

Hayward, D. C., Bastiani, M. J., Trueman, J. W., Truman, J. W., Riddiford, L. M., & Ball, E. E. (1999). The sequence of Locusta RXR, homologous to Drosophila Ultraspiracle, and its evolutionary implications. *Development Genes and Evolution*, *209*(9), 564–571.

Hayward, D. C., Dhadialla, T. S., Zhou, S., Kuiper, M. J., Ball, E. E., Wyatt, G. R., & Walker, V. K. (2003). Ligand specificity and developmental expression of RXR and ecdysone receptor in the migratory locust. *Journal of Insect Physiology*, *49*(12), 1135–1144. https://doi.org/10.1016/j.jinsphys.2003.08.007.

Heffer, A., Grubbs, N., Mahaffey, J., & Pick, L. (2013). The evolving role of the orphan nuclear receptor ftz-f1, a pair-rule segmentation gene. *Evolution & Development*, *15*(6), 406–417. https://doi.org/10.1111/ede.12050.

Henrich, V. C., Szekely, A. A., Kim, S. J., Brown, N. E., Antoniewski, C., Hayden, M. A., … Gilbert, L. I. (1994). Expression and function of the ultraspiracle (usp) gene during development of Drosophila melanogaster. *Developmental Biology*, *165*(1), 38–52. https://doi.org/10.1006/dbio.1994.1232.

Higashijima, S., Shishido, E., Matsuzaki, M., & Saigo, K. (1996). Eagle, a member of the steroid receptor gene superfamily, is expressed in a subset of neuroblasts and regulates the fate of their putative progeny in the Drosophila CNS. *Development (Cambridge, England)*, *122*(2), 527–536.

Horner, M. A., Chen, T., & Thummel, C. S. (1995). Ecdysteroid regulation and DNA binding properties of Drosophila nuclear hormone receptor superfamily members. *Developmental Biology*, *168*(2), 490–502. https://doi.org/10.1006/dbio.1995.1097.

Hoshizaki, D. K., Blackburn, T., Price, C., Ghosh, M., Miles, K., Ragucci, M., & Sweis, R. (1994). Embryonic fat-cell lineage in Drosophila melanogaster. *Development (Cambridge, England)*, *120*(9), 2489–2499.

Hou, H. Y., Heffer, A., Anderson, W. R., Liu, J., Bowler, T., & Pick, L. (2009). Stripy Ftz target genes are coordinately regulated by Ftz-F1. *Developmental Biology*, *335*(2), 442–453. https://doi.org/10.1016/j.ydbio.2009.08.004.

Hwang, D.-S., Lee, B.-Y., Kim, H.-S., Lee, M. C., Kyung, D.-H., Om, A.-S., … Lee, J.-S. (2014). Genome-wide identification of nuclear receptor (NR) superfamily genes in the copepod Tigriopus japonicus. *BMC Genomics*, *15*, 993. https://doi.org/10.1186/1471-2164-15-993.

Iwema, T., Billas, I. M. L., Beck, Y., Bonneton, F., Nierengarten, H., Chaumot, A., … Moras, D. (2007). Structural and functional characterization of a novel type of ligand-independent RXR-USP receptor. *The EMBO Journal*, *26*(16), 3770–3782. https://doi.org/10.1038/sj.emboj.7601810.

Jacob, A. L., Lund, J., Martinez, P., & Hedin, L. (2001). Acetylation of steroidogenic factor 1 protein regulates its transcriptional activity and recruits the coactivator GCN5. *The Journal of Biological Chemistry*, *276*(40), 37659–37664. https://doi.org/10.1074/jbc.M104427200.

Jürgens, G., Wieschaus, E., Nüsslein-Volhard, C., & Kluding, H. (1984). Mutations affecting the pattern of the larval cuticle in Drosophila melanogaster. *Wilhelm Roux's Archives of Developmental Biology*, *193*(5), 283–295. https://doi.org/10.1007/BF00848157.

Kanai, M. I., Okabe, M., & Hiromi, Y. (2005). seven-up controls switching of transcription factors that specify temporal identities of drosophila neuroblasts. *Developmental Cell*, *8*(2), 203–213. https://doi.org/10.1016/j.devcel.2004.12.014.

Kerber, B., Fellert, S., & Hoch, M. (1998). Seven-up, the Drosophila homolog of the COUP-TF orphan receptors, controls cell proliferation in the insect kidney. *Genes & Development*, *12*(12), 1781–1786.

King, N., Westbrook, M. J., Young, S. L., Kuo, A., Abedin, M., Chapman, J., … Rokhsar, D. (2008). The genome of the choanoflagellate Monosiga brevicollis and the origin of metazoans. *Nature*, *451*(7180), 783–788. https://doi.org/10.1038/nature06617.

King-Jones, K., Horner, M. A., Lam, G., & Thummel, C. S. (2006). The DHR96 nuclear receptor regulates xenobiotic responses in Drosophila. *Cell Metabolism*, *4*(1), 37–48. https://doi.org/10.1016/j.cmet.2006.06.006.

King-Jones, K., & Thummel, C. S. (2005). Nuclear receptors—A perspective from Drosophila. *Nature Reviews. Genetics*, *6*(4), 311–323.

Kozlova, T., & Thummel, C. S. (2003). Essential roles for ecdysone signaling during drosophila mid-embryonic development. *Science*, *301*(5641), 1911–1914. https://doi.org/10.1126/science.1087419.

Krylova, I. N., Sablin, E. P., Moore, J., Xu, R. X., Waitt, G. M., MacKay, J. A., … Ingraham, H. A. (2005). Structural analyses reveal phosphatidyl inositols as ligands for the NR5 orphan receptors SF-1 and LRH-1. *Cell*, *120*(3), 343–355. https://doi.org/10.1016/j.cell.2005.01.024.

Lamka, M. L., & Lipshitz, H. D. (1999). Role of the amnioserosa in germ band retraction of the Drosophila melanogaster embryo. *Developmental Biology*, *214*(1), 102–112. https://doi.org/10.1006/dbio.1999.9409.

Laudet, V. (1997). Evolution of the nuclear receptor superfamily: Early diversification from an ancestral orphan receptor. *Journal of Molecular Endocrinology*, *19*(3), 207–226.

Laudet, V., Hänni, C., Coll, J., Catzeflis, F., & Stéhelin, D. (1992). Evolution of the nuclear receptor gene superfamily. *The EMBO Journal*, *11*(3), 1003–1013.

Lemke, S., Busch, S. E., Antonopoulos, D. A., Meyer, F., Domanus, M. H., & Schmidt-Ott, U. (2010). Maternal activation of gap genes in the hover fly Episyrphus. *Development (Cambridge, England)*, *137*(10), 1709–1719. https://doi.org/10.1242/dev.046649.

Lo, P. C., & Frasch, M. (2001). A role for the COUP-TF-related gene seven-up in the diversification of cardioblast identities in the dorsal vessel of Drosophila. *Mechanisms of Development*, *104*(1–2), 49–60.

Lu, Y., Anderson, W. R., Zhang, H., Feng, S., & Pick, L. (2013). Functional conservation of Drosophila FTZ-F1 and its mammalian homologs suggests ligand-independent regulation of NR5A family transcriptional activity. *Development Genes and Evolution, 223*(3), 199–205. https://doi.org/10.1007/s00427-012-0435-3.

Lundell, M. J., & Hirsh, J. (1998). Eagle is required for the specification of serotonin neurons and other neuroblast 7–3 progeny in the Drosophila CNS. *Development, 125*(3), 463–472.

Lynch, J. A., Olesnicky, E. C., & Desplan, C. (2006). Regulation and function of tailless in the long germ wasp Nasonia vitripennis. *Development Genes and Evolution, 216*(7–8), 493–498. https://doi.org/10.1007/s00427-006-0076-5.

Maestro, O., Cruz, J., Pascual, N., Martín, D., & Bellés, X. (2005). Differential expression of two RXR/ultraspiracle isoforms during the life cycle of the hemimetabolous insect Blattella germanica (Dictyoptera, Blattellidae). *Molecular and Cellular Endocrinology, 238*(1–2), 27–37. https://doi.org/10.1016/j.mce.2005.04.004.

Mané-Padrós, D., Cruz, J., Vilaplana, L., Pascual, N., Bellés, X., & Martín, D. (2008). The nuclear hormone receptor BgE75 links molting and developmental progression in the direct-developing insect Blattella germanica. *Developmental Biology, 315*(1), 147–160. https://doi.org/10.1016/j.ydbio.2007.12.015.

MartÍn, D. (2010). Functions of nuclear receptors in insect development. In C. M. Bunce & M. J. Campbell (Eds.), *Nuclear receptors* (pp. 31–61). The Netherlands: Springer. https://doi.org/10.1007/978-90-481-3303-1_3.

Mishra, A. K., Tsachaki, M., Rister, J., Ng, J., Celik, A., & Sprecher, S. G. (2013). Binary cell fate decisions and fate transformation in the Drosophila larval eye. *PLoS Genetics, 9*(12). e1004027. https://doi.org/10.1371/journal.pgen.1004027

Misof, B., Liu, S., Meusemann, K., Peters, R. S., Donath, A., Mayer, C., … Zhou, X. (2014). Phylogenomics resolves the timing and pattern of insect evolution. *Science, 346*(6210), 763–767. https://doi.org/10.1126/science.1257570.

Mlodzik, M., Hiromi, Y., Weber, U., Goodman, C. S., & Rubin, G. M. (1990). The drosophila seven-up gene, a member of the steroid receptor gene superfamily, controls photoreceptor cell fates. *Cell, 60*(2), 211–224. https://doi.org/10.1016/0092-8674(90)90737-Y.

Molina, M. R., & Cripps, R. M. (2001). Ostia, the inflow tracts of the Drosophila heart, develop from a genetically distinct subset of cardial cells. *Mechanisms of Development, 109*(1), 51–59. https://doi.org/10.1016/S0925-4773(01)00509-3.

Mullican, S. E., DiSpirito, J. R., & Lazar, M. A. (2013). The orphan nuclear receptors at their 25-year reunion. *Journal of Molecular Endocrinology, 51*(3), T115–T140. https://doi.org/10.1530/JME-13-0212.

Nauber, U., Pankratz, M. J., Kienlin, A., Seifert, E., Klemm, U., & Jäckle, H. (1988). Abdominal segmentation of the Drosophila embryo requires a hormone receptor-like protein encoded by the gap gene knirps. *Nature, 336*(6198), 489–492. https://doi.org/10.1038/336489a0.

Nowickyj, S. M., Chithalen, J. V., Cameron, D., Tyshenko, M. G., Petkovich, M., Wyatt, G. R., … Walker, V. K. (2008). Locust retinoid X receptors: 9-cis-Retinoic acid in embryos from a primitive insect. *Proceedings of the National Academy of Sciences of the United States of America, 105*(28), 9540–9545. https://doi.org/10.1073/pnas.0712132105.

Nuclear Receptors Nomenclature Committee. (1999). A unified nomenclature system for the nuclear receptor superfamily. *Cell, 97*(2), 161–163.

Nüsslein-Volhard, C., Kluding, H., & Jürgens, G. (1985). Genes affecting the segmental subdivision of the Drosophila embryo. *Cold Spring Harbor Symposia on Quantitative Biology, 50*, 145–154.

Nüsslein-Volhard, C., & Wieschaus, E. (1980). Mutations affecting segment number and polarity in Drosophila. *Nature*, *287*(5785), 795–801. https://doi.org/10.1038/287795a0.

Ohno, C. K., & Petkovich, M. (1993). FTZ-F1 beta, a novel member of the Drosophila nuclear receptor family. *Mechanisms of Development*, *40*(1–2), 13–24.

Ohno, C. K., Ueda, H., & Petkovich, M. (1994). The Drosophila nuclear receptors FTZ-F1 alpha and FTZ-F1 beta compete as monomers for binding to a site in the fushi tarazu gene. *Molecular and Cellular Biology*, *14*(5), 3166–3175.

Oro, A. E., McKeown, M., & Evans, R. M. (1990). Relationship between the product of the Drosophila ultraspiracle locus and the vertebrate retinoid X receptor. *Nature*, *347*(6290), 298–301. https://doi.org/10.1038/347298a0.

Oro, A. E., McKeown, M., & Evans, R. M. (1992). The Drosophila retinoid X receptor homolog ultraspiracle functions in both female reproduction and eye morphogenesis. *Development (Cambridge, England)*, *115*(2), 449–462.

Palanker, L., Necakov, A. S., Sampson, H. M., Ni, R., Hu, C., Thummel, C. S., & Krause, H. M. (2006). Dynamic regulation of Drosophila nuclear receptor activity in vivo. *Development (Cambridge, England)*, *133*(18), 3549–3562. https://doi.org/10.1242/dev.02512.

Palanker, L., Tennessen, J. M., Lam, G., & Thummel, C. S. (2009). Drosophila HNF4 regulates lipid mobilization and β-oxidation. *Cell Metabolism*, *9*(3), 228–239. https:/doi.org/10.1016/j.cmet.2009.01.009.

Pawlak, M., Lefebvre, P., & Staels, B. (2012). General molecular biology and architecture of nuclear receptors. *Current Topics in Medicinal Chemistry*, *12*(6), 486–504.

Peel, A. D., Schanda, J., Grossmann, D., Ruge, F., Oberhofer, G., Gilles, A. F., … Bucher, G. (2013). Tc-knirps plays different roles in the specification of antennal and mandibular parasegment boundaries and is regulated by a pair-rule gene in the beetle Tribolium castaneum. *BMC Developmental Biology*, *13*, 25. https://doi.org/10.1186/1471-213X-13-25.

Perl, T. N., Schmid, B. G. M., Schwirz, J., & Chipman, A. D. (2013). The evolution of the knirps family of transcription factors in arthropods. *Molecular Biology and Evolution*, *30*, 1348–1357. mst046. https://doi.org/10.1093/molbev/mst046.

Perrimon, N., Engstrom, L., & Mahowald, A. P. (1985). Developmental genetics of the 2c-D region of the drosophila X chromosome. *Genetics*, *111*(1), 23–41.

Perrimon, N., Lanjuin, A., Arnold, C., & Noll, E. (1996). Zygotic lethal mutations with maternal effect phenotypes in drosophila melanogaster. II. Loci on the second and third chromosomes identified by P-element-induced mutations. *Genetics*, *144*(4), 1681–1692.

Pick, L. (2016). Hox genes, evo-devo, and the case of the ftz gene. *Chromosoma*, *125*(3), 535–551. https://doi.org/10.1007/s00412-015-0553-6.

Pick, L., Anderson, W. R., Shultz, J., & Woodard, C. T. (2006). The Ftz-F1 family: Orphan nuclear receptors regulated by novel protein–protein interactions. *Advances in Developmental Biology*, *16*, 255–296. Elsevier. http://www.sciencedirect.com/science/article/pii/S1574334906160081.

Pignoni, F., Baldarelli, R. M., Steingrimsson, E., Diaz, R. J., Patapoutian, A., Merriam, J. R., & Lengyel, J. A. (1990). The Drosophila gene tailless is expressed at the embryonic termini and is a member of the steroid receptor superfamily. *Cell*, *62*(1), 151–163. https://doi.org/10.1016/0092-8674(90)90249-E.

Poukka, H., Karvonen, U., Jänne, O. A., & Palvimo, J. J. (2000). Covalent modification of the androgen receptor by small ubiquitin-like modifier 1 (SUMO-1). *Proceedings of the National Academy of Sciences of the United States of America*, *97*(26), 14145–14150. https:/doi.org/10.1073/pnas.97.26.14145.

Regier, J. C., Shultz, J. W., Zwick, A., Hussey, A., Ball, B., Wetzer, R., ... Cunningham, C. W. (2010). Arthropod relationships revealed by phylogenomic analysis of nuclear protein-coding sequences. *Nature, 463*(7284), 1079–1083. https://doi.org/10.1038/nature08742.
Reinking, J., Lam, M. M. S., Pardee, K., Sampson, H. M., Liu, S., Yang, P., ... Krause, H. M. (2005). The Drosophila nuclear receptor e75 contains heme and is gas responsive. *Cell, 122*(2), 195–207. https://doi.org/10.1016/j.cell.2005.07.005.
Rochette-Egly, C. (2003). Nuclear receptors: Integration of multiple signalling pathways through phosphorylation. *Cellular Signalling, 15*(4), 355–366.
Rothe, M., Nauber, U., & Jäckle, H. (1989). Three hormone receptor-like Drosophila genes encode an identical DNA-binding finger. *The EMBO Journal, 8*(10), 3087–3094.
Rothe, M., Pehl, M., Taubert, H., & Jäckle, H. (1992). Loss of gene function through rapid mitotic cycles in the Drosophila embryo. *Nature, 359*(6391), 156–159. https://doi.org/10.1038/359156a0.
Ruaud, A.-F., Lam, G., & Thummel, C. S. (2010). The Drosophila nuclear receptors DHR3 and betaFTZ-F1 control overlapping developmental responses in late embryos. *Development (Cambridge, England), 137*(1), 123–131. https://doi.org/10.1242/dev.042036.
Rühl, R., Krzyżosiak, A., Niewiadomska-Cimicka, A., Rochel, N., Szeles, L., Vaz, B., ... Krężel, W. (2015). 9-cis-13,14-Dihydroretinoic acid is an endogenous retinoid acting as RXR ligand in mice. *PLoS Genetics, 11*(6). e1005213. https://doi.org/10.1371/journal.pgen.1005213.
Schroder, R., Eckert, C., Wolff, C., & Tautz, D. (2000). Conserved and divergent aspects of terminal patterning in the beetle Tribolium castaneum. *Proceedings of the National Academy of Sciences of the United States of America, 97*(12), 6591–6596. https://doi.org/10.1073/pnas.100005497.
Schwartz, C. J., Sampson, H. M., Hlousek, D., Percival-Smith, A., Copeland, J. W., Simmonds, A. J., & Krause, H. M. (2001). FTZ-Factor1 and Fushi tarazu interact via conserved nuclear receptor and coactivator motifs. *The EMBO Journal, 20*(3), 510–519. https://doi.org/10.1093/emboj/20.3.510.
Shiau, A. K., Barstad, D., Loria, P. M., Cheng, L., Kushner, P. J., Agard, D. A., & Greene, G. L. (1998). The structural basis of estrogen receptor/coactivator recognition and the antagonism of this interaction by tamoxifen. *Cell, 95*(7), 927–937.
Sladek, F. M. (2011). What are nuclear receptor ligands? *Molecular and Cellular Endocrinology, 334*(1–2), 3–13. https://doi.org/10.1016/j.mce.2010.06.018.
Sprecher, S. G., Pichaud, F., & Desplan, C. (2007). Adult and larval photoreceptors use different mechanisms to specify the same Rhodopsin fates. *Genes & Development, 21*(17), 2182–2195. https://doi.org/10.1101/gad.1565407.
Strecker, T. R., Merriam, J. R., & Lengyel, J. A. (1988). Graded requirement for the zygotic terminal gene, tailless, in the brain and tail region of the Drosophila embryo. *Development, 102*(4), 721–734.
Sullivan, A. A., & Thummel, C. S. (2003). Temporal profiles of nuclear receptor gene expression reveal coordinate transcriptional responses during drosophila development. *Molecular Endocrinology, 17*(11), 2125–2137. https://doi.org/10.1210/me.2002-0430.
Sung, C., Wong, L. E., Chang Sen, L. Q., Nguyen, E., Lazaga, N., Ganzer, G., ... Robinow, S. (2009). The unfulfilled/DHR51 gene of Drosophila melanogaster modulates wing expansion and fertility. *Developmental Dynamics, 238*(1), 171–182. https://doi.org/10.1002/dvdy.21817.
Tennessen, J. M., Baker, K. D., Lam, G., Evans, J., & Thummel, C. S. (2011). The drosophila estrogen-related receptor directs a metabolic switch that supports developmental growth. *Cell Metabolism, 13*(2), 139–148. https://doi.org/10.1016/j.cmet.2011.01.005.

Thomson, S. A., Baldwin, W. S., Wang, Y. H., Kwon, G., & LeBlanc, G. A. (2009). Annotation, phylogenetics, and expression of the nuclear receptors in Daphnia pulex. *BMC Genomics*, *10*, 500. https://doi.org/10.1186/1471-2164-10-500.

Thummel, C. S. (2001). Molecular mechanisms of developmental timing in C. elegans and Drosophila. *Developmental Cell*, *1*(4), 453–465. https://doi.org/10.1016/S1534-5807(01)00060-0.

Ueda, H., Sonoda, S., Brown, J. L., Scott, M. P., & Wu, C. (1990). A sequence-specific DNA-binding protein that activates fushi tarazu segmentation gene expression. *Genes & Development*, *4*(4), 624–635.

Velarde, R. A., Robinson, G. E., & Fahrbach, S. E. (2006). Nuclear receptors of the honey bee: Annotation and expression in the adult brain. *Insect Molecular Biology*, *15*(5), 583–595. https://doi.org/10.1111/j.1365-2583.2006.00679.x.

Wakimoto, B. T., Turner, F. R., & Kaufman, T. C. (1984). Defects in embryogenesis in mutants associated with the Antennapedia gene complex of Drosophila melanogaster. *Developmental Biology*, *102*, 147–172.

Wang, L., Rajan, H., Pitman, J. L., McKeown, M., & Tsai, C.-C. (2006). Histone deacetylase-associating Atrophin proteins are nuclear receptor corepressors. *Genes & Development*, *20*(5), 525–530. https://doi.org/10.1101/gad.1393506.

Weisbrod, A., Cohen, M., & Chipman, A. D. (2013). Evolution of the insect terminal patterning system—Insights from the milkweed bug, Oncopeltus fasciatus. *Developmental Biology*, *380*(1), 125–131. https://doi.org/10.1016/j.ydbio.2013.04.030.

Wilk, R., Hu, J., & Krause, H. M. (2013). Spatial profiling of nuclear receptor transcription patterns over the course of drosophila development. *G3 (Bethesda, MD)*, *3*(7), 1177–1189.

Wilson, M. J., & Dearden, P. K. (2009). Tailless patterning functions are conserved in the honeybee even in the absence of Torso signaling. *Developmental Biology*, *335*(1), 276–287. https://doi.org/10.1016/j.ydbio.2009.09.002.

Wisely, G. B., Miller, A. B., Davis, R. G., Thornquest, A. D., Johnson, R., Spitzer, T., … Williams, S. P. (2002). Hepatocyte nuclear factor 4 is a transcription factor that constitutively binds fatty acids. *Structure (London, England: 1993)*, *10*(9), 1225–1234.

Xu, H. E., Stanley, T. B., Montana, V. G., Lambert, M. H., Shearer, B. G., Cobb, J. E., … Stimmel, J. B. (2002). Structural basis for antagonist-mediated recruitment of nuclear co-repressors by PPARα. *Nature*, *415*(6873), 813–817. https://doi.org/10.1038/415813a.

Xu, J., Tan, A., & Palli, S. R. (2010). The function of nuclear receptors in regulation of female reproduction and embryogenesis in the red flour beetle, Tribolium castaneum. *Journal of Insect Physiology*, *56*(10), 1471–1480. https://doi.org/10.1016/j.jinsphys.2010.04.004.

Yamada, M., Murata, T., Hirose, S., Lavorgna, G., Suzuki, E., & Ueda, H. (2000). Temporally restricted expression of transcription factor betaFTZ-F1: Significance for embryogenesis, molting and metamorphosis in Drosophila melanogaster. *Development*, *127*(23), 5083–5092.

Yao, T.-P., Segraves, W. A., Oro, A. E., McKeown, M., & Evans, R. M. (1992). Drosophila ultraspiracle modulates ecdysone receptor function via heterodimer formation. *Cell*, *71*(1), 63–72. https://doi.org/10.1016/0092-8674(92)90266-F.

Yoo, J., Ko, S., Kim, H., Sampson, H., Yun, J.-H., Choe, K.-M., … Lee, W. (2011). Crystal structure of Fushi tarazu factor 1 ligand binding domain/Fushi tarazu peptide complex identifies new class of nuclear receptors. *The Journal of Biological Chemistry*, *286*(36), 31225–31231. https://doi.org/10.1074/jbc.M111.252916.

Younossi-Hartenstein, A., Green, P., Liaw, G. J., Rudolph, K., Lengyel, J., & Hartenstein, V. (1997). Control of early neurogenesis of the Drosophila brain by the head gap genes tll, otd, ems, and btd. *Developmental Biology*, *182*(2), 270–283. https://doi.org/10.1006/dbio.1996.8475.

Younossi-Hartenstein, A., Nassif, C., Green, P., & Hartenstein, V. (1996). Early neurogenesis of the Drosophila brain. *The Journal of Comparative Neurology*, *370*(3), 313–329. http://dx.doi.org/10.1002/(SICI)1096-9861(19960701)370:3<313::AID-CNE3>3.0.CO;2-7.

Yu, Y., Li, W., Su, K., Yussa, M., Han, W., Perrimon, N., & Pick, L. (1997). The nuclear hormone receptor Ftz-F1 is a cofactor for the Drosophila homeodomain protein Ftz. *Nature*, *385*(6616), 552–555. https://doi.org/10.1038/385552a0.

Yussa, M., Löhr, U., Su, K., & Pick, L. (2001). The nuclear receptor Ftz-F1 and homeodomain protein Ftz interact through evolutionarily conserved protein domains. *Mechanisms of Development*, *107*(1–2), 39–53.

Zhi, X., Zhou, X. E., He, Y., Searose-Xu, K., Zhang, C.-L., Tsai, C.-C., … Xu, H. E. (2015). Structural basis for corepressor assembly by the orphan nuclear receptor TLX. *Genes & Development*, *29*(4), 440–450. https://doi.org/10.1101/gad.254904.114.

Zhong, W., Sladek, F. M., & Darnell, J. E. (1993). The expression pattern of a Drosophila homolog to the mouse transcription factor HNF-4 suggests a determinative role in gut formation. *The EMBO Journal*, *12*(2), 537–544.

CHAPTER THREE

Nuclear Receptors in Skeletal Homeostasis

Hao Zuo, Yihong Wan[1]
The University of Texas Southwestern Medical Center, Dallas, TX, United States
[1]Corresponding author: e-mail address: yihong.wan@utsouthwestern.edu

Contents

Abstract

Nuclear receptors are a family of transcription factors that can be activated by lipophilic ligands. They are fundamental regulators of development, reproduction, and energy metabolism. In bone, nuclear receptors enable bone cells, including osteoblasts, osteoclasts, and osteocytes, to sense their dynamic microenvironment and maintain normal bone development and remodeling. Our views of the molecular mechanisms in this process have advanced greatly in the past decade. Drugs targeting nuclear receptors are widely used in the clinic for treating patients with bone disorders such as osteoporosis by modulating bone formation and resorption rates. Deficiency in the natural ligands of certain nuclear receptors can cause bone loss; for example, estrogen loss in postmenopausal women leads to osteoporosis and increases bone fracture risk. In contrast, excessive ligands of other nuclear receptors, such as glucocorticoids, can also be detrimental to bone health. Nonetheless, the ligand-induced osteoprotective effects of many other nuclear receptors, e.g., vitamin D receptor, are still in debate and require further characterizations. This review summarizes previous studies on the roles of

Current Topics in Developmental Biology, Volume 125
ISSN 0070-2153
http://dx.doi.org/10.1016/bs.ctdb.2017.01.002

nuclear receptors in bone homeostasis and incorporates the most recent findings. The advancement of our understanding in this field will help researchers improve the applications of agonists, antagonists, and selective modulators of nuclear receptors for therapeutic purposes; in particular, determining optimal pharmacological drug doses, preventing side effects, and designing new drugs that are more potent and specific.

1. INTRODUCTION

1.1 Bone and Bone Cells

Bone is a multifunctional organ that affects many aspects throughout the life. During embryonic and juvenile stages, bone development determines body size and provides support for skeletal muscle and organs. After bone growth is completed, bone stays in a homeostatic state, which is controlled by dynamic bone formation and bone resorption (Rodan, 1998). Bone also plays essential roles in endocrine functions, blood cell production, and mineral metabolism. Bone mainly consists of solid bone tissues and bone marrow. The solid bone tissues include cortical bone and cancellous bone (also called trabecular bone); both contain three major bone cell types, osteoblast, osteocyte, and osteoclast, which cooperate to maintain bone homeostasis.

Osteoblasts, the major bone-forming cells, produce type I collagen (Col1), osteocalcin, osteopontin (OPN), and alkaline phosphatase (ALP) that contribute to bone matrix formation and mineralization. Osteoblasts are derived from mesenchymal stem cells (MSCs). MSCs first go through a preosteoblast or immature osteoblast stage, and then a mature osteoblast stage, and at last become bone-lining cells and osteocytes (Fig. 1). Experimentally, differentiation stage-specific marker gene expression enables the detection of altered osteoblastogenesis and the development of conditional osteoblast stage-specific gene knockout tools. Paired-related homeobox gene 1 (Prx1) and transcription factor Sox9 are osteoblast progenitor markers (Akiyama et al., 2005; Nohno et al., 1993). Preosteoblasts express transcription factors Runt-related transcription factor 2 (Runx2) and osterix (Osx) (Nakashima et al., 2002; Otto et al., 1997), while Col1 and osteocalcin (Ocn) serve as protein makers for mature osteoblasts.

Osteocytes are derived from mature osteoblasts that are embedded into bone matrix. They sense mechanical loading and control bone adaptation to it. Osteocytes also have endocrine functions in regulating bone homeostasis by secreting proteins such as fibroblast growth factor 23 (FGF-23),

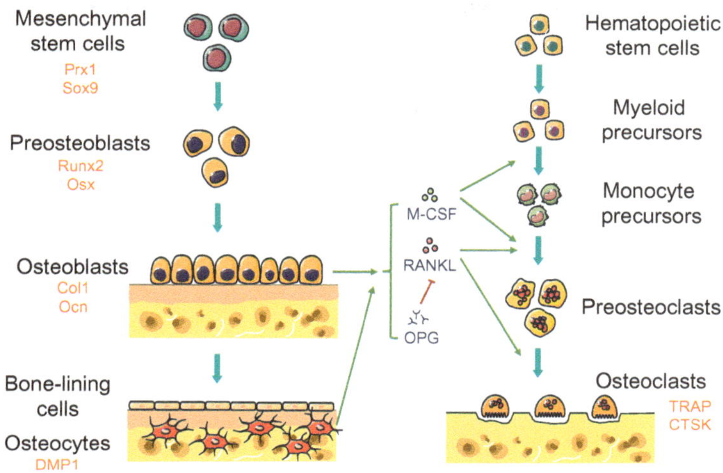

Fig. 1 Osteoblasts, osteocytes, and osteoclasts in bone. Bone-forming osteoblasts are differentiated from mesenchymal stem cells. Osteoblasts are then further matured into bone-lining cells and osteocytes that are embedded in bone matrix. Osteoblasts and osteocytes express M-CSF, RANKL, and OPG to control the differentiation of bone-resorbing osteoclasts from hematopoietic stem cells. Osteoclastogenesis from myeloid precursors is driven by M-CSF and RANKL, but counteracted by OPG, a RANKL decoy receptor. The orchestrated performance by osteoblasts, osteocytes, and osteoclasts maintains skeletal homeostasis. Cell type-specific protein markers are shown in *orange*.

Dickkopf-related protein 1 (DKK1), and sclerostin (SOST). Dentin matrix protein 1 (DMP1) is commonly used as an osteocyte marker gene (Toyosawa et al., 2001).

Osteoclasts are the only bone-resorbing cells. These multinucleated cells are differentiated from monocyte/macrophage precursors in the myeloid lineage, which originate from hematopoietic stem cells (Fig. 1). The process of osteoclast differentiation—osteoclastogenesis—is under the control of several cytokines produced by osteoblasts, osteocytes, and stromal cells. Two major cytokines, macrophage colony-stimulating factor and receptor activator of nuclear factor-κB ligand (RANKL), activate their receptors M-CSFR and RANK, respectively, to stimulate osteoclast differentiation (Boyle, Simonet, & Lacey, 2003; Fig. 1). Osteoprotegerin (OPG) is a RANKL decoy receptor that impedes osteoclast differentiation (Khosla, 2001; Fig. 1). Activated RANKL/RANK signaling upregulates nuclear factor of activated T cells, cytoplasmic 1 (NFATc1), a master transcription factor for osteoclast differentiation (Takayanagi, 2007), as well as enzymes for bone matrix degradation including tartrate-resistant acid phosphatase (TRAP) and cathepsin K (CTSK).

During bone development and remodeling, the differentiation and activity of osteoblasts, osteoclasts, and osteocytes are tightly controlled to maintain normal bone morphology. Bone cells sense environmental factors by expressing their respective receptors, among which nuclear receptor (NR) superfamily members play critical roles. Dysfunction of one or more of these bone cell types, resulted from genetic alterations of NR genes or environmental changes of NR ligands, may cause bone disorders such as osteoporosis, osteopetrosis, bone fracture, and rickets.

1.2 Nuclear Receptors

NRs are a superfamily of transcription factors that share similar structures containing a DNA-binding domain, a ligand-binding domain (LBD), a hinge region, and an N-terminal domain (Mangelsdorf et al., 1995). NR superfamily consists of 48 members in human (Evans & Mangelsdorf, 2014; Fig. 2). They are divided into steroid receptors, nonsteroid receptors, and orphan NRs that have no known endogenous ligand. NRs form complexes with coregulators, including corepressors and coactivators, and exhibit efficient transcriptional regulation by coregulator exchange upon ligand binding or dissociation (Glass & Rosenfeld, 2000; McKenna, Lanz, & O'Malley, 1999).

The important roles of NRs in many biological processes make them ideal drug targets for numerous metabolic diseases such as diabetes and osteoporosis. Thus, a number of NR-specific agonists, antagonists, and modulators have been designed to control aberrant NR functions in patients (Moore, Collins, & Pearce, 2006). A better understanding of the fundamental mechanisms of NR function and regulation will eventually benefit patients by the discoveries of new drugs and the improvement of currently utilized drugs. This review summarizes the current understanding of the roles of several NRs in regulating bone health, which may provide new insights for disease therapy to scientists in this field.

2. NRs AND BONE HOMEOSTASIS

2.1 Estrogen Receptor

2.1.1 Estrogens and Bone Health

The steroid hormone estrogens play important roles in regulating various physiological processes, such as reproductive organ development, energy

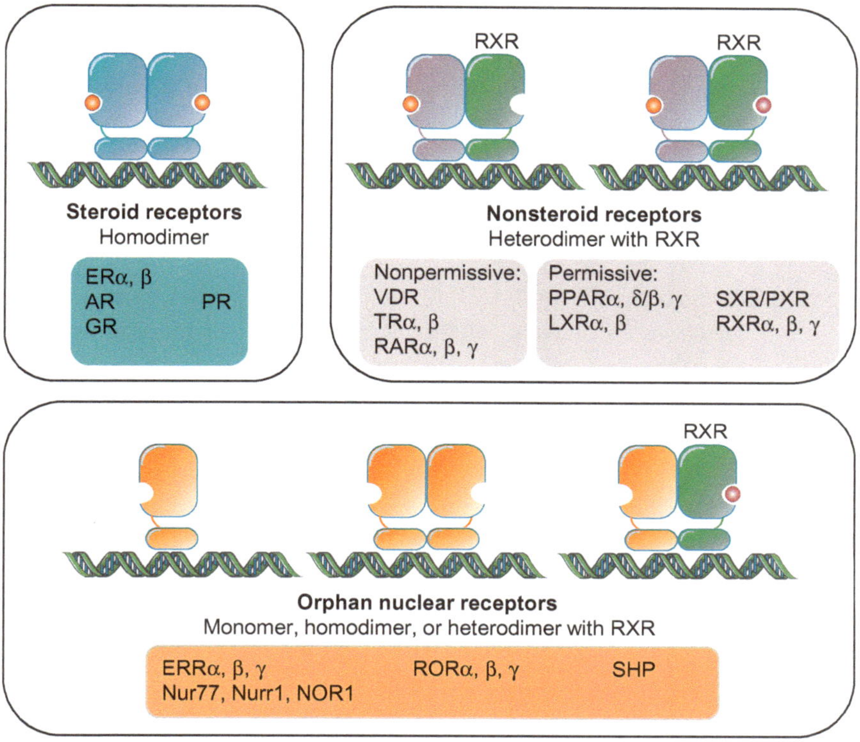

Fig. 2 Nuclear receptor classification and their members mentioned in this chapter. Nuclear receptor superfamily consists of 48 members. They are divided into three subfamilies, steroid receptors, nonsteroid receptors, and orphan receptors. Upon ligand binding, steroid receptors form homodimers to exert their transcriptional regulations. The members in this subfamily discussed here are estrogen receptor (ER) α and β, androgen receptor (AR), glucocorticoid receptor (GR), and progesterone receptor (PR). Nonsteroid receptors form heterodimers with retinoid X receptors (RXRs). Vitamin D receptor (VDR), thyroid hormone receptor (TR) α and β, and retinoic acid receptor (RAR) α, β, and γ are classified as nonpermissive receptors because their heterodimers are only activated by ligands of RXR partners. Other nonsteroid receptors, including peroxisome proliferator-activated receptor (PPAR) α, δ/β, and γ, liver X receptor (LXR) α and β, and steroid and xenobiotic receptor (SXR) (murine ortholog pregnane X receptor (PXR)), are permissive receptors, and their heterodimers can be activated by ligands of either RXR or RXR partners. Orphan nuclear receptors have no known endogenous ligand when initially identified. They regulate gene transcription as monomers, homodimers, or heterodimers with RXR. The orphan receptors mentioned in this chapter include estrogen-related receptor (ERR) α, β, and γ, nuclear receptor 77 (Nur77) (alternative name, nerve growth factor-induced gene B (NGFI-B)), nuclear receptor-related factor 1 (Nurr1), neuron-derived orphan receptor 1 (NOR1), RAR-related orphan receptor (ROR) α, β, and γ, and short heterodimeric partner (SHP).

metabolism, and bone homeostasis. Impaired estrogen levels in postmenopausal females or aged males result in decreased bone formation and increased bone resorption, leading to osteopenia and osteoporosis (Falahati-Nini et al., 2000). Bone loss observed in estrogen-deficient women and rodent models, such as ovariectomized (OVX) rats and mice (Thompson, Simmons, Pirie, & Ke, 1995), can be significantly diminished by estrogen treatment (Bain, Bailey, Celino, Lantry, & Edwards, 1993; Lindsay et al., 1976).

NRs for estrogen, ERα and ERβ, are the major mediators of estrogen actions (Barros & Gustafsson, 2011). Upon estrogen binding, nuclear ERα and ERβ form homodimers, associate with estrogen response elements (EREs), and then activate or repress target gene transcriptions (Cowley, Hoare, Mosselman, & Parker, 1997). In addition to the classical transcriptional regulations, estrogen receptors (ERs) also have nonclassical functions. They can modulate the activities of other transcription factors such as the activator protein-1 (AP-1) and Sp1 (Marino, Galluzzo, & Ascenzi, 2006). ERs on plasma membrane can also rapidly induce multiple signaling pathways upon ligand activation, such as ERK (Acconcia et al., 2005). Interestingly, although NRs have been viewed as the major ERs, a G protein-coupled estrogen receptor, also known as GPR30, has been reported to contribute to estrogen actions (Martensson et al., 2009).

2.1.2 ER Knockout Models

ERα and ERβ are expressed in osteoblasts, osteoclasts, osteocytes, as well as chondrocytes (Bord, Horner, Beavan, & Compston, 2001), suggesting that ER activation may regulate the functions of multiple bone and cartilage cells. Several animal models have been established to study the roles of ERs in bone cells and skeletal homeostasis. Global knockout of ERα led to increased trabecular bone mineral density (BMD), decreased cortical BMD, and lower bone turnover rate in both male and female mice (Sims et al., 2002). Complete deletion of ERβ resulted in increased trabecular BMD, unchanged cortical BMD, and attenuated bone resorption in female mice, while bone morphology was unaltered in male mice (Sims et al., 2002). In contrast, decreased trabecular and cortical BMD as well as elevated bone turnover rate were observed in estrogen-deficient women after menopause or ovariectomy, in aged men, and in men with a mutated ERα (Riggs, Khosla, & Melton, 1998; Smith et al., 1994). The reason for the differences between bone phenotypes of ER knockout mice and estrogen/ER-deficient human is still unclear.

Osteoblasts in a specific differentiation stage express several protein markers, e.g., Prx1 in MSCs, osterix1 (Osx1) in early differentiating osteoblasts, Ocn and collagen 1a1 (Col1a1) in mature osteoblasts, and Dmp1 in osteocytes—mature osteoblasts that migrate into bone matrix after finishing mineralization (Fig. 1). To determine the functions of an NR in osteoblast differentiation, maturation, mineralization, and mechanical response, Cre–Lox recombination technology is widely utilized. Specific deletion of an NR in a certain bone cell type at a specific developmental stage can be achieved via the expression of Cre recombinase driven by the gene promoters of these specific protein markers in engineered NR flox mice. Specific deletion of ERα in osteoblast progenitors with Prx1-Cre reduced cortical bone thickness and unchanged cancellous bone mass (Almeida et al., 2013). ERα deletion in immature osteoblasts with Osx1-Cre reduced the thickness of both cortical and trabecular bone (Almeida et al., 2013). ERα deletion in mature osteoblasts and osteocytes with Col1a1-Cre did not alter bone mass in either male or female mice (Almeida et al., 2013). However, osteocalcin-Cre-mediated ERα deletion in mature osteoblast and osteocytes reduced both trabecular and cortical bone volume in female mice (Maatta et al., 2013; Melville et al., 2014) and reduced trabecular bone volume in male (Maatta et al., 2013). Dmp1-Cre-induced osteocyte-specific ERα knockout showed decreased trabecular bone volume in male mice (Windahl, Borjesson, et al., 2013). Summarizing these observations, ERα in the osteoblast linage generally plays protective roles in bone mass accrual and maintenance.

In contrast, ERβ in osteoblasts may have an opposite role to ERα. Increased trabecular bone volume and unchanged cortical bone were observed in mice with Prx1-Cre-mediated ERβ deletion in osteoblast progenitors (Nicks et al., 2016). These opposite effects from loss of ERβ vs loss of ERα may be due to the differences in their transcriptional regulation and target genes, as well as their different affinities for transcription coregulators (An, Tzagarakis-Foster, Scharschmidt, Lomri, & Leitman, 2001; Katzenellenbogen & Katzenellenbogen, 2000). These findings suggest that selective ERα agonists may be better therapeutic choices than dual agonists.

ERα knockout in osteoclast precursors using LysM-Cre led to increased osteoclast number, decreased cancellous bone mass, and unchanged cortical bone mass (Martin-Millan et al., 2010). ERα deletion in mature osteoclasts by Ctsk-Cre resulted in decreased trabecular bone volume and unchanged cortical bone in female but not male mice (Nakamura et al., 2007). These findings indicate that ERα in osteoclasts suppresses osteoclast function in trabecular bone but not cortical bone.

In summary, ERα and ERβ have little effect on bone development during embryonic stage and in young mice (Sims et al., 2002) but modulate bone homeostasis in adult and old mice through both direct and indirect pathways. ERα defect in the osteoblast/osteocyte linage reduces bone mass of trabecular bone and/or cortical bone. This reduction is mostly due to attenuated osteoblast differentiation followed by decreased bone formation. ERα defect in the osteoclast linage dramatically enhances osteoclastogenesis and thus accelerates bone resorption. More evidence is required for drawing a conclusion on the direct effect of ERβ on bone cells. Besides ER in bone cells, ER in other organs also mediates the indirect regulation on bone development. Global ERα or ERα/β deletion significantly elevated circulating levels of estradiol and testosterone (Sims et al., 2002). On the other hand, estrogen deficiency in female reduced intestinal calcium absorption, increased renal calcium excretion, and increased the secretion of parathyroid hormone (PTH) (Riggs et al., 1998), which enhances calcium release from bone. These effects of ER on sex hormone production, calcium homeostasis, and PTH secretion are mostly not mediated through bone cells but through ovary, liver, parathyroid gland, and other organs. All these findings may explain the differences of the bone phenotypes between global and conditional knockout mice, and between knockout mouse models and human.

2.1.3 Mechanisms

Several mechanisms have been proposed for osteoblast regulation by estrogen and ERs. It was reported that estrogen treatment enhanced osteoblast differentiation from mouse bone marrow MSCs through upregulating osteoblastic makers such as ALP and osteocalcin (Qu et al., 1998). It was also shown that ERα stimulated osteoblast differentiation by directly binding to and activating the transcription factor Runx2 (McCarthy, Chang, Liu, & Centrella, 2003).

Multiple mechanisms may also account for osteoclast regulation by estrogen and ERs. First, estrogen treatment decreased the RANKL/OPG ratio in human osteoblasts (Bord, Ireland, Beavan, & Compston, 2003; Hofbauer et al., 1999), which could lead to less osteoclast differentiation and bone resorption. Second, osteoclast-specific ERα deletion inhibited osteoclast apoptosis by abolishing estrogen induction of Fas ligand (FasL) in osteoclasts, leading to increased osteoclast number (Nakamura et al., 2007). Third, FasL was reported to be a direct ERα target gene in osteoblast. Estrogen-induced

FasL expression in osteoblasts was sufficient to increase osteoclast apoptosis (Krum et al., 2008). Fourth, bone loss in osteoclast-specific ERα knockout mice was shown to be rescued by osteoclast-specific deletion of hypoxia-inducible factor 1 alpha (HIF1α), a proosteoclastogenic factor that was destabilized by estrogen but became stabilized upon estrogen deficiency (Miyauchi et al., 2013). In summary, evidence so far indicates that estrogen inhibits osteoclast differentiation by lowering RANKL/OPG ratio and destabilizing osteoclastic HIF1α and at the same time also inhibits osteoclast survival by inducing FasL and facilitating osteoclast apoptosis.

Ligand-activated nuclear ERs may regulate bone cell differentiation and activity by directly stimulating or repressing the transcription of target genes. A set of 3665 ERα-binding sites were found in human genome (Carroll et al., 2006), indicating a genome-wide regulation of gene transcription by estrogen, although different cell types may exhibit distinct ER targets. Future studies are required to systematically identify ERE and ER-binding sites, as well as functionally characterize ER target genes in each type of bone cells.

Besides direct promoter binding and target gene transcription regulation, ERs also physically interact with other transcription factors and modulate their transcriptional activities. For example, it was shown that the association of ERα with NF-κB and C/EBPβ repressed the transcription of interleukin-6 (IL-6) (Galien & Garcia, 1997; Stein & Yang, 1995), which promotes bone loss (Poli et al., 1994). Future studies are also needed to systemically identify ER-binding partners in bone cells, determine whether the binding is DNA dependent or DNA independent, and assess whether the binding is functionally significant.

It has been shown that ERs located on plasma membrane have similar ligand-binding affinities as nuclear ERs, and membrane ERs can rapidly transduce estrogen signaling via G proteins, ERK and c-Jun (Razandi, Pedram, Greene, & Levin, 1999). A series of studies have begun to determine the functional significance of membrane ERs in bone and delineate the mechanisms of gene transcription regulation via kinase-initiated regulation by membrane ERα vs classical genotropic regulation by nuclear ERα (Almeida, Han, O'Brien, Kousteni, & Manolagas, 2006; Almeida et al., 2010; Bartell et al., 2013; Kousteni et al., 2007, 2001, 2002, 2003).

ERs can also act in a ligand-independent manner during mechanical strain-induced bone response (Windahl, Saxon, et al., 2013). ERα mediates strain-induced osteoblast proliferation (Galea et al., 2013) and enhances loading-induced cortical bone increase in female mice (Saxon, Galea,

Meakin, Price, & Lanyon, 2012). ERβ mediates strain-induced Sost down-regulation (Galea et al., 2013), leading to enhanced Wnt/β-catenin signaling and osteoblast differentiation (Sharifi, Ereifej, & Lewiecki, 2015). However, ERβ was also shown to impair loading-induced cortical bone increase in both male and female mice (Saxon et al., 2012).

It should be noted that G protein-coupled estrogen receptor was reported to express in osteoblasts, osteocytes, and osteoclasts (Heino, Chagin, & Savendahl, 2008). Its deletion led to reduced bone growth (Martensson et al., 2009). Thus, G protein-coupled estrogen receptor should be considered during the design of estrogen-based drugs targeting nuclear ERs, as well as during the investigations of nuclear ER functions in bone cells upon estrogen treatment or depletion.

2.2 Androgen Receptor

Androgen is another steroid hormone that affects reproductive system, skeleton development, and many other aspects in both male and female (Chang, Lee, Wang, Yeh, & Chang, 2013). Upon androgen binding, androgen receptors (ARs) form homodimers and bind to a specific DNA sequence to regulate gene transcription. AR is expressed in osteoblasts (Colvard et al., 1989), osteoclasts (Mizuno et al., 1994), and osteocytes (Abu, Horner, Kusec, Triffitt, & Compston, 1997). Global knockout of AR in male mice reduced trabecular and cortical bone mass and upregulated RANKL expression in osteoblasts (Kawano et al., 2003). However, female mice with AR deficiency showed no bone abnormality (Kawano et al., 2003). Deletion of AR from osteoblast progenitors by Prx1-Cre in male mice led to decreased trabecular number, unchanged cortical bone, and increased osteoclast number (Ucer et al., 2015). Specific AR knockout in mature osteoblasts and osteocytes with osteocalcin-Cre triggered a reduction of trabecular bone volume and cortical bone thickness in young mice but not in adult mice (Chiang et al., 2009). Deletion of AR in mineralizing osteoblasts with Col2.3-Cre resulted in decreased trabecular bone volume and number, while no change was observed in cortical bone (Notini et al., 2007). Osteocyte-specific deletion of AR using Dmp1-Cre led to decreased trabecular bone mass and unchanged cortical bone in male mice (Sinnesael et al., 2012). Therefore, AR in all stages of the osteoblast/osteocyte differentiation prevents loss of trabecular bone and may have no effect on cortical bone mass. The reduced cortical bone thickness by osteocalcin-Cre-mediated AR deletion in young mice may need further investigations. Specific AR deletion in osteoclast precursors by LysM-Cre or in mature osteoclasts by Ctsk-Cre had no effect on bone mass or osteoclast number (Sinnesael et al., 2015; Ucer et al., 2015). Taking together, these results indicate that AR mainly acts in the osteoblast lineage to protect bone. Future studies are needed to elucidate AR direct target genes and downstream signaling events in osteoblasts, as well as explore whether AR beyond bone cells may indirectly modulate skeletal homeostasis via neuroendocrine mechanisms.

2.3 Glucocorticoid Receptor

Glucocorticoid receptor (GR) mediates actions of glucocorticoids (GCs) on gene transcriptional regulation. GCs are widely used in treating diseases such as autoimmune disorders. However, GC-induced osteoporosis is a severe side effect for these patients (den Uyl, Bultink, & Lems, 2011). High level of GCs impairs osteoblast differentiation and induces osteoblast apoptosis (Ishida & Heersche, 1998; Weinstein, Jilka, Parfitt, & Manolagas, 1998). Interestingly, a recent study shows that monomeric GR, rather than dimeric GR, is responsible for GC suppression of osteoblast differentiation and bone formation (Rauch et al., 2010). GR deletion in osteoblasts by Runx2-Cre abolishes the ability of GCs to repress bone formation because GCs can no longer induce osteoblast apoptosis or inhibit osteoblast proliferation and differentiation (Rauch et al., 2010). In contrast, mice carrying a mutation that only disrupts GR dimerization (GR^{dim} mice) are still sensitive to GC repression of bone formation. The antiosteoblastic effects of GCs may involve a suppression of cytokines, such as interleukin-11, via an interaction of the monomeric GR with AP-1, but not NF-κB (Rauch et al., 2010). GR deletion in osteoclast precursors by LysM-Cre diminishes the inhibitory effect of GCs on bone formation as well as osteoclastic bone resorption (Kim et al., 2006). These studies provide new mechanistic insights to how GC causes bone loss and how GR regulates bone.

2.4 Peroxisome Proliferator-Activated Receptor

Peroxisome proliferator-activated receptors (PPARs) are activated by various natural ligands such as unsaturated fatty acids and derivatives (Ahmadian et al., 2013; Evans & Mangelsdorf, 2014). They are key regulators of glucose and lipid metabolism; as such, their synthetic ligands such as rosiglitazone in the thiazolidinedione (TZD) class of drugs are widely used for the treatment of insulin resistance and type 2 diabetes (Ahmadian et al., 2013; Evans &

Mangelsdorf, 2014). Three PPAR isoforms have been discovered in human and mouse: PPARα, PPARδ (named PPARβ in mouse), and PPARγ. PPARα and PPARδ/β participate in energy combustion, while PPARγ facilitates energy storage. All three members form heterodimers with retinoid X receptor (RXR) and bind to peroxisome proliferator response element (PPRE) in target DNA. PPRE consists of a direct repeat AGGTCA that are separated by one base pair (Michalik et al., 2006). A number of coregulators participate in the transcriptional regulation by PPAR/RXR (Viswakarma et al., 2010), for example, PPARγ coactivator-1α (PGC-1α) (Vega, Huss, & Kelly, 2000). Ligand binding to PPAR/RXR heterodimer induces the exchange of corepressors for coactivators followed by transcriptional activation of target genes (Viswakarma et al., 2010).

2.4.1 Peroxisome Proliferator-Activated Receptor α

PPARα expression was found in osteoblasts, osteoclasts, and chondrocytes (Chan et al., 2007; Giaginis, Tsantili-Kakoulidou, & Theocharis, 2007). A global PPARα knockout mouse model was generated (Lee et al., 1995), but no significant change in bone was observed (Wu et al., 2000). However, treatment with PPARα agonists fenofibrate and Wyeth 14643 in rats increased whole body and femoral BMD and protected OVX rats from osteoporosis (Stunes et al., 2011). In vitro treatment with fenofibrate not only increased osteoblast differentiation (Still, Grabowski, Mackie, Perry, & Bishop, 2008) but also decreased osteoclast number (Chan et al., 2007). These findings suggest that activated PPARα protects bone through promoting bone formation while suppressing bone resorption.

2.4.2 Peroxisome Proliferator-Activated Receptor δ/β

PPARδ/β expression was detected in osteoblasts, osteoclasts, and chondrocytes (Giaginis et al., 2007). PPARδ/β knockout mice had smaller body size than wild-type control mice (Peters et al., 2000). In a mouse model with PPARδ/β deleted in all tissues except placenta by Sox2-Cre, reduced trabecular bone mass was observed (Scholtysek et al., 2013). At the same time, lower Wnt signaling activity, less serum OPG, and more osteoclasts were also detected in this mouse model (Scholtysek et al., 2013), suggesting that PPARδ/β deletion upregulates osteoblast-mediated osteoclastogenesis. Administration of PPARδ/β agonist GW501516 prevented bone loss in OVX mice (Scholtysek et al., 2013). Therefore, similar to PPARα, PPARδ/β also has a protective role in bone homeostasis.

2.4.3 Peroxisome Proliferator-Activated Receptor γ

There are two PPARγ isoforms, PPARγ1 and PPARγ2, due to alternative splicing and promoter usage. Because of the critical roles of PPARγ in regulating energy metabolism, several PPARγ agonists, such as TZD family members rosiglitazone and pioglitazone, have been developed to treat type 2 diabetes (Mayerson et al., 2002; Miyazaki et al., 2002). However, side effects of bone loss and increased risk of fracture arise from long-term TZD treatment (Bilezikian et al., 2013; Billington, Grey, & Bolland, 2015; Jin, Li, & Wan, 2015; Schwartz et al., 2015; Soccio, Chen, & Lazar, 2014; Wan, 2010), indicating a negative impact of activated PPARγ on bone.

PPARγ expression was found in osteoblasts, osteoclasts, and chondrocytes (Giaginis et al., 2007). Although homozygous PPARγ deficiency is lethal, higher bone mass, enhanced bone formation, and increased osteoblast differentiation were observed in heterozygous PPARγ-deficient mice (Akune et al., 2004). Deletion of PPARγ in mesenchymal progenitor cells with Col3.6-Cre increased BMD, bone volume, trabecular bone number, and osteoblast number (Cao et al., 2015). Deletion of PPARγ in osteoblast by Osx-Cre also led to increased trabecular bone number, unchanged cortical bone, and increased osteoblast differentiation (Sun et al., 2013). PPARγ deletion in osteoclast progenitor cells and endothelial cells with Tie2-Cre increased bone mass due to a reduction in osteoclast differentiation and bone resorption (Wan, Chong, & Evans, 2007).

Pharmacologically, treatment with TZDs such as rosiglitazone increases bone resorption and decreases bone formation by enhancing osteoclastogenesis and inhibiting osteoblastogenesis, in both humans and mice (Bilezikian et al., 2013; Billington et al., 2015; Jin et al., 2015; Schwartz et al., 2015; Wan, 2010). Moreover, it has been shown that rosiglitazone treatment also increases circulating osteoclast precursors but decreases circulating osteoblast precursors in postmenopausal women with type 2 diabetes mellitus (Rubin et al., 2014).

PPARγ cooperates with transcription coregulators to modulate bone cell differentiation. In osteoblasts, without ligand binding, PPARγ2 was shown to interact with small leucine zipper protein (sLZIP), which acted as a transcriptional corepressor by enhancing the association of PPARγ2 and HDAC3 to compete with PGC-1α recruitment (Kim & Ko, 2014). As a result of PPARγ2 suppression by sLZIP, osteogenesis was increased, while adipogenesis was decreased (Kim & Ko, 2014). In osteoclasts, PPARγ coactivator PGC-1β is indispensable during PPARγ stimulation of osteoclastogenesis, as PGC-1β knockout mice showed complete resistance to bone loss after rosiglitazone treatment (Wei et al., 2010). These findings further deepen our understanding of how PPARγ suppresses osteoblast differentiation and enhances osteoclastogenesis to induce bone loss.

In addition, chondrocyte-specific PPARγ knockout in mice by Col2-Cre increased chondrocyte apoptosis and cartilage degradation, resulting in accelerated osteoarthritis (OA) (Vasheghani et al., 2015). PPARγ agonists showed antiinflammatory, antimatrix metalloprotease, and antiapoptotic effects in vitro (Fahmi, Martel-Pelletier, Pelletier, & Kapoor, 2011). In vivo treatment of PPARγ agonists decreased cartilage lesions in OA animals (Fahmi et al., 2011). These results suggest that activated PPARγ may be protective in the cartilage. Future studies are required to examine whether this is the case in human patients taking the TZD drugs.

2.5 Vitamin D Receptor

Vitamin D is a key regulator of serum calcium homeostasis and skeletal development. Vitamin D deficiency in children results in rickets. In human physiology, vitamin D_3 can be obtained through ultraviolet-driven biosynthesis or diet intake and can be converted into its active form of metabolite, $1\alpha,25(OH)_2D_3$. Upon $1\alpha,25(OH)_2D_3$ binding, vitamin D receptor (VDR) forms a heterodimer with RXR to activate target gene transcription through binding to vitamin D response elements (VDREs) that consist of two direct hexameric repeats separated by three base pairs (Heikkinen et al., 2011). In addition to the roles of vitamin D in maintaining serum calcium levels, VDR was detected in osteoblasts, osteocytes, and chondrocytes, but not in osteoclasts (Wang, Zhu, & DeLuca, 2014). Thus, vitamin D regulates bone homeostasis through both indirect effects on calcium and direct effects on bone cells.

The roles of VDR in bone have been investigated using several animal models. In global VDR knockout mice, femoral length was reduced; growth plate, osteoid width, and trabecular bone volume were increased, while osteoclast number was decreased (Masuyama et al., 2003; Yoshizawa et al., 1997). The abnormal bone phenotypes in these mice were mainly due to impaired intestinal calcium absorption (Van Cromphaut et al., 2001). A diet supplement of higher calcium over phosphorus ratio or intestine-specific reexpression of VDR rescued the bone defects in VDR knockout mice (Masuyama et al., 2003; Xue & Fleet, 2009). Under normal serum calcium levels, VDR regulation of bone homeostasis is mainly mediated by controlling intestinal calcium absorption, whereas VDR signaling in osteoblasts and osteocytes is secondary (Lieben et al., 2012). However, low serum calcium triggers calcium release from bone to ensure a normal serum calcium level (Lieben et al., 2012). In this situation, vitamin D accelerates bone turnover, induces osteopenia, and suppresses bone mineralization (Lieben et al., 2012).

The negative role of VDR signaling in bone cells to decrease bone mass has been confirmed in conditional knockout mice. Col1a1-Cre-mediated VDR deletion in mature osteoblasts was reported to increase bone mass by reducing bone resorption and RANKL expression without affecting bone formation (Yamamoto et al., 2013). VDR deletion in osteocytes by Dmp1-Cre had no effect on bone (Lieben et al., 2012). VDR deletion in chondrocytes did not alter growth plate development, but increased trabecular bone mass through downregulating RANKL expression in chondrocytes and reducing osteoclastogenesis (Masuyama et al., 2006). In contrast to the observations in conditional VDR knockout mice, VDR transgenic overexpression driven by osteocalcin promoter in mature osteoblasts significantly increased trabecular bone volume by reducing bone resorption and osteoclastogenesis in mice (Baldock et al., 2006; Gardiner et al., 2000; Lam et al., 2014).

Besides the genetically modified animal models, conflicting data were also reported for the pharmacological effects of vitamin D on bone cells. High dose of vitamin D inhibits osteoblast mineralization in vitro (Yamaguchi & Weitzmann, 2012). Discontinuation of high vitamin D diet leads to gains of BMD in osteoporosis patients (Adams & Lee, 1997). It has been shown that activated VDR in osteoblasts directly upregulates the expression of OPN (Lieben et al., 2012), which is a negative regulator of osteoblast differentiation (Huang et al., 2004), by binding to the VDRE in *opn* promoter (Meyer, Goetsch, & Pike, 2010). It has also been reported that treatment with $1\alpha,25(OH)_2D_3$ downregulates Runx2 in osteoblasts to inhibit osteoblastogenesis and instead promote adipogenesis (Kim et al., 2016). Vitamin D also promotes osteoblast- and chondrocyte-driven osteoclast differentiation by increasing RANKL production (Lee, Kalinowski, Jastrzebski, & Lorenzo, 2002; Masuyama et al., 2006).

Nonetheless, opposite results were reported in other studies. Vitamin D-activated VDR in a human osteoblast cell line increased ALP activity and osteocalcin expression, as well as enhanced osteoblast differentiation and mineralization (van Driel et al., 2006). Vitamin D was also reported to inhibit osteoclast differentiation by suppressing expressions of NFATc1,

HIF1α, and c-Fos (Sakai et al., 2009; Sato et al., 2014; Takasu et al., 2006). Moreover, vitamin D was shown to block osteoclastogenesis by promoting migration of osteoclast precursors from bone to blood through suppressing S1PR2 expression in these cells (Kikuta et al., 2013).

The actions of vitamin D and VDR in maintaining serum calcium balance are very complicated involving their functions in the intestine, parathyroid gland, kidney, and bone. Thus, in order to elucidate tissue-specific and cell-specific VDR functions, multiple factors should be considered during in vivo and in vitro investigations, such as diet supplement, serum levels of calcium and vitamin D, tissue distribution of functional VDR, concentrations of calcium and vitamin D in cell culture medium, and cross talk among different organs and cell types.

2.6 Retinoid Acid Receptor and RXR

Retinoic acid (RA), including all-*trans*- and 9-*cis*-RA, belongs to vitamin A metabolites and fundamentally modulates vertebrate development especially organogenesis. RA functions as ligands for retinoid acid receptors (RARs) (RARα, RARβ, and RARγ) and RXRs (RXRα, RXRβ, and RXRγ). All-*trans*-RA only associates with RARs, while 9-*cis*-RA binds to both RARs and RXRs (Chambon, 1996). RARs and RXRs need to form RAR/RXR heterodimers to respond to RA and act as transcriptional factors. RAR/RXR heterodimers bind to specific RA response element and interact with corepressors in the absence of RA. RAR agonists are able to activate RAR/RXR complex by triggering exchange of corepressor with coactivator, while RXR agonists alone cannot (Germain, Iyer, Zechel, & Gronemeyer, 2002; Love et al., 2002).

In addition, 9-*cis*-RA induces the homodimer formation and transcriptional activation of RXRs (Zhang et al., 1992). Moreover, RXRs are also involved in the activation of other members of the NR superfamily by forming heterodimers with them. These heterodimers are divided into permissive and nonpermissive heterodimers. Activation of permissive heterodimers is induced by the ligand binding of either RXR or RXR partner (PPARs, LXRs, FXR, SXR/PXR, and CAR), while nonpermissive ones are only activated by the ligands of RXR partner (TRs, VDR, and RARs) (Evans & Mangelsdorf, 2014).

RARα and RARγ are expressed in osteoblasts and osteoclasts, whereas RARβ is detectable only after RA stimulation (Conaway et al., 2011; Inoue et al., 1996; Kindmark, Torma, Johansson, Ljunghall, & Melhus, 1993; van

Beek, Lowik, Karperien, & Papapoulos, 2006). Global RARα or RARβ knockout did not cause any significant change in bone (Li, Sucov, Lee, Evans, & Jaenisch, 1993; Lufkin et al., 1993; Mendelsohn et al., 1994). However, global RARγ knockout mice had reduced trabecular bone mass and increased osteoclastogenesis (Green et al., 2015), suggesting that RARγ is a negative regulator of bone resorption. In light of the potential functional redundancy within RARs and RXRs, several double global knockouts have been generated and characterized (Lohnes et al., 1994). In addition, chondrocyte-specific RAR double deletion by Col2a1-Cre demonstrated that $RAR\alpha^{-/-}$;$RAR\gamma^{-/-}$ mice and $RAR\beta^{-/-}$;$RAR\gamma^{-/-}$ mice showed growth plate defect (Williams et al., 2009), indicating a central role of RARγ in the cartilage.

High vitamin A diet and high serum retinol level were reported to be positively correlated with a higher risk of bone fracture in humans and animals (Feskanich, Singh, Willett, & Colditz, 2002; Lind et al., 2013; Melhus et al., 1998; Michaelsson, Lithell, Vessby, & Melhus, 2003; Whiting & Lemke, 1999). However, a number of reports showed no correlation between serum retinol level and fracture risk, or even a protective effect of retinol on bone (Ambrosini et al., 2014; Caire-Juvera, Ritenbaugh, Wactawski-Wende, Snetselaar, & Chen, 2009; Holvik et al., 2015; Ribaya-Mercado & Blumberg, 2007). Thus, the relationship between serum retinol level and fracture risk remains unclear.

Several in vitro studies demonstrated that RA negatively regulated osteoblast differentiation and mineralization (Iba, Chiba, Yamashita, Ishii, & Sawada, 2001; Lind et al., 2013; Ohishi et al., 1995). For example, RA treatment in vitro resulted in decreased mineralization and proliferation of osteoblasts accompanied by RAR-mediated reduced expressions of ALP, osteocalcin, Runx2, and Osx (Lind et al., 2013). However, RA treatment was also shown to enhance osteoblast differentiation (Gazit, Ebner, Kahn, & Derynck, 1993; Skillington, Choy, & Derynck, 2002; Song et al., 2005). Similar to the ambiguous roles of RA in osteoblast differentiation, RA treatment led to increased osteoclastogenesis in some reported studies (Conaway et al., 2011; Kneissel, Studer, Cortesi, & Susa, 2005; Saneshige et al., 1995) but was shown to decrease osteoclastogenesis in others (Balkan, Rodriguez-Gonzalez, Pang, Fernandez, & Troen, 2011; Hu, Lind, Sundqvist, Jacobson, & Melhus, 2010; Kneissel et al., 2005). RA-activated RARs were shown to negatively regulate longitudinal bone growth by suppressing growth plate chondrogenesis (Ballock et al., 1994; De Luca et al., 2000), but another group reported that RA increased chondrocyte maturation and mineralization (Iwamoto et al., 1993). To understand these differences and elucidate the bona fide roles of RARs/RXRs in bone will require more in depth in vitro and in vivo future studies.

2.7 Estrogen Receptor-Related Receptor

Estrogen receptor-related receptors (ERRs), including ERRα, ERRβ, and ERRγ, belong to the subfamily of orphan NR and share sequence homology with ERs. However, ERRs do not respond to ligands of ER such as estradiol (Horard & Vanacker, 2003). ERRα is the oldest orphan NR that has long been thought to not require an endogenous ligand, because it is active as long as the cells or tissues express its coactivators such as PGC-1α/β. Recently, our group has identified cholesterol as a potential endogenous ligand for ERRα that increased ERRα transcriptional activity (Wei et al., 2016). This discovery explains why ERRα is constitutively active because cholesterol is ubiquitous. Nonetheless, ERRα activities can be fine-tuned by altering cellular cholesterol levels. ERRα functions are diminished under cholesterol-depleted conditions such as following statin or bisphosphonate treatment, which are rescued by cholesterol add back (Wei et al., 2016). Moreover, ERRα is a key mediator of statin and bisphosphonate actions in bone, muscle, and macrophages (Wei et al., 2016). These findings deorphanize ERRα and provide new insights to the physiological regulation of ERRα activities, thereby revealing potential pharmacological strategies to control ERRα-related disorders.

In light of the multiple roles of ERRs in physiology and diseases, several synthetic small-molecule ligands have been developed. These ligands have been shown to bind to the LBD of ERRs (Wang et al., 2006) and function as agonists (GSK4716 and DY131) (Yu & Forman, 2005; Zuercher et al., 2005) or inverse agonists (4-hydroxytamoxifen and diethylstilbestrol) (Coward, Lee, Hull, & Lehmann, 2001; Tremblay et al., 2001). These findings not only provide useful chemical tools to probe ERR functions but also facilitate the development of ERR-targeting therapies.

Using mouse genetic models, ERRα was found to suppress osteoblastogenesis but promote osteoclastogenesis. Global ERRα knockout mice had increased femoral cancellous bone volume and density as well as enhanced osteoblast differentiation (Delhon et al., 2009). Female mice with specific deletion of ERRα in mature osteoblasts with Col1a1-Cre were completely resistant to bone loss induced by ovariectomy (Gallet et al., 2013). Global ERRα knockout mice also showed osteoclastogenesis defects with reduced mitochondrial biogenesis during osteoclast differentiation, leading to lower bone resorption and higher bone mass (Wei et al., 2016, 2010). Therefore, ERRα is a proosteoclastogenic NR and transcription factor. The identification of cholesterol as a natural ERRα agonist reveals that cholesterol-stimulated osteoclastogenesis, high cholesterol diet-induced bone loss, and bisphosphonate-mediated osteoprotection are ERRα dependent (Wei et al., 2016).

Homozygous ERRγ deletion is lethal and heterozygous ERRγ deletion elevates bone mass in male mice by increasing trabecular number, trabecular thickness, and osteoblast number (Cardelli & Aubin, 2014). This suggests that like ERRα, ERRγ also exerts negative effects on bone by suppressing bone formation. Homozygous ERRβ deletion was also lethal due to its high expression in placenta (Luo et al., 1997). Sox2-Cre-mediated deletion of ERRβ in all tissues except placenta altered body weight and physical activity (Byerly, Swanson, Wong, & Blackshaw, 2013), but no bone phenotype has been reported in this animal model. Future studies using conditional ERRγ/β knockout mice will further delineate their specific functions in each bone cell type and potentially also in nonbone cell types during skeletal maintenance.

2.8 NR4A Orphan NRs

Nuclear receptor subfamily 4 group A (NR4A) is a group of orphan NRs consisting of three members: Nur77 (NR4A1, NGFI-B), Nurr1 (NR4A2), and NOR1 (NR4A3). They exert their transcriptional activities as monomers or homodimers (Philips et al., 1997; Wilson, Fahrner, Johnston, & Milbrandt, 1991). In addition, Nur77 and Nurr1, but not NOR1, are able to form heterodimers with RXR and promote 9-*cis*-RA-induced RXR activation (Perlmann & Jansson, 1995; Zetterstrom, Solomin, Mitsiadis, Olson, & Perlmann, 1996). In the protein structures of Nur77 and Nurr1, there is no classical ligand-binding pocket as in other NRs (Flaig, Greschik, Peluso-Iltis, & Moras, 2005; Wang et al., 2003). However, cytosporone B (Zhan et al., 2008) and ethyl[2,3,4-trimethoxy-6-(1-octanoyl)phenyl] acetate (Liu et al., 2010) have been reported as ligands for Nur77. Unsaturated fatty acids, such as docosahexaenoic acid, have also been reported as ligands for both Nur77 and Nurr1 (de Vera et al., 2016; Vinayavekhin & Saghatelian, 2011). Although careful studies, for example, using knockout mice, are required in the future to examine the specificity of these potential NR4A ligands, these interesting findings provide the opportunities to deorphanize NR4A receptors and develop therapeutic drugs targeting NR4A in associated diseases.

Nur77, Nurr1, and NOR1 are expressed in osteoblasts (Lammi, Huppunen, & Aarnisalo, 2004; Pirih, Nervina, Pham, Aghaloo, & Tetradis, 2003; Tetradis, Bezouglaia, Tsingotjidou, & Vila, 2001). Their expression is stimulated by PTH and fibroblast growth factor 8b (FGF-8b) (Lammi & Aarnisalo, 2008; Pirih et al., 2003; Tetradis, Bezouglaia, & Tsingotjidou, 2001; Tetradis, Bezouglaia, Tsingotjidou, & Vila, 2001). In vitro Nurr1 knockdown in osteoblasts downregulated the expressions of osteocalcin and Col1a1, as well as ALP activity, while Nurr1 overexpression showed opposite effects (Lee, Choi, Gil, & Nikodem, 2006). These findings indicate that Nurr1 may play positive roles in promoting bone formation.

Opposite effects of NR4A on osteoblasts were also reported. Overexpression of Nur77, Nurr1, or NOR1 in osteoblasts was shown to activate OPN promoter (Lammi et al., 2004; Pirih, Tang, Ozkurt, Nervina, & Tetradis, 2004). OPN, one of major noncollagen proteins in bone matrix, is a negative regulator of osteoblast differentiation (Huang et al., 2004) and a mediator of bone resorption (Yoshitake, Rittling, Denhardt, & Noda, 1999). However, it remains unclear whether the NR4A-induced OPN expression is functionally significant for the overall impact on bone formation and bone resorption. Another study showed that overexpression of Nur77, Nurr1, or NOR1 in the U2-OS osteoblastic cell line suppressed β-catenin transcriptional activity and β-catenin in turn also inhibited the transcriptional activities of NR4As (Rajalin & Aarnisalo, 2011). Despite the important functions of β-catenin in osteoblasts, the impact of this potential cross talk between NR4As and β-catenin on osteoblast activity and bone formation has not been demonstrated. Future studies are required to determine the physiological relevance and functional significance of these interesting in vitro observations.

Nur77 expression is elevated during osteoclast differentiation, while the expression of Nurr1 and NOR1 is very low in osteoclasts (Li et al., 2015). Work from our lab shows that global Nur77 knockout reduced bone mass and increased osteoclastogenesis (Li et al., 2015), indicating a resorption-suppressive and osteoprotective effect of Nur77. Mechanistically, we uncover that Nur77 prevents excessive osteoclastogenesis by mediating an NFATc1 self-limiting regulatory loop. NFATc1 induces Nur77 expression at late stage of osteoclast differentiation; in turn, Nur77 transcriptionally upregulates E3 ubiquitin ligase Cbl-b, which triggers NFATc1 protein degradation (Li et al., 2015). These findings not only identify Nur77 as a key player in bone turnover and a new therapeutic target for bone diseases but also elucidate a previously unrecognized NFATc1 → Nur77 → Cbl-b–•NFATc1 feedback mechanism that confers NFATc1 signaling autoresolution.

3. CONCLUSIONS

Generally, activation of ERα, AR, PPARα, PPARδ/β, Nur77, and Nurr1 protects bone, while activation of ERβ, GR, PPARγ, and ERRα/γ causes bone loss. Conflicting roles of VDR and RAR in bone have been reported. Among other NRs that are not discussed here, SXR/PXR (Azuma et al., 2010), RORα (Benderdour, Fahmi, Beaudet, Fernandes, & Shi, 2011; Lyashenko et al., 2010; Meyer, Kneissel, Mariani, & Fournier, 2000), and SHP (Jeong et al., 2010) may play positive protective roles in bone, while PR (Rickard et al., 2008) negatively regulates bone homeostasis. In vivo analysis of global knockout mice indicates that LXRs may exert negative detrimental effects on bone, in part by elevating osteoclast functions (Robertson et al., 2006); but in vitro data show that activation of LXRs inhibits both osteoblastogenesis (Prawitt et al., 2011) and osteoclastogenesis (Kim et al., 2013; Kleyer et al., 2012; Remen, Henning, Lerner, Gustafsson, & Andersson, 2011; Robertson Remen, Gustafsson, & Andersson, 2013; Robertson Remen, Lerner, Gustafsson, & Andersson, 2013). The effect of thyroid hormone and thyroid hormone receptors, TRα and TRβ, on bone development and remodeling has been reviewed in detail recently (Bassett & Williams, 2016). Hypothyroid delays bone growth and maturation in children, while thyrotoxicosis accelerates these processes in children and causes osteoporosis in adult (Bassett & Williams, 2016). Global TRα deletion resulted in delayed bone development of young mice and increased bone mass with reduced osteoclast differentiation in adult mice (Bassett et al., 2007; Bassett & Williams, 2016). Global TRβ deletion led to elevated thyroid hormone levels, accelerated ossification and short stature in young mice, and reduced bone mass with increased osteoclast differentiation in adult mice, all of which may be due to hyperactivation of TRα by elevated thyroid hormone levels in these mice (Bassett et al., 2007; Bassett & Williams, 2016). Therefore, normal levels of thyroid hormone and functions of its receptors are necessary for both early bone development and adult bone maintenance. We have summarized the reported bone phenotypes in NR knockout animals (Table 1).

Table 1 Summary of Bone Phenotypes of NR Knockout Animal Models

NR	Cell Type	Cre	Phenotype				References
			T.b.	C.b.	Ob	Oc	
ERα	Global		↑	↓	↓	↓	Sims et al. (2002)
	Ob progenitor	Prx1	↓	→	↓	→	Almeida et al. (2013)
	Pre-Ob	Osx1	↓	↓	↓		Almeida et al. (2013)
	Mature Ob/Ocy	Col1a1	→	→	↓		Almeida et al. (2013)
	Mature Ob/Ocy	Ocn	↓	↓	↓	↓	Maatta et al. (2013) and Melville et al. (2014)
	Ocy	Dmp1	↓	→	↓	→	Windahl, Borjesson, et al. (2013)
	Oc precursor	LysM	↓	→	→	↑	Martin-Millan et al. (2010)
	Mature Oc	Ctsk	↓	→	→	↑	Nakamura et al. (2007)
ERβ	Global		↑	→	→	→	Sims et al. (2002)
	Ob progenitors	Prx1	↑	→	↑	→	Nicks et al. (2016)
AR	Global		↓	↓	↑	↑	Kawano et al. (2003)
	Ob progenitors	Prx1	↓	→	→	↑	Ucer et al. (2015)
	Mature Ob/Ocy	Ocn	↓	↓	→	↑	Chiang et al. (2009)
	Mature Ob/Ocy	Col2.3	↓	→		→	Notini et al. (2007)
	Ocy	Dmp1	↓	→		→	Sinnesael et al. (2012)
	Oc precursor	LysM	→	→		→	Ucer et al. (2015)
	Mature Oc	Ctsk	→	→		→	Sinnesael et al. (2012)
GR	Pre-Ob	Runx2	↓		↓	→	Rauch et al. (2010)
	Oc precursor	LysM			↑	↑	Kim et al. (2006)
PPARα	Global		→	→			Wu et al. (2000)
PPARβ	Except placenta	Sox2	↓		→	↑	Scholtysek et al. (2013)
PPARγ	Global		↑		↑	→	Akune et al. (2004)
	Ob progenitor	Col3.6	↑		↑	→	Cao et al. (2015)
	Pre-Ob	Osx	↑	→	↑		Sun et al. (2013)
	Oc precursor	Tie2	↑		→	↓	Wan et al. (2007)

Table 1 Summary of Bone Phenotypes of NR Knockout Animal Models—cont'd

NR	Cell Type	Cre	Phenotype T.b.	Phenotype C.b.	Phenotype Ob	Phenotype Oc	References
VDR	Global		↑			↓	Masuyama et al. (2003) and Yoshizawa et al. (1997)
	Mature Ob/Ocy	Col1a1	↑	→	→	↓	Yamamoto et al. (2013)
	Ocy	Dmp1	→	→	→	→	Lieben et al. (2012)
	Chondrocyte	Col2	↑	→	→	↓	Masuyama et al. (2006)
TRα	Global		↑	↑	↓	↓	Bassett et al. (2007)
TRβ	Global		↓	↓	↑	↑	Bassett et al. (2007)
RARα	Global		→	→			Li et al. (1993) and Lufkin et al. (1993)
RARβ	Global		→	→			Mendelsohn et al. (1994)
RARγ	Global		↓	→	→	↑	Green et al. (2015)
ERRα	Global		↑	→	↑	→	Delhon et al. (2009)
	Global		↑	→	↑	↓	Wei et al. (2016) and Wei et al. (2010)
	Mature Ob/Ocy	Col1a1	→		↑		Gallet et al. (2013)
ERRγ	Global		↑	→	↑	→	Cardelli and Aubin (2014)
Nur77	Global		↓	↓	→	↑	Li et al. (2015)
PR	Global		↑	↑	↑	↓	Rickard et al. (2008)
LXRα	Global		→	↑	→	↓	Robertson et al. (2006)
LXRβ	Global		→	→	↑	↓	Robertson et al. (2006)
PXR	Global		↓	↓	↓	↑	Azuma et al. (2010)
RORα	Global		↓	↓			Lyashenko et al. (2010) and Meyer et al. (2000)
SHP	Global		↓	↓	↓	↓	Jeong et al. (2010)

C.b., cortical bone; *Ob*, osteoblast; *Oc*, osteoclast; *Ocy*, osteocyte; *Ocn*, osteocalcin; *T.b.*, trabecular bone; ↑, increase; ↓, decrease; →, no change.

NRs are essential in controlling gene expressions during bone development and remodeling in response to the rise and fall of ligand concentration. NRs can act as constitutive transcriptional activators when agonists are abundant and/or ubiquitous, or can be induced to activation by binding to steroid/thyroid hormones, biological metabolites, vitamins, and other lipophilic natural or synthetic small molecules. Alterations of NR activities, resulted from abnormal levels of either receptors or ligands, can lead to bone disorders; for example, estrogen deficiency or GC treatment causes osteoporosis. NR modulation of skeletal homeostasis may involve mechanisms beyond direct transcriptional regulation, for example, membrane ERα-mediated kinase-initiated regulation. Moreover, NR control of bone remodeling may involve anatomic sites outside of bone that mediates neuronal, endocrine, and metabolic regulation on bone; for example, VDR acts in intestine to regulate calcium absorption. Therefore, the multifaceted roles of each NR in regulating bone homeostasis and the bone beneficial concentration ranges for their agonists, antagonists, or modulators need to be determined in future studies. NRs are excellent targets for drug design due to their broad functions in various biological processes and their functional tunability through small-molecule ligands. This review provides a timely update on our latest understanding of NR regulation of bone health. This new knowledge will facilitate the development of novel osteoprotective medicine and effective strategies to eliminate the bone loss side effects of current drugs.

ACKNOWLEDGMENTS

We thank all our colleagues whose studies have contributed to our understating of NRs in bone but could not be cited here due to space limitation. Y.W. is a Virginia Murchison Linthicum Scholar in Medical Research and Lawrence Raisz Endowed Professor in Bone Cell Metabolism. This work was in part supported by NIH (R01 DK089113, Y.W.), CPRIT (RP130145, Y.W.), DOD (W81XWH-13-1-0318, Y.W.), Welch Foundation (I-1751, Y.W.), March of Dimes (#6-FY13-137, Y.W.), Mary Kay Foundation (#073.14, Y.W.), and UTSW Endowed Scholar Startup Fund (Y.W.). The authors declare that they have no financial conflict of interest.

REFERENCES

Abu, E. O., Horner, A., Kusec, V., Triffitt, J. T., & Compston, J. E. (1997). The localization of androgen receptors in human bone. *The Journal of Clinical Endocrinology and Metabolism, 82*(10), 3493–3497.

Acconcia, F., Totta, P., Ogawa, S., Cardillo, I., Inoue, S., Leone, S., ... Marino, M. (2005). Survival versus apoptotic 17beta-estradiol effect: Role of ER alpha and ER beta activated non-genomic signaling. *Journal of Cellular Physiology, 203*(1), 193–201.

Adams, J. S., & Lee, G. (1997). Gains in bone mineral density with resolution of vitamin D intoxication. *Annals of Internal Medicine, 127*(3), 203–206.

Ahmadian, M., Suh, J. M., Hah, N., Liddle, C., Atkins, A. R., Downes, M., & Evans, R. M. (2013). PPARgamma signaling and metabolism: The good, the bad and the future. *Nature Medicine, 19*(5), 557–566.

Akiyama, H., Kim, J. E., Nakashima, K., Balmes, G., Iwai, N., Deng, J. M., ... de Crombrugghe, B. (2005). Osteo-chondroprogenitor cells are derived from Sox9 expressing precursors. *Proceedings of the National Academy of Sciences of the United States of America*, *102*(41), 14665–14670.

Akune, T., Ohba, S., Kamekura, S., Yamaguchi, M., Chung, U. I., Kubota, N., ... Kawaguchi, H. (2004). PPARgamma insufficiency enhances osteogenesis through osteoblast formation from bone marrow progenitors. *The Journal of Clinical Investigation*, *113*(6), 846–855.

Almeida, M., Han, L., O'Brien, C. A., Kousteni, S., & Manolagas, S. C. (2006). Classical genotropic versus kinase-initiated regulation of gene transcription by the estrogen receptor alpha. *Endocrinology*, *147*(4), 1986–1996.

Almeida, M., Iyer, S., Martin-Millan, M., Bartell, S. M., Han, L., Ambrogini, E., ... Manolagas, S. C. (2013). Estrogen receptor-alpha signaling in osteoblast progenitors stimulates cortical bone accrual. *The Journal of Clinical Investigation*, *123*(1), 394–404.

Almeida, M., Martin-Millan, M., Ambrogini, E., Bradsher, R., 3rd, Han, L., Chen, X. D., ... Manolagas, S. C. (2010). Estrogens attenuate oxidative stress and the differentiation and apoptosis of osteoblasts by DNA-binding-independent actions of the ERalpha. *Journal of Bone and Mineral Research*, *25*(4), 769–781.

Ambrosini, G. L., Alfonso, H., Reid, A., Mackerras, D., Bremner, A. P., Beilby, J., ... de Klerk, N. H. (2014). Plasma retinol and total carotenes and fracture risk after long-term supplementation with high doses of retinol. *Nutrition*, *30*(5), 551–556.

An, J., Tzagarakis-Foster, C., Scharschmidt, T. C., Lomri, N., & Leitman, D. C. (2001). Estrogen receptor beta-selective transcriptional activity and recruitment of coregulators by phytoestrogens. *The Journal of Biological Chemistry*, *276*(21), 17808–17814.

Azuma, K., Casey, S. C., Ito, M., Urano, T., Horie, K., Ouchi, Y., ... Inoue, S. (2010). Pregnane X receptor knockout mice display osteopenia with reduced bone formation and enhanced bone resorption. *The Journal of Endocrinology*, *207*(3), 257–263.

Bain, S. D., Bailey, M. C., Celino, D. L., Lantry, M. M., & Edwards, M. W. (1993). High-dose estrogen inhibits bone resorption and stimulates bone formation in the ovariectomized mouse. *Journal of Bone and Mineral Research*, *8*(4), 435–442.

Baldock, P. A., Thomas, G. P., Hodge, J. M., Baker, S. U., Dressel, U., O'Loughlin, P. D., ... Gardiner, E. M. (2006). Vitamin D action and regulation of bone remodeling: Suppression of osteoclastogenesis by the mature osteoblast. *Journal of Bone and Mineral Research*, *21*(10), 1618–1626.

Balkan, W., Rodriguez-Gonzalez, M., Pang, M., Fernandez, I., & Troen, B. R. (2011). Retinoic acid inhibits NFATc1 expression and osteoclast differentiation. *Journal of Bone and Mineral Metabolism*, *29*(6), 652–661.

Ballock, R. T., Heydemann, A., Wakefield, L. M., Flanders, K. C., Roberts, A. B., & Sporn, M. B. (1994). Inhibition of the chondrocyte phenotype by retinoic acid involves upregulation of metalloprotease genes independent of TGF-beta. *Journal of Cellular Physiology*, *159*(2), 340–346.

Barros, R. P., & Gustafsson, J. A. (2011). Estrogen receptors and the metabolic network. *Cell Metabolism*, *14*(3), 289–299.

Bartell, S. M., Han, L., Kim, H. N., Kim, S. H., Katzenellenbogen, J. A., Katzenellenbogen, B. S., ... Manolagas, S. C. (2013). Non-nuclear-initiated actions of the estrogen receptor protect cortical bone mass. *Molecular Endocrinology*, *27*(4), 649–656.

Bassett, J. H., O'Shea, P. J., Sriskantharajah, S., Rabier, B., Boyde, A., Howell, P. G., ... Williams, G. R. (2007). Thyroid hormone excess rather than thyrotropin deficiency induces osteoporosis in hyperthyroidism. *Molecular Endocrinology*, *21*(5), 1095–1107.

Bassett, J. H., & Williams, G. R. (2016). Role of thyroid hormones in skeletal development and bone maintenance. *Endocrine Reviews*, *37*(2), 135–187.

Benderdour, M., Fahmi, H., Beaudet, F., Fernandes, J. C., & Shi, Q. (2011). Nuclear receptor retinoid-related orphan receptor alpha1 modulates the metabolic activity of human osteoblasts. *Journal of Cellular Biochemistry, 112*(8), 2160–2169.
Bilezikian, J. P., Josse, R. G., Eastell, R., Lewiecki, E. M., Miller, C. G., Wooddell, M., … Fitzpatrick, L. A. (2013). Rosiglitazone decreases bone mineral density and increases bone turnover in postmenopausal women with type 2 diabetes mellitus. *The Journal of Clinical Endocrinology and Metabolism, 98*(4), 1519–1528.
Billington, E. O., Grey, A., & Bolland, M. J. (2015). The effect of thiazolidinediones on bone mineral density and bone turnover: Systematic review and meta-analysis. *Diabetologia, 58*(10), 2238–2246.
Bord, S., Horner, A., Beavan, S., & Compston, J. (2001). Estrogen receptors alpha and beta are differentially expressed in developing human bone. *The Journal of Clinical Endocrinology and Metabolism, 86*(5), 2309–2314.
Bord, S., Ireland, D. C., Beavan, S. R., & Compston, J. E. (2003). The effects of estrogen on osteoprotegerin, RANKL, and estrogen receptor expression in human osteoblasts. *Bone, 32*(2), 136–141.
Boyle, W. J., Simonet, W. S., & Lacey, D. L. (2003). Osteoclast differentiation and activation. *Nature, 423*(6937), 337–342.
Byerly, M. S., Swanson, R. D., Wong, G. W., & Blackshaw, S. (2013). Estrogen-related receptor beta deficiency alters body composition and response to restraint stress. *BMC Physiology, 13*, 10.
Caire-Juvera, G., Ritenbaugh, C., Wactawski-Wende, J., Snetselaar, L. G., & Chen, Z. (2009). Vitamin A and retinol intakes and the risk of fractures among participants of the Women's Health Initiative Observational Study. *The American Journal of Clinical Nutrition, 89*(1), 323–330.
Cao, J., Ou, G., Yang, N., Ding, K., Kream, B. E., Hamrick, M. W., … Shi, X. M. (2015). Impact of targeted PPARgamma disruption on bone remodeling. *Molecular and Cellular Endocrinology, 410*, 27–34.
Cardelli, M., & Aubin, J. E. (2014). ERRgamma is not required for skeletal development but is a RUNX2-dependent negative regulator of postnatal bone formation in male mice. *PLoS One, 9*(10). e109592.
Carroll, J. S., Meyer, C. A., Song, J., Li, W., Geistlinger, T. R., Eeckhoute, J., … Brown, M. (2006). Genome-wide analysis of estrogen receptor binding sites. *Nature Genetics, 38*(11), 1289–1297.
Chambon, P. (1996). A decade of molecular biology of retinoic acid receptors. *The FASEB Journal, 10*(9), 940–954.
Chan, B. Y., Gartland, A., Wilson, P. J., Buckley, K. A., Dillon, J. P., Fraser, W. D., & Gallagher, J. A. (2007). PPAR agonists modulate human osteoclast formation and activity in vitro. *Bone, 40*(1), 149–159.
Chang, C., Lee, S. O., Wang, R. S., Yeh, S., & Chang, T. M. (2013). Androgen receptor (AR) physiological roles in male and female reproductive systems: Lessons learned from AR-knockout mice lacking AR in selective cells. *Biology of Reproduction, 89*(1), 21.
Chiang, C., Chiu, M., Moore, A. J., Anderson, P. H., Ghasem-Zadeh, A., McManus, J. F., … Davey, R. A. (2009). Mineralization and bone resorption are regulated by the androgen receptor in male mice. *Journal of Bone and Mineral Research, 24*(4), 621–631.
Colvard, D. S., Eriksen, E. F., Keeting, P. E., Wilson, E. M., Lubahn, D. B., French, F. S., … Spelsberg, T. C. (1989). Identification of androgen receptors in normal human osteoblast-like cells. *Proceedings of the National Academy of Sciences of the United States of America, 86*(3), 854–857.
Conaway, H. H., Pirhayati, A., Persson, E., Pettersson, U., Svensson, O., Lindholm, C., … Lerner, U. H. (2011). Retinoids stimulate periosteal bone resorption by enhancing the protein RANKL, a response inhibited by monomeric glucocorticoid receptor. *The Journal of Biological Chemistry, 286*(36), 31425–31436.

Coward, P., Lee, D., Hull, M. V., & Lehmann, J. M. (2001). 4-Hydroxytamoxifen binds to and deactivates the estrogen-related receptor gamma. *Proceedings of the National Academy of Sciences of the United States of America*, *98*(15), 8880–8884.

Cowley, S. M., Hoare, S., Mosselman, S., & Parker, M. G. (1997). Estrogen receptors alpha and beta form heterodimers on DNA. *The Journal of Biological Chemistry*, *272*(32), 19858–19862.

Delhon, I., Gutzwiller, S., Morvan, F., Rangwala, S., Wyder, L., Evans, G., … Fournier, B. (2009). Absence of estrogen receptor-related-alpha increases osteoblastic differentiation and cancellous bone mineral density. *Endocrinology*, *150*(10), 4463–4472.

De Luca, F., Uyeda, J. A., Mericq, V., Mancilla, E. E., Yanovski, J. A., Barnes, K. M., … Baron, J. (2000). Retinoic acid is a potent regulator of growth plate chondrogenesis. *Endocrinology*, *141*(1), 346–353.

den Uyl, D., Bultink, I. E., & Lems, W. F. (2011). Advances in glucocorticoid-induced osteoporosis. *Current Rheumatology Reports*, *13*(3), 233–240.

de Vera, I. M., Giri, P. K., Munoz-Tello, P., Brust, R., Fuhrmann, J., Matta-Camacho, E., … Kojetin, D. J. (2016). Identification of a binding site for unsaturated fatty acids in the orphan nuclear receptor Nurr1. *ACS Chemical Biology*, *11*(7), 1795–1799.

Evans, R. M., & Mangelsdorf, D. J. (2014). Nuclear receptors, RXR, and the big bang. *Cell*, *157*(1), 255–266.

Fahmi, H., Martel-Pelletier, J., Pelletier, J. P., & Kapoor, M. (2011). Peroxisome proliferator-activated receptor gamma in osteoarthritis. *Modern Rheumatology*, *21*(1), 1–9.

Falahati-Nini, A., Riggs, B. L., Atkinson, E. J., O'Fallon, W. M., Eastell, R., & Khosla, S. (2000). Relative contributions of testosterone and estrogen in regulating bone resorption and formation in normal elderly men. *The Journal of Clinical Investigation*, *106*(12), 1553–1560.

Feskanich, D., Singh, V., Willett, W. C., & Colditz, G. A. (2002). Vitamin A intake and hip fractures among postmenopausal women. *JAMA*, *287*(1), 47–54.

Flaig, R., Greschik, H., Peluso-Iltis, C., & Moras, D. (2005). Structural basis for the cell-specific activities of the NGFI-B and the Nurr1 ligand-binding domain. *The Journal of Biological Chemistry*, *280*(19), 19250–19258.

Galea, G. L., Meakin, L. B., Sugiyama, T., Zebda, N., Sunters, A., Taipaleenmaki, H., … Price, J. S. (2013). Estrogen receptor alpha mediates proliferation of osteoblastic cells stimulated by estrogen and mechanical strain, but their acute down-regulation of the Wnt antagonist Sost is mediated by estrogen receptor beta. *The Journal of Biological Chemistry*, *288*(13), 9035–9048.

Galien, R., & Garcia, T. (1997). Estrogen receptor impairs interleukin-6 expression by preventing protein binding on the NF-kappaB site. *Nucleic Acids Research*, *25*(12), 2424–2429.

Gallet, M., Saidi, S., Hay, E., Photsavang, J., Marty, C., Sailland, J., … Vanacker, J. M. (2013). Repression of osteoblast maturation by ERRalpha accounts for bone loss induced by estrogen deficiency. *PLoS One*, *8*(1). e54837.

Gardiner, E. M., Baldock, P. A., Thomas, G. P., Sims, N. A., Henderson, N. K., Hollis, B., … Eisman, J. A. (2000). Increased formation and decreased resorption of bone in mice with elevated vitamin D receptor in mature cells of the osteoblastic lineage. *The FASEB Journal*, *14*(13), 1908–1916.

Gazit, D., Ebner, R., Kahn, A. J., & Derynck, R. (1993). Modulation of expression and cell surface binding of members of the transforming growth factor-beta superfamily during retinoic acid-induced osteoblastic differentiation of multipotential mesenchymal cells. *Molecular Endocrinology*, 7(2), 189–198.

Germain, P., Iyer, J., Zechel, C., & Gronemeyer, H. (2002). Co-regulator recruitment and the mechanism of retinoic acid receptor synergy. *Nature*, *415*(6868), 187–192.

Giaginis, C., Tsantili-Kakoulidou, A., & Theocharis, S. (2007). Peroxisome proliferator-activated receptors (PPARs) in the control of bone metabolism. *Fundamental and Clinical Pharmacology*, *21*(3), 231–244.

Glass, C. K., & Rosenfeld, M. G. (2000). The coregulator exchange in transcriptional functions of nuclear receptors. *Genes and Development*, *14*(2), 121–141.
Green, A. C., Poulton, I. J., Vrahnas, C., Hausler, K. D., Walkley, C. R., Wu, J. Y., ... Purton, L. E. (2015). RARgamma is a negative regulator of osteoclastogenesis. *The Journal of Steroid Biochemistry and Molecular Biology*, *150*, 46–53.
Heikkinen, S., Vaisanen, S., Pehkonen, P., Seuter, S., Benes, V., & Carlberg, C. (2011). Nuclear hormone 1alpha,25-dihydroxyvitamin D3 elicits a genome-wide shift in the locations of VDR chromatin occupancy. *Nucleic Acids Research*, *39*(21), 9181–9193.
Heino, T. J., Chagin, A. S., & Savendahl, L. (2008). The novel estrogen receptor G-protein-coupled receptor 30 is expressed in human bone. *The Journal of Endocrinology*, *197*(2), R1–R6.
Hofbauer, L. C., Khosla, S., Dunstan, C. R., Lacey, D. L., Spelsberg, T. C., & Riggs, B. L. (1999). Estrogen stimulates gene expression and protein production of osteoprotegerin in human osteoblastic cells. *Endocrinology*, *140*(9), 4367–4370.
Holvik, K., Ahmed, L. A., Forsmo, S., Gjesdal, C. G., Grimnes, G., Samuelsen, S. O., ... Meyer, H. E. (2015). No increase in risk of hip fracture at high serum retinol concentrations in community-dwelling older Norwegians: The Norwegian Epidemiologic Osteoporosis Studies. *The American Journal of Clinical Nutrition*, *102*(5), 1289–1296.
Horard, B., & Vanacker, J. M. (2003). Estrogen receptor-related receptors: Orphan receptors desperately seeking a ligand. *Journal of Molecular Endocrinology*, *31*(3), 349–357.
Hu, L., Lind, T., Sundqvist, A., Jacobson, A., & Melhus, H. (2010). Retinoic acid increases proliferation of human osteoclast progenitors and inhibits RANKL-stimulated osteoclast differentiation by suppressing RANK. *PLoS One*, *5*(10). e13305.
Huang, W., Carlsen, B., Rudkin, G., Berry, M., Ishida, K., Yamaguchi, D. T., & Miller, T. A. (2004). Osteopontin is a negative regulator of proliferation and differentiation in MC3T3-E1 pre-osteoblastic cells. *Bone*, *34*(5), 799–808.
Iba, K., Chiba, H., Yamashita, T., Ishii, S., & Sawada, N. (2001). Phase-independent inhibition by retinoic acid of mineralization correlated with loss of tetranectin expression in a human osteoblastic cell line. *Cell Structure and Function*, *26*(4), 227–233.
Inoue, A., Otsuka, E., Hiruma, Y., Hirose, S., Furuya, M., Tanaka, S., & Hagiwara, H. (1996). Stimulation by retinoids of the natriuretic peptide system of osteoblastic MC3T3-E1 cells. *Biochemical and Biophysical Research Communications*, *228*(1), 182–186.
Ishida, Y., & Heersche, J. N. (1998). Glucocorticoid-induced osteoporosis: Both in vivo and in vitro concentrations of glucocorticoids higher than physiological levels attenuate osteoblast differentiation. *Journal of Bone and Mineral Research*, *13*(12), 1822–1826.
Iwamoto, M., Shapiro, I. M., Yagami, K., Boskey, A. L., Leboy, P. S., Adams, S. L., & Pacifici, M. (1993). Retinoic acid induces rapid mineralization and expression of mineralization-related genes in chondrocytes. *Experimental Cell Research*, *207*(2), 413–420.
Jeong, B. C., Lee, Y. S., Bae, I. H., Lee, C. H., Shin, H. I., Ha, H. J., ... Koh, J. T. (2010). The orphan nuclear receptor SHP is a positive regulator of osteoblastic bone formation. *Journal of Bone and Mineral Research*, *25*(2), 262–274.
Jin, Z., Li, X., & Wan, Y. (2015). Minireview: Nuclear receptor regulation of osteoclast and bone remodeling. *Molecular Endocrinology*, *29*(2), 172–186.
Katzenellenbogen, B. S., & Katzenellenbogen, J. A. (2000). Estrogen receptor transcription and transactivation: Estrogen receptor alpha and estrogen receptor beta: Regulation by selective estrogen receptor modulators and importance in breast cancer. *Breast Cancer Research*, *2*(5), 335–344.
Kawano, H., Sato, T., Yamada, T., Matsumoto, T., Sekine, K., Watanabe, T., ... Kato, S. (2003). Suppressive function of androgen receptor in bone resorption. *Proceedings of the National Academy of Sciences of the United States of America*, *100*(16), 9416–9421.

Khosla, S. (2001). Minireview: The OPG/RANKL/RANK system. *Endocrinology*, *142*(12), 5050–5055.

Kikuta, J., Kawamura, S., Okiji, F., Shirazaki, M., Sakai, S., Saito, H., & Ishii, M. (2013). Sphingosine-1-phosphate-mediated osteoclast precursor monocyte migration is a critical point of control in antibone-resorptive action of active vitamin D. *Proceedings of the National Academy of Sciences of the United States of America*, *110*(17), 7009–7013.

Kim, J., & Ko, J. (2014). A novel PPARgamma2 modulator sLZIP controls the balance between adipogenesis and osteogenesis during mesenchymal stem cell differentiation. *Cell Death and Differentiation*, *21*(10), 1642–1655.

Kim, J. H., Seong, S., Kim, K., Kim, I., Jeong, B. C., & Kim, N. (2016). Downregulation of Runx2 by 1,25-dihydroxyvitamin D(3) induces the transdifferentiation of osteoblasts to adipocytes. *International Journal of Molecular Sciences*, *17*(5), 770.

Kim, H. J., Yoon, K. A., Yoon, H. J., Hong, J. M., Lee, M. J., Lee, I. K., & Kim, S. Y. (2013). Liver X receptor activation inhibits osteoclastogenesis by suppressing NF-kappaB activity and c-Fos induction and prevents inflammatory bone loss in mice. *Journal of Leukocyte Biology*, *94*(1), 99–107.

Kim, H. J., Zhao, H., Kitaura, H., Bhattacharyya, S., Brewer, J. A., Muglia, L. J., … Teitelbaum, S. L. (2006). Glucocorticoids suppress bone formation via the osteoclast. *The Journal of Clinical Investigation*, *116*(8), 2152–2160.

Kindmark, A., Torma, H., Johansson, A., Ljunghall, S., & Melhus, H. (1993). Reverse transcription-polymerase chain reaction assay demonstrates that the 9-cis retinoic acid receptor alpha is expressed in human osteoblasts. *Biochemical and Biophysical Research Communications*, *192*(3), 1367–1372.

Kleyer, A., Scholtysek, C., Bottesch, E., Hillienhof, U., Beyer, C., Distler, J. H., … Kronke, G. (2012). Liver X receptors orchestrate osteoblast/osteoclast crosstalk and counteract pathologic bone loss. *Journal of Bone and Mineral Research*, *27*(12), 2442–2451.

Kneissel, M., Studer, A., Cortesi, R., & Susa, M. (2005). Retinoid-induced bone thinning is caused by subperiosteal osteoclast activity in adult rodents. *Bone*, *36*(2), 202–214.

Kousteni, S., Almeida, M., Han, L., Bellido, T., Jilka, R. L., & Manolagas, S. C. (2007). Induction of osteoblast differentiation by selective activation of kinase-mediated actions of the estrogen receptor. *Molecular and Cellular Biology*, *27*(4), 1516–1530.

Kousteni, S., Bellido, T., Plotkin, L. I., O'Brien, C. A., Bodenner, D. L., Han, L., … Manolagas, S. C. (2001). Nongenotropic, sex-nonspecific signaling through the estrogen or androgen receptors: Dissociation from transcriptional activity. *Cell*, *104*(5), 719–730.

Kousteni, S., Chen, J. R., Bellido, T., Han, L., Ali, A. A., O'Brien, C. A., … Manolagas, S. C. (2002). Reversal of bone loss in mice by nongenotropic signaling of sex steroids. *Science*, *298*(5594), 843–846.

Kousteni, S., Han, L., Chen, J. R., Almeida, M., Plotkin, L. I., Bellido, T., & Manolagas, S. C. (2003). Kinase-mediated regulation of common transcription factors accounts for the bone-protective effects of sex steroids. *The Journal of Clinical Investigation*, *111*(11), 1651–1664.

Krum, S. A., Miranda-Carboni, G. A., Hauschka, P. V., Carroll, J. S., Lane, T. F., Freedman, L. P., & Brown, M. (2008). Estrogen protects bone by inducing Fas ligand in osteoblasts to regulate osteoclast survival. *The EMBO Journal*, *27*(3), 535–545.

Lam, N. N., Triliana, R., Sawyer, R. K., Atkins, G. J., Morris, H. A., O'Loughlin, P. D., & Anderson, P. H. (2014). Vitamin D receptor overexpression in osteoblasts and osteocytes prevents bone loss during vitamin D-deficiency. *The Journal of Steroid Biochemistry and Molecular Biology*, *144*(Pt. A), 128–131.

Lammi, J., & Aarnisalo, P. (2008). FGF-8 stimulates the expression of NR4A orphan nuclear receptors in osteoblasts. *Molecular and Cellular Endocrinology*, *295*(1–2), 87–93.

Lammi, J., Huppunen, J., & Aarnisalo, P. (2004). Regulation of the osteopontin gene by the orphan nuclear receptor NURR1 in osteoblasts. *Molecular Endocrinology*, *18*(6), 1546–1557.

Lee, M. K., Choi, H., Gil, M., & Nikodem, V. M. (2006). Regulation of osteoblast differentiation by Nurr1 in MC3T3-E1 cell line and mouse calvarial osteoblasts. *Journal of Cellular Biochemistry, 99*(3), 986–994.
Lee, S. K., Kalinowski, J., Jastrzebski, S., & Lorenzo, J. A. (2002). 1,25(OH)2 vitamin D3-stimulated osteoclast formation in spleen-osteoblast cocultures is mediated in part by enhanced IL-1 alpha and receptor activator of NF-kappa B ligand production in osteoblasts. *The Journal of Immunology, 169*(5), 2374–2380.
Lee, S. S., Pineau, T., Drago, J., Lee, E. J., Owens, J. W., Kroetz, D. L., … Gonzalez, F. J. (1995). Targeted disruption of the alpha isoform of the peroxisome proliferator-activated receptor gene in mice results in abolishment of the pleiotropic effects of peroxisome proliferators. *Molecular and Cellular Biology, 15*(6), 3012–3022.
Li, E., Sucov, H. M., Lee, K. F., Evans, R. M., & Jaenisch, R. (1993). Normal development and growth of mice carrying a targeted disruption of the alpha 1 retinoic acid receptor gene. *Proceedings of the National Academy of Sciences of the United States of America, 90*(4), 1590–1594.
Li, X., Wei, W., Huynh, H., Zuo, H., Wang, X., & Wan, Y. (2015). Nur77 prevents excessive osteoclastogenesis by inducing ubiquitin ligase Cbl-b to mediate NFATc1 self-limitation. *eLife, 4*. e07217.
Lieben, L., Masuyama, R., Torrekens, S., Van Looveren, R., Schrooten, J., Baatsen, P., … Carmeliet, G. (2012). Normocalcemia is maintained in mice under conditions of calcium malabsorption by vitamin D-induced inhibition of bone mineralization. *The Journal of Clinical Investigation, 122*(5), 1803–1815.
Lind, T., Sundqvist, A., Hu, L., Pejler, G., Andersson, G., Jacobson, A., & Melhus, H. (2013). Vitamin A is a negative regulator of osteoblast mineralization. *PLoS One, 8*(12). e82388.
Lindsay, R., Hart, D. M., Aitken, J. M., MacDonald, E. B., Anderson, J. B., & Clarke, A. C. (1976). Long-term prevention of postmenopausal osteoporosis by oestrogen. Evidence for an increased bone mass after delayed onset of oestrogen treatment. *Lancet, 1*(7968), 1038–1041.
Liu, J. J., Zeng, H. N., Zhang, L. R., Zhan, Y. Y., Chen, Y., Wang, Y., … Wu, Q. (2010). A unique pharmacophore for activation of the nuclear orphan receptor Nur77 in vivo and in vitro. *Cancer Research, 70*(9), 3628–3637.
Lohnes, D., Mark, M., Mendelsohn, C., Dolle, P., Dierich, A., Gorry, P., … Chambon, P. (1994). Function of the retinoic acid receptors (RARs) during development (I). Craniofacial and skeletal abnormalities in RAR double mutants. *Development, 120*(10), 2723–2748.
Love, J. D., Gooch, J. T., Benko, S., Li, C., Nagy, L., Chatterjee, V. K., … Schwabe, J. W. (2002). The structural basis for the specificity of retinoid-X receptor-selective agonists: New insights into the role of helix H12. *The Journal of Biological Chemistry, 277*(13), 11385–11391.
Lufkin, T., Lohnes, D., Mark, M., Dierich, A., Gorry, P., Gaub, M. P., … Chambon, P. (1993). High postnatal lethality and testis degeneration in retinoic acid receptor alpha mutant mice. *Proceedings of the National Academy of Sciences of the United States of America, 90*(15), 7225–7229.
Luo, J., Sladek, R., Bader, J. A., Matthyssen, A., Rossant, J., & Giguere, V. (1997). Placental abnormalities in mouse embryos lacking the orphan nuclear receptor ERR-beta. *Nature, 388*(6644), 778–782.
Lyashenko, N., Weissenbock, M., Sharir, A., Erben, R. G., Minami, Y., & Hartmann, C. (2010). Mice lacking the orphan receptor ror1 have distinct skeletal abnormalities and are growth retarded. *Developmental Dynamics, 239*(8), 2266–2277.
Maatta, J. A., Buki, K. G., Gu, G., Alanne, M. H., Vaaraniemi, J., Liljenback, H., … Vaananen, K. (2013). Inactivation of estrogen receptor alpha in bone-forming cells induces bone loss in female mice. *The FASEB Journal, 27*(2), 478–488.

Mangelsdorf, D. J., Thummel, C., Beato, M., Herrlich, P., Schutz, G., Umesono, K., ... Evans, R. M. (1995). The nuclear receptor superfamily: The second decade. *Cell*, *83*(6), 835–839.
Marino, M., Galluzzo, P., & Ascenzi, P. (2006). Estrogen signaling multiple pathways to impact gene transcription. *Current Genomics*, 7(8), 497–508.
Martensson, U. E., Salehi, S. A., Windahl, S., Gomez, M. F., Sward, K., Daszkiewicz-Nilsson, J., ... Leeb-Lundberg, L. M. (2009). Deletion of the G protein-coupled receptor 30 impairs glucose tolerance, reduces bone growth, increases blood pressure, and eliminates estradiol-stimulated insulin release in female mice. *Endocrinology*, *150*(2), 687–698.
Martin-Millan, M., Almeida, M., Ambrogini, E., Han, L., Zhao, H., Weinstein, R. S., ... Manolagas, S. C. (2010). The estrogen receptor-alpha in osteoclasts mediates the protective effects of estrogens on cancellous but not cortical bone. *Molecular Endocrinology*, *24*(2), 323–334.
Masuyama, R., Nakaya, Y., Katsumata, S., Kajita, Y., Uehara, M., Tanaka, S., ... Suzuki, K. (2003). Dietary calcium and phosphorus ratio regulates bone mineralization and turnover in vitamin D receptor knockout mice by affecting intestinal calcium and phosphorus absorption. *Journal of Bone and Mineral Research*, *18*(7), 1217–1226.
Masuyama, R., Stockmans, I., Torrekens, S., Van Looveren, R., Maes, C., Carmeliet, P., ... Carmeliet, G. (2006). Vitamin D receptor in chondrocytes promotes osteoclastogenesis and regulates FGF23 production in osteoblasts. *The Journal of Clinical Investigation*, *116*(12), 3150–3159.
Mayerson, A. B., Hundal, R. S., Dufour, S., Lebon, V., Befroy, D., Cline, G. W., ... Petersen, K. F. (2002). The effects of rosiglitazone on insulin sensitivity, lipolysis, and hepatic and skeletal muscle triglyceride content in patients with type 2 diabetes. *Diabetes*, *51*(3), 797–802.
McCarthy, T. L., Chang, W. Z., Liu, Y., & Centrella, M. (2003). Runx2 integrates estrogen activity in osteoblasts. *The Journal of Biological Chemistry*, *278*(44), 43121–43129.
McKenna, N. J., Lanz, R. B., & O'Malley, B. W. (1999). Nuclear receptor coregulators: Cellular and molecular biology. *Endocrine Reviews*, *20*(3), 321–344.
Melhus, H., Michaelsson, K., Kindmark, A., Bergstrom, R., Holmberg, L., Mallmin, H., ... Ljunghall, S. (1998). Excessive dietary intake of vitamin A is associated with reduced bone mineral density and increased risk for hip fracture. *Annals of Internal Medicine*, *129*(10), 770–778.
Melville, K. M., Kelly, N. H., Khan, S. A., Schimenti, J. C., Ross, F. P., Main, R. P., & van der Meulen, M. C. (2014). Female mice lacking estrogen receptor-alpha in osteoblasts have compromised bone mass and strength. *Journal of Bone and Mineral Research*, *29*(2), 370–379.
Mendelsohn, C., Mark, M., Dolle, P., Dierich, A., Gaub, M. P., Krust, A., ... Chambon, P. (1994). Retinoic acid receptor beta 2 (RAR beta 2) null mutant mice appear normal. *Developmental Biology*, *166*(1), 246–258.
Meyer, M. B., Goetsch, P. D., & Pike, J. W. (2010). Genome-wide analysis of the VDR/RXR cistrome in osteoblast cells provides new mechanistic insight into the actions of the vitamin D hormone. *The Journal of Steroid Biochemistry and Molecular Biology*, *121*(1–2), 136–141.
Meyer, T., Kneissel, M., Mariani, J., & Fournier, B. (2000). In vitro and in vivo evidence for orphan nuclear receptor RORalpha function in bone metabolism. *Proceedings of the National Academy of Sciences of the United States of America*, *97*(16), 9197–9202.
Michaelsson, K., Lithell, H., Vessby, B., & Melhus, H. (2003). Serum retinol levels and the risk of fracture. *The New England Journal of Medicine*, *348*(4), 287–294.
Michalik, L., Auwerx, J., Berger, J. P., Chatterjee, V. K., Glass, C. K., Gonzalez, F. J., ... Wahli, W. (2006). International union of pharmacology. LXI. Peroxisome proliferator-activated receptors. *Pharmacological Reviews*, *58*(4), 726–741.

Miyauchi, Y., Sato, Y., Kobayashi, T., Yoshida, S., Mori, T., Kanagawa, H., ... Miyamoto, T. (2013). HIF1alpha is required for osteoclast activation by estrogen deficiency in postmenopausal osteoporosis. *Proceedings of the National Academy of Sciences of the United States of America*, *110*(41), 16568–16573.
Miyazaki, Y., Mahankali, A., Matsuda, M., Mahankali, S., Hardies, J., Cusi, K., ... DeFronzo, R. A. (2002). Effect of pioglitazone on abdominal fat distribution and insulin sensitivity in type 2 diabetic patients. *The Journal of Clinical Endocrinology and Metabolism*, *87*(6), 2784–2791.
Mizuno, Y., Hosoi, T., Inoue, S., Ikegami, A., Kaneki, M., Akedo, Y., ... Orimo, H. (1994). Immunocytochemical identification of androgen receptor in mouse osteoclast-like multinucleated cells. *Calcified Tissue International*, *54*(4), 325–326.
Moore, J. T., Collins, J. L., & Pearce, K. H. (2006). The nuclear receptor superfamily and drug discovery. *ChemMedChem*, *1*(5), 504–523.
Nakamura, T., Imai, Y., Matsumoto, T., Sato, S., Takeuchi, K., Igarashi, K., ... Kato, S. (2007). Estrogen prevents bone loss via estrogen receptor alpha and induction of Fas ligand in osteoclasts. *Cell*, *130*(5), 811–823.
Nakashima, K., Zhou, X., Kunkel, G., Zhang, Z., Deng, J. M., Behringer, R. R., & de Crombrugghe, B. (2002). The novel zinc finger-containing transcription factor osterix is required for osteoblast differentiation and bone formation. *Cell*, *108*(1), 17–29.
Nicks, K. M., Fujita, K., Fraser, D., McGregor, U., Drake, M. T., McGee-Lawrence, M. E., ... Khosla, S. (2016). Deletion of estrogen receptor beta in osteoprogenitor cells increases trabecular but not cortical bone mass in female mice. *Journal of Bone and Mineral Research*, *31*(3), 606–614.
Nohno, T., Koyama, E., Myokai, F., Taniguchi, S., Ohuchi, H., Saito, T., & Noji, S. (1993). A chicken homeobox gene related to Drosophila paired is predominantly expressed in the developing limb. *Developmental Biology*, *158*(1), 254–264.
Notini, A. J., McManus, J. F., Moore, A., Bouxsein, M., Jimenez, M., Chiu, W. S., ... Davey, R. A. (2007). Osteoblast deletion of exon 3 of the androgen receptor gene results in trabecular bone loss in adult male mice. *Journal of Bone and Mineral Research*, *22*(3), 347–356.
Ohishi, K., Nishikawa, S., Nagata, T., Yamauchi, N., Shinohara, H., Kido, J., & Ishida, H. (1995). Physiological concentrations of retinoic acid suppress the osteoblastic differentiation of fetal rat calvaria cells in vitro. *European Journal of Endocrinology*, *133*(3), 335–341.
Otto, F., Thornell, A. P., Crompton, T., Denzel, A., Gilmour, K. C., Rosewell, I. R., ... Owen, M. J. (1997). Cbfa1, a candidate gene for cleidocranial dysplasia syndrome, is essential for osteoblast differentiation and bone development. *Cell*, *89*(5), 765–771.
Perlmann, T., & Jansson, L. (1995). A novel pathway for vitamin A signaling mediated by RXR heterodimerization with NGFI-B and NURR1. *Genes and Development*, *9*(7), 769–782.
Peters, J. M., Lee, S. S., Li, W., Ward, J. M., Gavrilova, O., Everett, C., ... Gonzalez, F. J. (2000). Growth, adipose, brain, and skin alterations resulting from targeted disruption of the mouse peroxisome proliferator-activated receptor beta(delta). *Molecular and Cellular Biology*, *20*(14), 5119–5128.
Philips, A., Lesage, S., Gingras, R., Maira, M. H., Gauthier, Y., Hugo, P., & Drouin, J. (1997). Novel dimeric Nur77 signaling mechanism in endocrine and lymphoid cells. *Molecular and Cellular Biology*, *17*(10), 5946–5951.
Pirih, F. Q., Nervina, J. M., Pham, L., Aghaloo, T., & Tetradis, S. (2003). Parathyroid hormone induces the nuclear orphan receptor NOR-1 in osteoblasts. *Biochemical and Biophysical Research Communications*, *306*(1), 144–150.
Pirih, F. Q., Tang, A., Ozkurt, I. C., Nervina, J. M., & Tetradis, S. (2004). Nuclear orphan receptor Nurr1 directly transactivates the osteocalcin gene in osteoblasts. *The Journal of Biological Chemistry*, *279*(51), 53167–53174.

Poli, V., Balena, R., Fattori, E., Markatos, A., Yamamoto, M., Tanaka, H., ... Costantini, F. (1994). Interleukin-6 deficient mice are protected from bone loss caused by estrogen depletion. *The EMBO Journal*, *13*(5), 1189–1196.
Prawitt, J., Beil, F. T., Marshall, R. P., Bartelt, A., Ruether, W., Heeren, J., ... Niemeier, A. (2011). Short-term activation of liver X receptors inhibits osteoblasts but long-term activation does not have an impact on murine bone in vivo. *Bone*, *48*(2), 339–346.
Qu, Q., Perala-Heape, M., Kapanen, A., Dahllund, J., Salo, J., Vaananen, H. K., & Harkonen, P. (1998). Estrogen enhances differentiation of osteoblasts in mouse bone marrow culture. *Bone*, *22*(3), 201–209.
Rajalin, A. M., & Aarnisalo, P. (2011). Cross-talk between NR4A orphan nuclear receptors and beta-catenin signaling pathway in osteoblasts. *Archives of Biochemistry and Biophysics*, *509*(1), 44–51.
Rauch, A., Seitz, S., Baschant, U., Schilling, A. F., Illing, A., Stride, B., ... Tuckermann, J. P. (2010). Glucocorticoids suppress bone formation by attenuating osteoblast differentiation via the monomeric glucocorticoid receptor. *Cell Metabolism*, *11*(6), 517–531.
Razandi, M., Pedram, A., Greene, G. L., & Levin, E. R. (1999). Cell membrane and nuclear estrogen receptors (ERs) originate from a single transcript: Studies of ERalpha and ERbeta expressed in Chinese hamster ovary cells. *Molecular Endocrinology*, *13*(2), 307–319.
Remen, K. M., Henning, P., Lerner, U. H., Gustafsson, J. A., & Andersson, G. (2011). Activation of liver X receptor (LXR) inhibits receptor activator of nuclear factor kappaB ligand (RANKL)-induced osteoclast differentiation in an LXRbeta-dependent mechanism. *The Journal of Biological Chemistry*, *286*(38), 33084–33094.
Ribaya-Mercado, J. D., & Blumberg, J. B. (2007). Vitamin A: Is it a risk factor for osteoporosis and bone fracture? *Nutrition Reviews*, *65*(10), 425–438.
Rickard, D. J., Iwaniec, U. T., Evans, G., Hefferan, T. E., Hunter, J. C., Waters, K. M., ... Turner, R. T. (2008). Bone growth and turnover in progesterone receptor knockout mice. *Endocrinology*, *149*(5), 2383–2390.
Riggs, B. L., Khosla, S., & Melton, L. J., 3rd. (1998). A unitary model for involutional osteoporosis: Estrogen deficiency causes both type I and type II osteoporosis in postmenopausal women and contributes to bone loss in aging men. *Journal of Bone and Mineral Research*, *13*(5), 763–773.
Robertson Remen, K. M., Gustafsson, J. A., & Andersson, G. (2013). The liver X receptor promotes macrophage differentiation and suppresses osteoclast formation in mouse RAW264.7 promyelocytic leukemia cells exposed to bacterial lipopolysaccharide. *Biochemical and Biophysical Research Communications*, *430*(1), 375–380.
Robertson Remen, K. M., Lerner, U. H., Gustafsson, J. A., & Andersson, G. (2013). Activation of the liver X receptor-beta potently inhibits osteoclastogenesis from lipopolysaccharide-exposed bone marrow-derived macrophages. *Journal of Leukocyte Biology*, *93*(1), 71–82.
Robertson, K. M., Norgard, M., Windahl, S. H., Hultenby, K., Ohlsson, C., Andersson, G., & Gustafsson, J. A. (2006). Cholesterol-sensing receptors, liver X receptor alpha and beta, have novel and distinct roles in osteoclast differentiation and activation. *Journal of Bone and Mineral Research*, *21*(8), 1276–1287.
Rodan, G. A. (1998). Bone homeostasis. *Proceedings of the National Academy of Sciences of the United States of America*, *95*(23), 13361–13362.
Rubin, M. R., Manavalan, J. S., Agarwal, S., McMahon, D. J., Nino, A., Fitzpatrick, L. A., & Bilezikian, J. P. (2014). Effects of rosiglitazone vs metformin on circulating osteoclast and osteogenic precursor cells in postmenopausal women with type 2 diabetes mellitus. *The Journal of Clinical Endocrinology and Metabolism*, *99*(10), E1933–E1942.
Sakai, S., Takaishi, H., Matsuzaki, K., Kaneko, H., Furukawa, M., Miyauchi, Y., ... Toyama, Y. (2009). 1-Alpha, 25-dihydroxy vitamin D3 inhibits osteoclastogenesis through IFN-beta-dependent NFATc1 suppression. *Journal of Bone and Mineral Metabolism*, *27*(6), 643–652.

Saneshige, S., Mano, H., Tezuka, K., Kakudo, S., Mori, Y., Honda, Y., et al. (1995). Retinoic acid directly stimulates osteoclastic bone resorption and gene expression of cathepsin K/OC-2. *The Biochemical Journal, 309*(Pt. 3), 721–724.

Sato, Y., Miyauchi, Y., Yoshida, S., Morita, M., Kobayashi, T., Kanagawa, H., … Miyamoto, T. (2014). The vitamin D analogue ED71 but not 1,25(OH)2D3 targets HIF1alpha protein in osteoclasts. *PLoS One, 9*(11). e111845.

Saxon, L. K., Galea, G., Meakin, L., Price, J., & Lanyon, L. E. (2012). Estrogen receptors alpha and beta have different gender-dependent effects on the adaptive responses to load bearing in cancellous and cortical bone. *Endocrinology, 153*(5), 2254–2266.

Scholtysek, C., Katzenbeisser, J., Fu, H., Uderhardt, S., Ipseiz, N., Stoll, C., … Kronke, G. (2013). PPARbeta/delta governs Wnt signaling and bone turnover. *Nature Medicine, 19*(5), 608–613.

Schwartz, A. V., Chen, H., Ambrosius, W. T., Sood, A., Josse, R. G., Bonds, D. E., … Margolis, K. L. (2015). Effects of TZD use and discontinuation on fracture rates in ACCORD bone study. *The Journal of Clinical Endocrinology and Metabolism, 100*(11), 4059–4066.

Sharifi, M., Ereifej, L., & Lewiecki, E. M. (2015). Sclerostin and skeletal health. *Reviews in Endocrine & Metabolic Disorders, 16*(2), 149–156.

Sims, N. A., Dupont, S., Krust, A., Clement-Lacroix, P., Minet, D., Resche-Rigon, M., … Baron, R. (2002). Deletion of estrogen receptors reveals a regulatory role for estrogen receptors-beta in bone remodeling in females but not in males. *Bone, 30*(1), 18–25.

Sinnesael, M., Claessens, F., Laurent, M., Dubois, V., Boonen, S., Deboel, L., & Vanderschueren, D. (2012). Androgen receptor (AR) in osteocytes is important for the maintenance of male skeletal integrity: Evidence from targeted AR disruption in mouse osteocytes. *Journal of Bone and Mineral Research, 27*(12), 2535–2543.

Sinnesael, M., Jardi, F., Deboel, L., Laurent, M. R., Dubois, V., Zajac, J. D., … Vanderschueren, D. (2015). The androgen receptor has no direct antiresorptive actions in mouse osteoclasts. *Molecular and Cellular Endocrinology, 411*, 198–206.

Skillington, J., Choy, L., & Derynck, R. (2002). Bone morphogenetic protein and retinoic acid signaling cooperate to induce osteoblast differentiation of preadipocytes. *The Journal of Cell Biology, 159*(1), 135–146.

Smith, E. P., Boyd, J., Frank, G. R., Takahashi, H., Cohen, R. M., Specker, B., … Korach, K. S. (1994). Estrogen resistance caused by a mutation in the estrogen-receptor gene in a man. *The New England Journal of Medicine, 331*(16), 1056–1061.

Soccio, R. E., Chen, E. R., & Lazar, M. A. (2014). Thiazolidinediones and the promise of insulin sensitization in type 2 diabetes. *Cell Metabolism, 20*(4), 573–591.

Song, H. M., Nacamuli, R. P., Xia, W., Bari, A. S., Shi, Y. Y., Fang, T. D., & Longaker, M. T. (2005). High-dose retinoic acid modulates rat calvarial osteoblast biology. *Journal of Cellular Physiology, 202*(1), 255–262.

Stein, B., & Yang, M. X. (1995). Repression of the interleukin-6 promoter by estrogen receptor is mediated by NF-kappa B and C/EBP beta. *Molecular and Cellular Biology, 15*(9), 4971–4979.

Still, K., Grabowski, P., Mackie, I., Perry, M., & Bishop, N. (2008). The peroxisome proliferator activator receptor alpha/delta agonists linoleic acid and bezafibrate upregulate osteoblast differentiation and induce periosteal bone formation in vivo. *Calcified Tissue International, 83*(4), 285–292.

Stunes, A. K., Westbroek, I., Gustafsson, B. I., Fossmark, R., Waarsing, J. H., Eriksen, E. F., … Syversen, U. (2011). The peroxisome proliferator-activated receptor (PPAR) alpha agonist fenofibrate maintains bone mass, while the PPAR gamma agonist pioglitazone exaggerates bone loss, in ovariectomized rats. *BMC Endocrine Disorders, 11*, 11.

Sun, H., Kim, J. K., Mortensen, R., Mutyaba, L. P., Hankenson, K. D., & Krebsbach, P. H. (2013). Osteoblast-targeted suppression of PPARgamma increases osteogenesis through activation of mTOR signaling. *Stem Cells, 31*(10), 2183–2192.

Takasu, H., Sugita, A., Uchiyama, Y., Katagiri, N., Okazaki, M., Ogata, E., & Ikeda, K. (2006). c-Fos protein as a target of anti-osteoclastogenic action of vitamin D, and synthesis of new analogs. *The Journal of Clinical Investigation, 116*(2), 528–535.

Takayanagi, H. (2007). The role of NFAT in osteoclast formation. *The Annals of the New York Academy of Sciences, 1116*, 227–237.

Tetradis, S., Bezouglaia, O., & Tsingotjidou, A. (2001). Parathyroid hormone induces expression of the nuclear orphan receptor Nurr1 in bone cells. *Endocrinology, 142*(2), 663–670.

Tetradis, S., Bezouglaia, O., Tsingotjidou, A., & Vila, A. (2001). Regulation of the nuclear orphan receptor Nur77 in bone by parathyroid hormone. *Biochemical and Biophysical Research Communications, 281*(4), 913–916.

Thompson, D. D., Simmons, H. A., Pirie, C. M., & Ke, H. Z. (1995). FDA guidelines and animal models for osteoporosis. *Bone, 17*(4 Suppl.), 125S–133S.

Toyosawa, S., Shintani, S., Fujiwara, T., Ooshima, T., Sato, A., Ijuhin, N., & Komori, T. (2001). Dentin matrix protein 1 is predominantly expressed in chicken and rat osteocytes but not in osteoblasts. *Journal of Bone and Mineral Research, 16*(11), 2017–2026.

Tremblay, G. B., Kunath, T., Bergeron, D., Lapointe, L., Champigny, C., Bader, J. A., … Giguere, V. (2001). Diethylstilbestrol regulates trophoblast stem cell differentiation as a ligand of orphan nuclear receptor ERR beta. *Genes and Development, 15*(7), 833–838.

Ucer, S., Iyer, S., Bartell, S. M., Martin-Millan, M., Han, L., Kim, H. N., … Manolagas, S. C. (2015). The effects of androgens on murine cortical bone do not require AR or ERalpha signaling in osteoblasts and osteoclasts. *Journal of Bone and Mineral Research, 30*(7), 1138–1149.

van Beek, E., Lowik, C., Karperien, M., & Papapoulos, S. (2006). Independent pathways in the modulation of osteoclastic resorption by intermediates of the mevalonate biosynthetic pathway: The role of the retinoic acid receptor. *Bone, 38*(2), 167–171.

Van Cromphaut, S. J., Dewerchin, M., Hoenderop, J. G., Stockmans, I., Van Herck, E., Kato, S., … Carmeliet, G. (2001). Duodenal calcium absorption in vitamin D receptor-knockout mice: Functional and molecular aspects. *Proceedings of the National Academy of Sciences of the United States of America, 98*(23), 13324–13329.

van Driel, M., Koedam, M., Buurman, C. J., Roelse, M., Weyts, F., Chiba, H., … van Leeuwen, J. P. (2006). Evidence that both 1alpha,25-dihydroxyvitamin D3 and 24-hydroxylated D3 enhance human osteoblast differentiation and mineralization. *Journal of Cellular Biochemistry, 99*(3), 922–935.

Vasheghani, F., Zhang, Y., Li, Y. H., Blati, M., Fahmi, H., Lussier, B., … Kapoor, M. (2015). PPARgamma deficiency results in severe, accelerated osteoarthritis associated with aberrant mTOR signalling in the articular cartilage. *Annals of the Rheumatic Diseases, 74*(3), 569–578.

Vega, R. B., Huss, J. M., & Kelly, D. P. (2000). The coactivator PGC-1 cooperates with peroxisome proliferator-activated receptor alpha in transcriptional control of nuclear genes encoding mitochondrial fatty acid oxidation enzymes. *Molecular and Cellular Biology, 20*(5), 1868–1876.

Vinayavekhin, N., & Saghatelian, A. (2011). Discovery of a protein-metabolite interaction between unsaturated fatty acids and the nuclear receptor Nur77 using a metabolomics approach. *Journal of the American Chemical Society, 133*(43), 17168–17171.

Viswakarma, N., Jia, Y., Bai, L., Vluggens, A., Borensztajn, J., Xu, J., & Reddy, J. K. (2010). Coactivators in PPAR-regulated gene expression. *PPAR Research, 2010*, 250126.

Wan, Y. (2010). PPARgamma in bone homeostasis. *Trends in Endocrinology and Metabolism, 21*(12), 722–728.

Wan, Y., Chong, L. W., & Evans, R. M. (2007). PPAR-gamma regulates osteoclastogenesis in mice. *Nature Medicine, 13*(12), 1496–1503.
Wang, Z., Benoit, G., Liu, J., Prasad, S., Aarnisalo, P., Liu, X., … Perlmann, T. (2003). Structure and function of Nurr1 identifies a class of ligand-independent nuclear receptors. *Nature, 423*(6939), 555–560.
Wang, Y., Zhu, J., & DeLuca, H. F. (2014). Identification of the vitamin D receptor in osteoblasts and chondrocytes but not osteoclasts in mouse bone. *Journal of Bone and Mineral Research, 29*(3), 685–692.
Wang, L., Zuercher, W. J., Consler, T. G., Lambert, M. H., Miller, A. B., Orband-Miller, L. A., … Nolte, R. T. (2006). X-ray crystal structures of the estrogen-related receptor-gamma ligand binding domain in three functional states reveal the molecular basis of small molecule regulation. *The Journal of Biological Chemistry, 281*(49), 37773–37781.
Wei, W., Schwaid, A. G., Wang, X., Wang, X., Chen, S., Chu, Q., … Wan, Y. (2016). Ligand activation of ERRalpha by cholesterol mediates statin and bisphosphonate effects. *Cell Metabolism, 23*(3), 479–491.
Wei, W., Wang, X., Yang, M., Smith, L. C., Dechow, P. C., Sonoda, J., … Wan, Y. (2010). PGC1beta mediates PPARgamma activation of osteoclastogenesis and rosiglitazone-induced bone loss. *Cell Metabolism, 11*(6), 503–516.
Weinstein, R. S., Jilka, R. L., Parfitt, A. M., & Manolagas, S. C. (1998). Inhibition of osteoblastogenesis and promotion of apoptosis of osteoblasts and osteocytes by glucocorticoids. Potential mechanisms of their deleterious effects on bone. *The Journal of Clinical Investigation, 102*(2), 274–282.
Whiting, S. J., & Lemke, B. (1999). Excess retinol intake may explain the high incidence of osteoporosis in northern Europe. *Nutrition Reviews, 57*(6), 192–195.
Williams, J. A., Kondo, N., Okabe, T., Takeshita, N., Pilchak, D. M., Koyama, E., … Iwamoto, M. (2009). Retinoic acid receptors are required for skeletal growth, matrix homeostasis and growth plate function in postnatal mouse. *Developmental Biology, 328*(2), 315–327.
Wilson, T. E., Fahrner, T. J., Johnston, M., & Milbrandt, J. (1991). Identification of the DNA binding site for NGFI-B by genetic selection in yeast. *Science, 252*(5010), 1296–1300.
Windahl, S. H., Borjesson, A. E., Farman, H. H., Engdahl, C., Moverare-Skrtic, S., Sjogren, K., … Ohlsson, C. (2013). Estrogen receptor-alpha in osteocytes is important for trabecular bone formation in male mice. *Proceedings of the National Academy of Sciences of the United States of America, 110*(6), 2294–2299.
Windahl, S. H., Saxon, L., Borjesson, A. E., Lagerquist, M. K., Frenkel, B., Henning, P., … Ohlsson, C. (2013). Estrogen receptor-alpha is required for the osteogenic response to mechanical loading in a ligand-independent manner involving its activation function 1 but not 2. *Journal of Bone and Mineral Research, 28*(2), 291–301.
Wu, X., Peters, J. M., Gonzalez, F. J., Prasad, H. S., Rohrer, M. D., & Gimble, J. M. (2000). Frequency of stromal lineage colony forming units in bone marrow of peroxisome proliferator-activated receptor-alpha-null mice. *Bone, 26*(1), 21–26.
Xue, Y., & Fleet, J. C. (2009). Intestinal vitamin D receptor is required for normal calcium and bone metabolism in mice. *Gastroenterology, 136*(4), 1317–1327. e1311–1312.
Yamaguchi, M., & Weitzmann, M. N. (2012). High dose 1,25(OH)2D3 inhibits osteoblast mineralization in vitro. *International Journal of Molecular Medicine, 29*(5), 934–938.
Yamamoto, Y., Yoshizawa, T., Fukuda, T., Shirode-Fukuda, Y., Yu, T., Sekine, K., … Imai, Y. (2013). Vitamin D receptor in osteoblasts is a negative regulator of bone mass control. *Endocrinology, 154*(3), 1008–1020.

Yoshitake, H., Rittling, S. R., Denhardt, D. T., & Noda, M. (1999). Osteopontin-deficient mice are resistant to ovariectomy-induced bone resorption. *Proceedings of the National Academy of Sciences of the United States of America, 96*(14), 8156–8160.

Yoshizawa, T., Handa, Y., Uematsu, Y., Takeda, S., Sekine, K., Yoshihara, Y., ... Kato, S. (1997). Mice lacking the vitamin D receptor exhibit impaired bone formation, uterine hypoplasia and growth retardation after weaning. *Nature Genetics, 16*(4), 391–396.

Yu, D. D., & Forman, B. M. (2005). Identification of an agonist ligand for estrogen-related receptors ERRbeta/gamma. *Bioorganic and Medicinal Chemistry Letters, 15*(5), 1311–1313.

Zetterstrom, R. H., Solomin, L., Mitsiadis, T., Olson, L., & Perlmann, T. (1996). Retinoid X receptor heterodimerization and developmental expression distinguish the orphan nuclear receptors NGFI-B, Nurr1, and Nor1. *Molecular Endocrinology, 10*(12), 1656–1666.

Zhan, Y., Du, X., Chen, H., Liu, J., Zhao, B., Huang, D., ... Wu, Q. (2008). Cytosporone B is an agonist for nuclear orphan receptor Nur77. *Nature Chemical Biology, 4*(9), 548–556.

Zhang, X. K., Lehmann, J., Hoffmann, B., Dawson, M. I., Cameron, J., Graupner, G., ... Pfahl, M. (1992). Homodimer formation of retinoid X receptor induced by 9-cis retinoic acid. *Nature, 358*(6387), 587–591.

Zuercher, W. J., Gaillard, S., Orband-Miller, L. A., Chao, E. Y., Shearer, B. G., Jones, D. G., ... Willson, T. M. (2005). Identification and structure-activity relationship of phenolic acyl hydrazones as selective agonists for the estrogen-related orphan nuclear receptors ERRbeta and ERRgamma. *Journal of Medicinal Chemistry, 48*(9), 3107–3109.

CHAPTER FOUR

Estrogen Hormone Biology

Katherine J. Hamilton, Sylvia C. Hewitt, Yukitomo Arao, Kenneth S. Korach[1]

Reproductive and Developmental Biology Laboratory, National Institute of Environmental Health Sciences/NIH, Research Triangle Park, NC, United States

[1]Corresponding author: e-mail address: korach@niehs.nih.gov

Contents

Abstract

The hormone estrogen is involved in both female and male reproduction, as well as numerous other biological systems including the neuroendocrine, vascular, skeletal, and immune systems. Therefore, it is also implicated in many different diseases and conditions such as infertility, obesity, osteoporosis, endometriosis, and a variety of cancers. Estrogen works through its two distinct nuclear receptors, estrogen receptor alpha (ERα) and estrogen receptor beta (ERβ). Various transcriptional regulation mechanisms have been identified as the mode of action for estrogen, mainly the classical mechanism with direct DNA binding but also a nongenomic mode of action and one using tethered or

Current Topics in Developmental Biology, Volume 125
ISSN 0070-2153
http://dx.doi.org/10.1016/bs.ctdb.2016.12.005

indirect binding. The expression profiles of ERα and ERβ are unique with the primary sites of ERα expression being the uterus and pituitary gland and the main site of ERβ expression being the granulosa cells of the ovary. Mouse models with knockout or mutation of Esr1 and Esr2 have furthered our understanding of the role of each individual receptor plays in physiology. From these studies, it is known that the primary roles for ERα are in the uterus and neuroendocrine system, as female mice lacking ERα are infertile due to impaired ovarian and uterine function, whereas female mice lacking ERβ are subfertile due to ovarian defects. The development of effective therapies for estrogen-related diseases has relied on an understanding of the physiological roles and mechanistic functionalities of ERα and ERβ in human health and disease.

1. INTRODUCTION

Our understanding of mechanisms by which hormones act has evolved since the first description over 100 years ago; especially, regarding the role of nuclear receptors and receptor-mediated signaling first proposed by Jensen over 50 years ago (Jensen, 1962; Jensen & Jacobson, 1960, 1962). Our knowledge of cellular mechanisms from the initial concepts of ligand receptor binding, activation, direct DNA binding, and resulting gene regulation now includes non-DNA binding or tethering, cellular nongenomic signaling, and receptor-mediated nonligand hormone activities (Hewitt, Winuthayanon, & Korach, 2016). Estrogen, one of the first hormone substances identified, was thought to have only female-selective activities important in female reproduction. We now know, however, that estrogen is also involved in male reproduction and in numerous other systems including the neuroendocrine, vascular, skeletal, and immune systems of both males and females. Estrogen influences many physiological processes, as it is also implicated in many different diseases including obesity, metabolic disorder, and a variety of cancers, osteoporosis, lupus, endometriosis, and uterine fibroids (Burns & Korach, 2012; Deroo & Korach, 2006).

2. CELL MECHANISMS

It is now accepted that the predominant mechanism of estrogen action is through nuclear estrogen receptor (ER) expression in estrogen target organs (Mangelsdorf et al., 1995). For many years, it was thought there was only a single estrogen receptor (ERα), but a second form of estrogen receptor (ERβ), was later discovered (Kuiper, Enmark, Pelto-Huikko, Nilsson, & Gustafsson, 1996). The biological effects of estrogen, described later in this chapter, are mediated through these two distinct ER proteins, ERα and ERβ. Separate genes on nonhomologous chromosomes encode these receptors; the expression profiles of which are quite different across tissues and cell types. The predominant expression tissues for ERα include: uterus and pituitary gland with the highest levels, liver, hypothalamus, bone, mammary gland, cervix, and vagina. ERβ expression on the other hand is expressed in fewer tissues by most analyses, but tissues with predominant levels include ovary, lung, and prostate (Couse, Lindzey, Grandien, Gustafsson, & Korach, 1997). ERβ expression is especially high in the ovary and is found exclusively in the granulosa cells. Many therapeutic interventions to estrogen-related diseases target the functions of ERα and ERβ. Such therapeutic approaches highlight the importance of understanding the physiological role of ERα and ERβ in tissues and their in vivo mechanistic functionality to identify effective treatments and minimize side effects.

The ERs are members of the nuclear receptor superfamily of hormone receptors and are composed of several main structural features that are consistent in these proteins (Aagaard, Siersbaek, & Mandrup, 2011). All members of this superfamily are comprised of four structural and functional domains: an amino-terminal domain (A/B-domain), a DNA-binding domain (DBD; C-domain), a hinge region (D-domain), and a ligand-binding domain (LBD; E-domain). The ERs have an additional fifth domain: the carboxyl-terminal domain (F-domain) whose function is still unknown (Mangelsdorf et al., 1995). In the case of ERα and ERβ, the C and E domains carry a high degree of homology between the two forms; however, the A/B, D, and F domains are divergent (Germain, Staels, Dacquet, Spedding, & Laudet, 2006; Mangelsdorf et al., 1995). The A/B-domain contains the transcription activation function 1 (AF-1) which is reported to be important for ligand-independent transactivation (Bourguet, Germain, & Gronemeyer, 2000). The LBD or E-domain of ERs contains the transcription activation function 2 (AF-2) that is important for ligand-dependent transcriptional regulation (Bourguet et al., 2000). Helix 12 is a highly conserved region within the LBD and is the core of the AF-2 functionality. The structural configuration of helix 12 is changed by ligand binding resulting in either an active (agonist bound) or inactive (antagonist bound) form for the transcription regulation (Green & Carroll, 2007; Klinge, 2000).

Estrogen works through several possible cellular mechanisms to mediate its biological responses as shown in Fig. 1. These include two major cellular actions involving the receptors: rapid nongenomic effects and genomic

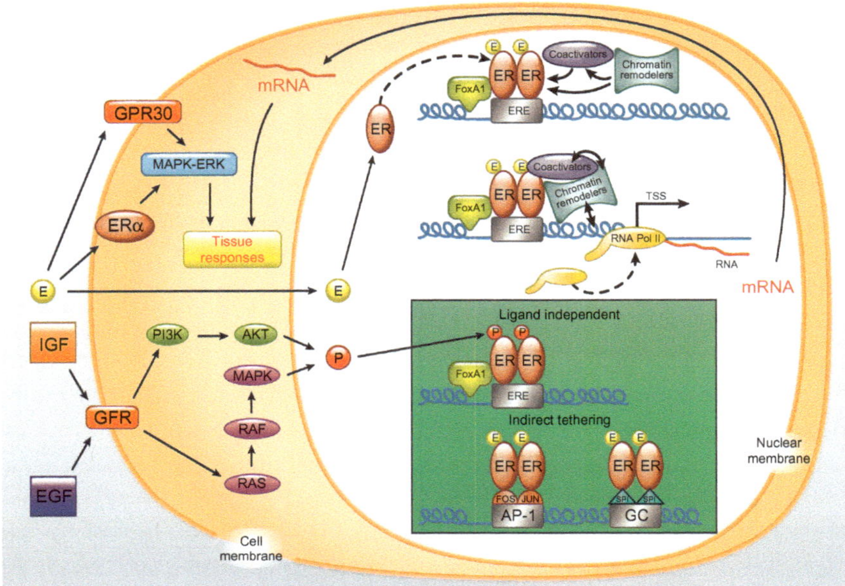

Fig. 1 Cellular mechanisms of estrogen action. Model of nuclear and nonnuclear estrogen receptor (ER) action. Estrogen (E circles) and ER complex binds directly to the regulatory DNA elements (estrogen-responsive element (ERE)) recruiting additional factors involved in transcriptional regulation. ER can also bind indirectly through a tethering mechanism to AP1- or Sp1-binding sites (GC) to regulate transcription. Growth factors (IGF and EGF) can phosphorylate ER through membrane growth factor receptor (GFR)-mediated intracellular signaling pathways (P circles) to regulate gene expression in the absence of ligand (nuclear action). Estrogen also binds and activates membrane ERα or GPR30, inducing the intracellular signaling pathway (nonnuclear action) that is rapid.

activities (Hewitt et al., 2016). Several studies, primarily in cell culture, have shown these rapid actions occur within minutes of hormone treatment and can be silenced by inhibition of either the MAPK/ERK or AKT signaling pathways (Clark et al., 2014; Kelly & Levin, 2001). Activation of these intracellular signaling pathways has been shown to involve a plasma membrane-associated process that is mediated by either a G protein-coupled receptor, GPER1 (originally designated GPR30), or a caveolin-associated form of ERα (Levin, 2015; Prossnitz & Hathaway, 2015). Estrogen signaling at the membrane solely involves ERα and is shown to require palmitoylation at cysteine 447 (human) or cysteine 451 (mouse) (Levin, 2015). Mutation of this cysteine in mouse models results in differing phenotypes, effects, and mechanistic interpretations (Adlanmerini et al., 2014; Pedram, Razandi, Lewis, Hammes, & Levin, 2014). How significantly nongenomic signaling contributes to genomic actions of ERα and the biology of hormone responsiveness is still not totally resolved and requires continued study.

Three major genomic ER-mediated transcriptional regulation mechanisms have been characterized. These include the direct binding to regulatory elements of DNA (classical), indirect binding to other existing transcription factors which are bound to DNA (tethering), and ligand-independent receptor activation, which is proposed to involve altered phosphorylation of sites on the receptor protein. Examples of each of these genomic modes of action for estrogen and ER have been published, although most studies have investigated the activity primarily involving ERα. In the first example, hormone–ER binding causes a conformational change in the LBD, allowing helix 12 to accept coactivator interactions. Coactivator binding is required for the resulting genomic response and is directly proportional to the amplitude of this response. In the absence of hormone, ERα is bound to DNA in an inactive state, as shown in both cell culture and in vivo mouse studies by ChIP-Seq (Carroll et al., 2005; Hewitt et al., 2012). Hormone binding increases the number of binding peaks in the genome. Mouse models in which the DBD of ERα is mutated indicate that direct DNA binding is required to elicit hormone responses and biological activity. Further research will determine whether it is the only activity that is required or if complementary actions of other signaling mechanisms (Ahlbory-Dieker et al., 2009; Hewitt et al., 2014). As shown in Fig. 1, other nuclear factors influence direct binding such as a pioneering factor FoxA1, which is bound at sites to allow the recruitment of chromatin remodeling proteins, opening the chromatin to give ER accessibility to its regulatory DNA sites. Following the assembly of the ER transcription complex, composed of a multitude of components (Carroll & Brown, 2006) gene transcription is initiated by recruitment of polymerase II. A second method (shown in Fig. 1, Box) primarily described in cell culture is the indirect or tethered mechanism of hormone receptor action in which the hormone receptor modulates gene expression by protein–protein interactions with existing transcription factors (e.g., *Fos/jun*), which bind directly to their respective response elements (AP1) (Jakacka et al., 2001). Other examples of regulatory elements have included binding to factors in Sp1 sites in GC-rich regions of DNA (Kushner et al., 2000). Lastly, ER can have regulatory activities and drive hormone responses in the absence of hormone through ligand-independent activation by growth factors or other intracellular signaling pathways, thought to involve phosphorylation of certain serine residues on the receptor (Smith, 1998). Such a coupling of nongenomic and genomic signaling may be an explanation for the complementation of different cellular signaling pathways, which elicit the broad spectrum of hormone responses of estrogen action.

3. UTERINE ESTROGEN RESPONSE

The ovariectomized (ovexed) mouse uterus is an estrogen-responsive organ and therefore a valuable model to study ER-mediated responses. The rodent uterus is a bicornuate tube made up of outer muscle cell layers (myometria), an inner lumen lined with luminal epithelial cells, and a layer of stroma cells between the lumen and myometrium. The uterus also contains glandular structures that are lined with epithelial cells. The endogenous stimulation of the uterus occurs during the transient proestrus surge of the reproductive cycle. Experimentally, a single injection of estrogen can mimic this stimulation by being administered to an ovexed animal, and the uterus will then undergo a series of ordered, well-characterized events that can be divided into an initial (early) phase and subsequent (late) responses culminating specifically in waves of mitosis restricted to only uterine epithelial cells. These responses are mediated by ERα, which is expressed in all uterine cells (luminal and glandular epithelial, stromal, and myometrial cells).

Studies using the ovexed mouse uterine model have examined the regulation of endogenous uterine genes over a 24-h stimulatory time course and shown that the gene regulation pattern follows the progression of early (within 2 h after estrogen was administered) or late (occurring 12–24 h after estrogen was administered) events. Some gene regulation was seen at intervening time points, but most fall within either the early or late clusters (Hewitt et al., 2003). In experimental analysis of estrogen-responsive uterine genes, it is apparent that samplings at 2-h or 24-h time points will represent most of the observed gene responses correlating with tissue physiological actions.

Interaction of ERα and RNA polymerase II (Pol II) was analyzed using ChIP-seq in uterine tissue (Hewitt et al., 2012) to understand ERα DNA binding within an in vivo system. In vehicle-treated unstimulated samples, more than 5000 peaks were mapped indicating ERα was already bound to the chromatin in the absence of hormone (Hewitt et al., 2012). Estrogen treatment increased the amount of ERα binding at these sites and also led to ERα binding to additional regions (Hewitt et al., 2012) with a total of more than 17,000 sites. The number of active annotated genes (with Pol II at the transcription start site) within 100 KB of ERα peaks increased from

4672 (1.1 ERα peaks/active gene) to 6519 (2.6 ERα peaks per active gene) thereby showing that in the absence of hormone, ERα is bound to DNA sites, and that hormone treatment increased the number 2.4-fold. Analysis of the ER binding sites for transcription factor binding motifs revealed that ERE motifs were present in 35% of the vehicle sites and were more abundant (59%) in the estrogen-treated sites (Hewitt et al., 2012). Thus, ERE motifs are important for estrogen-dependent ER recruitment. The computed consensus motif for the sequences bound to ERα in uterine chromatin matched the experimentally derived ERE (GGTCAnnnTGACC) (Hewitt et al., 2012), indicating preference for this motif in a biological system. Interestingly, at the sites that were not enriched for ERE motifs, numerous other motifs were seen; notably homeobox (Hox) motifs were highly enriched (Hewitt et al., 2012). Many Hox family members are expressed in the uterus (Hewitt et al., 2012), and Hoxa10 and Hoxa11 have been demonstrated to play key roles in uterine function (Eun Kwon & Taylor, 2004). ERα binding in the uterine tissue was primarily distal from promoters (Hewitt et al., 2012), which has similarly been observed in ERα ChIP-seq in MCF-7 cells. When comparing ChIP-seq data to microarray profiles, upregulated transcripts at early time points (2 and 6 h) were significantly more likely to have ERα binding at their promoters (0–10 kb 5′) than downregulated genes (Hewitt et al., 2012).

3.1 Genetic Control of Estrogen Responses

Differences in uterine estrogen sensitivity of two mouse strains, C57Bl6 (more responsive uterus) and C3H (less sensitive uterus) have been mapped to associated quantitative trait loci (QTL) (Roper et al., 1999). Uterine transcriptional profiles of C57Bl6 and C3H mice (basal or 2 h or 24 h after estrogen treatment) include response differences correlated with the QTL on chromosomes 4, 5, 11, and 16 (Wall et al., 2013). For example, *Ngfr* is in the chromosome 11 QT locus, and its transcript is expressed at a threefold higher level in ovexed untreated C57Bl6 than in ovexed untreated C3H uterine samples. Uterine NGF signaling has been shown to impact pregnancy (Hah & Kraus, 2014). *Runx1*, which was within the chromosome 16 QT locus, and can enhance estrogen responses (Chimge & Frenkel, 2013), was shown to be present at higher levels in C57Bl6 than C3H uterine epithelial cells (Wall et al., 2013). Transcripts that showed strain-selective differences indicated C3H-selective enrichment of apoptosis, consistent with increase in the apoptosis indicator *Casp3*, and decrease in the apoptosis inhibitor *Naip1* (*Birc1a*) in C3H vs C57Bl6 following treatment with estrogen (Wall et al., 2013). Mammary gland response differences were also examined (Wall et al., 2014), where an opposite strain sensitivity observation was reported (C3H more sensitive than C57Bl6). Strain-selective transcripts were identified in the mammary samples as well. Most interesting was the opposite pattern of *Runx1* expression, with higher levels in C3H than C57Bl6 mammary epithelia, a pattern consistent with higher estrogen sensitivity of C3H mammary glands (Wall et al., 2014). Understanding differences in sensitivity to estrogen is important in understanding genetic contributions to the impact of xenoestrogens on populations of exposed humans and wildlife.

3.2 ERα Mutations Demonstrate Uterine Mechanisms

Since ERα is detected in all uterine cells, deletion or mutation of the receptor is expected to have a profound impact on 17β-estradiol (E2)-mediated responses. Mice with ERα deletion or mutation are therefore an optimal biological model in which to dissect details of ERα mechanisms (Table 1). Mice that lack ERα ($Esr1^{-/-}$; a.k.a. αERKO) develop a hypoplastic uterus that includes all uterine cell types and structures; however, there is no uterine maturation or growth at puberty, and no response to E2 (Couse & Korach, 1999). Since the female reproductive tract is composed of several tissue types, all expressing ERα, identifying the cell type selectivity of ERα activity related to biological responses is critical. This cell specificity becomes of particular significance when comparing the varying physiological responses of the uterus to its inherent functions involving proliferative to luteal secretory responses and implantation (Wang & Dey, 2006) and the development of diseases, including endometriosis, cystic endometrial hyperplasia, fibroids, and endometrial cancer. More specifically, in contrast to adults, neonates, and prepubertal experimental animals exhibit both stromal and epithelial tissue proliferation under E2 stimulation, while in adults, E2 selectively stimulates growth only in epithelial cells with the stroma remaining quiescent (Quarmby & Korach, 1984). Two major concepts have been published to explain the mitogenic mechanisms for the E2 activity; one involves direct estrogen action through ER in the epithelial cell and the second regards a paracrine mechanism of direct stimulation of E2/ER in stromal cells to induce a mitogenic signal (e.g., growth factor) on the epithelium. To identify the specific epithelial responses to E2, uterine

Table 1 Uterine Phenotypes of Estrogen Receptor Mutants

Gene	Mutation	Nick-Names	Uterine Phenotypes	References
Esr1	Homozygous null for ERα	αERKO, Ex3αERKO	Normal uterine development but exhibits hypoplastic uteri	Antonson, Omoto, Humire, and Gustafsson (2012), Curtis Hewitt, Goulding, Eddy, and Korach (2002), Curtis and Korach (1999), Dupont et al. (2000), Hewitt et al. (2010), and Lubahn et al. (1993)
			Insensitive to the proliferative and differentiating effects of endogenous E2, growth factors, and exogenous E2	
			Implantation defect	
			Lack decidualization	
			Infertile	
Esr1	One mutated allele of two-point mutation in ERα DBD (E207A and G208A) and one WT allele	$NERKI^{+/-}$ $ER^{AA/+}$	Normal uterine development but exhibits hyperplastic uteri	Jakacka et al. (2002)
			Hypersensitive to estrogen	
			Infertile	
Esr1	One mutated allele of two-point mutation in DNA-binding domain of ERα (E207A and G208A) and one ERα null allele	*ERα KIKO*, $ER^{AA/-}$	Normal uterine development	Hewitt, Li, Li, and Korach (2010) and O'Brien et al. (2006)
			Insensitive to the proliferative effects of exogenous E2 treatment	
			Infertile	

Continued

Table 1 Uterine Phenotypes of Estrogen Receptor Mutants—cont'd

Gene	Mutation	Nick-Names	Uterine Phenotypes	References
Esr1	Four-point mutation of DBD ERα (Y201E, K210S, K214A, and R215A)	$ER\alpha^{EAAE/EAAE}$	Normal uterine development but exhibits hypoplastic uteri	Ahlbory-Dieker et al. (2009)
			Loss of E2-induced uterine transcripts	
			Infertile	
Esr1	Deletion of amino acids 2–128 including AF2 domain of ERα	$ER\alpha AF\text{-}1^0$	Normal uterine development and architecture	Abot et al. (2013) and Billon-Gales et al. (2009)
			Blunted E2 response	
			Infertile	
Esr1	Deletion of amino acids 543–549 in LBD/AF-2 of ERα	$ER\alpha AF\text{-}2^0$	Normal uterine development but exhibits hypoplastic uteri	Billon-Gales et al. (2011)
			Insensitive to E2 treatment	
			Infertile	
Esr1	Two-point mutation in LBD/AF-2 of ERα (L543A and L544A)	$AF2ER^{KI/KI}$	Normal uterine development but exhibits hypoplastic uteri	Arao et al. (2011)
			Insensitive to E2 treatment	
			ER antagonists and partial agonist (ICI 182, 780, and TAM) induced uterine epithelial proliferation	
			Growth factor did not induce the uterine epithelial cell proliferation	
			Infertile	

Esr1	Point mutation in LBD of ERα (G525L)	*ENERKI* $ER\alpha^{G525L}$	Normal uterine development but exhibits hypoplastic uteri	Sinkevicius et al. (2008)
			Insensitive to E2 treatment. Synthetic estrogens PPT and DES induce uterine growth	
			IGF-1-induced patchy uterine epithelial growth	
			Infertile	
Esr1	Female reproductive tract epithelial cell-specific deletion of ERα	$Wnt7a^{Cre+}$; $Esr1^{f/f}$ Epi ERα cKO $WE^{d/d}$	Normal uterine development	Pawar, Laws, Bagchi, and Bagchi (2015), Winuthayanon, Hewitt, and Korach (2014), and Winuthayanon, Hewitt, Orvis, Behringer, and Korach (2010)
			Sensitive to E2- and growth factors-induced epithelial cell proliferation	
			Selective loss of E2-target gene response	
			Implantation defect	
			Decidualization defect	
			Infertile	
Esr1	Uterine-specific deletion of ERα	$PgrCre^{Cre+}$; $Esr1^{f/f}$ERα Ut cKO $Esr^{d/d}$	Normal uterine development	Pawar et al. (2015)
			Insensitive to E2	
			Decidualization defect	
			Infertile	

Continued

Table 1 Uterine Phenotypes of Estrogen Receptor Mutants—cont'd

Gene	Mutation	Nick-Names	Uterine Phenotypes	References
Esr1	Point mutation of ERα palmitoylation site (C541A)	C451A-ERα, NOER (nuclear-only ERa)	**C451A-ERα:** normal uterine development, E2 growth response **NOER:** hypoplastic ERα-null like uterus	Adlanmerini et al. (2014) and Pedram et al. (2014)
Esr1	LBD of ERα fused with multiple palmitoylation sites from the neuromodulin protein	MOER	Normal uterine development but exhibits hypoplastic uteri	
$Esr2^{-/-}$	Homozygous null for ERβ	$Esr2^{-/-}$ (βERKO, Ex3βERKO, $ER\beta_{ST}{}^{L-/L-}$)	Exhibit grossly normal uterine development and function Sensitive to E2 treatment Some $Esr2^{-/-}$ lines reported elevated uterine epithelial proliferation after E treatment	Antal, Krust, Chambon, and Mark (2008a), Dupont et al. (2000), Krege et al. (1998), Wada-Hiraike et al. (2006), and Binder et al. (2013)
Esr1 and *Esr2*	Homozygous null for both ERα and ERβ	αβERKO	Normal uterine development but exhibit hypoplastic uteri, similar αERKO. Insensitive to E2, infertile	Couse et al. (1999) and Dupont et al. (2000)

epithelial-specific ERα knockout mice (*Wnt7a*$^{Cre/+}$; *Esr1*$^{f/f}$ or referred to as "epithelial ER cKO") were generated by crossing *Esr1*-floxed mice (Hewitt, Kissling, et al., 2010) with *Wnt7a*$^{Cre/+}$ mice (Huang, Orvis, Wang, & Behringer, 2012). Using the epithelial ER cKO mouse model, it has been shown that ERα in uterine epithelium is dispensable for epithelial growth response to E2. This finding supports mediation by stromal mitogenic paracrine factors, such as IGF-1. However, epithelial ERα is required for a full growth response of endometrial hyperplasia by actively inhibiting epithelial apoptosis in the uterus (Winuthayanon et al., 2010). To dissect the mediators of epithelial ERα response during uterine transcription, microarray analysis was performed to evaluate the differentially expressed genes in the presence (WT control littermates, referred to as WT) or absence (epithelial ER cKO) of epithelial ERα after E2 treatment for 2 h or 24 h in ovexed adult females (Winuthayanon et al., 2014). RNA microarray analysis revealed approximately 20% of the genes differentially expressed at 2 h were epithelial ERα independent, as they were preserved in the epithelial ER cKO uteri. This indicates that regulation of the early uterine transcripts mediated by stromal ERα is sufficient to promote initial proliferative responses. However, more than 80% of the differentially expressed transcripts at 24 h were not regulated in the epithelial ER cKO uteri, indicating most late transcriptional regulation required epithelial ERα, especially those involved in mitosis. This shows that loss of regulation of these later transcripts results in the blunted subsequent uterine growth after 3 days of E2 treatment. These transcriptional profiles at 2 and 24 h of E2 treatment correlate with previously observed biological responses, in which the initial proliferative response (at 24 h E2 treatment) is independent of epithelial ERα and thus dependent on stromal ERα, yet epithelial ERα is essential for subsequent maintenance of tissue responsiveness during 3 days of E2 treatment.

In addition to uterine response to E2, epithelial ER cKO females were infertile partly due to an implantation defect (Winuthayanon et al., 2010). In addition, they fail to decidualize (Pawar et al., 2015). The role of epithelial ERα during implantation was examined using a uterine receptivity model that has been previously published (Tong & Pollard, 1999) by treating the mice with a series of E2 and P4 injections to mimic the hormonal profile during implantation. E + Pe treatment significantly increases uterine weight in wild-type (WT) females, as well as proliferation of stromal cells, but not epithelial cells. In epithelial ER cKO uteri, treatment with E + Pe showed a dampened uterine weight increase when compared to the WT-treated group (Winuthayanon et al., 2014), and a slight decrease in stromal cell proliferation. However, epithelial cell proliferation was significantly higher in epithelial ER cKO compared to WT uteri. This suggests that lack of uterine epithelial ERα does not affect stromal cell proliferation but leads to an inability to appropriately arrest epithelial cell proliferation, a key requirement for embryo attachment and implantation. Additionally, leukemia inhibitory factor (*Lif*) (Stewart et al., 1992) and Indian hedgehog (*Ihh*) (Lee et al., 2006), both required for uterine receptivity and induced in WT uteri, were not induced in the epithelial ER cKO samples, confirming induction of these factors requires epithelial ERα. Comparable expression of PR is seen in WT and epithelial ER cKO uteri using immunohistochemical analysis. Moreover, expression of HAND2, a PR-regulated transcription factor expressed in uterine stromal cells during implantation (Li et al., 2011), showed a similar pattern in the epithelial ER cKO uteri and WT. This indicates that the expression of PR and its downstream effector (HAND2) in the stromal cells were not disrupted by a lack of epithelial ERα. In summary, loss of epithelial ERα disrupted progesterone's ability to inhibit E_2-induced epithelial cell proliferation, but did not affect uterine stromal cell proliferation. Understanding the ERα epithelial cell-specific mechanisms and gene responses for controlling cell growth in the uterus is informative toward understanding a basis for uterine diseases such as endometriosis and endometrial cancer.

3.3 Tethered Pathway Analysis Using DNA-Binding Deficient ERα Mutants

Studies using in vitro cell culture-based models have indicated that estrogen-responsive genes that lack the canonical ERE sequence can interact with ERs via a tethering mechanism whereby ER is recruited by AP1 or SP1 bound to their respective response elements (Jakacka et al., 2001; Kushner et al., 2000; O'Lone, Frith, Karlsson, & Hansen, 2004; Safe, 2001) (Fig. 1). To study the relative biological roles for tethered and ERE DNA-binding mechanisms in vivo, two different ERα "knock in" mouse models have been created that have mutations of the first zinc finger of the ERα DBD. Both DNA-binding mutations were designed to prevent ER–ERE binding, while retaining the ability to regulate genes via the tethered pathway (Ahlbory-Dieker et al., 2009; Jakacka et al., 2001). The first mouse model was referred to as the "nonclassical ER knock in" (Nerki) (Jakacka et al., 2002). Female mice heterozygous for this mutation are infertile due to ovarian and uterine pathologies (Jakacka et al., 2002); however, by intercrossing with the *Esr1*$^{-/-}$ global knockout line, a mouse possessing one copy of the Nerki ERα allele and one copy of the null ERα allele has been generated (O'Brien et al., 2006) thereby expressing the Nerki mutant as its only ERα protein. The Nerki/αERKO (*Esr1*$^{AA/-}$; a.k.a. KIKO) uterus is not hypoplastic, however it resembles the *Esr1*$^{-/-}$ in that estrogen fails to elicit uterine weight increase or cell proliferation (Hewitt, Li, et al., 2010; O'Brien et al., 2006). Microarray comparison of transcripts after estrogen treatment indicated the KIKO uterus retains some of the gene regulation (~24%) also seen in the WT uterus (Hewitt, O'Brien, Jameson, Kissling, & Korach, 2009). WT vs KIKO differentially regulated genes in this microarray profile were enriched for components of the Wnt/Ctnnb signaling pathway as transcripts for Wnt ligands, receptors, transducers, and targets were misregulated by E2 in KIKO vs WT uteri (Hewitt et al., 2009).

Microarray and later ChIP-seq analyses (Hewitt et al., 2014) also showed unexpected results, which were the appearance of numerous estrogen-regulated responses in the KIKO that were not observed in normal WT uteri. Evaluation of the KIKO uterine cistrome by ERα ChIP-seq revealed that these transcripts result from an unanticipated "gain of function" of the KIKO DNA-binding mutation. Analyses of the sequences bound to KIKO ERα revealed enrichment of hormone response element (HRE) DNA motifs, which typically bind androgen, progesterone, and glucocorticoid receptors (AR, PR, and GR). Further in vivo and in vitro analyses have shown that the KIKO ERα binds HRE DNA and regulates uterine genes that are normally PR targets (Hewitt et al., 2014), indicating this particular ERα mutation has an aberrant binding activity with loss of ERE binding but a gain of HRE binding and gene regulation (Hewitt et al., 2014). The KIKO ERα was created by introducing two-point mutations (E207A and G207A) at the base of the first zinc finger of the DBD. These positions in the PR, AR, and GR are occupied by glycine–serine. However, modeling the interaction between ERα and ERE based on structural studies revealed the critical importance of E207 in forming a hydrogen bond with the G/C nucleotides in an ERE at position 2/12 (G**G**TCAnnnTGA**C**C). Additionally, attempting to interact with the T/A found at the equivalent position of an HRE (G**A**ACAnnnTGT**T**C) results in significant steric clash, thus excluding ERα/HRE binding. However, by replacing the critical E207 residue with A, the hydrogen binding with ERE and the exclusion of HRE are lost, leading to an ability to interact with HRE (Hewitt et al., 2014).

The second DNA-binding mutant ERα mouse model (EAAE) has an ERα that does not bind to HRE or ERE motifs either in vitro or in vivo

(Hewitt et al., 2014) but retains AP1-mediated gene induction (Hewitt et al., 2014). It was created by introducing four-point mutations (Y201A, K210A, K214A, and R215E) in the first zinc finger and between the first and second zinc finger. Since EAAE ERα maintains the critical E207 residue, it lacks aberrant HRE binding seen with the KIKO ERα. The EAAE mouse uterus is hypoplastic and refractory to estrogen responses, like the αERKO, indicating that the tethering HRE mechanism is not a major physiological regulatory response in the uterus. Microarray profiling of uterine RNA indicated a lack of estrogen-responsive transcripts (Hewitt et al., 2014). Altogether, this shows that DNA-binding activity of the ERα is critical for uterine function and estrogen response, and DNA-binding-independent activity appears to have little role on its own, but may complement the direct DNA-binding activity in eliciting the full uterine hormone response.

Induction of IGF1 signaling by E2 is known to be a major mediator of uterine growth in a paracrine manner, whereby uterine stromal cells secrete IGF1, which then stimulates epithelial cell growth (Adesanya, Zhou, Samathanam, Powell-Braxton, & Bondy, 1999; Cunha, Cooke, & Kurita, 2004; Zhu & Pollard, 2007). One major surprising observation in the KIKO and EAAE models was the inability of estrogen to increase the transcript of insulin-like growth factor 1 (*Igf1*), which had been reported to be regulated via interaction between ERα to AP1 motifs in the Igf1 promoter via tethering. Analysis of *Igf1* genomic sequences indicated that the AP1 motif previously identified using the chicken Igf1 gene is absent in mammals. Several potential ERE sequences were identified and tested for ERα binding by ChIP and gel shift (Hewitt et al., 2012; Hewitt, Li, et al., 2010). ChIP-PCR analysis confirmed ER bound to specific ERE sequences of the *Igf1* in WT but not KIKO uteri. Interestingly, exogenous treatment with IGF-1 did not restore KIKO uterine growth, indicating additional ERE-mediated responses are needed to modulate the stimulatory action of IGF-1 in the full uterine response. Additionally, analysis of our uterine ERα ChIP-seq data revealed an enhancer 50 kb 5′ of the *Igf1* promoter that has more estrogen-dependent ERα enrichment than the previously tested EREs (Hewitt et al., 2012).

3.4 Analysis of AF-1- and AF-2-Mediated Responses

As described in the Section 2 "Cell Mechanisms", ERα activity requires interaction with coregulators through AF-1 and AF-2. To understand how these impact uterine responses, mice with mutations in AF-1

(ERαAF-1^0) or AF-2 (ERαAF-2^0 and AF2ER) have been created. ERαAF-1^0 mice were made by deleting amino acids 2–128, which includes the AF-1. The uterus develops normally, but exhibits a blunted response to E2 (Abot et al., 2013; Billon-Gales et al., 2009). In contrast, mice lacking AF-2 exhibit more severe uterine phenotypes, with the development of a hypoplastic uterus and complete insensitivity to E2 treatments (Arao et al., 2011; Billon-Gales et al., 2011). ERαAF-2^0 mice were made by deleting amino acids 543–549, whereas AF2ER mice were made using two-point mutations in the LBD. Both experimental approaches resulted in inactivation of the AF-2 function and lack of response to E2. One advantage of the AF2ER model is that ER antagonists such as tamoxifen and fulvestrant (ICI182780) exhibit agonist activity, resulting in an antagonist–agonist reversal, thus allowing reactivation of some ERα-mediated responses, including uterine epithelial cell proliferation, presumably through the AF-1 function of the mutant ERα. Overall, studies with AF-1 and AF-2 mutant mice demonstrate that AF-2 activity is critical for uterine E2 response, but that AF-1 activity can promote uterine response. Further characterization of these mice has also uncovered the tissue selectivity of ER actions through either AF-1 or AF-2. AF-1 activity is sufficient in promoting uterine and male efferent duct responses; in contrast, AF-2 activity is necessary for pituitary and mammary responses (Arao et al., 2012, 2011).

A mouse with G525L mutation in the LBD called the estrogen-nonresponsive ER knock-in (ENERKI), shows lack of response to E2 (Sinkevicius et al., 2008). Like the AF2ER mice, high doses of the synthetic ERα selective agonist propyl pyrazole triol (PPT) and the ER agonist diethylstilbestrol (DES) are able to induce uterine growth (Sinkevicius et al., 2008). These observations support the findings from mice with mutations in AF-1 or AF-2 activities, that full ER function is needed for optimal uterine response.

3.5 Analysis of Biological Impact of Membrane-Initiated Signaling

To address the impact of cell membrane-associated signals, two mouse models with an identical mutation of the palmitoylation site (C451A) have been made which prevent membrane localization of ERα: nuclear-only ER (NOER) (Pedram et al., 2014) and C451A-ERα (Adlanmerini et al., 2014). The two models differed in their uterine phenotypes: the NOER having a hypoplastic uterus that lacks E2 responses and the C451A-ERα having normal uterine development and E2 responses. The C451A-ERα model only had ~55% reduction in membrane ERα (measured only in hepatocytes) which perhaps explains the different phenotypes (Pedram et al., 2014). Conversely, another mouse model was created that prevented nuclear localization (membrane-only ER, MOER) (Pedram et al., 2009) by expressing the LBD fused with multiple palmitoylation sites from the neuromodulin protein, resulting in a hypoplastic, E2-insensitive uterus (Pedram et al., 2009). Clearly, nuclear ERα is critical to uterine function, however the role of membrane-localized ERα is uncertain and the focus of ongoing investigations is to identify the intracellular signaling pathways involved.

3.6 ERβ Does Not Impact Uterine Responses

Observation from ERβ-null females indicates their subfertility is due to diminished ovarian responses, with normal uterine development, function, and responses to E2 (Hewitt et al., 2003; Krege et al., 1998). Although females with deletion of both ERα and ERβ have more severe ovarian defects, the uterine phenotypes are similar to that observed in ERα-null mice (Couse, 1999).

3.7 Importance of ERα to Uterine Function Informs Mechanisms of Disease

Owing to the biological and molecular events that require ERα, the mouse uterus has enabled advancement of our understanding of the details underlying estrogen-initiated responses. Studies in the many mouse models developed with deletion or mutation of ERα have highlighted the essential role of DNA binding and AF-1 and -2 functions in achieving optimal development and response. Additionally, ERα has cell type-dependent roles. Our increased understanding of these molecular details and their roles in normal uterine function are critical to understanding perturbation that leads to impaired embryo implantation or endometrial diseases including endometriosis, endometrial cancer, and leiomyoma.

4. ER IN THE OVARY

Both known forms of nuclear ER, ERα and ERβ, are expressed in mammalian ovaries but localized to distinct functional compartments. ERβ is highly expressed but limited to the granulosa cells of growing follicles, while ERα is generally localized to the interstitium and theca cells. This expression pattern is highly conserved among several mammalian species. Adult ER alpha null (αERKO) females are anovulatory, possessing pre- and small antral follicles but lacking corpora lutea, resulting in infertility. By 50 days of age, αERKO mice are infertile and have ovaries that exhibit multiple enlarged, hemorrhagic, and cystic follicles, with increased gonadotropin and gonadotropin receptor levels, elevated steroid synthesis, and hypertrophied theca cells (Couse & Korach, 1999, 2001; Schomberg et al., 1999). Adult estrogen receptor beta null (βERKO) females are subfertile, as evidenced by reduced litter number and size (Couse, Yates, Deroo, & Korach, 2005). Despite speculated roles of ER in granulosa cells, βERKO ovaries appear relatively normal and possess follicles at all stages of growth and are not overtly impaired by losing ERβ (Couse et al., 2005; Krege et al., 1998). Consistent with the subfertility, superovulatory treatments in βERKO females result in significantly fewer ovulations and observation of trapped oocyte follicles (Couse et al., 2005; Krege et al., 1998). In addition, reduced expression of PR and Cox2, increased rates of follicle atresia, and a paucity of corpora lutea in βERKO ovaries indicate that the subfertility is likely due to a reduced ovulatory frequency (Emmen et al., 2005; Krege et al., 1998).

4.1 Ovarian Phenotypes of ERα Mutant Mice

Although αERKO females are anovulatory, it is generally thought to be due to the cystic ovarian phenotype seen in these mice (Couse & Korach, 1999, 2001; Schomberg et al., 1999). Accompanying the pathology is a severe disruption in steroid hormone levels; mainly, αERKO mice have chronically elevated luteinizing hormone (LH), E2, and testosterone (T) due to a disruption of negative feedback (Couse, Bunch, Lindzey, Schomberg, & Korach, 1999; Couse, Yates, Walker, & Korach, 2003). Treatment of αERKO females with a gonadotropin-releasing hormone (GnRH) antagonist (antide) corrected the cystic follicle and elevated steroidogenesis levels and is therefore thought to be the primary cause of the ovarian phenotypes seen in αERKO females (Couse, Bunch, et al., 1999; Couse et al., 2005). One distinct phenotype of the ERα null females is the aberrant expression and extremely high levels of the enzymes involved in androgen biosynthesis (Couse et al., 2003). Aberrant expression of *Hsd17b3*, a testis specific gene, is observed in the αERKO female ovary. HSD17B3 catalyzes the conversion of androstendione to testosterone (T), is expressed in theca cells, and contributes to the high serum T in the αERKO female. T produced by

HSD17B3 in theca cells is not converted to E2 by granulosa Cyp19 (aromatase), even though higher expression of Cyp19 is observed in αERKO female ovary. The expression of *Hsd17b3* in the αERKO ovary is regulated by LH and treatment with a GnRH-antagonist can normalize T levels in the αERKO female (Couse et al., 2003). *Cyp17*, the enzyme necessary for androstendione synthesis, is increased threefold in αERKO females when compared to WT (Couse et al., 2003), resulting in increased serum levels of E2 and T. *Cyp17* is found in the theca cells, where ERα is also localized, and therefore it is speculated that ERα mediates thecal cell steroidogenesis. Additionally, data show a very modest increase in *Cyp17* in WT females with chronic LH expression (Couse et al., 2003). ERα null follicles grown in culture produce more androgens relative to wild-type follicles under a controlled gonadotropin environment (Emmen et al., 2005), further supporting a role for ovarian ERα in T production. Taken together, these data suggest that the role of ERα in ovulation is through regulation of androgen biosynthesis by way of a short loop feedback mechanism in theca cells within the ovary.

It is also important to note the ovarian phenotypes of various ERα mutants that have been described (Table 2). As discussed earlier, there are a group of mutants that were developed to limit the genomic activity of ERα and their mutations only allow estrogen signaling through the nonclassical mechanism. One such model is the Nerki. Female mice that are heterozygous for this mutation are infertile and the ovaries contain follicles of all stages but no corpora lutea and lipid-filled cells in the ovarian stroma (Jakacka et al., 2002). When superovulated, Nerki heterozygous females develop large hemorrhagic cysts like those seen in the *Esr1*$^{-/-}$ and ovulation does not reach the level seen in WT mice; however, they do not display the altered steroidogenic enzyme or hormone profile (Jakacka et al., 2002). Jackacka et al. speculated that the ovarian phenotypes result from the Nerki ERα acting in a dominant-negative matter (Jakacka et al., 2002). As described earlier, more recent research revealed that Nerki ERα has aberrant DNA-binding activity and binds to steroid HRE sequences including progesterone receptor responsive element (PRE) (Hewitt et al., 2014). Unexpected PRE-mediated gene regulation together with the ERE-mediated gene regulation by WT and heterozygote mice may be a cause of the disrupted phenotype described in the Nerki ovary.

As described in the uterine section of this chapter, the ERα$^{AA/-}$ (KIKO) and ERα$^{EAAE/EAAE}$ (EAAE) mouse models were developed to further understand the role of ERα in nonclassical estrogen signaling

Table 2 Ovarian Phenotypes of Estrogen Receptor Mutants

Gene	Mutation	Nickname	Ovarian Phenotypes	Hormone Levels	References
Esr1	Homozygous null for ERα	αERKO or Ex3αERKO	Anovulatory and infertile hemorrhagic and cystic ovaries with no CLs present in histological sections	Elevated T, E2, and LH Normal FSH & P	Antonson et al. (2012), Curtis Hewitt et al. (2002), Curtis, Clark, Myers, and Korach (1999), Dupont et al. (2000), Hewitt, Kissling, et al. (2010), and Lubahn et al. (1993)
			Increased expression of steroidogenic enzymes		
			Lack of response to superovulation		
Esr1	One mutated allele of two-point mutation in ERα DBD and one WT allele	$NERKI^{+/-}$	Anovulatory and infertile	Normal LH, FSH, and E2 Reduced P	Jakacka et al. (2002)
			Lack of plugs in NERKI females after superovulation and natural mating		
			Superovulation partially restored ovulation while increasing cyst presence		
Esr1	One mutated allele of two-point mutation in DNA-binding domain of ERα and one ERαKO allele	*KIKO* ($ER^{AA/-}$)	Anovulatory and infertile	Normal E2 and P	Hewitt, Li, et al. (2010) and O'Brien et al. (2006)
			No CLs present in histological sections		
Esr1	Homozygous animal of four-point mutation of DBD ERα	$ER\alpha^{EAAE/EAAE}$	Infertile	Not reported	Ahlbory-Dieker et al. (2009)
			Hemorrhagic ovaries		

Continued

Table 2 Ovarian Phenotypes of Estrogen Receptor Mutants—cont'd

Gene	Mutation	Nickname	Ovarian Phenotypes	Hormone Levels	References
Esr1	Homozygous animal of one-point mutation in LBD of ERα	*ENERKI* ($ER\alpha^{G525L}$)	Anovulatory	Elevated serum E2, T, and LH Normal FSH	Sinkevicius et al. (2008)
			Hemorrhagic and cystic ovaries with increased atretic antral follicles and No CLs in histological sections		
			Hyperplastic theca cells in response to LH (data not shown)		
Esr1	Homozygous knock-in of two-point mutation in LBD of ERα	$AF2ER^{KI/KI}$	Anovulatory and infertile	Elevated serum LH and E2	Arao et al. (2011)
			Hemorrhagic and cystic ovaries with no CLs present in histological sections		
			Lack of reseponse to superovulation		
Esr2	Homozygous null alleles for ERβ	$Esr2^{-/-}$ βERKO, Ex3βERKO, and $ER\beta_{ST}{}^{L-/L}$	Subfertile–infertile (lines vary)	Normal LH and FSH	Antal, Krust, Chambon, and Mark (2008b), Dupont et al. (2000), Krege et al. (1998), Wada-Hiraike et al. (2006), and Binder et al. (2013)
			Reduction or failure to respond to superovulation		
			Lack of COC expansion		

Esr1 and Esr2	Homozygous null for both ERα and ERβ	αβERKO	Anovulatory and infertile	Elevated LH and T, normal FSH, and P	Couse, Hewitt, et al. (1999) and Dupont et al. (2000)
			No CLs and few large follicles		
			Ovarian transdifferentiation to Sertoli-like cells		
			Altered expression of steroidogenic enzymes		
Cyp19a1	Homozygous null aromatase: ArKO Unable to synthesize endogenous E2.	*Cyp19a1*$^{-/-}$	Anovulatory and infertile	No E2 Elevated LH, FSH, and T	Fisher, Graves, Parlow, and Simpson (1998), Honda, Harada, Ito, Takagi, and Maeda (1998), Toda et al. (2001), Britt et al. (2000, 2002), Fisher et al. (1998), Honda et al. (1998), Toda, Hayashi, Ono, and Saibara (2012), and Toda et al. (2001)
			Hemorrhagic and cystic ovaries with no CLs present in histological sections		
			Failure to respond to superovulation with partial rescue with E2 treatment. Ovarian transdifferentiation to Sertoli-like cells that express *Sox9*		

Continued

Table 2 Ovarian Phenotypes of Estrogen Receptor Mutants—cont'd

Gene	Mutation	Nickname	Ovarian Phenotypes	Hormone Levels	References
Esr1	Theca cell-specific ERα knockout	*Cyp17cre*; $ERa^{flox/flox}$	Fertility normal in young mice, but 6-month-old animals have reduced fertility and longer estrous cycle	Elevated T at both 2 and 6-months	Bridges et al. (2008) and Lee et al. (2009)
				Normal FSH	
				Decreased LH at 2 months with further decrease at 6 months	
Esr1	Palmitoylation deficient mutants	Esr1 C541ANOER	Cystic ovaries	LH elevated, E2 normal	Adlanmerini et al. (2014) and Pedram et al. (2014)
			Cystic ovaries	LH and E2 elevated	

(Ahlbory-Dieker et al., 2009; O'Brien et al., 2006; Sinkevicius et al., 2008). These models both express ERα that is unable to bind to ERE sequences. The ovaries of KIKO mice have follicles with most stages of development, but lack corpora lutea (O'Brien et al., 2006). EAAE mice have hemorrhagic and cystic ovaries and are infertile, similar to the *Esr1*$^{-/-}$ mice (Ahlbory-Dieker et al., 2009). Taken together, these data suggest that direct ERE binding by ERα is critical for ovarian ERα functionality and regulation required for proper ovulation and therefore fertility.

There are also two notable ERα mouse lines that have mutations in the LBD of ERα. In one model, the ENERKI, a single point mutation was created by switching a glycine to a leucine at residue 525 creating an altered ligand-binding pocket which prevents ligand binding (Sinkevicius et al., 2008). ENERKI has hemorrhagic and cystic ovaries and does not ovulate based on the lack of corpora lutea found in ovarian sections (Sinkevicius et al., 2008). The ovarian defects and lack of ovulation in the ENERKI reveal the importance of estrogen hormone binding for normal ovarian function and neuroendocrine negative feedback (Sinkevicius et al., 2008). A second line, the AF2ER, has two-point mutations in the AF-2 region of ERα and, while E2 can be bound by this mutant, it is not able to engage transcriptional machinery due to the inability to interact with coactivators (Arao et al., 2011). The ovarian phenotype of AF2ER females looks very similar to the ERα null mice, as they have hemorrhagic and cystic ovaries and ovarian sections have no corpora lutea (Arao et al., 2011). AF2ER female mice also have a disruption in negative feedback and have elevated E, T, and LH (Arao et al., 2011). This suggests that the AF-2 region of ERα is critical for regulating ovarian function, neuroendocrine control, and ovulation.

4.2 Ovary-Specific ERα Knockouts

The global ERα knockout mouse has shown that loss of estrogen is detrimental to ovarian function, as evidenced by large hemorrhagic cysts and infertility. However, this approach does not allow us to dissect the role of ERα in the ovary due to confounding factors, such as the high persistent steroid and gonadotropin serum levels following disruption of negative feedback. However, theca cell-specific ERα knockout mice (thEsr1KO) were generated by crossing the floxed ERα strain with *Cyp17* cre mice (Bridges et al., 2008). At 2 months of age, thEsr1KO mice had comparable fertility to WT mice and displayed a normal estrus cycle despite a reduction of serum LH. After superovulation, the same numbers of oocytes were found in the oviduct of the cKO and WT suggesting there was no defect in ovulation (Lee et al., 2009). However, by 6 months of age the thEsr1KO ovaries displayed hemorrhagic cysts and superovulation resulted in significantly fewer oocytes (Lee et al., 2009). Females had an even further reduction of serum LH, suggesting that ERα expression in the theca cells of the ovary is important for feedback regulation and control of LH expression. At both time points, testosterone was elevated, confirming the original role of ovarian ERα regulation of androgen production in the ovary as proposed in the *Esr1*$^{-/-}$ mice (Couse et al., 2006; Lee et al., 2009). These data suggest that thecal cell ERα is associated with an age-related reduction in ovarian function, but is not required for ovulation and normal ovarian function in young mice.

4.3 Ovarian Phenotypes of ERβ Mutant Mice

The role of ERβ in the maintenance of normal reproduction and fertility has not been fully elucidated. It has been reported that female βERKO mice are subfertile compared to WT, as characterized by fewer pregnancies, fewer litters, and smaller litter sizes (Dupont et al., 2000; Krege et al., 1998). Ovaries of adult βERKO mice contain a reduced number of corpora lutea, indicating fewer ovulations, and ovulation rates cannot be rescued by exogenous gonadotropins (Couse et al., 2005; Emmen et al., 2005; Krege et al., 1998). ERβ deletion results in impaired follicular maturation and a reduced number of follicles responsive to LH, which could explain why βERKO mice are poor responders to ovulatory stimulation and have smaller litters. The fact that βERKO mice have fewer pregnancies and produce fewer litters may also be due to fewer adequate ovulatory signals (i.e., LH surges). βERKO ovaries and granulosa cells isolated from βERKO mice have an attenuated cAMP accumulation in response to FSH and altered expression of several genes including *Lhcgr* (LH receptor) and *Cyp19a1* (aromatase) (Couse et al., 2005; Emmen et al., 2005). Cultured follicles from βERKO follicles produce less estrogen than WT follicles (Emmen et al., 2005; Rodriguez et al., 2010) and may fail to provide a sufficient stimulus required to trigger physiologically relevant LH surges. The amplitude and timing of the naturally occurring LH surge was measured in individual intact βERKO and WT mice (Jayes, Burns, Rodriguez, Kissling, & Korach, 2014) and it was determined that while the pituitary levels of LH revealed no differences, the amplitude of the LH surge was severely blunted in βERKO mice compared to WT. The βERKO mice did not produce an adequate preovulatory E2 surge. To determine if the smaller LH surges and the reduced number of litters in βERKO were due to the lack of ERβ in the hypothalamic–pituitary axis or ovary, ovaries were transplanted from WT into βERKO mice and vice versa (Jayes et al., 2014). The size of the LH surge was reduced only in mice lacking ERβ within the ovary, and these mice had fewer litters. Fertility and size of the LH surge were rescued in βERKO mice receiving a WT ovary. These data provided the first experimental evidence that the LH surge is impaired in βERKO females and may be another aspect of their overall reduced fertility. This study shows that ERβ is not necessary within the pituitary and hypothalamus for the generation of a normal LH surge and for normal fertility, but ERβ is essential within the ovary to provide proper hormonal signals for the ovulatory cycle.

4.4 Role of ERβ Signaling in Granulosa Cells

In the ovary, there is a heterogeneous cell population, and ERβ is specifically expressed in granulosa cells, while ERα is predominately expressed in the theca cells (Binder et al., 2013; Krege et al., 1998). To isolate pure populations of granulosa cells, laser capture microdissection (LCM) was performed using ovaries from WT and βERKO mice after stimulation with FSH alone or FSH in combination with LH. Use of LCM allowed for targeted isolation of granulosa cells from large antral follicles (FSH) or preovulatory follicles (FSH + LH) in both WT and βERKO mice so that cells at the similar stages of development could be compared. Microarray analysis demonstrated that granulosa cells isolated at similar stages of follicular maturation have altered gene expression in βERKO mice compared to WT mice. While a subset of follicles can grow and respond to FSH and LH in βERKO mice, the transcriptional profile differs in these cells from that observed in WT cells from similar sized follicles, suggesting ERβ-null granulosa cells are not properly differentiated to respond to hormonal stimulation (Binder et al., 2013).

Examination of the genes from large antral follicles after FSH stimulation revealed 414 genes differentially expressed in βERKO granulosa cells compared to WT. These genes included several implicated in E2 biosynthesis, including *Adcyap1* and *Runx2*, which correlates with reduced E2 concentrations in βERKO follicles (Emmen et al., 2005; Rodriguez et al., 2010). While a subset of granulosa cells were able to respond to LH and differentiate into preovulatory granulosa cells in βERKO mice, these cells showed 1258 genes differentially expressed compared to WT preovulatory granulosa cells. These genes included members of several signaling pathways, including *Akap12* and other members of the cAMP/PKA signaling pathway.

These findings indicate that ERβ is necessary for proper differentiation of ovarian granulosa cells in response to gonadotropins during folliculogenesis and provides a novel list of ERβ-dependent estrogen-regulated genes that may contribute to proper follicle maturation and ovulation in vivo.

4.5 Ovarian Phenotypes of ERα and ERβ Compound Mutant Mice

Mice that possess neither estrogen receptors alpha nor estrogen receptors beta (αβERKO) are anovulatory and infertile similar to the αERKO (Couse, Hewitt, et al., 1999; Dupont et al., 2000). In ovarian sections, there are normal follicular stages found but upon aging to 6–12 months the antrum is underdeveloped, granulosa cell number is small, and the theca is thin and disorganized. Furthermore, there are cystic follicles present and the mice have the same hormonal disruption as seen in the αERKO mice, with even higher LH levels (Couse, Hewitt, et al., 1999; Dupont et al., 2000). The main difference seen in the αβERKO when compared to the single ER KOs is the presence of seminiferous tubule-like structures in the post-pubertal ovary that appear to arise from atretic follicles and have cells with characteristics of Sertoli-like cells found in the male testis (Couse, Hewitt, et al., 1999; Dupont et al., 2000). This phenotype of transdifferentiation in the αβERKO, while absent in each individual KO (αERKO or βERKO), suggests that estrogen signaling involving both ERα and ERβ is necessary for proper ovarian formation and function. There is possible compensation by one or the other ER present in the individual knockout lines or an even more overt effect of the even higher levels of LH in combination with the improper cellular tissue differentiation.

4.6 Ovarian Phenotypes in Mice Lacking Estradiol Synthesis

When considering the role of E2 in various body functions, it is helpful to not only look at the ER mutants but also mice that lack the ability to synthesize E2. Taking this approach, it is possible to dissect hormone ligand-dependent responses from nonligand dependent. Cyp19-null mice (ArKO) were developed for this purpose. Initially these mice had no noticeable phenotype; however, transition to a soy-free diet resulted in noticeable differences between ArKO and WT mice (Fisher et al., 1998). The discrepancy appears to be due to hormonally active components of the feed. The ovaries of ArKO mice have follicles of all stages of development, but no corpora lutea (Britt et al., 2000; Fisher et al., 1998; Toda et al., 2001). Additionally, ArKO mice develop hemorrhagic and cystic follicles with age due, in part, to disrupted negative feedback that can be corrected with E2 treatment (Toda et al., 2001). With carefully timed exogenous hormone treatment, ovulation can be partially rescued, showing that the ovulation defect is estrogen dependent (Toda et al., 2012). When fed a soy-free diet, ArKO ovaries develop the transdifferentiation phenotype seen in the αβERKO ovaries (Britt et al., 2001). The steroid hormone serum composition of the ArKO female is disrupted (E is undetectable, androgens are elevated) and the mice have elevated serum LH (Britt et al., 2000; Fisher et al., 1998). When data from the ArKO mice (lacking ligand) is compiled with mice lacking ERα or ERβ (lack of receptors), it is clear that estrogen plays a major role in ovarian physiology. This is a compound issue that can be attributed to estrogen signaling not only in the ovary, but also in the hypothalamus and pituitary gland, which are involved in negative feedback controlling the trophic hormone levels.

5. ER IN METABOLISM

Estrogen regulates multiple physiological functions, including reproduction, bone density, and metabolic regulations. As a consequence of pleiotropic effects of estrogen, the decline of endogenous estrogen production by the ovaries at menopause often leads to functional disorders including dyslipidemia, impaired glucose tolerance (IGT), and type 2 diabetes mellitus, which increase cardiovascular disease risk in postmenopausal women and directly affect the quality of life (Munoz, Derstine, & Gower, 2002). Several animal models have been developed to further explore the clinical findings of estrogen-dependent metabolic regulation. Indeed, ovexed mice lacking intact estrogen signaling display obesity and IGT; these effects are reversible with the reintroduction of estrogen (E2) (Zhu et al., 2013). Similar results have been seen in *Cyp19* (aromatase) knockout (KO) mice, which are unable to synthesize E2 from testosterone. Treatment of *Cyp19*KO mice with exogenous E2 restores the E2 protective effect against the development of metabolic syndrome in both male and female mice (Hewitt, Pratis, Jones, & Simpson, 2004; Jones et al., 2000). Studies using the ER (ERα and ERβ) knockout mice have demonstrated that ERα plays the essential role in estrogen-mediated metabolic regulation, whereas ERβ does not (Bryzgalova et al., 2006).

5.1 Metabolic Phenotype of ERα Knockout Mice

Metabolic phenotypes of αERKO have been described previously. αERKO females present with obesity, IGT, and insulin resistance (Heine, Taylor, Iwamoto, Lubahn, & Cooke, 2000). Fat deposition of parametrial and inguinal white adipose tissues (WAT) were higher in regular diet fed 3-month-old αERKO females than in wild-type (WT) littermates. No difference in WT vs αERKO perirenal WAT or brown adipose tissue (BAT) was observed. Increased adipocyte volume in parametrial and inguinal WAT was accompanied by increased adipocyte number (Heine et al., 2000). These observations suggested that ERα is involved in the adipogenesis; however, the mechanisms responsible for ERα-dependent regulation of adipogenesis remain unclear. Energy intake of WT and αERKO was equal, indicating that obesity was not induced by hyperphagia. In contrast, energy expenditure was reduced in αERKO compared with WT, indicating that altered energy expenditure may contribute to the observed obesity (Heine et al., 2000). Recent reports suggest that decreased locomotion is a cause of reduction of energy expenditure in αERKO mice (Park et al., 2011; Xu et al., 2011).

5.2 Physiological Role of ERα Transactivation Domains in Metabolism

As described previously, ERα has two transcription activation domains, named AF-1 and AF-2. Physiological roles of ERα AF-1 and AF-2 have been reported using the mouse models, which deleted ERα AF-1 (ERaAF-1^0) or ERα AF-2 (ERaAF-2^0) (Handgraaf et al., 2013). ERaAF-2^0 females present with obesity, IGT, and insulin resistance, that mimics that seen in αERKO females. In striking contrast, metabolic phenotypes were lacking in ERaAF-1^0 mice being identical to WT. HFD-induced metabolic disturbances in ovexed ERaAF-1^0 and WT mice were prevented by E2 administration, whereas an E2-mediated protective effect was totally abrogated in ERaAF-2^0 and αERKO mice. Thus, the report concluded that the protective effect of E2 toward obesity and insulin resistance is ERα AF-2 dependent but does not require AF-1 (Handgraaf et al., 2013). The molecular mechanism of AF-1 or AF-2 activation or cooperative regulation of ERα AF-1 by AF-2 is still unresolved (Arao, Coons, Zuercher, & Korach, 2015). Although the ERα AF-2 mutations (ERaAF-2^0 and AF2ER) disrupt E2-mediated physiological responses, antagonistic ligands such as fulvestrant and tamoxifen activate AF-1-mediated physiological functions in these mutant mice (Arao et al., 2012, 2011; Moverare-Skrtic et al., 2014). AF2ER females present with disrupted metabolic phenotypes similar to ERaAF-2^0 and αERKO mice. Treatment with Tamoxifen to AF2ER females rescued the metabolic phenotypes (Arao, Hamilton, Lierz, & Korach, 2016). This result suggested that ERα AF-1 is able to modulate metabolic regulation, even though it is in contrast to the previous report using a different model system (ERaAF-1^0) (Handgraaf et al., 2013). Understanding the mechanism of ligand-dependent ERα AF-1 and AF-2 cooperative regulation will be necessary to delineate new therapeutic options for selective modulation of ERα-mediated metabolic regulation.

5.3 Phenotype of ERα DNA-Binding Domain Mutant Mice in Metabolism

ERα DBD mutant mice (KIKO) were analyzed to characterize the role for nongenomic and indirect DNA-binding transcription (nonclassical ERα signaling) toward mediating metabolic regulation (Park et al., 2011). KIKO mice restored metabolic parameters dysregulated in αERKO mice to normal values, suggesting that the nonclassical ERα signaling rescues body weight and metabolic function. The normalization of energy expenditure, including voluntary locomotor activity leads to nonclassical ERα signaling-mediated normalization of metabolic regulation (Park et al., 2011). The phenotype of KIKO mice suggested that the nonclassical ERα signaling is a potential target for selective modulation of ERα-mediated metabolic regulation. Based on the aberrant DNA-binding activity of the KIKO mouse model, further consideration of the metabolic phenotype of the EAAE mouse model with no DNA-binding activity will provide a more accurate assessment of the signaling mechanisms involved in metabolic regulation. Additionally, development of other knock-in mutation mouse models will facilitate further evaluation of nongenomic extranuclear ERα action. The H2NES ERα mutation which is a cytosol-only form of ERα mutant, even in the presence of hormone, might be useful for such purposes (Burns, Li, Liu, & Korach, 2014). We have currently developed such a mouse model and are characterizing the phenotypes to assess the role of nongenomic ERα signaling.

As described earlier, various functional domains of ERα contribute to differential estrogen-mediated metabolic regulations. Development of ligands that selectively regulate specific ERα functional domains and ER cellular signaling mechanisms may be useful for developing more effective targeted therapies for postmenopausal women without undesirable side effects.

REFERENCES

Aagaard, M. M., Siersbaek, R., & Mandrup, S. (2011). Molecular basis for gene-specific transactivation by nuclear receptors. *Biochimica et Biophysica Acta, 1812*, 824–835.

Abot, A., Fontaine, C., Raymond-Letron, I., Flouriot, G., Adlanmerini, M., Buscato, M., et al. (2013). The AF-1 activation function of estrogen receptor alpha is necessary and sufficient for uterine epithelial cell proliferation in vivo. *Endocrinology, 154*, 2222–2233.

Adesanya, O. O., Zhou, J., Samathanam, C., Powell-Braxton, L., & Bondy, C. A. (1999). Insulin-like growth factor 1 is required for G2 progression in the estradiol-induced mitotic cycle. *Proceedings of the National Academy of Sciences of the United States of America, 96*, 3287–3291.

Adlanmerini, M., Solinhac, R., Abot, A., Fabre, A., Raymond-Letron, I., Guihot, A. L., et al. (2014). Mutation of the palmitoylation site of estrogen receptor alpha in vivo reveals tissue-specific roles for membrane versus nuclear actions. *Proceedings of the National Academy of Sciences of the United States of America, 111*, E283–E290.

Ahlbory-Dieker, D. L., Stride, B. D., Leder, G., Schkoldow, J., Trolenberg, S., Seidel, H., et al. (2009). DNA binding by estrogen receptor-alpha is essential for the transcriptional response to estrogen in the liver and the uterus. *Molecular Endocrinology, 23*, 1544–1555.

Antal, M. C., Krust, A., Chambon, P., & Mark, M. (2008a). Sterility and absence of histopathological defects in nonreproductive organs of a mouse ER beta-null mutant. *Proceedings of the National Academy of Sciences of the United States of America, 105*, 2433–2438.

Antal, M. C., Krust, A., Chambon, P., & Mark, M. (2008b). Sterility and absence of histopathological defects in nonreproductive organs of a mouse ERbeta-null mutant. *Proceedings of the National Academy of Sciences of the United States of America, 105*, 2433–2438.

Antonson, P., Omoto, Y., Humire, P., & Gustafsson, J. A. (2012). Generation of ERalpha-floxed and knockout mice using the Cre/LoxP system. *Biochemical and Biophysical Research Communications, 424*, 710–716.

Arao, Y., Coons, L. A., Zuercher, W. J., & Korach, K. S. (2015). Transactivation function-2 of estrogen receptor alpha contains transactivation function-1-regulating element. *The Journal of Biological Chemistry, 290*, 17611–17627.

Arao, Y., Hamilton, K. J., Goulding, E. H., Janardhan, K. S., Eddy, E. M., & Korach, K. S. (2012). Transactivating function (AF) 2-mediated AF-1 activity of estrogen receptor alpha is crucial to maintain male reproductive tract function. *Proceedings of the National Academy of Sciences of the United States of America, 109*, 21140–21145.

Arao, Y., Hamilton, K. J., Lierz, S. L., & Korach, K. S. (2016). Differential function of estrogen receptor α in fat accumulation of intra-abdominal and inguinal adipose tissues. *Endocrine Reviews, 37*(Suppl. 2), 2016 Adipocyte Biology (Posters). Endocrine Society; 2016.

Arao, Y., Hamilton, K. J., Ray, M. K., Scott, G., Mishina, Y., & Korach, K. S. (2011). Estrogen receptor alpha AF-2 mutation results in antagonist reversal and reveals tissue selective function of estrogen receptor modulators. *Proceedings of the National Academy of Sciences of the United States of America, 108*, 14986–14991.

Billon-Gales, A., Fontaine, C., Filipe, C., Douin-Echinard, V., Fouque, M. J., Flouriot, G., et al. (2009). The transactivating function 1 of estrogen receptor alpha is dispensable for the vasculoprotective actions of 17 beta-estradiol. *Proceedings of the National Academy of Sciences of the United States of America, 106*, 2053–2058.

Billon-Gales, A., Krust, A., Fontaine, C., Abot, A., Flouriot, G., Toutain, C., et al. (2011). Activation function 2 (AF2) of estrogen receptor-alpha is required for the atheroprotective action of estradiol but not to accelerate endothelial healing. *Proceedings of the National Academy of Sciences of the United States of America, 108*, 13311–13316.

Binder, A. K., Rodriguez, K. F., Hamilton, K. J., Stockton, P. S., Reed, C. E., & Korach, K. S. (2013). The absence of ER-beta results in altered gene expression in ovarian granulosa cells isolated from in vivo preovulatory follicles. *Endocrinology, 154*, 2174–2187.

Bourguet, W., Germain, P., & Gronemeyer, H. (2000). Nuclear receptor ligand-binding domains: Three-dimensional structures, molecular interactions and pharmacological implications. *Trends in Pharmacological Sciences, 21*, 381–388.

Bridges, P. J., Koo, Y., Kang, D.-W., Hudgins-Spivey, S., Lan, Z.-J., Xu, X., et al. (2008). Generation of Cyp17iCre transgenic mice and their application to conditionally delete estrogen receptor alpha (Esr1) from the ovary and testis. *Genesis, 46*, 499–505.

Britt, K. L., Drummond, A. E., Cox, V. A., Dyson, M., Wreford, N. G., Jones, M. E., et al. (2000). An age-related ovarian phenotype in mice with targeted disruption of the Cyp 19 (aromatase) gene. *Endocrinology, 141*, 2614–2623.

Britt, K. L., Drummond, A. E., Dyson, M., Wreford, N. G., Jones, M. E. E., Simpson, E. R., et al. (2001). The ovarian phenotype of the aromatase knockout (ArKO) mouse. *Journal of Steroid Biochemistry and Molecular Biology, 79*, 181–185.

Britt, K. L., Kerr, J., O'Donnell, L., Jones, M. E., Drummond, A. E., Davis, S. R., et al. (2002). Estrogen regulates development of the somatic cell phenotype in the eutherian ovary. *FASEB Journal: Official Publication of the Federation of American Societies for Experimental Biology, 16*, 1389–1397.

Bryzgalova, G., Gao, H., Ahren, B., Zierath, J. R., Galuska, D., Steiler, T. L., et al. (2006). Evidence that oestrogen receptor-alpha plays an important role in the regulation of glucose homeostasis in mice: Insulin sensitivity in the liver. *Diabetologia, 49*, 588–597.

Burns, K. A., & Korach, K. S. (2012). Estrogen receptors and human disease: An update. *Archives of Toxicology, 86*, 1491–1504.

Burns, K. A., Li, Y., Liu, L., & Korach, K. S. (2014). Research resource: Comparison of gene profiles from wild-type ERalpha and ERalpha hinge region mutants. *Molecular Endocrinology, 28*, 1352–1361.

Carroll, J. S., & Brown, M. (2006). Estrogen receptor target gene: An evolving concept. *Molecular Endocrinology, 20*, 1707–1714.

Carroll, J. S., Liu, X. S., Brodsky, A. S., Li, W., Meyer, C. A., Szary, A. J., et al. (2005). Chromosome-wide mapping of estrogen receptor binding reveals long-range regulation requiring the forkhead protein FoxA1. *Cell, 122*, 33–43.

Chimge, N. O., & Frenkel, B. (2013). The RUNX family in breast cancer: Relationships with estrogen signaling. *Oncogene, 32*, 2121–2130.

Clark, S., Rainville, J., Zhao, X., Katzenellenbogen, B. S., Pfaff, D., & Vasudevan, N. (2014). Estrogen receptor-mediated transcription involves the activation of multiple kinase pathways in neuroblastoma cells. *Journal of Steroid Biochemistry and Molecular Biology, 139*, 45–53.

Couse, J. (1999). Reproductive phenotypes in estrogen receptor knockout mice: Contrasting roles for estrogen receptor-alpha and estrogen receptor-beta. *Biology of Reproduction, 60*, 88.

Couse, J. F., Bunch, D. O., Lindzey, J., Schomberg, D. W., & Korach, K. S. (1999). Prevention of the polycystic ovarian phenotype and characterization of ovulatory capacity in the estrogen receptor-alpha knockout mouse. *Endocrinology, 140*, 5855–5865.

Couse, J. F., Hewitt, S. C., Bunch, D. O., Sar, M., Walker, V. R., Davis, B. J., et al. (1999). Postnatal sex reversal of the ovaries in mice lacking estrogen receptors alpha and beta. *Science*, *286*, 2328–2331.
Couse, J. F., & Korach, K. S. (1999). Estrogen receptor null mice: What have we learned and where will they lead us? *Endocrine Reviews*, *20*, 358–417.
Couse, J. F., & Korach, K. S. (2001). Contrasting phenotypes in reproductive tissues of female estrogen receptor null mice. *Annals of the New York Academy of Sciences*, *948*, 1–8.
Couse, J. F., Lindzey, J., Grandien, K., Gustafsson, J. A., & Korach, K. S. (1997). Tissue distribution and quantitative analysis of estrogen receptor-alpha (ERalpha) and estrogen receptor-beta (ERbeta) messenger ribonucleic acid in the wild-type and ERalpha-knockout mouse. *Endocrinology*, *138*, 4613–4621.
Couse, J. F., Yates, M. M., Deroo, B. J., & Korach, K. S. (2005). Estrogen receptor-beta is critical to granulosa cell differentiation and the ovulatory response to gonadotropins. *Endocrinology*, *146*, 3247–3262.
Couse, J. F., Yates, M. M., Rodriguez, K. F., Johnson, J. A., Poirier, D., & Korach, K. S. (2006). The intraovarian actions of estrogen receptor-alpha are necessary to repress the formation of morphological and functional Leydig-like cells in the female gonad. *Endocrinology*, *147*, 3666–3678.
Couse, J. F., Yates, M. M., Walker, V. R., & Korach, K. S. (2003). Characterization of the hypothalamic–pituitary–gonadal axis in estrogen receptor (ER) null mice reveals hypergonadism and endocrine sex reversal in females lacking ER alpha but not ER beta. *Molecular Endocrinology*, *17*, 1039–1053.
Cunha, G. R., Cooke, P. S., & Kurita, T. (2004). Role of stromal–epithelial interactions in hormonal responses. *Archives of Histology and Cytology*, *67*, 417–434.
Curtis Hewitt, S., Goulding, E. H., Eddy, E. M., & Korach, K. S. (2002). Studies using the estrogen receptor alpha knockout uterus demonstrate that implantation but not decidualization-associated signaling is estrogen dependent. *Biology of Reproduction*, *67*, 1268–1277.
Curtis, S. W., Clark, J., Myers, P., & Korach, K. S. (1999). Disruption of estrogen signaling does not prevent progesterone action in the estrogen receptor or knockout mouse uterus. *Proceedings of the National Academy of Sciences of the United States of America*, *96*, 3646–3651.
Curtis, S. H., & Korach, K. S. (1999). Steroid receptor knockout models: Phenotypes and responses illustrate interactions between receptor signaling pathways in vivo. In B. W. O'Malley (Ed.), *Hormones and signaling* (pp. 357–380). San Diego: Academic Press.
Deroo, B. J., & Korach, K. S. (2006). Estrogen receptors and human disease. *The Journal of Clinical Investigation*, *116*, 561–570.
Dupont, S., Krust, A., Gansmuller, A., Dierich, A., Chambon, P., & Mark, M. (2000). Effect of single and compound knockouts of estrogen receptors alpha (ERalpha) and beta (ERbeta) on mouse reproductive phenotypes. *Development*, *127*, 4277–4291.
Emmen, J. M., Couse, J. F., Elmore, S. A., Yates, M. M., Kissling, G. E., & Korach, K. S. (2005). In vitro growth and ovulation of follicles from ovaries of estrogen receptor (ER) {alpha} and ER{beta} null mice indicate a role for ER{beta} in follicular maturation. *Endocrinology*, *146*, 2817–2826.
Eun Kwon, H., & Taylor, H. S. (2004). The role of HOX genes in human implantation. *Annals of the New York Academy of Sciences*, *1034*, 1–18.
Fisher, C. R., Graves, K. H., Parlow, A. F., & Simpson, E. R. (1998). Characterization of mice deficient in aromatase (ArKO) because of targeted disruption of the cyp19 gene. *Proceedings of the National Academy of Sciences of the United States of America*, *95*, 6965–6970.
Germain, P., Staels, B., Dacquet, C., Spedding, M., & Laudet, V. (2006). Overview of nomenclature of nuclear receptors. *Pharmacological Reviews*, *58*, 685–704.

Green, K. A., & Carroll, J. S. (2007). Oestrogen-receptor-mediated transcription and the influence of co-factors and chromatin state. *Nature Reviews. Cancer*, *7*, 713–722.

Hah, N., & Kraus, W. L. (2014). Hormone-regulated transcriptomes: Lessons learned from estrogen signaling pathways in breast cancer cells. *Molecular and Cellular Endocrinology*, *382*, 652–664.

Handgraaf, S., Riant, E., Fabre, A., Waget, A., Burcelin, R., Liere, P., et al. (2013). Prevention of obesity and insulin resistance by estrogens requires ERalpha activation function-2 (ERalphaAF-2), whereas ERalphaAF-1 is dispensable. *Diabetes*, *62*, 4098–4108.

Heine, P. A., Taylor, J. A., Iwamoto, G. A., Lubahn, D. B., & Cooke, P. S. (2000). Increased adipose tissue in male and female estrogen receptor- alpha knockout mice. *Proceedings of the National Academy of Sciences of the United States of America*, *97*, 12729–12734.

Hewitt, S. C., Deroo, B. J., Hansen, K., Collins, J., Grissom, S., Afshari, C. A., et al. (2003). Estrogen receptor-dependent genomic responses in the uterus mirror the biphasic physiological response to estrogen. *Molecular Endocrinology (Baltimore, Md.)*, *17*, 2070–2083.

Hewitt, S. C., Kissling, G. E., Fieselman, K. E., Jayes, F. L., Gerrish, K. E., & Korach, K. S. (2010). Biological and biochemical consequences of global deletion of exon 3 from the ER alpha gene. *FASEB Journal: Official Publication of the Federation of American Societies for Experimental Biology*, *24*, 4660–4667.

Hewitt, S. C., Li, L., Grimm, S. A., Chen, Y., Liu, L., Li, Y., et al. (2012). Research resource: Whole-genome estrogen receptor alpha binding in mouse uterine tissue revealed by ChIP-Seq. *Molecular Endocrinology (Baltimore, Md.)*, *26*, 887–898.

Hewitt, S. C., Li, L., Grimm, S. A., Winuthayanon, W., Hamilton, K. J., Pockette, B., et al. (2014). Novel DNA motif binding activity observed in vivo with an estrogen receptor alpha mutant mouse. *Molecular Endocrinology*, *28*, 899–911.

Hewitt, S. C., Li, Y., Li, L., & Korach, K. S. (2010). Estrogen-mediated regulation of Igf1 transcription and uterine growth involves direct binding of estrogen receptor alpha to estrogen-responsive elements. *The Journal of Biological Chemistry*, *285*, 2676–2685.

Hewitt, S. C., O'Brien, J. E., Jameson, J. L., Kissling, G. E., & Korach, K. S. (2009). Selective disruption of ER alpha DNA-binding activity alters uterine responsiveness to estradiol. *Molecular Endocrinology*, *23*, 2111–2116.

Hewitt, K. N., Pratis, K., Jones, M. E., & Simpson, E. R. (2004). Estrogen replacement reverses the hepatic steatosis phenotype in the male aromatase knockout mouse. *Endocrinology*, *145*, 1842–1848.

Hewitt, S. C., Winuthayanon, W., & Korach, K. S. (2016). What's new in estrogen receptor action in the female reproductive tract. *Journal of Molecular Endocrinology*, *56*, R55–R71.

Honda, S., Harada, N., Ito, S., Takagi, Y., & Maeda, S. (1998). Disruption of sexual behavior in male aromatase-deficient mice lacking exons 1 and 2 of the cyp19 gene. *Biochemical and Biophysical Research Communications*, *252*, 445–449.

Huang, C. C., Orvis, G. D., Wang, Y., & Behringer, R. R. (2012). Stromal-to-epithelial transition during postpartum endometrial regeneration. *PLoS One*, *7*, e44285.

Jakacka, M., Ito, M., Martinson, F., Ishikawa, T., Lee, E. J., & Jameson, J. L. (2002). An estrogen receptor (ER)alpha deoxyribonucleic acid-binding domain knock-in mutation provides evidence for nonclassical ER pathway signaling in vivo. *Molecular Endocrinology*, *16*, 2188–2201.

Jakacka, M., Ito, M., Weiss, J., Chien, P. Y., Gehm, B. D., & Jameson, J. L. (2001). Estrogen receptor binding to DNA is not required for its activity through the nonclassical AP1 pathway. *Journal of Biological Chemistry*, *276*, 13615–13621.

Jayes, F. L., Burns, K. A., Rodriguez, K. F., Kissling, G. E., & Korach, K. S. (2014). The naturally occurring luteinizing hormone surge is diminished in mice lacking estrogen receptor Beta in the ovary. *Biology of Reproduction*, *90*, 24.

Jensen, E. V. (1962). On the mechanism of estrogen action. *Perspectives in Biology and Medicine*, *6*, 47–59.

Jensen, E. V., & Jacobson, H. I. (1960). Fate of steroidal estorgen in target tissues. In G. Pincus & E. P. Vollmer (Eds.), *Biological activities of steroids in relation to cancer* (pp. 161–174). New York: Academic Press.

Jensen, E. V., & Jacobson, H. I. (1962). Basic guides to the mechanism of estrogen action. *Recent Progress in Hormone Research, 18*, 387–414.

Jones, M. E., Thorburn, A. W., Britt, K. L., Hewitt, K. N., Wreford, N. G., Proietto, J., et al. (2000). Aromatase-deficient (ArKO) mice have a phenotype of increased adiposity. *Proceedings of the National Academy of Sciences of the United States of America, 97*, 12735–12740.

Kelly, M. J., & Levin, E. R. (2001). Rapid actions of plasma membrane estrogen receptors. *Trends in Endocrinology and Metabolism, 12*, 152–156.

Klinge, C. M. (2000). Estrogen receptor interaction with co-activators and co-repressors. *Steroids, 65*, 227–251.

Krege, J. H., Hodgin, J. B., Couse, J. F., Enmark, E., Warner, M., Mahler, J. F., et al. (1998). Generation and reproductive phenotypes of mice lacking estrogen receptor beta. *Proceedings of the National Academy of Sciences of the United States of America, 95*, 15677–15682.

Kuiper, G. G., Enmark, E., Pelto-Huikko, M., Nilsson, S., & Gustafsson, J. A. (1996). Cloning of a novel receptor expressed in rat prostate and ovary. *Proceedings of the National Academy of Sciences of the United States of America, 93*, 5925–5930.

Kushner, P. J., Agard, D. A., Greene, G. L., Scanlan, T. S., Shiau, A. K., Uht, R. M., et al. (2000). Estrogen receptor pathways to AP-1. *Journal of Steroid Biochemistry and Molecular Biology, 74*, 311–317.

Lee, K., Jeong, J., Kwak, I., Yu, C. T., Lanske, B., Soegiarto, D. W., et al. (2006). Indian hedgehog is a major mediator of progesterone signaling in the mouse uterus. *Nature Genetics, 38*, 1204–1209.

Lee, S., Kang, D. W., Hudgins-Spivey, S., Krust, A., Lee, E. Y., Koo, Y., et al. (2009). Theca-specific estrogen receptor-alpha knockout mice lose fertility prematurely. *Endocrinology, 150*, 3855–3862.

Levin, E. R. (2015). Extranuclear steroid receptors are essential for steroid hormone actions. *Annual Review of Medicine, 66*(66), 271–280.

Li, Q., Kannan, A., DeMayo, F. J., Lydon, J. P., Cooke, P. S., Yamagishi, H., et al. (2011). The antiproliferative action of progesterone in uterine epithelium is mediated by Hand2. *Science, 331*, 912–916.

Lubahn, D. B., Moyer, J. S., Golding, T. S., Couse, J. F., Korach, K. S., & Smithies, O. (1993). Alteration of reproductive function but not prenatal sexual development after insertional disruption of the mouse estrogen receptor gene. *Proceedings of the National Academy of Sciences of the United States of America, 90*, 11162–11166.

Mangelsdorf, D. J., Thummel, C., Beato, M., Herrlich, P., Schutz, G., Umesono, K., et al. (1995). The nuclear receptor superfamily: The second decade. *Cell, 83*, 835–839.

Moverare-Skrtic, S., Borjesson, A. E., Farman, H. H., Sjogren, K., Windahl, S. H., Lagerquist, M. K., et al. (2014). The estrogen receptor antagonist ICI 182,780 can act both as an agonist and an inverse agonist when estrogen receptor alpha AF-2 is modified. *Proceedings of the National Academy of Sciences of the United States of America, 111*, 1180–1185.

Munoz, J., Derstine, A., & Gower, B. A. (2002). Fat distribution and insulin sensitivity in postmenopausal women: Influence of hormone replacement. *Obesity Research, 10*, 424–431.

O'Brien, J. E., Peterson, T. J., Tong, M. H., Lee, E. J., Pfaff, L. E., Hewitt, S. C., et al. (2006). Estrogen-induced proliferation of uterine epithelial cells is independent of estrogen receptor alpha binding to classical estrogen response elements. *The Journal of Biological Chemistry, 281*, 26683–26692.

O'Lone, R., Frith, M. C., Karlsson, E. K., & Hansen, U. (2004). Genomic targets of nuclear estrogen receptors. *Molecular Endocrinology (Baltimore, Md.), 18*, 1859–1875.

Park, C. J., Zhao, Z., Glidewell-Kenney, C., Lazic, M., Chambon, P., Krust, A., et al. (2011). Genetic rescue of nonclassical ERalpha signaling normalizes energy balance in obese Eralpha-null mutant mice. *The Journal of Clinical Investigation*, *121*, 604–612.

Pawar, S., Laws, M. J., Bagchi, I. C., & Bagchi, M. K. (2015). Uterine epithelial estrogen receptor-alpha controls decidualization via a paracrine mechanism. *Molecular Endocrinology (Baltimore, Md.)*, *29*, 1362–1374.

Pedram, A., Razandi, M., Kim, J. K., O'Mahony, F., Lee, E. Y. H. P., Luderer, U., et al. (2009). Developmental phenotype of a membrane only estrogen receptor alpha(MOER) mouse. *Journal of Biological Chemistry*, *284*, 3488–3495.

Pedram, A., Razandi, M., Lewis, M., Hammes, S., & Levin, E. R. (2014). Membrane-localized estrogen receptor alpha is required for normal organ development and function. *Developmental Cell*, *29*, 482–490.

Prossnitz, E. R., & Hathaway, H. J. (2015). What have we learned about GPER function in physiology and disease from knockout mice? *Journal of Steroid Biochemistry and Molecular Biology*, *153*, 114–126.

Quarmby, V. E., & Korach, K. S. (1984). The influence of 17 beta-estradiol on patterns of cell division in the uterus. *Endocrinology*, *114*, 694–702.

Rodriguez, K. F., Couse, J. F., Jayes, F. L., Hamilton, K. J., Burns, K. A., Taniguchi, F., et al. (2010). Insufficient luteinizing hormone-induced intracellular signaling disrupts ovulation in preovulatory follicles lacking estrogen receptor-{beta}. *Endocrinology*, *151*, 2826–2834.

Roper, R. J., Griffith, J. S., Lyttle, C. R., Doerge, R. W., McNabb, A. W., Broadbent, R. E., et al. (1999). Interacting quantitative trait loci control phenotypic variation in murine estradiol-regulated responses. *Endocrinology*, *140*, 556–561.

Safe, S. (2001). Transcriptional activation of genes by 17 beta-estradiol through estrogen receptor-Sp1 interactions. In G. Litwack & T. Begley (Eds.), *Vitamins and hormones—Advances in research and applications: Vol. 62*(pp. 231–252). San Diego, CA: Academic Press.

Schomberg, D. W., Couse, J. F., Mukherjee, A., Lubahn, D. B., Sar, M., Mayo, K. E., et al. (1999). Targeted disruption of the estrogen receptor-alpha gene in female mice: Characterization of ovarian responses and phenotype in the adult. *Endocrinology*, *140*, 2733–2744.

Sinkevicius, K. W., Burdette, J. E., Woloszyn, K., Hewitt, S. C., Hamilton, K., Sugg, S. L., et al. (2008). An estrogen receptor-alpha knock-in mutation provides evidence of ligand-independent signaling and allows modulation of ligand-induced pathways in vivo. *Endocrinology*, *149*, 2970–2979.

Smith, C. L. (1998). Cross-talk between peptide growth factor and estrogen receptor signaling pathways. *Biology of Reproduction*, *58*, 627–632.

Stewart, C. L., Kaspar, P., Brunet, L. J., Bhatt, H., Gadi, I., Kontgen, F., et al. (1992). Blastocyst implantation depends on maternal expression of leukaemia inhibitory factor. *Nature*, *359*, 76–79.

Toda, K., Hayashi, Y., Ono, M., & Saibara, T. (2012). Impact of ovarian sex steroids on ovulation and ovulatory gene induction in aromatase-null mice. *Endocrinology*, *153*, 386–394.

Toda, K., Takeda, K., Okada, T., Akira, S., Saibara, T., Kaname, T., et al. (2001). Targeted disruption of the aromatase P450 gene (Cyp19) in mice and their ovarian and uterine responses to 17beta-oestradiol. *The Journal of Endocrinology*, *170*, 99–111.

Tong, W., & Pollard, J. W. (1999). Progesterone inhibits estrogen-induced cyclin D1 and cdk4 nuclear translocation, cyclin E- and cyclin A-cdk2 kinase activation, and cell proliferation in uterine epithelial cells in mice. *Molecular and Cellular Biology*, *19*, 2251–2264.

Wada-Hiraike, O., Hiraike, H., Okinaga, H., Imamov, O., Barros, R. P., Morani, A., et al. (2006). Role of estrogen receptor beta in uterine stroma and epithelium: Insights from

estrogen receptor beta-/- mice. *Proceedings of the National Academy of Sciences of the United States of America, 103*, 18350–18355.
Wall, E. H., Case, L. K., Hewitt, S. C., Nguyen-Vu, T., Candelaria, N. R., Teuscher, C., et al. (2014). Genetic control of ductal morphology, estrogen-induced ductal growth, and gene expression in female mouse mammary gland. *Endocrinology, 155*, 3025–3035.
Wall, E. H., Hewitt, S. C., Liu, L., del Rio, R., Case, L. K., Lin, C. Y., et al. (2013). Genetic control of estrogen-regulated transcriptional and cellular responses in mouse uterus. *FASEB Journal: Official Publication of the Federation of American Societies for Experimental Biology, 27*, 1874–1886.
Wang, H., & Dey, S. K. (2006). Roadmap to embryo implantation: Clues from mouse models. *Nature Reviews. Genetics, 7*, 185–199.
Winuthayanon, W., Hewitt, S. C., & Korach, K. S. (2014). Uterine epithelial cell estrogen receptor alpha-dependent and -independent genomic profiles that underlie estrogen responses in mice. *Biology of Reproduction, 91*, 110.
Winuthayanon, W., Hewitt, S. C., Orvis, G. D., Behringer, R. R., & Korach, K. S. (2010). Uterine epithelial estrogen receptor alpha is dispensable for proliferation but essential for complete biological and biochemical responses. *Proceedings of the National Academy of Sciences of the United States of America, 107*, 19272–19277.
Xu, Y., Nedungadi, T. P., Zhu, L., Sobhani, N., Irani, B. G., Davis, K. E., et al. (2011). Distinct hypothalamic neurons mediate estrogenic effects on energy homeostasis and reproduction. *Cell Metabolism, 14*, 453–465.
Zhu, L., Brown, W. C., Cai, Q., Krust, A., Chambon, P., McGuinness, O. P., et al. (2013). Estrogen treatment after ovariectomy protects against fatty liver and may improve pathway-selective insulin resistance. *Diabetes, 62*, 424–434.
Zhu, L., & Pollard, J. W. (2007). Estradiol-17 beta regulates mouse uterine epithelial cell proliferation through insulin-like growth factor 1 signaling. *Proceedings of the National Academy of Sciences of the United States of America, 104*, 15847–15851.

CHAPTER FIVE

Mechanisms of Glucocorticoid Action During Development

Jonathan T. Busada, John A. Cidlowski[1]
Molecular Endocrinology Group, Signal Transduction Laboratory, National Institute of Environmental Health Sciences, National Institutes of Health, Research Triangle Park, NC, United States
[1]Corresponding author: e-mail address: cidlows1@niehs.nih.gov

Contents

Abstract

Glucocorticoids are primary stress hormones produced by the adrenal cortex. The concentration of serum glucocorticoids in the fetus is low throughout most of gestation but surge in the weeks prior to birth. While their most well-known function is to stimulate differentiation and functional development of the lungs, glucocorticoids also play crucial roles in the development of several other organ systems. Mothers at risk of preterm delivery are administered glucocorticoids to accelerate fetal lung development and prevent respiratory distress. Conversely, excessive glucocorticoid signaling is detrimental for fetal development; slowing fetal and placental growth and programming the individual for disease later in adult life. This review explores the mechanisms that control glucocorticoid signaling during pregnancy and provides an overview of the impact of glucocorticoid signaling on fetal development.

1. INTRODUCTION

Glucocorticoids are steroid hormones produced by the adrenal cortex in a circadian manner and in response to environmental or biological stress. Their synthesis and secretion are controlled by the hypothalamic–pituitary–adrenal (HPA) axis. In the adult, glucocorticoids regulate a wide variety of

Current Topics in Developmental Biology, Volume 125
ISSN 0070-2153
http://dx.doi.org/10.1016/bs.ctdb.2016.12.004

biological processes including energy metabolism, cardiac output, inflammation, and immunity. Glucocorticoids also serve crucial and unique roles in pregnancy and fetal development. In the fetus, serum glucocorticoids rise dramatically in the final weeks of pregnancy and are required to prepare the fetus for life after birth. Insufficient glucocorticoid signaling can be fatal primarily due to impaired lung development. Conversely, excessive glucocorticoid signaling from chronic maternal stress or antenatal treatment with synthetic glucocorticoids may suppress fetal growth and program the fetus for life-long diseases. In this review, we will discuss the mechanisms controlling fetal glucocorticoid exposure and provide an overview of the various impacts of glucocorticoid signaling on fetal development.

2. ADRENAL GLAND MORPHOLOGY AND EMBRYOLOGY

The adult adrenal gland is broadly subdivided into two zones, the outer adrenal cortex which produces steroid hormones and the inner medulla which produces the catecholamines: epinephrine and norepinephrine. Initiation of glucocorticoid production by the fetus is closely tied to adrenal gland development. During development the intermediate mesoderm gives rise to the adrenal cortex and the ectoderm gives rise to the adrenal medulla. In humans, at 4–5 weeks postcoitum (WPC) coelomic epithelia cells and mesonephric mesenchymal cells migrate from the adjacent mesonephros to form the primitive adrenal primordia posteromedially to the genital ridge (Bridgham, Carroll, & Thornton, 2006; Parker et al., 2002; Sucheston & Cannon, 1968). The adrenal primordia separate from the gonads by 7–8 WPC and become encapsulated by 9 WPC. The primitive adrenal medulla forms at 6 WPC as neural crest cell-derived pheochromoblasts invade the fetal adrenal cortex and form isolated clusters of sympathetic neurons (Cooper, Hutchins, & Israel, 1990; Ehrhart-Bornstein et al., 1997; Ishimoto & Jaffe, 2011).

The adult adrenal cortex is divided into three functionally distinct zones. The outer zona glomerulosa produces the mineralocorticoid aldosterone, the middle zona fasciculata produces the glucocorticoid corticosterone in rodents and cortisol in humans, and the inner zona reticularis produces the androgens testosterone and dehydroepiandrosterone (DHEA) that is utilized by various tissues for extra gonadal production of testosterone and estrogen (Nakamura et al., 2009; Penning et al., 2000; Simpson, 2003). Hormone production is tightly controlled by regional expression of a cascade of steroidogenic enzymes (Fig. 1). Similar to the adult adrenal cortex, the fetal

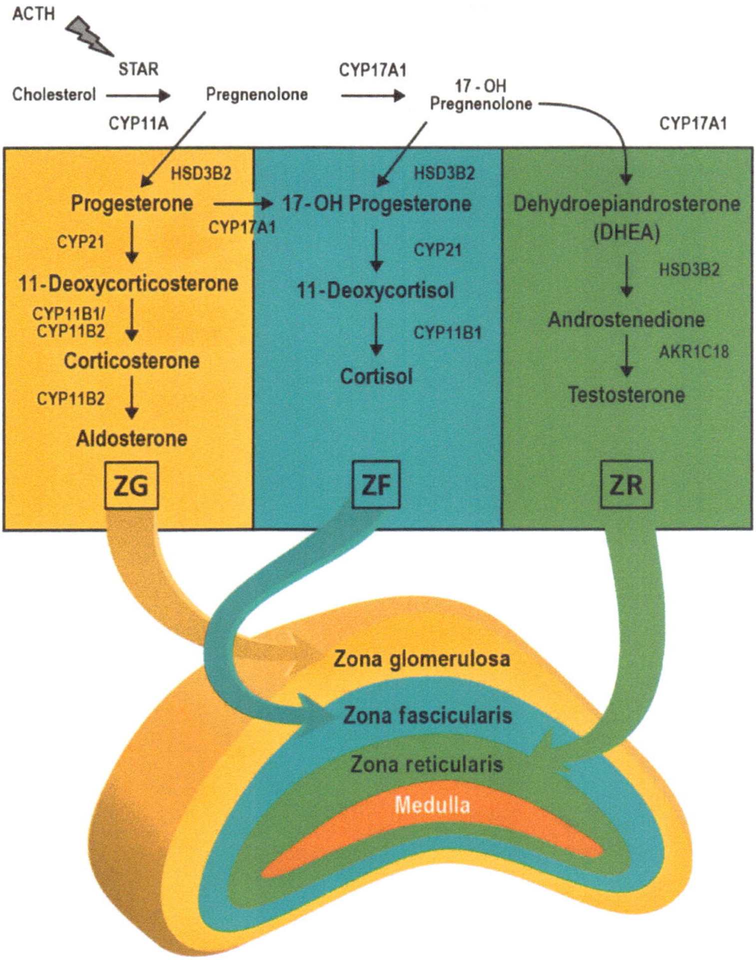

Fig. 1 Steroidogenesis in the adult adrenal gland. Steroidogenesis is segregated in distinct regions of the adrenal gland by spacial-specific expression of a cascade of enzymes. Synthesis is stimulated by ACTH which directs STAR to mobilize cholesterol to the inner mitochondrial membrane. Mineralocorticoids are synthesized by the zona glomerulosa (ZG), glucocorticoids are synthesized by the zona fascicularis (ZF), testosterone is synthesized by the zona reticularis (ZR), and catacolamines are synthesized by the medulla.

adrenal cortex is comprised of three regions: an outer definitive zone, middle transitional zone, and inner fetal zone (Fig. 2A). Within a few months after birth, the definitive zone and transitional zone develop into the zona glomerulosa and zona fasciculata, respectively. However, the fetal zone

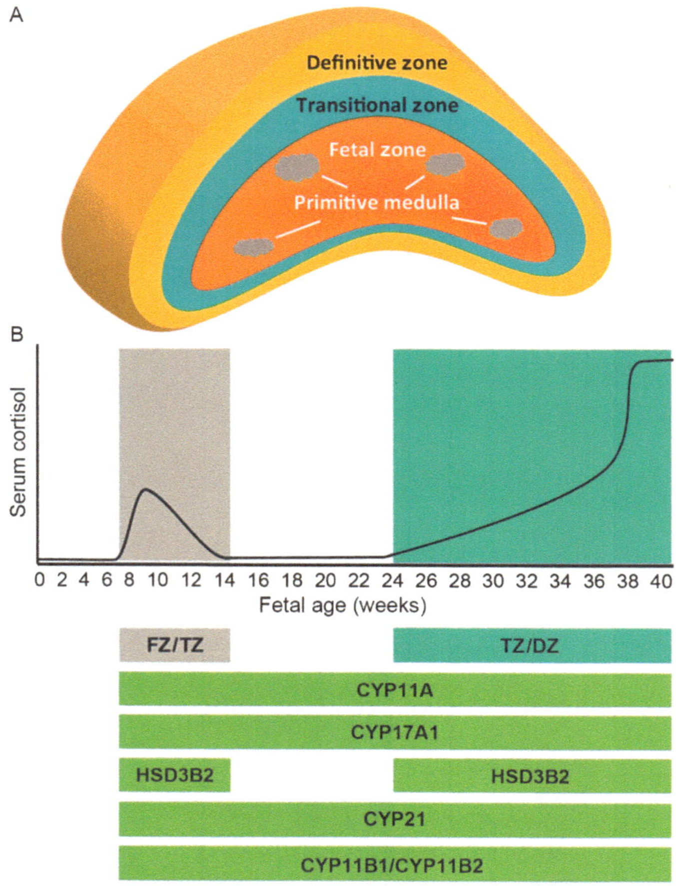

Fig. 2 (A) Depiction of the fetal adrenal gland. (B) In the human fetus, cortisol is produced early in pregnancy from 7 to 14 weeks postcoitum (WPC). Synthesis is dependent on expression of steroidogenic enzymes by the adrenal fetal zone (FZ) and transitional zone (TZ). Serum cortisol levels are low through midpregnancy and then slowly begin to increase toward the end of the second trimester as HSD3B2 expression resumes in the adrenal TZ and definitive zone (DZ). There is a surge in serum glucocorticoids during the last weeks of fetal life due to increased fetal production and from the ability of maternal cortisol to pass through the placenta.

rapidly involutes and degenerates. The adult zona reticularis has no known counterpart in the fetal adrenal gland and does not arise until approximately 6 years postpartum. The medulla is not recognized as a distinct structure during fetal development but rather consists of isolated islands of chromaffin cells spread throughout the cortex. After birth, these islands coalesce to form a primitive medulla at the center of the adrenal gland. By 1 year postpartum, the medulla has fully differentiated and closely resembles the adult medulla (Wilburn, Goldsmith, Chang, & Jaffe, 1986; Wilburn & Jaffe, 1988; Yon et al., 1998). Significant effort has been made to understand adrenal gland development and several review articles have been written describing this process (Ishimoto & Jaffe, 2011; Kempna & Fluck, 2008; Mesiano & Jaffe, 1997).

3. PRODUCTION AND METABOLISM OF GLUCOCORTICOIDS IN THE ADULT AND THE FETUS

Glucocorticoid synthesis is differentially regulated in the pre- and postnatal adrenal glands. In the adult, physiological cues signal the hypothalamus to release corticotrophin-releasing hormone (CRH) which stimulates neurons in the anterior pituitary to release adrenocorticotropic hormone (ACTH) which directs the adrenal glands to synthesize glucocorticoids from cholesterol. Circulating glucocorticoids exert negative feedback on the hypothalamus and the pituitary to inhibit the release of CRH and ACTH, respectively. Regulation of fetal glucocorticoid signaling is much more complex than in the adult, as maternal glucocorticoids can potentially cross the placenta. In addition, fetal glucocorticoid synthesis is only partially directed by the HPA axis and it is primarily regulated by differential expression of the enzymes required for glucocorticoid synthesis.

Two of the primary functions of the fetal adrenal glands in humans are to produce androgens and glucocorticoids. These two processes are functionally linked as the enzymes necessary for androgen production are also required for glucocorticoid synthesis. The fetal adrenal glands are capable of steroidogenesis shortly after they form. At 6 WPC in humans, the newly formed adrenal glands do not express enzymes necessary for steroidogenesis. However, by 7 WPC steroidogenic acute regulatory protein (STAR) and the enzymes CYP11A, CYP17A1, HSD3B2, CYP21, and CYP11B1 and 2 are expressed in the fetal zone and the transitional zone (Fig. 2B). In addition, there is a concomitant increase in HPA activity as ACTH is released by the pituitary gland. Glucocorticoid production can be attenuated by administration of the synthetic glucocorticoid dexamethasone (Goto et al., 2006). Glucocorticoid production peaks at 8–9 WPC and then ceases by 14 WPC. However, androgen production continues throughout fetal life and is an important source of DHEA which is utilized by the placenta for estrogen synthesis (Kaludjerovic & Ward, 2012; Siiteri & MacDonald, 1966). While glucocorticoid production is stimulated by ACTH, temporal expression of the enzyme HSD3B2, which is required for glucocorticoid production, controls the window of synthesis early in gestation. Analysis by immunohistochemistry has demonstrated that HSD3B2 levels peak at 9 WPC and gradually decrease thereafter and is undetectable by 14 WPC (Goto et al., 2006; Parker, Faye-Petersen, Stankovic, Mason, & Grizzle, 1995). The serum cortisol concentration reflects HSD3B2 expression and peaks at 9 WPC and then decreases until 14–15 WPC (Goto et al., 2006) (Fig. 2B). HSD3B2 levels remain undetectable throughout most of the second trimester and serum glucocorticoid levels are low despite persistent ACTH release from the pituitary (Goto et al., 2006; Mesiano, Coulter, & Jaffe, 1993; Narasaka, Suzuki, Moriya, & Sasano, 2001; Parker et al., 1995). Interestingly, ex vivo cultures of midgestation fetal adrenal glands induce HSD3B2 and readily synthesize glucocorticoids in response to ACTH. Several potential regulators of HSD3B2 expression have been identified including SF1, GATA4, GATA6, LRH1, NUR77, and STAT5 (Feltus, Groner, & Melner, 1999; Leers-Sucheta, Morohashi, Mason, & Melner, 1997; Martin & Tremblay, 2005; Martin et al., 2005) HSD3B2 expression resumes at approximately 24 WPC and corresponds to an increase in serum glucocorticoids (Narasaka et al., 2001; Parker et al., 1995). A possible explanation for this unusual mechanism regulating glucocorticoid synthesis by differential HSD3B2 expression is that it allows ACTH to support DHEA production without stimulating glucocorticoid synthesis.

Regulation of glucocorticoid synthesis is a critical mechanism that dictates the timing of fetal glucocorticoid exposure. However, corticosteroid metabolism also plays an equally important role in regulating fetal glucocorticoid signaling. Glucocorticoid metabolism in the fetus and adult is primarily performed by the isozymes HSD11B1 and HSD11B2. HSD11B1 is a bidirectional reductase that predominately catalyzes the conversion of the biologically inactive glucocorticoids 11-dehydrocorticosterone or cortisone to corticosterone or cortisol, which are the endogenous ligands for the glucocorticoid receptor (NR3C1, hereafter GR) in rodents and humans, respectively (Jamieson, Chapman, Edwards, & Seckl, 1995; Ricketts, Shoesmith, et al., 1998). HSD11B2 catalyzes the reverse reaction by catabolizing glucocorticoids to biologically inactive 11-dehydrocorticosterone or cortisone (Brown, Chapman, Edwards, & Seckl, 1993; Stewart, Murry, & Mason, 1994). In the adult, HSD11B1 is widely expressed in rodent and human tissues. High expression has been observed in the adult liver, lung, brain, stomach, adipose tissue, adrenal cortex, uterus, ovaries, and testes (Monder & Lakshmi, 1990; Paulsen, Pedersen, Fisker, & Richelsen, 2007; Ricketts, Verhaeg, et al., 1998). The putative function of HSD11B1 in these tissues is to generate high local glucocorticoid concentrations. Conversely, HSD11B2 is specifically expressed in tissues that require protection from glucocorticoids such as the kidney, colon, and salivary glands (Monder & Lakshmi, 1990; Smith et al., 1996). These tissues are dependent on signaling by the mineralocorticoid receptor (NR3C2; hereafter MR) which binds to glucocorticoids or its other endogenous ligand aldosterone with equal affinity. Circulating glucocorticoids are up to 1000 times higher than mineralocorticoids. HSD11B2 functions to protect the MR from glucocorticoid occupancy and promotes specific activation by mineralocorticoids (Edwards et al., 1988).

The HSD11B isozymes are highly expressed in the uterus, placenta, and fetal tissues. However, their exact role in establishing pregnancy and directing fetal glucocorticoid signaling is not fully understood. HSD11B1 is highly expressed in the decidua and the trophoblast where its putative function is to increase the local glucocorticoid concentration. This local increase in glucocorticoid concentration acts in concert with circulating glucocorticoids to suppress maternal immune rejection of the embryo and to promote uterine and vascular remodeling necessary for embryo implantation (Burton, Krozowski, & Waddell, 1998; Ricketts, Verhaeg, et al., 1998; Whirledge et al., 2015). Despite these functions attributed to HSD11B1, *Hsd11b1* null mice are born at normal mendelian ratios and are nearly phenotypically indistinguishable from wild type littermates (Kotelevtsev et al., 1997). Furthermore, development occurs normally in null pups born to homozygous null mothers indicating that HSD11B1 is dispensable for normal mouse development. In contrast, inhibition of HSD11B1 in sheep prevents conceptus elongation indicating that a high local glucocorticoid concentration gradient is necessary for embryonic development (Brooks, Burns, & Spencer, 2015). However, it is important to note that both HSD11B isozymes are promiscuous and may impact development independent of glucocorticoid metabolism (Balazs, Nashev, Chandsawangbhuwana, Baker, & Odermatt, 2009; Muller, Pompon, Urban, & Morfin, 2006).

Maternal glucocorticoid levels increase over the course of pregnancy and are much higher than serum glucocorticoid levels in the fetus. Glucocorticoid access to the placenta and fetus is primarily blocked by HSD11B2 which is robustly expressed by the syncytiotrophoblast and serves as an enzymatic barrier to maternal glucocorticoids (Krozowski et al., 1995). Overwhelming or circumventing the HSD11B2 barrier, such as during chronic maternal stress or caloric restriction or by administration of synthetic glucocorticoids, restricts fetal and placental growth and programs the fetus for greater risk of disease later in life (Bingham, Sheela Rani, Frazer, Strong, & Morilak, 2013; Reynolds, 2013). *Hsd11b2* null mice pups exhibit motor weakness and reduced suckling behavior and 50% of homozygous null pups die within 48 h of birth (Kotelevtsev et al., 1999). However, pups that survived for more than 48 h after birth generally lived to adulthood. A separate study found that when *Hsd11b2* null mice were backcrossed to a pure C57BL/6 background, null pups did not die but exhibited reduced birth weight and adrenal hypotrophy at birth and increased anxiety as adults (Holmes et al., 2006). In addition, null mice exhibit symptoms of mineralocorticoid excess syndrome such as: sodium retention, hypokalemia, and hypertension, but it is unknown if these symptoms are caused by aberrant embryonic development. Reduced fetal birth weight in *Hsd11b2* null mice is potentially caused by increased glucocorticoid signaling in the placenta which restricts placental growth, decreases placental nutrient and amino acid transport, and decreases placental vascularization (Wyrwoll, Seckl, & Holmes, 2009). In humans, congenital mutation of HSD11B2 is associated with reduced fetal birth weight, failure to thrive, and may be associated with fetal death (Ferrari et al., 1996; Krozowski, Stewart, Obeyesekere, Li, & Ferrari, 1997).

In addition to being expressed in the placenta, HSD11B2 is also expressed in several other fetal tissues including the central nervous system, kidney, hindgut, testes, bile ducts, and lung through midgestation (Brown et al., 1996; Diaz, Brown, & Seckl, 1998). The putative function in these tissues is to serve as a secondary barrier to maternal glucocorticoids where exposure may cause premature differentiation and growth arrest. Alternatively, HSD11B2 may be expressed in discrete fetal organs to protect them from glucocorticoids synthesized by the fetus. Fetal and placental HSD11B2 expression decreases at approximately E16 in rats and mice and there is a concurrent increase in HSD11B1, promoting glucocorticoid signaling in fetal tissues (Mark, Augustus, Lewis, Hewitt, & Waddell, 2009; Thompson, Han, & Yang, 2002). Similarly, placental HSD11B2 activity significantly decreases at 38 WPC in humans and allows maternal glucocorticoids to cross into fetal circulation and contribute to the surge in fetal glucocorticoid production prior to birth (Murphy & Clifton, 2003).

4. SIGNALING AND FUNCTION OF THE GLUCOCORTICOID RECEPTOR

At the cellular level, glucocorticoids function by binding the GR which is a member of the superfamily of ligand-dependent transcription factors. In adult tissues, the GR is nearly ubiquitously expressed and is responsible for a wide variety of cellular responses and tissue functions. The GR gene is composed of nine different exons and the GR protein is divided into three major domains. Exon 2 encodes an N-terminal transactivation domain, exons 3 and 4 encode a dimerization domain, exon 5 primarily encodes a hinge region, and exons 6–8 and a portion of exon 9 encode a ligand-binding domain (Fig. 3A). When unbound by ligand, the GR resides in the cytoplasm in a large protein complex which, in addition to the GR, is comprised of the chaperones HSP90, HSPA1B, and PTGES3 as well as the immunophilins FKBP5 and FKBP4. Endogenous glucocorticoids travel through the blood bound by corticosteroid-binding globulin (SERPINA6) which facilitates transport and regulates bioavailability. Unbound glucocorticoids freely diffuse through the cellular plasma membrane and bind to the cytosolic GR. Upon ligand binding, the GR undergoes a conformational change causing the protein complex to dissociate and exposing the nuclear localization signal allowing the receptor to translocate into the nucleus. Once in the nucleus, two GRs homodimerize and interact with a wide variety of coactivators, corepressors, and transcription factors to regulate the transcription of several thousand genes. The classic method by which the GR regulates gene expression is by binding to glucocorticoid response elements (GREs) in the promoters of target genes. Once the GR is bound to a GRE, it recruits transcription machinery and chromatin modulators to activate gene transcription. Recent studies have found many GREs in distant sites from the target gene rather than in the promotor (Burd & Archer, 2013). Therefore, the absence of a GRE in any given promoter does not exclude it from regulation by the GR. The GR functions to both induce or suppress transcription. Transcriptional activation by the GR may be achieved by multiple mechanisms such as by directly interacting with GREs or by tethering to other transcription factors and modifying their activity. Similar mechanisms are utilized to suppress gene transcription. For example, the GR binds to RELA (the NFKB p65 subunit) and inhibits its ability to induce transcription of cytokines (Nissen & Yamamoto, 2000). The GR can also bind as a monomer to

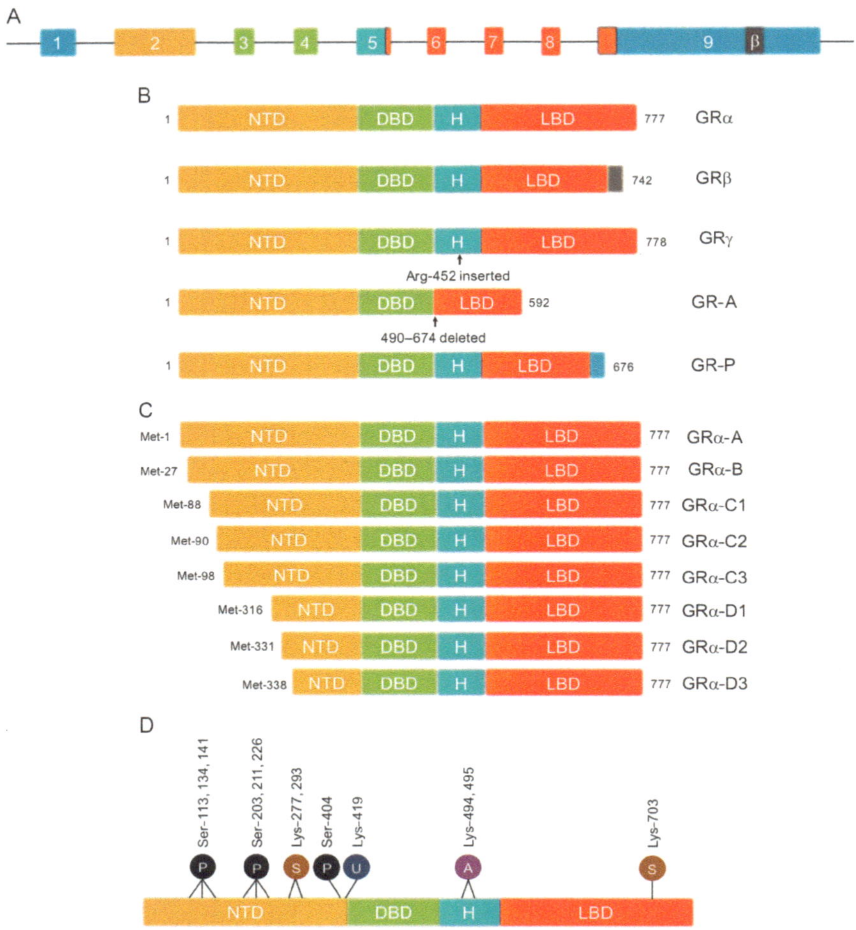

Fig. 3 (A) Depiction nascent GR RNA. Exon 2 comprises the N-terminal transactivation domain (NTD; *brown color*), exons 3 and 4 comprise the DNA-binding domain (DBD; *green*), the majority of exon 5 encodes the hinge region (H; *light blue*), and ligand-binding domain (LBD; *red*) is encoded by a portion of exon 5, all of exons 6, 7, 8, and part of exon 9. The LBD of GRβ utilizes an alternative portion of exon 9 (*gray*). (B) Alternative splicing of the GR mRNA yields five different GR isoforms. (C) Alternative AUG translation start sites in exon 2 generate eight different translational isoforms of GRα with progressively shorter transactivation domains. (D) Posttranslational modifications of the GR. *P*, phosphorylation sites; *S*, sumoylation sites; *U*, a ubiquitination site; and *A*, an acetylation site. The numbers for each modification are for the human GR.

negative GREs (nGRE) which function primary to suppress transcription (Surjit et al., 2011). The molecular mechanisms controlling glucocorticoid action in the adult are quite extensive and have been the focus of several recent review articles (Kadmiel & Cidlowski, 2013; Oakley & Cidlowski, 2013, 2015; Ramamoorthy & Cidlowski, 2016).

Responses to glucocorticoids can differ from tissue to tissue, over the course of development, or even within the same cell during different stages of the cell cycle. The various cellular responses are achieved by the GR interacting with different cofactors, expression of multiple GR transcriptional and translational isoforms (Kino, Su, & Chrousos, 2009; Lu & Cidlowski, 2004, 2005), and through posttranslational modification of the GR (Galliher-Beckley, Williams, & Cidlowski, 2011; Oakley & Cidlowski, 2013). Alternative splicing of the nascent GR RNA yields several different GR protein isoforms (Fig. 3B). The classic GR protein is termed GRα while alternative splicing of exon 9 generates GRβ which are the two most common GR isoforms (Oakley, Sar, & Cidlowski, 1996). Alternative splicing also generates three other GR isoforms termed GRγ, GR-A, and GR-P (Krett, Pillay, Moalli, Greipp, & Rosen, 1995; Rivers, Levy, Hancock, Lightman, & Norman, 1999). Only GRα and GRγ are able to bind to endogenous glucocorticoids while GRβ, GR-A, and GR-P are thought to modify GRα activity. However, GRγ and GRβ each have unique transcriptional profiles independent of GRα (Kino, Manoli, et al., 2009; Lewis-Tuffin, Jewell, Bienstock, Collins, & Cidlowski, 2007).

Alternative translation initiation generates additional GR isoforms (Fig. 3C). Exon 2 of the GR contains eight different AUG translation start sites which in combination with leaky ribosome scanning generate eight different translational isoforms of GRα (Lu & Cidlowski, 2005). These isoforms are termed: GRα-A, GRα-B, GRα-C1, GRα-C2, GRα-C3, Grα-D1, GRα-D2, and GRα-D3. Each of the isoforms has a progressively truncated transactivation domain. GRα-A, B, and C isoforms reside in the cytoplasm while unbound by ligand but GRα-D resides constitutively in the nucleus (Oakley & Cidlowski, 2013). Each of the isoforms exhibits similar affinity for glucocorticoids and can bind to GREs to modulate transcription. However, the transcriptional profiles and resultant cellular functions are distinct for each isoform (Lu, Collins, Grissom, & Cidlowski, 2007). Recently, it was found that eight different transcriptional and translational GR isoforms, GRα, GRβ, GR-P, GR-P, GRα-C, and GRα-D1–3, are expressed in the rodent and human placenta (Saif et al., 2016, 2014, 2015). Expression of the GR isoforms varied over the course of development and GRα-C was higher in the preterm placenta relative to placentas from individuals delivered at term. GRα-D3 was expressed higher in male placenta relative to female placentas. However, little is known concerning differential expression of GR isoforms in the fetus or how they function during development.

Additional variability and tissue specificity of glucocorticoid signaling is achieved through posttranslational modification of the GR. To date, 13 different sites of posttranslational modification have been identified including 7 phosphorylated serine residues in the transactivation domain, 2 sumoylation sites in the transactivation domain, 1 sumoylation site in the ligand-binding domain, 1 ubiquitination site in the transactivation domain, and 2 acetylation sites in the hinge region (Fig. 3D) (Oakley & Cidlowski, 2013). Phosphorylation is the best understood of these modifications and functions to modify the transcriptional activity of the GR by recruiting various coactivators and corepressors (Galliher-Beckley & Cidlowski, 2009). Sumoylation occurs on three specific lysine residues on the GR and recruits various coregulators which direct changes in the GR transcriptional activity (Druker et al., 2013). The GR is polyubiquitinated on a conserved lysine residue and which leads to degradation by the proteasome (Wallace & Cidlowski, 2001). Turnover of the GR by the proteasome is an important mechanism to control cellular sensitivity to glucocorticoid signaling (Wang & DeFranco, 2005). Finally, acetylation of two lysine residues in the hinge region occurs in response to ligand binding and modulates GR interactions with coregulators. For instance, deacetylation of the GR is required for efficient suppression of RELA-induced proinflammatory cytokines (Ito et al., 2006).

In addition to the classical transcriptional (genomic) glucocorticoid response, the GR can also elicit rapid nongenomic action which can occur within seconds. These nongenomic responses are independent of changes in message abundance or translation. There are conceivably multiple mechanisms whereby the GR may exert nongenomic actions but activation of kinase cascades is the best characterized. Activation of kinase cascades is performed by members of the protein complex associated with the unliganded GR. Upon ligand binding, the protein complex dissociates and the GR translocates to the nucleus liberating the binding proteins to mediate nongenomic actions associated with GR signaling. A recent study which utilizing cultures of primary rat neurons isolated from embryonic brains found that the GR associates with lipid rafts and activates the MAPK pathway, leading to phosphorylation of GJA1 and inhibition of neuron proliferation (Samarasinghe et al., 2011). Inhibition of proliferation depended the GR interacting with caveolin (CAV1) and SRC and occurred independent of transcription and translation. Inhibition of neuron proliferation mediated by nongenomic actions of the GR may contribute to the smaller brain size and neurological defects observed in infants after chronic maternal stress or after antenatal treatment with synthetic glucocorticoids. However, more studies are needed to understand how nongenomic actions of the GR affect fetal development.

5. THE IMPACT OF GLUCOCORTICOID SIGNALING ON FETAL DEVELOPMENT

Glucocorticoid signaling over the course of pregnancy is complex. Timing, intensity, and duration of the glucocorticoid signal are crucial to ensure proper development. In humans, glucocorticoid signaling occurs in three separate windows during embryonic and fetal development with intervening intervals of low or absent serum glucocorticoid concentration. Excessive glucocorticoid signaling or prolonged administration of synthetic glucocorticoids can negatively impact fetal development and predispose individuals to develop diseases later in juvenile and adult life. Conversely, loss of glucocorticoid signaling can cause widespread developmental defects that may cumulate in neonatal lethality.

The first window of glucocorticoid signaling occurs early in pregnancy and promotes embryo implantation, decidualization of the uterine wall, and to suppress the maternal immune system to prevent embryo rejection (Mastorakos et al., 1996). Deletion of the GR in the mouse uterus resulted in fewer embryo implantations (Whirledge et al., 2015). Furthermore, the same study reported progressive loss of embryos after E5.5 indicating that glucocorticoid signaling is also required for decidualization and continuation of pregnancy. Uterine GR knockout mice fail to induce expression of *Itga4* which is required to induce decidualization after implantation (Basak, Dhar, & Das, 2002). Furthermore, GR deletion reduced endometrial cell proliferation and prevented recruitment of macrophages which remodel the uterine wall after implantation (Brown, von Chamier, Allam, & Reyes, 2014; Care et al., 2013; Jasper et al., 2011; Whirledge et al., 2015). Similar studies in sheep have found that HSD11B1 is highly expressed in the endometrium and in the conceptus. Global inhibition or deletion of HSD11B1 in the conceptus trophectoderm prevents conceptus elongation but GR deletion in the conceptus did not affect elongation

(Brooks et al., 2015; Dorniak, Welsh, Bazer, & Spencer, 2013). These studies support the notion that glucocorticoid signaling in the endometrium is required for initiation and continuation of pregnancy.

The second window of glucocorticoid signaling during fetal development occurs from 7 to 14 WPC in humans and does not occur in rodents (Goto et al., 2006). As discussed earlier, primary control of fetal glucocorticoid production is through expression of HSD3B2. At 7 WPC, glucocorticoids are produced in response to the newly developed HPA axis and are necessary to suppress adrenal production of DHEA and direct adrenal development. Infants who carry homozygous mutation of *CYP21* cannot synthesize cortisol and are unable to exert negative feedback on the hypothalamus and pituitary, which causes excessive ACTH release (Ishimoto & Jaffe, 2011). Elevated ACTH stimulates adrenal hyperplasia and increases production of DHEA leading to elevated circulating androgens and virilization of the female external genitalia (Mendonca et al., 2002). CYP21 deficiency is treated by administering dexamethasone to susceptible fetuses based on the parental genotype. Treatment is initiated at 7–8 WPC to reflect endogenous cortisol synthesis and protect female genital development which occurs between 7 and 12 WPC (David & Forest, 1984; Lajic, Wedell, Bui, Ritzen, & Holst, 1998). At approximately 12 WPC, the fetus is genotyped and only female fetuses with homozygous mutation of *CYP21* continue to receive treatment (Lajic, Nordenstrom, & Hirvikoski, 2011). While long-term glucocorticoid treatment does successfully prevent virilization of female genitalia, it is also controversial as only female individuals with homozygous *CYP21* mutation, or 1 in 8 pregnancies from susceptible parents, will benefit from dexamethasone treatment, and therefore, 7 of 8 pregnancies will be needlessly treated. Some studies have linked early dexamethasone treatment of individuals with CYP21 deficiency with a decrease in intelligence and impaired social interaction (Hauser et al., 2008). Conversely, recent studies have found increased intelligence and improved social behavior following dexamethasone treatment compared to untreated CYP21-deficient individuals (Maryniak, Ginalska-Malinowska, Bielawska, & Ondruch, 2014; Meyer-Bahlburg, Dolezal, Haggerty, Silverman, & New, 2012). Future studies should examine the effects of a transient glucocorticoid regimen beginning during the 8th week and ending at the 14th week of gestation to more closely mimic endogenous glucocorticoid synthesis by the fetus.

The third window of glucocorticoid signaling occurs during the third trimester. The fetal adrenals begin to synthesize glucocorticoids at

24 WPC in humans and E15 in mice, and there is a surge in circulating fetal glucocorticoids from 38 to 40 WPC (Lockwood et al., 1996; Mesiano & Jaffe, 1997). The late stage surge in glucocorticoids is required for the maturation of several organ systems in preparation for life after birth. Several mouse models of glucocorticoid deficiency have provided valuable insight into how glucocorticoids direct organ development and differentiation prior to birth. GR null mice die shortly after birth due to respiratory distress (Cole et al., 1995). Lungs from GR null neonatal mice exhibit increased cellularity, a thickened alveolar surface, and reduced surfactant production. However, while the lung pathology was the most notable feature following deletion of the GR, recent studies using the GR null mice as well as conditional GR knockout mice have found several other developmental defects. At birth, the hearts of GR null mice are smaller and underdeveloped, exhibiting altered heart electrical activity and impaired cardiac function (Rog-Zielinska et al., 2013). GR knockout in fetal hepatocytes causes dysregulation of genes required for glucose metabolism and gluconeogenesis and 50% of knockout mice die within 48 h of birth (Opherk et al., 2004). Numerous studies have examined the role of glucocorticoid signaling on brain development and there are several conditional brain GR knockout models. Surprisingly, aberrant glucocorticoid signaling or ablation of the GR in the brain generally does not result in fetal or postnatal lethality except in the case of simultaneous GR deletion from the brain and pituitary gland which results in excessive serum glucocorticoids culminating in death by 1 week after birth (Erdmann, Schutz, & Berger, 2008). Deletion of the GR in the central nervous system resulted in decreased anxiety while deletion in the forebrain resulted in increased depression (Boyle et al., 2005; Boyle, Kolber, Vogt, Wozniak, & Muglia, 2006; Tronche et al., 1999). However, it is difficult to know if these behavioral phenotypes result from aberrant glucocorticoid signaling during development or if they occur postnatally.

While the predominant mode of glucocorticoid signaling is through the GR, glucocorticoids also bind with similar affinity to the MR. Late in fetal development, the adrenal glands begin to produce the endogenous MR ligand aldosterone. As a result, MR bound by either glucocorticoids or mineralocorticoids could impact fetal development. However, MR null mice are born at expected mendelian ratios and appear normal at birth. Homozygous null mice die 10 days after birth due to dehydration and excessive renal sodium loss (Berger et al., 1998). Postnatal death can be prevented by administration of sodium and adult MR null mice appear healthy and are able reproduce. While serum aldosterone was elevated in neonatal MR null mice and renin-producing cells were hyperplastic in adult mice, these are likely not caused by aberrant development because low blood volume caused by sodium wasting and potassium retention are well-known consequences of adrenalectomy (Bleich et al., 1999).

For more than 40 years, glucocorticoids have been used as a frontline treatment option for women at risk of preterm delivery to improve viability of preterm infants (Liggins & Howie, 1972). Antenatal glucocorticoid treatment presumably functions similarly to the endogenous glucocorticoid surge that occurs during the final weeks of pregnancy. A single injection of the synthetic glucocorticoids dexamethasone or betamethasone 48 h prior to birth accelerates fetal lung development and reduces respiratory distress which is the most common cause of death in preterm infants. However, antenatal glucocorticoid treatment also accelerates the differentiation of several other tissues including the heart to increase contractility and cardiac output as well as accelerate closure of the foramen ovale and ductus arteriosus (Eronen, Kari, Pesonen, & Hallman, 1993; Rog-Zielinska, Richardson, Denvir, & Chapman, 2014). Glucocorticoid treatment has profoundly increased survival of preterm infants. Considerable effort has been dedicated to improve antenatal glucocorticoid treatment and to determine if multiple doses or long-term glucocorticoid treatment of mothers at risk of preterm delivery could further increase fetal survival (Murphy et al., 2008). Multiple glucocorticoid doses have been shown to further stimulate lung development and increase fetal survival. However, the increased rate of survival is minimal and prolonged exposure to high levels of glucocorticoids suppresses fetal and placental growth and has been linked to neurological and learning defects, cardiovascular disease, and metabolic disease later in life (Reynolds, 2013).

A potential shortcoming in the current treatment regimen of preterm infants is that exogenous glucocorticoid treatment is usually not continued after birth. In infants carried to term, cortisol synthesis continues after birth and supports the function and postnatal development of the lungs, heart, liver, and intestines. However, preterm infants have low serum cortisol levels and exhibit transient adrenal insufficiency likely due to immaturity of various components of their HPA axis (Ng et al., 2004). Transient adrenal insufficiency of preterm infants is associated with increased infant mortality and may increase disease susceptibility later in life (Fernandez & Watterberg, 2009; Nykanen, Anttila, Heinonen, Hallman, & Voutilainen, 2007). Clinical trials have demonstrated that postnatal glucocorticoid treatment of preterm infants accelerates lung development and extubation, prevents chronic lung disease, facilitates closure of the ductus arteriosium, and prevents chronic cardiovascular diseases as adults (Halliday, Ehrenkranz, & Doyle, 2009; Rademaker et al., 2007; Watterberg et al., 2007). However, postnatal glucocorticoid treatment is also associated with several negative effects including gastrointestinal bleeding, intestinal perforation, hyperglycemia, hypertension, hypertrophic cardiomyopathy, and temporary reduction of weight gain (Halliday et al., 2009) Future studies should examine if alternative glucocorticoid doses may promote continued neonatal development while minimizing risks of adverse effects.

6. CONCLUDING REMARKS

It is clear that glucocorticoids play a crucial role in directing fetal development; the timing and extent of fetal glucocorticoid exposure is crucial for survival, development, and to prevent fetal programming for diseases later in life. We have made tremendous advances in understanding how glucocorticoid signaling prepares the fetus for extrauterine life. However, there remains a large knowledge gap concerning how glucocorticoids actually direct fetal development and the molecular mechanism of glucocorticoid action in the fetus. Particularly, little is known about glucocorticoid action early in fetal development. By increasing our understanding of the impact of glucocorticoid action in development, we may be able to improve treatment of mothers at risk for preterm delivery without the risk of long-term adverse effects.

ACKNOWLEDGMENT

Support provided by the Intramural Research Program of the National Institute of Environment Health Sciences/National Institutes of Health.

REFERENCES

Balazs, Z., Nashev, L. G., Chandsawangbhuwana, C., Baker, M. E., & Odermatt, A. (2009). Hexose-6-phosphate dehydrogenase modulates the effect of inhibitors and alternative substrates of 11beta-hydroxysteroid dehydrogenase 1. *Molecular and Cellular Endocrinology, 301*, 117–122.

Basak, S., Dhar, R., & Das, C. (2002). Steroids modulate the expression of alpha4 integrin in mouse blastocysts and uterus during implantation. *Biology of Reproduction, 66*, 1784–1789.

Berger, S., Bleich, M., Schmid, W., Cole, T. J., Peters, J., Watanabe, H., et al. (1998). Mineralocorticoid receptor knockout mice: Pathophysiology of Na + metabolism. *Proceedings of the National Academy of Sciences of the United States of America, 95*, 9424–9429.

Bingham, B. C., Sheela Rani, C. S., Frazer, A., Strong, R., & Morilak, D. A. (2013). Exogenous prenatal corticosterone exposure mimics the effects of prenatal stress on adult brain stress response systems and fear extinction behavior. *Psychoneuroendocrinology, 38*, 2746–2757.

Bleich, M., Warth, R., Schmidt-Hieber, M., Schulz-Baldes, A., Hasselblatt, P., Fisch, D., et al. (1999). Rescue of the mineralocorticoid receptor knock-out mouse. *Pflügers Archiv/European Journal of Physiology, 438*, 245–254.

Boyle, M. P., Brewer, J. A., Funatsu, M., Wozniak, D. F., Tsien, J. Z., Izumi, Y., et al. (2005). Acquired deficit of forebrain glucocorticoid receptor produces depression-like changes in adrenal axis regulation and behavior. *Proceedings of the National Academy of Sciences of the United States of America, 102*, 473–478.

Boyle, M. P., Kolber, B. J., Vogt, S. K., Wozniak, D. F., & Muglia, L. J. (2006). Forebrain glucocorticoid receptors modulate anxiety-associated locomotor activation and adrenal responsiveness. *The Journal of Neuroscience: The Official Journal of the Society for Neuroscience, 26*, 1971–1978.

Bridgham, J. T., Carroll, S. M., & Thornton, J. W. (2006). Evolution of hormone-receptor complexity by molecular exploitation. *Science (New York, N.Y.), 312*, 97–101.

Brooks, K., Burns, G., & Spencer, T. E. (2015). Biological roles of hydroxysteroid (11-beta) dehydrogenase 1 (HSD11B1), HSD11B2, and glucocorticoid receptor (NR3C1) in sheep conceptus elongation. *Biology of Reproduction, 93*, 38.

Brown, R. W., Chapman, K. E., Edwards, C. R., & Seckl, J. R. (1993). Human placental 11 beta-hydroxysteroid dehydrogenase: Evidence for and partial purification of a distinct NAD-dependent isoform. *Endocrinology, 132*, 2614–2621.

Brown, R. W., Diaz, R., Robson, A. C., Kotelevtsev, Y. V., Mullins, J. J., Kaufman, M. H., et al. (1996). The ontogeny of 11 beta-hydroxysteroid dehydrogenase type 2 and mineralocorticoid receptor gene expression reveal intricate control of glucocorticoid action in development. *Endocrinology, 137*, 794–797.

Brown, M. B., von Chamier, M., Allam, A. B., & Reyes, L. (2014). M1/M2 macrophage polarity in normal and complicated pregnancy. *Frontiers in Immunology, 5*, 606.

Burd, C. J., & Archer, T. K. (2013). Chromatin architecture defines the glucocorticoid response. *Molecular and Cellular Endocrinology, 380*, 25–31.

Burton, P. J., Krozowski, Z. S., & Waddell, B. J. (1998). Immunolocalization of 11beta-hydroxysteroid dehydrogenase types 1 and 2 in rat uterus: Variation across the estrous cycle and regulation by estrogen and progesterone. *Endocrinology, 139*, 376–382.

Care, A. S., Diener, K. R., Jasper, M. J., Brown, H. M., Ingman, W. V., & Robertson, S. A. (2013). Macrophages regulate corpus luteum development during embryo implantation in mice. *The Journal of Clinical Investigation, 123*, 3472–3487.

Cole, T. J., Blendy, J. A., Monaghan, A. P., Krieglstein, K., Schmid, W., Aguzzi, A., et al. (1995). Targeted disruption of the glucocorticoid receptor gene blocks adrenergic chromaffin cell development and severely retards lung maturation. *Genes & Development, 9*, 1608–1621.

Cooper, M. J., Hutchins, G. M., & Israel, M. A. (1990). Histogenesis of the human adrenal medulla. An evaluation of the ontogeny of chromaffin and nonchromaffin lineages. *The American Journal of Pathology, 137*, 605–615.

David, M., & Forest, M. G. (1984). Prenatal treatment of congenital adrenal hyperplasia resulting from 21-hydroxylase deficiency. *The Journal of Pediatrics, 105*, 799–803.

Diaz, R., Brown, R. W., & Seckl, J. R. (1998). Distinct ontogeny of glucocorticoid and mineralocorticoid receptor and 11beta-hydroxysteroid dehydrogenase types I and II mRNAs in the fetal rat brain suggest a complex control of glucocorticoid actions. *The Journal of Neuroscience: The Official Journal of the Society for Neuroscience, 18*, 2570–2580.

Dorniak, P., Welsh, T. H., Jr., Bazer, F. W., & Spencer, T. E. (2013). Cortisol and interferon tau regulation of endometrial function and conceptus development in female sheep. *Endocrinology, 154*, 931–941.

Druker, J., Liberman, A. C., Antunica-Noguerol, M., Gerez, J., Paez-Pereda, M., Rein, T., et al. (2013). RSUME enhances glucocorticoid receptor SUMOylation and transcriptional activity. *Molecular and Cellular Biology, 33*, 2116–2127.

Edwards, C. R., Stewart, P. M., Burt, D., Brett, L., McIntyre, M. A., Sutanto, W. S., et al. (1988). Localisation of 11 beta-hydroxysteroid dehydrogenase—Tissue specific protector of the mineralocorticoid receptor. *Lancet (London, England), 2*, 986–989.

Ehrhart-Bornstein, M., Breidert, M., Guadanucci, P., Wozniak, W., Bocian-Sobkowska, J., Malendowicz, L. K., et al. (1997). 17 alpha-Hydroxylase and chromogranin a in 6th week human fetal adrenals. *Hormone and Metabolic Research = Hormon- und Stoffwechselforschung = Hormones et metabolisme, 29*, 30–32.

Erdmann, G., Schutz, G., & Berger, S. (2008). Loss of glucocorticoid receptor function in the pituitary results in early postnatal lethality. *Endocrinology, 149*, 3446–3451.

Eronen, M., Kari, A., Pesonen, E., & Hallman, M. (1993). The effect of antenatal dexamethasone administration on the fetal and neonatal ductus arteriosus. A randomized double-blind study. *American Journal of Diseases of Children (1960), 147*, 187–192.

Feltus, F. A., Groner, B., & Melner, M. H. (1999). Stat5-mediated regulation of the human type II 3beta-hydroxysteroid dehydrogenase/delta5-delta4 isomerase gene: Activation by prolactin. *Molecular Endocrinology (Baltimore, Md), 13*, 1084–1093.

Fernandez, E. F., & Watterberg, K. L. (2009). Relative adrenal insufficiency in the preterm and term infant. *Journal of Perinatology: Official Journal of the California Perinatal Association, 29*(Suppl. 2), S44–S49.

Ferrari, P., Obeyesekere, V. R., Li, K., Wilson, R. C., New, M. I., Funder, J. W., et al. (1996). Point mutations abolish 11 beta-hydroxysteroid dehydrogenase type II activity in three families with the congenital syndrome of apparent mineralocorticoid excess. *Molecular and Cellular Endocrinology, 119*, 21–24.

Galliher-Beckley, A. J., & Cidlowski, J. A. (2009). Emerging roles of glucocorticoid receptor phosphorylation in modulating glucocorticoid hormone action in health and disease. *IUBMB Life, 61*, 979–986.

Galliher-Beckley, A. J., Williams, J. G., & Cidlowski, J. A. (2011). Ligand-independent phosphorylation of the glucocorticoid receptor integrates cellular stress pathways with nuclear receptor signaling. *Molecular and Cellular Biology, 31*, 4663–4675.

Goto, M., Piper Hanley, K., Marcos, J., Wood, P. J., Wright, S., Postle, A. D., et al. (2006). In humans, early cortisol biosynthesis provides a mechanism to safeguard female sexual development. *The Journal of Clinical Investigation, 116*, 953–960.

Halliday, H. L., Ehrenkranz, R. A., & Doyle, L. W. (2009). Early (<8 days) postnatal corticosteroids for preventing chronic lung disease in preterm infants. *The Cochrane Database of Systematic Reviews, 20*(1), CD001146.

Hauser, J., Knapman, A., Zurcher, N. R., Pilloud, S., Maier, C., Diaz-Heijtz, R., et al. (2008). Effects of prenatal dexamethasone treatment on physical growth, pituitary-adrenal hormones, and performance of motor, motivational, and cognitive tasks in juvenile and adolescent common marmoset monkeys. *Endocrinology, 149*, 6343–6355.

Holmes, M. C., Abrahamsen, C. T., French, K. L., Paterson, J. M., Mullins, J. J., & Seckl, J. R. (2006). The mother or the fetus? 11beta-Hydroxysteroid dehydrogenase type 2 null mice provide evidence for direct fetal programming of behavior by endogenous glucocorticoids. *The Journal of Neuroscience: The Official Journal of the Society for Neuroscience, 26*, 3840–3844.

Ishimoto, H., & Jaffe, R. B. (2011). Development and function of the human fetal adrenal cortex: A key component in the feto-placental unit. *Endocrine Reviews, 32*, 317–355.

Ito, K., Yamamura, S., Essilfie-Quaye, S., Cosio, B., Ito, M., Barnes, P. J., et al. (2006). Histone deacetylase 2-mediated deacetylation of the glucocorticoid receptor enables NF-kappaB suppression. *The Journal of Experimental Medicine, 203*, 7–13.

Jamieson, P. M., Chapman, K. E., Edwards, C. R., & Seckl, J. R. (1995). 11 beta-hydroxysteroid dehydrogenase is an exclusive 11 beta-reductase in primary cultures of rat hepatocytes: Effect of physicochemical and hormonal manipulations. *Endocrinology, 136*, 4754–4761.

Jasper, M. J., Care, A. S., Sullivan, B., Ingman, W. V., Aplin, J. D., & Robertson, S. A. (2011). Macrophage-derived LIF and IL1B regulate alpha(1,2)fucosyltransferase 2 (Fut2) expression in mouse uterine epithelial cells during early pregnancy. *Biology of Reproduction, 84*, 179–188.

Kadmiel, M., & Cidlowski, J. A. (2013). Glucocorticoid receptor signaling in health and disease. *Trends in Pharmacological Sciences, 34*, 518–530.

Kaludjerovic, J., & Ward, W. E. (2012). The interplay between estrogen and fetal adrenal cortex. *Journal of Nutrition and Metabolism, 2012*, 837901.

Kempna, P., & Fluck, C. E. (2008). Adrenal gland development and defects. *Best Practice & Research Clinical Endocrinology & Metabolism, 22*, 77–93.

Kino, T., Manoli, I., Kelkar, S., Wang, Y., Su, Y. A., & Chrousos, G. P. (2009). Glucocorticoid receptor (GR) beta has intrinsic, GRalpha-independent transcriptional activity. *Biochemical and Biophysical Research Communications, 381*, 671–675.

Kino, T., Su, Y. A., & Chrousos, G. P. (2009). Human glucocorticoid receptor isoform beta: Recent understanding of its potential implications in physiology and pathophysiology. *Cellular and Molecular Life Sciences, 66*, 3435–3448.

Kotelevtsev, Y., Brown, R. W., Fleming, S., Kenyon, C., Edwards, C. R., Seckl, J. R., et al. (1999). Hypertension in mice lacking 11beta-hydroxysteroid dehydrogenase type 2. *The Journal of Clinical Investigation, 103*, 683–689.

Kotelevtsev, Y., Holmes, M. C., Burchell, A., Houston, P. M., Schmoll, D., Jamieson, P., et al. (1997). 11beta-Hydroxysteroid dehydrogenase type 1 knockout mice show attenuated glucocorticoid-inducible responses and resist hyperglycemia on obesity or stress. *Proceedings of the National Academy of Sciences of the United States of America, 94*, 14924–14929.

Krett, N. L., Pillay, S., Moalli, P. A., Greipp, P. R., & Rosen, S. T. (1995). A variant glucocorticoid receptor messenger RNA is expressed in multiple myeloma patients. *Cancer Research, 55*, 2727–2729.

Krozowski, Z., MaGuire, J. A., Stein-Oakley, A. N., Dowling, J., Smith, R. E., & Andrews, R. K. (1995). Immunohistochemical localization of the 11 beta-hydroxysteroid dehydrogenase type II enzyme in human kidney and placenta. *The Journal of Clinical Endocrinology and Metabolism, 80*, 2203–2209.

Krozowski, Z. S., Stewart, P. M., Obeyesekere, V. R., Li, K., & Ferrari, P. (1997). Mutations in the 11 beta-hydroxysteroid dehydrogenase type II enzyme associated with hypertension and possibly stillbirth. *Clinical and Experimental Hypertension (New York, N.Y.: 1993), 19*, 519–529.

Lajic, S., Nordenstrom, A., & Hirvikoski, T. (2011). Long-term outcome of prenatal dexamethasone treatment of 21-hydroxylase deficiency. *Endocrine Development, 20*, 96–105.

Lajic, S., Wedell, A., Bui, T. H., Ritzen, E. M., & Holst, M. (1998). Long-term somatic follow-up of prenatally treated children with congenital adrenal hyperplasia. *The Journal of Clinical Endocrinology and Metabolism, 83*, 3872–3880.

Leers-Sucheta, S., Morohashi, K., Mason, J. I., & Melner, M. H. (1997). Synergistic activation of the human type II 3beta-hydroxysteroid dehydrogenase/delta5-delta4 isomerase promoter by the transcription factor steroidogenic factor-1/adrenal 4-binding protein and phorbol ester. *The Journal of Biological Chemistry, 272*, 7960–7967.

Lewis-Tuffin, L. J., Jewell, C. M., Bienstock, R. J., Collins, J. B., & Cidlowski, J. A. (2007). Human glucocorticoid receptor beta binds RU-486 and is transcriptionally active. *Molecular and Cellular Biology, 27*, 2266–2282.

Liggins, G. C., & Howie, R. N. (1972). A controlled trial of antepartum glucocorticoid treatment for prevention of the respiratory distress syndrome in premature infants. *Pediatrics*, *50*, 515–525.

Lockwood, C. J., Radunovic, N., Nastic, D., Petkovic, S., Aigner, S., & Berkowitz, G. S. (1996). Corticotropin-releasing hormone and related pituitary-adrenal axis hormones in fetal and maternal blood during the second half of pregnancy. *Journal of Perinatal Medicine*, *24*, 243–251.

Lu, N. Z., & Cidlowski, J. A. (2004). The origin and functions of multiple human glucocorticoid receptor isoforms. *Annals of the New York Academy of Sciences*, *1024*, 102–123.

Lu, N. Z., & Cidlowski, J. A. (2005). Translational regulatory mechanisms generate N-terminal glucocorticoid receptor isoforms with unique transcriptional target genes. *Molecular Cell*, *18*, 331–342.

Lu, N. Z., Collins, J. B., Grissom, S. F., & Cidlowski, J. A. (2007). Selective regulation of bone cell apoptosis by translational isoforms of the glucocorticoid receptor. *Molecular and Cellular Biology*, *27*, 7143–7160.

Mark, P. J., Augustus, S., Lewis, J. L., Hewitt, D. P., & Waddell, B. J. (2009). Changes in the placental glucocorticoid barrier during rat pregnancy: Impact on placental corticosterone levels and regulation by progesterone. *Biology of Reproduction*, *80*, 1209–1215.

Martin, L. J., Taniguchi, H., Robert, N. M., Simard, J., Tremblay, J. J., & Viger, R. S. (2005). GATA factors and the nuclear receptors, steroidogenic factor 1/liver receptor homolog 1, are key mutual partners in the regulation of the human 3beta-hydroxysteroid dehydrogenase type 2 promoter. *Molecular Endocrinology (Baltimore, Md)*, *19*, 2358–2370.

Martin, L. J., & Tremblay, J. J. (2005). The human 3beta-hydroxysteroid dehydrogenase/Delta5-Delta4 isomerase type 2 promoter is a novel target for the immediate early orphan nuclear receptor Nur77 in steroidogenic cells. *Endocrinology*, *146*, 861–869.

Maryniak, A., Ginalska-Malinowska, M., Bielawska, A., & Ondruch, A. (2014). Cognitive and social function in girls with congenital adrenal hyperplasia—Influence of prenatally administered dexamethasone. *Child Neuropsychology: A Journal on Normal and Abnormal Development in Childhood and Adolescence*, *20*, 60–70.

Mastorakos, G., Scopa, C. D., Kao, L. C., Vryonidou, A., Friedman, T. C., Kattis, D., et al. (1996). Presence of immunoreactive corticotropin-releasing hormone in human endometrium. *The Journal of Clinical Endocrinology and Metabolism*, *81*, 1046–1050.

Mendonca, B. B., Leite, M. V., de Castro, M., Kino, T., Elias, L. L., Bachega, T. A., et al. (2002). Female pseudohermaphroditism caused by a novel homozygous missense mutation of the GR gene. *The Journal of Clinical Endocrinology and Metabolism*, *87*, 1805–1809.

Mesiano, S., Coulter, C. L., & Jaffe, R. B. (1993). Localization of cytochrome P450 cholesterol side-chain cleavage, cytochrome P450 17 alpha-hydroxylase/17, 20-lyase, and 3 beta-hydroxysteroid dehydrogenase isomerase steroidogenic enzymes in human and rhesus monkey fetal adrenal glands: Reappraisal of functional zonation. *The Journal of Clinical Endocrinology and Metabolism*, 77, 1184–1189.

Mesiano, S., & Jaffe, R. B. (1997). Developmental and functional biology of the primate fetal adrenal cortex. *Endocrine Reviews*, *18*, 378–403.

Meyer-Bahlburg, H. F., Dolezal, C., Haggerty, R., Silverman, M., & New, M. I. (2012). Cognitive outcome of offspring from dexamethasone-treated pregnancies at risk for congenital adrenal hyperplasia due to 21-hydroxylase deficiency. *European Journal of Endocrinology*, *167*, 103–110.

Monder, C., & Lakshmi, V. (1990). Corticosteroid 11 beta-dehydrogenase of rat tissues: Immunological studies. *Endocrinology*, *126*, 2435–2443.

Muller, C., Pompon, D., Urban, P., & Morfin, R. (2006). Inter-conversion of 7alpha- and 7beta-hydroxy-dehydroepiandrosterone by the human 11beta-hydroxysteroid dehydrogenase type 1. *The Journal of Steroid Biochemistry and Molecular Biology*, *99*, 215–222.

Murphy, V. E., & Clifton, V. L. (2003). Alterations in human placental 11beta-hydroxysteroid dehydrogenase type 1 and 2 with gestational age and labour. *Placenta, 24*, 739–744.

Murphy, K. E., Hannah, M. E., Willan, A. R., Hewson, S. A., Ohlsson, A., Kelly, E. N., et al. (2008). Multiple courses of antenatal corticosteroids for preterm birth (MACS): A randomised controlled trial. *Lancet (London, England), 372*, 2143–2151.

Nakamura, Y., Hornsby, P. J., Casson, P., Morimoto, R., Satoh, F., Xing, Y., et al. (2009). Type 5 17beta-hydroxysteroid dehydrogenase (AKR1C3) contributes to testosterone production in the adrenal reticularis. *The Journal of Clinical Endocrinology and Metabolism, 94*, 2192–2198.

Narasaka, T., Suzuki, T., Moriya, T., & Sasano, H. (2001). Temporal and spatial distribution of corticosteroidogenic enzymes immunoreactivity in developing human adrenal. *Molecular and Cellular Endocrinology, 174*, 111–120.

Ng, P. C., Lee, C. H., Lam, C. W., Ma, K. C., Fok, T. F., Chan, I. H., et al. (2004). Transient adrenocortical insufficiency of prematurity and systemic hypotension in very low birthweight infants. *Archives of Disease in Childhood. Fetal and Neonatal Edition, 89*, F119–F126.

Nissen, R. M., & Yamamoto, K. R. (2000). The glucocorticoid receptor inhibits NFkappaB by interfering with serine-2 phosphorylation of the RNA polymerase II carboxy-terminal domain. *Genes & Development, 14*, 2314–2329.

Nykanen, P., Anttila, E., Heinonen, K., Hallman, M., & Voutilainen, R. (2007). Early hypoadrenalism in premature infants at risk for bronchopulmonary dysplasia or death. *Acta Paediatrica (Oslo, Norway: 1992), 96*, 1600–1605.

Oakley, R. H., & Cidlowski, J. A. (2013). The biology of the glucocorticoid receptor: New signaling mechanisms in health and disease. *The Journal of Allergy and Clinical Immunology, 132*, 1033–1044.

Oakley, R. H., & Cidlowski, J. A. (2015). Glucocorticoid signaling in the heart: A cardiomyocyte perspective. *The Journal of Steroid Biochemistry and Molecular Biology, 153*, 27–34.

Oakley, R. H., Sar, M., & Cidlowski, J. A. (1996). The human glucocorticoid receptor beta isoform. Expression, biochemical properties, and putative function. *The Journal of Biological Chemistry, 271*, 9550–9559.

Opherk, C., Tronche, F., Kellendonk, C., Kohlmuller, D., Schulze, A., Schmid, W., et al. (2004). Inactivation of the glucocorticoid receptor in hepatocytes leads to fasting hypoglycemia and ameliorates hyperglycemia in streptozotocin-induced diabetes mellitus. *Molecular Endocrinology (Baltimore, Md.), 18*, 1346–1353.

Parker, C. R., Jr., Faye-Petersen, O., Stankovic, A. K., Mason, J. I., & Grizzle, W. E. (1995). Immunohistochemical evaluation of the cellular localization and ontogeny of 3 beta-hydroxysteroid dehydrogenase/delta 5–4 isomerase in the human fetal adrenal gland. *Endocrine Research, 21*, 69–80.

Parker, K. L., Rice, D. A., Lala, D. S., Ikeda, Y., Luo, X., Wong, M., et al. (2002). Steroidogenic factor 1: An essential mediator of endocrine development. *Recent Progress in Hormone Research, 57*, 19–36.

Paulsen, S. K., Pedersen, S. B., Fisker, S., & Richelsen, B. (2007). 11Beta-HSD type 1 expression in human adipose tissue: Impact of gender, obesity, and fat localization. *Obesity (Silver Spring, Md.), 15*, 1954–1960.

Penning, T. M., Burczynski, M. E., Jez, J. M., Hung, C. F., Lin, H. K., Ma, H., et al. (2000). Human 3alpha-hydroxysteroid dehydrogenase isoforms (AKR1C1-AKR1C4) of the aldo-keto reductase superfamily: Functional plasticity and tissue distribution reveals roles in the inactivation and formation of male and female sex hormones. *The Biochemical Journal, 351*, 67–77.

Rademaker, K. J., Uiterwaal, C. S., Groenendaal, F., Venema, M. M., van Bel, F., Beek, F. J., et al. (2007). Neonatal hydrocortisone treatment: Neurodevelopmental outcome and MRI at school age in preterm-born children. *The Journal of Pediatrics*, *150*, 351–357.

Ramamoorthy, S., & Cidlowski, J. A. (2016). Corticosteroids: Mechanisms of action in health and disease. *Rheumatic Diseases Clinics of North America*, *42*, 15–31, vii.

Reynolds, R. M. (2013). Programming effects of glucocorticoids. *Clinical Obstetrics and Gynecology*, *56*, 602–609.

Ricketts, M. L., Shoesmith, K. J., Hewison, M., Strain, A., Eggo, M. C., & Stewart, P. M. (1998). Regulation of 11 beta-hydroxysteroid dehydrogenase type 1 in primary cultures of rat and human hepatocytes. *The Journal of Endocrinology*, *156*, 159–168.

Ricketts, M. L., Verhaeg, J. M., Bujalska, I., Howie, A. J., Rainey, W. E., & Stewart, P. M. (1998). Immunohistochemical localization of type 1 11beta-hydroxysteroid dehydrogenase in human tissues. *The Journal of Clinical Endocrinology and Metabolism*, *83*, 1325–1335.

Rivers, C., Levy, A., Hancock, J., Lightman, S., & Norman, M. (1999). Insertion of an amino acid in the DNA-binding domain of the glucocorticoid receptor as a result of alternative splicing. *The Journal of Clinical Endocrinology and Metabolism*, *84*, 4283–4286.

Rog-Zielinska, E. A., Richardson, R. V., Denvir, M. A., & Chapman, K. E. (2014). Glucocorticoids and foetal heart maturation; implications for prematurity and foetal programming. *Journal of Molecular Endocrinology*, *52*, R125–R135.

Rog-Zielinska, E. A., Thomson, A., Kenyon, C. J., Brownstein, D. G., Moran, C. M., Szumska, D., et al. (2013). Glucocorticoid receptor is required for foetal heart maturation. *Human Molecular Genetics*, *22*, 3269–3282.

Saif, Z., Dyson, R. M., Palliser, H. K., Wright, I. M., Lu, N., & Clifton, V. L. (2016). Identification of eight different isoforms of the glucocorticoid receptor in guinea pig placenta: Relationship to preterm delivery, sex and betamethasone exposure. *PLoS One*, *11*, e0148226.

Saif, Z., Hodyl, N. A., Hobbs, E., Tuck, A. R., Butler, M. S., Osei-Kumah, A., et al. (2014). The human placenta expresses multiple glucocorticoid receptor isoforms that are altered by fetal sex, growth restriction and maternal asthma. *Placenta*, *35*, 260–268.

Saif, Z., Hodyl, N. A., Stark, M. J., Fuller, P. J., Cole, T., Lu, N., et al. (2015). Expression of eight glucocorticoid receptor isoforms in the human preterm placenta vary with fetal sex and birthweight. *Placenta*, *36*, 723–730.

Samarasinghe, R. A., Di Maio, R., Volonte, D., Galbiati, F., Lewis, M., Romero, G., et al. (2011). Nongenomic glucocorticoid receptor action regulates gap junction intercellular communication and neural progenitor cell proliferation. *Proceedings of the National Academy of Sciences of the United States of America*, *108*, 16657–16662.

Siiteri, P. K., & MacDonald, P. C. (1966). Placental estrogen biosynthesis during human pregnancy. *The Journal of Clinical Endocrinology and Metabolism*, *26*, 751–761.

Simpson, E. R. (2003). Sources of estrogen and their importance. *The Journal of Steroid Biochemistry and Molecular Biology*, *86*, 225–230.

Smith, R. E., Maguire, J. A., Stein-Oakley, A. N., Sasano, H., Takahashi, K., Fukushima, K., et al. (1996). Localization of 11 beta-hydroxysteroid dehydrogenase type II in human epithelial tissues. *The Journal of Clinical Endocrinology and Metabolism*, *81*, 3244–3248.

Stewart, P. M., Murry, B. A., & Mason, J. I. (1994). Human kidney 11 beta-hydroxysteroid dehydrogenase is a high affinity nicotinamide adenine dinucleotide-dependent enzyme and differs from the cloned type I isoform. *The Journal of Clinical Endocrinology and Metabolism*, *79*, 480–484.

Sucheston, M. E., & Cannon, M. S. (1968). Development of zonular patterns in the human adrenal gland. *Journal of Morphology*, *126*, 477–491.

Surjit, M., Ganti, K. P., Mukherji, A., Ye, T., Hua, G., Metzger, D., et al. (2011). Widespread negative response elements mediate direct repression by agonist-liganded glucocorticoid receptor. *Cell*, *145*, 224–241.

Thompson, A., Han, V. K., & Yang, K. (2002). Spatial and temporal patterns of expression of 11beta-hydroxysteroid dehydrogenase types 1 and 2 messenger RNA and glucocorticoid receptor protein in the murine placenta and uterus during late pregnancy. *Biology of Reproduction*, *67*, 1708–1718.

Tronche, F., Kellendonk, C., Kretz, O., Gass, P., Anlag, K., Orban, P. C., et al. (1999). Disruption of the glucocorticoid receptor gene in the nervous system results in reduced anxiety. *Nature Genetics*, *23*, 99–103.

Wallace, A. D., & Cidlowski, J. A. (2001). Proteasome-mediated glucocorticoid receptor degradation restricts transcriptional signaling by glucocorticoids. *The Journal of Biological Chemistry*, *276*, 42714–42721.

Wang, X., & DeFranco, D. B. (2005). Alternative effects of the ubiquitin-proteasome pathway on glucocorticoid receptor down-regulation and transactivation are mediated by CHIP, an E3 ligase. *Molecular Endocrinology (Baltimore, Md.)*, *19*, 1474–1482.

Watterberg, K. L., Shaffer, M. L., Mishefske, M. J., Leach, C. L., Mammel, M. C., Couser, R. J., et al. (2007). Growth and neurodevelopmental outcomes after early low-dose hydrocortisone treatment in extremely low birth weight infants. *Pediatrics*, *120*, 40–48.

Whirledge, S. D., Oakley, R. H., Myers, P. H., Lydon, J. P., DeMayo, F., & Cidlowski, J. A. (2015). Uterine glucocorticoid receptors are critical for fertility in mice through control of embryo implantation and decidualization. *Proceedings of the National Academy of Sciences of the United States of America*, *112*, 15166–15171.

Wilburn, L. A., Goldsmith, P. C., Chang, K. J., & Jaffe, R. B. (1986). Ontogeny of enkephalin and catecholamine-synthesizing enzymes in the primate fetal adrenal medulla. *The Journal of Clinical Endocrinology and Metabolism*, *63*, 974–980.

Wilburn, L. A., & Jaffe, R. B. (1988). Quantitative assessment of the ontogeny of met-enkephalin, norepinephrine and epinephrine in the human fetal adrenal medulla. *Acta Endocrinologica*, *118*, 453–459.

Wyrwoll, C. S., Seckl, J. R., & Holmes, M. C. (2009). Altered placental function of 11beta-hydroxysteroid dehydrogenase 2 knockout mice. *Endocrinology*, *150*, 1287–1293.

Yon, L., Breault, L., Contesse, V., Bellancourt, G., Delarue, C., Fournier, A., et al. (1998). Localization, characterization, and second messenger coupling of pituitary adenylate cyclase-activating polypeptide receptors in the fetal human adrenal gland during the second trimester of gestation. *The Journal of Clinical Endocrinology and Metabolism*, *83*, 1299–1305.

CHAPTER SIX

Progesterone Receptor Signaling in Uterine Myometrial Physiology and Preterm Birth

San-Pin Wu, Francesco J. DeMayo[1]
Reproductive and Developmental Biology Laboratory, National Institute of Environmental Health Sciences, National Institute of Health, Research Triangle Park, NC, United States
[1]Corresponding author: e-mail address: francesco.demayo@nih.gov

Contents

Abstract

Myometrium holds the structural integrity for the uterus and generates force for parturition with its primary component, the smooth muscle cells. The progesterone receptor mediates progesterone-dependent signaling and connects to a network of pathways for regulation of contractility and inflammatory responses in myometrium. Dysfunctional progesterone signaling has been linked to pregnancy complications including preterm birth. In the present review, we summarize recent findings on modifiers and effectors of the progesterone receptor signaling. Discussions include novel conceptual discoveries and new development in legacy pathways such as the signal transducers NF-κB, ZEB, microRNA, and the unfolded protein response pathways. We also discuss the impact of progesterone receptor isoform composition and ligand accessibility in modification of the progesterone receptor genomic actions.

1. INTRODUCTION

The uterus goes through extensive structural remodeling and functional adaptation over the course of pregnancy. To accommodate fetal

Current Topics in Developmental Biology, Volume 125 2017 Published by Elsevier Inc.
ISSN 0070-2153
http://dx.doi.org/10.1016/bs.ctdb.2017.03.001

growth, myometrium expands its size through hyperplasia and hypertrophy of smooth muscle cells at distinct phases (Shynlova, Kwong, & Lye, 2010; Shynlova et al., 2006). Functionally, while myometrium remains quiescent throughout gestation, at term, inflammation responses stimulate coordinated contraction for laboring and parturition. At the postpartum stage, uterine involution occurs and its size returns to the prepregnancy stage ready for pregnancy again (Hsu et al., 2014). Dysregulated functional adaptation may result in premature laboring and subsequent preterm birth.

Preterm birth is defined as delivery before 37 weeks of pregnancy and is the leading cause of perinatal lethality. Premature infants resulted from preterm birth often suffer complications of retinopathy, jaundice, anemia, infections, respiratory distress, neurological disorders, cardiovascular defects, and necrotizing enterocolitis. According to data from the Centers for Disease Control and Prevention, the US preterm birth rate raises to 9.63% in 2015 after a 7-year decrease from 2007 to 2014 with higher occurrences in African American and American Indian or Alaska Native women. While many pharmacological agents for treatment of premature uterine contraction have beneficial effects in the first 48 h, the steroid hormone progesterone (P4) serves as the most promising agent to maintain uterine quiescence and prevent preterm birth beyond the acute phase (Navathe & Berghella, 2016).

P4 is produced by ovaries, placenta, and adrenal glands, and its actions are primarily mediated by the progesterone receptor (PGR). PGR is expressed in the central nervous system, ovaries, breasts, and the female reproductive tracts, including the vagina, cervix, fallopian tubes, and uterine endometrium and myometrium. PGR is important for female sexual behavior, ovulation, the establishment and maintenance of pregnancy, and development of mammary gland (Bhurke, Bagchi, & Bagchi, 2016; Diep, Daniel, Mauro, Knutson, & Lange, 2015). In uterine myometrium, P4 signaling maintains uterine quiescence by suppressing prostaglandin and oxytocin-dependent inflammatory responses and contractile activities before term. At term, in a species-dependent manner, either withdrawal of P4 by a decrease in hormone levels or alteration of PGR signaling relieves the suppression on inflammation and contraction, which allows the myometrium to adopt a contractile phenotype for laboring. Multiple mechanisms, including P4 metabolism, regulation of PGR gene expression, PGR posttranslational modifications, and PGR coregulators, that mediate or regulate uterine P4/PGR signaling have been identified and reviewed extensively (Bhurke et al., 2016; Grimm, Hartig, & Edwards, 2016; Patel et al., 2015; Renthal et al., 2015; Rubel et al., 2016; Szwarc, Kommagani, Lessey, & Lydon, 2014). Here we focus on discussing uterine P4/PGR signaling modifiers and effectors in myometrial physiology and diseases.

2. LIGAND AVAILABILITY

Withdrawal of P4 signaling may occur at the level of ligand availability, which can be achieved by reduction of P4 synthesis and/or an increase of P4 degradation. Indeed, P4 degradation has been reported at the source (ovary) and target (myometrium) tissues. Prostaglandin signaling is required for parturition at term (Sugimoto et al., 1997). In ovaries, prostaglandin signaling attenuates progesterone synthesis by lowering the levels of steroidogenic acute regulatory protein (STAR) (Fiedler, Plouffe, Hales, Hales, & Khan, 1999). Emerging evidence suggests that the prostaglandin-STAR pathway is modulated in part by fetal–maternal cross talk via surfactant protein-A and platelet-activating factor that are secreted by fetal lungs under the control of steroid receptor coactivators 1 and 2 (SRC-1 and SRC-2) (Gao et al., 2015). It is proposed that, at maturation, signals from fetal lungs downregulate ovarian P4 synthesis that leads to reduction of P4 signaling, and subsequent initiation of parturition (Mendelson, Montalbano, & Gao, 2016). Moreover, prostaglandin also elevates transcription factor *Nur77* (also known as *nuclear receptor subfamily 4, group A, member 1, Nr4a1*) levels to promote expression of 20α-hydroxysteroid dehydrogenase (20α-HSD, also known as *aldo-keto reductase family 1, member C18, Akr1c18*) for P4 degradation (Stocco et al., 2000). 20α-HSD is essential for P4 removal and timely induction of parturition, as evidenced by higher serum progesterone levels and prolonged gestation in 20α-HSD knockout mice (Piekorz, Gingras, Hoffmeyer, Ihle, & Weinstein, 2005). Ovarian 20α-HSD expression is also repressed by *signal transducer and activator of transcription 5b* (*Stat5b*) (Piekorz et al., 2005) and positively regulated by *mastermind-like domain-containing 1* (*Mamld1*) (Miyado et al., 2015). *Stat5b* null mice exhibit elevated ovarian 20α-HSD expression, low serum P4 levels, and early abortion of pregnancy (Piekorz et al., 2005; Udy et al., 1997). In contrast, ablation of *Mamld1* leads to reduced 20α-HSD levels, higher serum P4 concentration, and delayed parturition (Miyado et al., 2015). Interestingly, *Mamld1* is required to suppress *Stat5b* expression in ovaries at 18.5 days post coitum, while *Mamld1* deficiency does not alter expression of *prostaglandin F receptor* and *Nur77* (Miyado et al., 2015). These findings suggest that *Mamld1* and other regulators jointly suppress *Stat5b* expression to increase 20α-HSD levels in a prostaglandin-independent manner.

Increasing evidence indicates that regulation of ligand availability at the local microenvironment contributes, in part, to functional P4 withdrawal. In addition to ovaries, 20α-HSD is also expressed in uterine myometrium, adrenal glands, thymocytes, and kidney (Akinola, Poutanen, Vihko, & Vihko, 1997; Choi, Ishida, Matsuwaki, Yamanouchi, & Nishihara, 2008; Egert & Maass, 1973; Hirabayashi, Suzuki, Takahashi, & Nishihara, 2001; Nadeem et al., 2016). In laboring human myometrium, increased 20α-HSD expression is associated with a reduction of nuclear P4 levels (Nadeem et al., 2016; Walsh, Stanczyk, & Novy, 1984), suggesting that myometrium 20α-HSD may play a role in regulating local P4 levels. During gestation, P4/PGR signaling promotes expression of the zinc finger E-box-binding homeobox 1 (*Zeb1*, see more discussion in Section 5) transcription factor in the myometrium (Renthal et al., 2010). ZEB1 then increases STAT5B protein levels by directly suppressing expression of *miR-200a*, a STAT5B upstream repressor. Decreased *miR-200a* also relieves the repression on zinc finger E-box-binding homeobox 2 (ZEB2) whose increasing levels further suppress *miR-200a* expression. As a result, elevated STAT5B subsequently decreases levels of 20α-HSD to prevent P4 removal (Williams, Renthal, Condon, Gerard, & Mendelson, 2012). In the presence of ligand, PGR isoform B (PGR-B) moves into nuclei to represses expression of contractile genes reducing myometrium contraction (Nadeem et al., 2016). On the other hand, when P4 levels are lower, reduced P4/PGR signaling decreases ZEB1/2 and subsequently increases *miR-200a* expression, leading to reduced STAT5B levels, elevated 20α-HSD expression, and further reduction of ligand availability (Renthal et al., 2010; Williams, Renthal, Condon, et al., 2012). Without ligand binding, PGR-B resides in cytosol, while unliganded PGR isoform A (PGR-A) translocates to nuclei and promotes expression of contractile genes for subsequent increase of uterine contraction (Nadeem et al., 2016). Taken together, P4/PGR signaling feeds forward to maintain ligand availability by reducing the P4 metabolizing rate through activation of the ZEB1/2-*miR-200*-STAT5B pathway. At the onset of parturition, prostaglandin signaling decreases P4 levels by reducing synthesis and increasing degradation of P4, eventually leading to withdrawal of P4 signaling (Fig. 1). Importantly, this mechanism provides a plausible

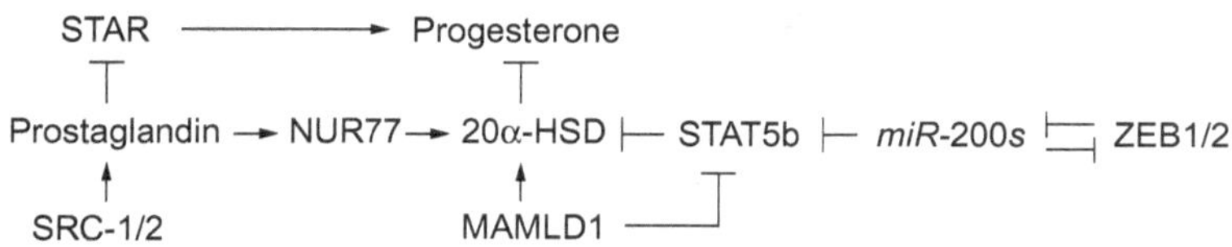

Fig. 1 Modulation of the progesterone availability.

explanation on the association between myometrial reduction of liganded PGR and preterm labor (Nadeem et al., 2016). Moreover, regulation of local ligand availability may also partially explain the P4 functional withdrawal in human when circulating P4 levels remain unchanged around parturition.

3. COMPOSITION OF PGR ISOFORMS

The two PGR isoforms PGR-A and PGR-B are transcribed from the same genomic locus by two different promoters, resulting in a 114-kDa PGR-B and a 94-kDa PGR-A proteins (Kastner et al., 1990). Upon ligand stimulation, PGR-A and PGR-B have the capacity to modulate distinct sets of downstream target genes, suggesting that the P4 signaling is mediated by a combined action of both PGR isoforms (Khan, Bellance, Guiochon-Mantel, Lombes, & Loosfelt, 2012; Tan, Yi, Rote, Hurd, & Mesiano, 2012). Therefore, it is logical to assume that changes in relative abundance of PGR-A and PGR-B may result in different translation of P4 signaling and subsequent physiological responses. This notion finds support from observations that alterations of PGR-A and PGR-B relative abundance are associated with various physiological and pathological conditions, including breast cancer metastasis and uterine myometrium contractility (Chai, Smith, Zakar, Mitchell, & Madsen, 2012; Khan et al., 2012; Merlino et al., 2007). In human myometrium, PGR-B is the more abundant isoform at the preterm stage and maintains its levels at both nonlaboring and laboring term stages. In contrast, PGR-A levels rise over the course of pregnancy and eventually exceed PGR-B in the laboring myometrium (Chai et al., 2012; Merlino et al., 2007). Because the resulting increase of PGR-A to PGR-B ratio occurs concomitant with the P4 functional withdrawal and the transition from quiescent to laboring state, such correlation implicates that regulation of the PGR-A/PGR-B ratio may contribute to preparation and switch of myometrium into a contractile state.

PGR isoforms can be either activators or repressors through preferential interaction with different coregulators (Giangrande, Kimbrel, Edwards, & McDonnell, 2000; Jacobsen & Horwitz, 2012; Nadeem et al., 2016; Yore et al., 2010). With respect to the myometrium, PGR-A preferentially interacts with AP-1 subunits FOSL1 and FOSL2 in comparison to PGR-B, while PGR-B exhibits a stronger interaction with JUN in nuclei of hTERT-HM human myometrial cells (Nadeem et al., 2016). The functional significance of such preferential interactions is exemplified by an observation that PGR-A works as an activator in the presence of FOSL2 and JUND while exhibits a repressive function on the *GJA1* promoter in Syrian hamster myometrium cells (Nadeem et al., 2016). In contrast, PGR-B only suppresses *GJA1* promoter activities in the presence of JUNB and JUND, but not in the context of FOSL2 and JUND (Nadeem et al., 2016). These findings implicate that, during gestation, the more abundant PGR-B interacts with JUN–JUN dimers to recruit p54/mSin3A/HDAC repressors to suppress expression of contractile genes. During transition to the laboring stage, elevated PGR-A and FOS levels reach a threshold to change the myometrial genomic profile in favor of contraction via the interaction between PGR-A and JUN–FOS heterodimers (Mitchell & Lye, 2002; Nadeem et al., 2016). Additionally, PGR-A can also attenuate PGR-B transcription activities in hTERT-HM cells (Peters et al., 2017), possibly through a dominant-negative effect (Vegeto et al., 1993). Moreover, interleukin-1 β (IL-1β)-stimulated phosphorylation of PGR-A at serine 345 (S345) is required for the PGR-A-dependent inhibition of PGR-B actions on the inflammatory genes (Amini et al., 2016). This mechanism may further tilt the functional impact of rising PGR-A/PGR-B ratio in favor of a PGR-A-dependent genomic regulation.

Transcriptional regulation of the *PGR-A* expression serves as one of the mechanisms to mediate the rise of PGR-A levels. Studies on human tissues and mouse models reveal that *KRUPPEL-LIKE FACTOR 9* (*KLF9*) promotes PGR-A expression in myometrium. *Klf9* knockout mice have reduced myometrium PGR-A levels at term, delayed parturition, and resistance to RU486-induced pregnancy termination (Zeng, Velarde, Simmen, & Simmen, 2008). *Klf9* ablation lowers PGR-A protein levels, reduces expression of contractile genes, including *Gja1* and *Oxtr*, and decreases the binding capacity of NF-κB proteins p65 and p50 to their target genes. In humans, compared to term patients, myometrium of patients with late-term pregnancy exhibits lower levels of KLF9, PGR-A, and PGR-B, as well as a lower PGR-A/PGR-B ratio and an increased antiinflammatory and decreased proinflammation profile of gene expression. (Pabona, Zhang, Ginsburg, Simmen, & Simmen, 2015). Moreover, knocking down KLF9 attenuates RU486-dependent suppression of proinflammatory gene expression. Collectively, these findings indicate that KLF9 promotes *PGR-A* expression resulting in a molecular profile favorable for inflammation and contraction in myometrium.

Emerging evidence indicates that *PGR-A* expression is subject to epigenetic regulation. In human myometrium, the *PGR-A* promoter exhibits higher levels of active histone marks such as histone H3 lysine 4 trimethylation (H3K4me3), acetylated histone 3, and acetylated histone 4, compared with the *PGR-B* promoter (Chai et al., 2012). This finding suggests that the *PGR-A* promoter is relatively permissible to transcriptional regulation by other transcription factors such as KLF9. This is also in line with the observation that *PGR-A* mRNA levels increase over the course of pregnancy, while the *PGR-B* mRNA abundance has minimal change (Chai et al., 2012). On the *PGR-A* promoter, laboring human myometrium exhibits higher H3K4me3 levels and reduced occupancy of the H3K4 demethylase, Jumonji AT-rich interactive domain 1A (JARID1A, also known as KDM5A), compared with quiescent myometrium (Chai et al., 2014). Because JARID1A can repress *PGR* expression in MCF-7 cells (Stratmann & Haendler, 2011), it is proposed that the reduced JARID1A occupancy may result in higher levels of H3K4me3, increased enhancer activities, and subsequent further induction of *PGR-A* transcription in laboring myometrium. Additionally, another inverse correlation between *PGR-A* and the *histone deacetylase 1* (*HDAC1*) mRNA levels has also been reported in human myometrium at the transition from quiescent to laboring stage. Further experimentations show that HDAC1 occupies the *PGR-A* promoter and suppresses *PGR-A* expression in cultured primary myometrial cells (Ke et al., 2016). These observations suggest that *PGR-A* expression is regulated by HDAC1-dependent histone modifications. In summary, in response to as yet unidentified signals, the epigenetic regulators JARID1A and HDAC1 may modulate *PGR-A* promoter activities through modifying the histone methylation and acetylation of myometrium during pregnancy.

PGR-A levels can also be regulated at the posttranslational level as evidenced by the stabilization of that myometrial PGR-A protein in response to inflammatory stimuli (Peters et al., 2017). In human myometrial tissue explants and myometrial cell lines, IL-1β attenuates progesterone-dependent reduction of PGR-A protein levels. IL-1β does not alter *PGR-A* mRNA abundance, but blockage of proteasome activities can

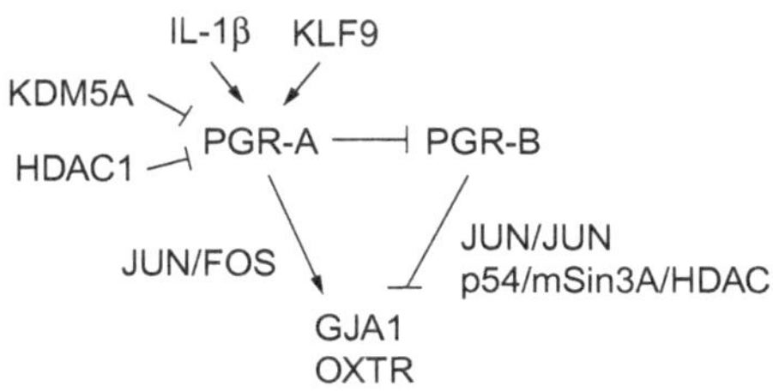

Fig. 2 The progesterone receptor isoform regulatory network.

sustain PGR-A protein levels in the presence of progesterone. Thus, IL-1β may regulate PGR-A stability through posttranslational modifications, possibly via the p38 mitogen-activated protein kinase-dependent phosphorylation (Jung et al., 2002; Khan et al., 2011). In summary, multiple mechanisms regulate the increase of PGR-A protein levels, subsequently leading to change of P4 signaling profile via switching from PGR-B to PGR-A dominance to stimulate uterine contraction and parturition (Fig. 2).

4. THE PGR–NF-κB AXIS

P4 acts through PGR to suppress the inflammatory response in uterus (Lydon et al., 1995; Tibbetts, Conneely, & O'Malley, 1999). In human myometrial cells, P4 treatment attenuates IL-1β-induced *prostaglandin-endoperoxide synthase 2* (*PTGS2*, also known as COX-2) expression via blocking the occupancy of the protein complex NF-κB (nuclear factor kappa-light-chain-enhancer of activated B cells) on the *PTGS2* promoter (Hardy, Janowski, Corey, & Mendelson, 2006). P4 signaling also induces expression of kappa light polypeptide gene enhancer in B cells inhibitor, alpha (IκBα) to sequester and keep NF-κB inactive (Hardy et al., 2006). In addition, three other mechanisms have been proposed for regulation of NF-κB activities by P4/PGR signaling. (1) P4/PGR signaling may block IL-1β induced and proteasome-dependent IκBα protein degradation (Renthal et al., 2015). (2) PGR may physically interact with NF-κB and inhibit NF-κB activities (Kalkhoven, Wissink, van der Saag, & van der Burg, 1996; Renthal et al., 2015). (3) PGR induces dual-specificity phosphatase 1 (DUSP1) expression to reduce phosphorylation of mitogen-activated protein kinase 1 and 3 (ERK1/2) and subsequently decrease NF-κB p65 subunit nuclear localization, as evidenced in the T47D breast cancer cell line (Chen, Hardy, & Mendelson, 2011). In summary,

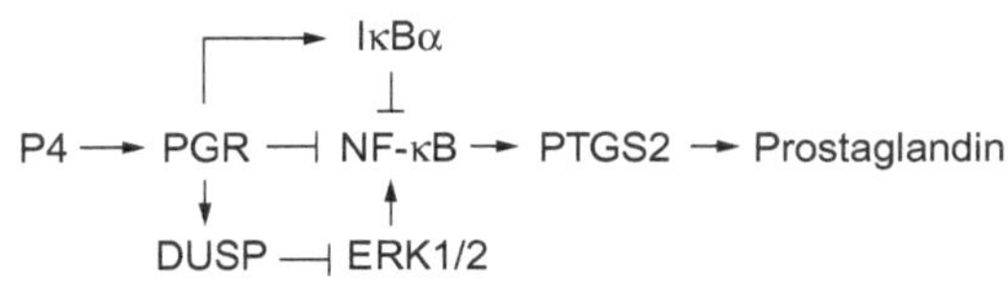

Fig. 3 Regulation of NF-κB signaling by P4/PGR.

P4/PGR signaling targets multiple points of the NF-κB pathway to suppress NF-κB-dependent induction of proinflammatory gene expression (Fig. 3).

Mounting evidence indicates sirtuins as a regulator of the NF-κB pathway. Sirtuins, known as inflammation and metabolic regulators, are histone deacetylases capable of epigenetically modulating transcription. SIRT1 and SIRT2 suppress NF-κB activities by deacetylating p65 and subsequently decrease proinflammatory cytokine production (Rothgiesser, Erener, Waibel, Luscher, & Hottiger, 2010; Yang et al., 2012). SIRT6 deacetylates histone H3 lysine 9 (H3K9) at promoters of NF-κB target genes, decreases RELA promoter occupancy, and suppresses expression of NF-κB downstream targets (Kawahara et al., 2009). In uteri, laboring myometrium exhibits a lower level of SIRT3 compared with nonlaboring myometrium at term (Lim, Barker, Menon, & Lappas, 2016), coincided with the withdrawal of P4 signaling. In cultured primary myometrial cells, SIRT3 levels are reduced by IL-1β and tumor necrosis factor α (TNFα) treatment, while knocking down SIRT3 expression augments NF-κB activities and increases expression of downstream cytokines and chemokines (Lim et al., 2016). These findings suggest a role of SIRT3 in maintaining uterine quiescence by suppressing the inflammatory response during gestation. Although it is unclear how SIRT3 interacts with NF-κB and whether PGR is involved, given the role of sirtuin family proteins in modulating activities of nuclear hormone receptors such as estrogen receptor and liver X receptor (Li et al., 2007; Moore & Faller, 2013), it would be interesting to examine potential interactions between PGR and SIRT3 in the myometrium.

5. THE PGR-ZEB-MicroRNA REGULATORY CIRCUIT

ZEB1 and ZEB2 belong to the ZEB family transcription factors that are pivotal for embryonic development and tumor progression (Vandewalle, Van Roy, & Berx, 2009; Zhang, Sun, & Ma, 2015). Studies in mouse models reveal that *Zeb1* mediates transforming growth factor β (TGFβ)-dependent expression of smooth muscle genes and is required for differentiation of vascular smooth muscle (Nishimura et al., 2006), while *Zeb2* is essential for normal vagina development (Miyoshi et al., 2006). In both human and mouse uteri, *ZEB1* and *ZEB2* are prominently expressed in quiescent myometrium and their levels are reduced at laboring (Renthal et al., 2010), concomitant with the temporal change of P4 signaling. The functional relationship between P4/PGR signaling and ZEB1/2 expression finds support from the requirement of P4 signaling in sustaining *Zeb1* and *Zeb2* expression with *Zeb1* being a potential direct downstream target of PGR (Renthal et al., 2010). Results from mouse and cell models further suggest that ZEB1 occupies the promoters of gap junction protein, alpha 1 (*Gja1*) and oxytocin receptor (*Oxtr*) genes and directly represses transcription of these classic marker genes for uterine contraction (Renthal et al., 2010). These findings collectively indicate that ZEB1/2 mediates P4/PGR signaling in silencing the myometrium activities through regulation of contractile gene expression.

Results from a series of studies indicate that a regulatory network, consisting of ZEB1/2 and microRNAs (miRs), modulates myometrial molecular profiles during the transition from quiescent to laboring. These interactions include reciprocal repression between ZEB1/2 and *miR-200* family, positive regulation of *miR-199a/214* cluster expression by ZEB1 and repression of ZEB1 by *miR-10b* (Guo et al., 2015; Renthal et al., 2010; Tang et al., 2015; Williams, Renthal, Condon, et al., 2012; Williams, Renthal, Gerard, & Mendelson, 2012). Expression of ZEB1/2 and *miR-200b/429* correlates inversely in myometrium (Renthal et al., 2010). ZEB1 occupancy is observed on the genomic locus of *miR-200b/429* cluster, while such occupancy is stronger in quiescent than in laboring myometrium. Also, the 3′-untranslated regions of both *ZEB1* and *ZEB2* mRNA contain binding sites for *miR-200b/429*. Functionally, P4 elevates *Zeb1/2* levels and suppresses *miR-200b/429* expressions. On the other hand, *miR-200b/429* overexpression reduces ZEB1 and ZEB2 expression in human myometrial cells. These findings together suggest that ZEB1/2 and *miR-200b/429* exert a mutually repressive function. During gestation, P4/PGR signaling sustains a high level of ZEB1/2 that suppresses expression of *miR-200* family to maintain their own abundance and ligand availability in myometrium (discussed in Section 2). At the laboring stage, P4 withdrawal reduces ZEB1 expression and relives ZEB1/2's repression effect on *miR-200b/429*, whose elevated expression further decrease ZEB1/2 levels.

P4 signaling also promotes expression of microRNAs to reduce levels of contractile genes. In both human and mice, *miR-199a-3p* and *miR-214* are more abundant in the quiescent myometrium than during labor (Williams, Renthal, Gerard, et al., 2012). P4 treatment and ZEB1 over-expression both can increase *miR-199a-3p* and *miR-214* expressions, while reduced ZEB1 levels lead to lower abundance of *miR-199a-3p* and *miR-214*. In addition, ZEB1 occupancy is observed on the genomic locus of *miR-199a/214* cluster and ZEB1 exhibits a much stronger occupancy pattern in quiescent myometrium. Moreover, *miR-199a-3p* and *miR-214* both target the mRNA of *PTGS2*, while overexpression of either microRNA attenuates IL-1β-induced increase of PTGS2 protein abundance and TNFα-stimulated myometrial cell contraction. Taken together, these results reveal that ZEB1-*miR-199a/214* mediated a P4/PGR-dependent suppression of inflammatory gene expression, in addition to the aforementioned NF-κB pathway. Of note, estrogen treatment is able to reduce the abundance of myometrial *miR-199a-3p* and *miR-214*. Similarly, *miR-181a*, another antiinflammatory microRNA that shows decreased expression near term, is also negatively regulated by estrogen (Gao et al., 2016). Given that estrogen receptor α directly represses *miR-181a* transcription, it would be interesting to examine whether the *miR-199a/214* cluster is regulated by estrogen with a similar mechanism.

Emerging data show increased *miR-10b* expression in the human preterm myometrium compared with myometrium not in labor (Tang et al., 2015). Studies in endometrial epithelial cells reveal that *miR-10b* targets *ZEB1* and *PIK3CA* mRNAs and reduces their protein abundance (Guo et al., 2015). Collectively, although more investigations are needed, it is tempting to speculate that the elevated *miR-10b* could be a contributing factor to tilt the balance between ZEB1 and *miR-200* family toward the contractile profile, subsequently leading to early P4 withdrawal and preterm birth. In summary, P4/PGR signaling via the ZEB transcription factors controls multiple miRs with either pro- or anticontractile/inflammatory roles to regulate the myometrial function (Fig. 4).

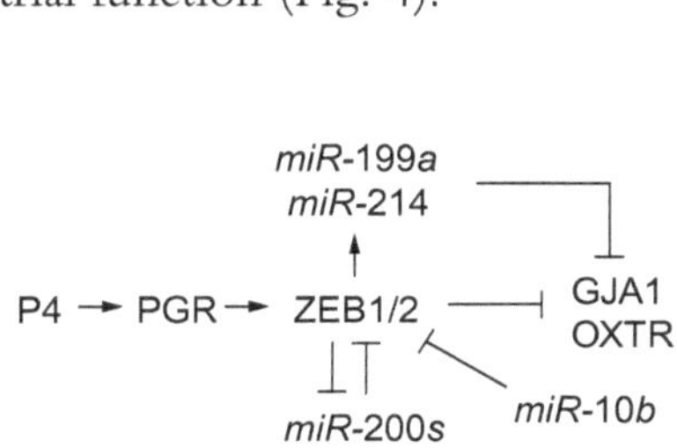

Fig. 4 Interaction among P4/PGR, ZEB1/2, and microRNAs for regulation of contractile genes.

6. ENDOPLASMIC RETICULUM STRESS AND UNFOLDED PROTEIN RESPONSE

Protein production, including polypeptide synthesis, protein folding, and some posttranslational modifications, takes place in the organelle endoplasmic reticulum (ER). When accumulation of unfolded proteins stresses ER, the unfolded protein response (UPR) elicits prosurvival signaling to reduce translation rates, facilitate protein degradation, and induce ER chaperones rectifying the ER stress. Emerging evidence suggests smooth muscle UPR as a regulator of gestational length and P4 signaling as a UPR modulator (Kawakami et al., 2014; Kyathanahalli et al., 2015). Physiological UPR is present in uteri of pregnant mice and is thought to sustain enough active caspase 3 and 7 that keeps GJA1 protein at low level in order to maintain uterine quiescence (Kyathanahalli et al., 2015). At late gestation, decreased UPR results in a reduction of active caspase 3/7, which increases GJA1 protein levels by decreasing its degradation and leads to subsequent increase of muscle contraction (Kyathanahalli et al., 2015). P4 treatment delays UPR reduction, maintains levels of active caspase 3/7, and keeps GJA1 at a low level. In contrast, RU486 suppresses UPR, decreases abundance of active caspase 3/7, and leads to elevated levels of GJA1 (Kyathanahalli et al., 2015). Moreover, P4 also promotes uterine expression of *caspase 3*, possibly through PGR binding motifs in the promoter region (Jeyasuria, Wetzel, Bradley, Subedi, & Condon, 2009). Taken together, these findings indicate an additional layer of P4 regulation through post-translational modulation of GJA1 levels via UPR and caspase 3/7-dependent mechanisms (Fig. 5).

In uteri, P4 signaling interacts with another ER stress transducer, cAMP-response element-binding 3-like protein 1 (CREB3L1, also known as OASIS), that mediates cross talk among subcellular organelles to enhance the protein processing capacity of ER (Kondo, Saito, Asada, Kanemoto, & Imaizumi, 2011). CREB3L1 is a transcription factor with a transmembrane domain that allows its association with ER. Once activated by regulated intramembrane proteolysis, CREB3L1 translocates to the

P4 → PGR → UPR → Caspase 3/7 ⊣ GJA1

Fig. 5 Interplay between P4/PGR signaling and the unfolded protein response in control of contractile marker gene expression.

nucleus and exerts genomic actions (Mellor et al., 2013; Murakami et al., 2009). CREB3L1 also acts downstream of PGR to promote ERK signaling for stromal cell decidualization (Ahn et al., 2016). In human uteri, detection of CREB3L1 protein is reported in both epithelial and stromal compartments of endometrium with the strongest expression in the decidual cells at the late-secretory stage when the P4 signaling is prominent (Ahn et al., 2016). Although it is not clear whether CREB3L1 is expressed in the myometrium, its mRNA and protein can be detected in various smooth muscle tissues including bladder, stomach, and intestine. Based on the mouse model, endometrial CREB3L1 expression depends on PGR-mediated P4 signaling through mechanisms yet to be identified (Ahn et al., 2016). Functionally, CREB3L1 is required for the decidualization process as evidenced by the failure of decidualization and reduced expression of decidualization marker genes in human endometrial stromal cells after knocking down the CREB3L1 in vitro (Ahn et al., 2016). Decidual cells exhibit distinct features compared with stromal cells they derived from, including larger cell size, enlarged and dilated rough ER, increased in size of the Golgi complexes, and increased secretion of prolactin and IGFBP-1 (Lane, Oxberry, Mazella, & Tseng, 1994). These changes are indicative of major structural and functional alterations of ER in response to P4 stimulation. In light of the roles of CREB3L1 in the decidualization response and in ER–nucleus interaction, it is tempting to speculate that the CREB3L1 may utilize similar mechanisms as those in ER stress response to mediate P4-induced ER and Golgi changes during decidualization.

Notably, excessive ER stress results in preterm delivery with the most impact at late gestation in the tunicamycin-treated mouse model (Kawakami et al., 2014; Kyathanahalli et al., 2015). Moreover, ER stress is associated with fetal growth restriction and endometriosis (Ahn et al., 2016; Lian et al., 2011; Yung et al., 2008). These findings are significant because ER stress has been linked to numerus environmental insults (Kitamura, 2013), and pregnant women are exposed to various environmental stimulations over the course of pregnancy.

7. CONCLUDING REMARKS AND FUTURE PERSPECTIVES

In myometrium, P4/PGR signaling is connected to a complex regulatory network through the NF-κB, ZEB-microRNAs, and UPR pathways as well as direct transcriptional regulation, which together modulate the

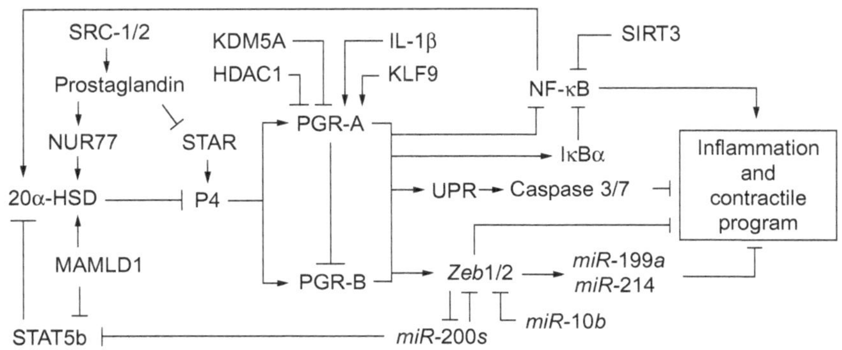

Fig. 6 A genetic network of PGR signaling for regulation of uterine myometrium states in adult.

levels and activities of downstream effectors. On the other hand, the overall P4/PGR signaling-dependent molecular profiles are modified by the combined activities of PGR isoforms, the concerted efforts of coregulators, and the ligand availability. P4/PGR signaling mediates and utilizes these interconnected pathways, as summarized in Fig. 6, to determine the state of myometrium over the entire course of pregnancy. Although the proof of concept has been performed on a few model genes, a better understanding on the genomic impact of PGR-A and PGR-B isoforms in myometrium at the global level and in a physiological context would benefit discovery of novel and distinct pathways specific for myometrium. In addition, modifiers that result in preterm birth and fetal growth restriction and are associated with the P4/PGR signaling network (e.g., microRNAs, ER stress, and premature loss of liganded PGR in nuclei) are of particular interest in understanding mechanisms that elicit and mediate such associations and the extent of the individual modifier's pathological impact.

P4 supplement is currently in use for prevention of preterm birth and can prevent about one-third of recurrent preterm birth (Norwitz & Caughey, 2011). The fact that P4 fails to benefit many more patients highlights not only the complex etiology behind the preterm birth but also a few emerging questions in the myometrial biology. It is not clear whether all myometrial smooth muscle cells are the same and can respond equally to P4. Heterogeneity of smooth muscle cells are present in the arteries with respect to response to growth stimuli (Topouzis & Majesky, 1996). It is also believed that predisposed subpopulations of smooth muscle cells may contribute to pathological alterations in atherosclerosis and restenosis (Hao, Gabbiani, & Bochaton-Piallat, 2003). Examining the molecular signature of individual myometrial smooth muscle cells with single cell transcriptome profiling may help to address these questions. Meanwhile, at the laboring stage, coordinated contraction of each individual smooth muscle cells is essential to generate effective force for parturition. Emerging evidence indicates the presence of a group of specialized cells named "interstitial cells of Cajal" or "telocytes" in uterine myometrium (Allix et al., 2008; Cretoiu, Cretoiu, Marin, Radu, & Popescu, 2013). This type of cells serves as pacemaker cells in various several smooth muscle tissues generating electric waves to trigger coordinated muscle contraction. What factors these telocytes interact with and whether P4 signaling affects telocytes' development and functions remain to be defined.

ACKNOWLEDGMENT

This work is supported by the Intramural Research Program of the National Institute of Environmental Health Sciences, National Institute of Health: Project Z1AES103311-01.

REFERENCES

Ahn, J. I., Yoo, J. Y., Kim, T. H., Kim, Y. I., Ferguson, S. D., Fazleabas, A. T., et al. (2016). cAMP-response element-binding 3-like protein 1 (CREB3L1) is required for decidualization and its expression is decreased in women with endometriosis. *Current Molecular Medicine*, *16*, 276–287.

Akinola, L. A., Poutanen, M., Vihko, R., & Vihko, P. (1997). Expression of 17beta-hydroxysteroid dehydrogenase type 1 and type 2, P450 aromatase, and 20alpha-hydroxysteroid dehydrogenase enzymes in immature, mature, and pregnant rats. *Endocrinology*, *138*, 2886–2892.

Allix, S., Reyes-Gomez, E., Aubin-Houzelstein, G., Noel, D., Tiret, L., Panthier, J. J., et al. (2008). Uterine contractions depend on KIT-positive interstitial cells in the mouse: Genetic and pharmacological evidence. *Biology of Reproduction*, *79*, 510–517.

Amini, P., Michniuk, D., Kuo, K., Yi, L., Skomorovska-Prokvolit, Y., Peters, G. A., et al. (2016). Human parturition involves phosphorylation of progesterone receptor-A at serine-345 in myometrial cells. *Endocrinology*, *157*, 4434–4445.

Bhurke, A. S., Bagchi, I. C., & Bagchi, M. K. (2016). Progesterone-regulated endometrial factors controlling implantation. *American Journal of Reproductive Immunology*, *75*, 237–245.

Chai, S. Y., Smith, R., Fitter, J. T., Mitchell, C., Pan, X., Ilicic, M., et al. (2014). Increased progesterone receptor A expression in laboring human myometrium is associated with decreased promoter occupancy by the histone demethylase JARID1A. *Molecular Human Reproduction*, *20*, 442–453.

Chai, S. Y., Smith, R., Zakar, T., Mitchell, C., & Madsen, G. (2012). Term myometrium is characterized by increased activating epigenetic modifications at the progesterone receptor-A promoter. *Molecular Human Reproduction*, *18*, 401–409.

Chen, C. C., Hardy, D. B., & Mendelson, C. R. (2011). Progesterone receptor inhibits proliferation of human breast cancer cells via induction of MAPK phosphatase 1 (MKP-1/DUSP1). *The Journal of Biological Chemistry*, *286*, 43091–43102.

Choi, J. H., Ishida, M., Matsuwaki, T., Yamanouchi, K., & Nishihara, M. (2008). Involvement of 20alpha-hydroxysteroid dehydrogenase in the maintenance of pregnancy in mice. *The Journal of Reproduction and Development*, *54*, 408–412.

Cretoiu, S. M., Cretoiu, D., Marin, A., Radu, B. M., & Popescu, L. M. (2013). Telocytes: Ultrastructural, immunohistochemical and electrophysiological characteristics in human myometrium. *Reproduction*, *145*, 357–370.

Diep, C. H., Daniel, A. R., Mauro, L. J., Knutson, T. P., & Lange, C. A. (2015). Progesterone action in breast, uterine, and ovarian cancers. *Journal of Molecular Endocrinology*, *54*, R31–R53.

Egert, D., & Maass, H. (1973). On the fate of progesterone in the uterus of pregnant and nonpregnant rats. *Acta Endocrinologica. Supplementum*, *173*, 47.

Fiedler, E. P., Plouffe, L., Jr., Hales, D. B., Hales, K. H., & Khan, I. (1999). Prostaglandin F(2alpha) induces a rapid decline in progesterone production and steroidogenic acute regulatory protein expression in isolated rat corpus luteum without altering messenger ribonucleic acid expression. *Biology of Reproduction*, *61*, 643–650.

Gao, L., Rabbitt, E. H., Condon, J. C., Renthal, N. E., Johnston, J. M., Mitsche, M. A., et al. (2015). Steroid receptor coactivators 1 and 2 mediate fetal-to-maternal signaling that initiates parturition. *The Journal of Clinical Investigation*, *125*, 2808–2824.

Gao, L., Wang, G., Liu, W. N., Kinser, H., Franco, H. L., & Mendelson, C. R. (2016). Reciprocal feedback between miR-181a and E2/ERalpha in myometrium enhances inflammation leading to labor. *The Journal of Clinical Endocrinology and Metabolism*, *101*, 3646–3656.

Giangrande, P. H., Kimbrel, E. A., Edwards, D. P., & McDonnell, D. P. (2000). The opposing transcriptional activities of the two isoforms of the human progesterone receptor are due to differential cofactor binding. *Molecular and Cellular Biology*, *20*, 3102–3115.

Grimm, S. L., Hartig, S. M., & Edwards, D. P. (2016). Progesterone receptor signaling mechanisms. *Journal of Molecular Biology*, *428*, 3831–3849.

Guo, Y., Lang, X., Lu, Z., Wang, J., Li, T., Liao, Y., et al. (2015). MiR-10b directly targets ZEB1 and PIK3CA to curb adenomyotic epithelial cell invasiveness via upregulation of E-cadherin and inhibition of Akt phosphorylation. *Cellular Physiology and Biochemistry*, *35*, 2169–2180.

Hao, H., Gabbiani, G., & Bochaton-Piallat, M. L. (2003). Arterial smooth muscle cell heterogeneity: Implications for atherosclerosis and restenosis development. *Arteriosclerosis, Thrombosis, and Vascular Biology*, *23*, 1510–1520.

Hardy, D. B., Janowski, B. A., Corey, D. R., & Mendelson, C. R. (2006). Progesterone receptor plays a major antiinflammatory role in human myometrial cells by antagonism of nuclear factor-kappaB activation of cyclooxygenase 2 expression. *Molecular Endocrinology*, *20*, 2724–2733.

Hirabayashi, K., Suzuki, M., Takahashi, M., & Nishihara, M. (2001). Expression of ovarian 20alpha-hydroxysteroid dehydrogenase in rat thymus. *Endocrine Journal*, *48*, 557–563.

Hsu, K. F., Pan, H. A., Hsu, Y. Y., Wu, C. M., Chung, W. J., & Huang, S. C. (2014). Enhanced myometrial autophagy in postpartum uterine involution. *Taiwanese Journal of Obstetrics & Gynecology*, *53*, 293–302.

Jacobsen, B. M., & Horwitz, K. B. (2012). Progesterone receptors, their isoforms and progesterone regulated transcription. *Molecular and Cellular Endocrinology*, *357*, 18–29.

Jeyasuria, P., Wetzel, J., Bradley, M., Subedi, K., & Condon, J. C. (2009). Progesterone-regulated caspase 3 action in the mouse may play a role in uterine quiescence during pregnancy through fragmentation of uterine myocyte contractile proteins. *Biology of Reproduction*, *80*, 928–934.

Jung, Y. D., Fan, F., McConkey, D. J., Jean, M. E., Liu, W., Reinmuth, N., et al. (2002). Role of P38 MAPK, AP-1, and NF-kappaB in interleukin-1beta-induced IL-8 expression in human vascular smooth muscle cells. *Cytokine*, *18*, 206–213.

Kalkhoven, E., Wissink, S., van der Saag, P. T., & van der Burg, B. (1996). Negative interaction between the RelA(p65) subunit of NF-kappaB and the progesterone receptor. *The Journal of Biological Chemistry*, *271*, 6217–6224.

Kastner, P., Krust, A., Turcotte, B., Stropp, U., Tora, L., Gronemeyer, H., et al. (1990). Two distinct estrogen-regulated promoters generate transcripts encoding the two functionally different human progesterone receptor forms A and B. *The EMBO Journal*, *9*, 1603–1614.

Kawahara, T. L., Michishita, E., Adler, A. S., Damian, M., Berber, E., Lin, M., et al. (2009). SIRT6 links histone H3 lysine 9 deacetylation to NF-kappaB-dependent gene expression and organismal life span. *Cell*, *136*, 62–74.

Kawakami, T., Yoshimi, M., Kadota, Y., Inoue, M., Sato, M., & Suzuki, S. (2014). Prolonged endoplasmic reticulum stress alters placental morphology and causes low birth weight. *Toxicology and Applied Pharmacology*, *275*, 134–144.

Ke, W., Chen, C., Luo, H., Tang, J., Zhang, Y., Gao, W., et al. (2016). Histone deacetylase 1 regulates the expression of progesterone receptor A during human parturition by occupying the progesterone receptor A promoter. *Reproductive Sciences*, *23*, 955–964.

Khan, J. A., Amazit, L., Bellance, C., Guiochon-Mantel, A., Lombes, M., & Loosfelt, H. (2011). p38 and p42/44 MAPKs differentially regulate progesterone receptor A and B isoform stabilization. *Molecular Endocrinology*, *25*, 1710–1724.

Khan, J. A., Bellance, C., Guiochon-Mantel, A., Lombes, M., & Loosfelt, H. (2012). Differential regulation of breast cancer-associated genes by progesterone receptor isoforms PRA and PRB in a new bi-inducible breast cancer cell line. *PloS One*, 7, e45993.

Kitamura, M. (2013). The unfolded protein response triggered by environmental factors. *Seminars in Immunopathology*, *35*, 259–275.

Kondo, S., Saito, A., Asada, R., Kanemoto, S., & Imaizumi, K. (2011). Physiological unfolded protein response regulated by OASIS family members, transmembrane bZIP transcription factors. *IUBMB Life*, *63*, 233–239.

Kyathanahalli, C., Organ, K., Moreci, R. S., Anamthathmakula, P., Hassan, S. S., Caritis, S. N., et al. (2015). Uterine endoplasmic reticulum stress-unfolded protein response regulation of gestational length is caspase-3 and -7-dependent. *Proceedings of the National Academy of Sciences of the United States of America*, *112*, 14090–14095.

Lane, B., Oxberry, W., Mazella, J., & Tseng, L. (1994). Decidualization of human endometrial stromal cells in vitro: Effects of progestin and relaxin on the ultrastructure and production of decidual secretory proteins. *Human Reproduction*, *9*, 259–266.

Li, X., Zhang, S., Blander, G., Tse, J. G., Krieger, M., & Guarente, L. (2007). SIRT1 deacetylates and positively regulates the nuclear receptor LXR. *Molecular Cell*, *28*, 91–106.

Lian, I. A., Loset, M., Mundal, S. B., Fenstad, M. H., Johnson, M. P., Eide, I. P., et al. (2011). Increased endoplasmic reticulum stress in decidual tissue from pregnancies complicated by fetal growth restriction with and without pre-eclampsia. *Placenta*, *32*, 823–829.

Lim, R., Barker, G., Menon, R., & Lappas, M. (2016). A novel role for SIRT3 in regulating mediators involved in the terminal pathways of human labor and delivery. *Biology of Reproduction*, *95*, 95.

Lydon, J. P., DeMayo, F. J., Funk, C. R., Mani, S. K., Hughes, A. R., Montgomery, C. A., Jr., et al. (1995). Mice lacking progesterone receptor exhibit pleiotropic reproductive abnormalities. *Genes & Development*, *9*, 2266–2278.

Mellor, P., Deibert, L., Calvert, B., Bonham, K., Carlsen, S. A., & Anderson, D. H. (2013). CREB3L1 is a metastasis suppressor that represses expression of genes regulating metastasis, invasion, and angiogenesis. *Molecular and Cellular Biology*, *33*, 4985–4995.

Mendelson C.R., Montalbano A.P., & Gao L. (2016). Fetal-to-maternal signaling in the timing of birth, *The Journal of Steroid Biochemistry and Molecular Biology*, (in press). http://doi.org/10.1016/j.jsbmb.2016.09.006.

Merlino, A. A., Welsh, T. N., Tan, H., Yi, L. J., Cannon, V., Mercer, B. M., et al. (2007). Nuclear progesterone receptors in the human pregnancy myometrium: Evidence that parturition involves functional progesterone withdrawal mediated by increased expression of progesterone receptor-A. *The Journal of Clinical Endocrinology and Metabolism*, *92*, 1927–1933.

Mitchell, J. A., & Lye, S. J. (2002). Differential expression of activator protein-1 transcription factors in pregnant rat myometrium. *Biology of Reproduction*, *67*, 240–246.

Miyado, M., Miyado, K., Katsumi, M., Saito, K., Nakamura, A., Shihara, D., et al. (2015). Parturition failure in mice lacking Mamld1. *Scientific Reports*, *5*, 14705.

Miyoshi, T., Maruhashi, M., Van De Putte, T., Kondoh, H., Huylebroeck, D., & Higashi, Y. (2006). Complementary expression pattern of Zfhx1 genes Sip1 and deltaEF1 in the mouse embryo and their genetic interaction revealed by compound mutants. *Developmental Dynamics*, *235*, 1941–1952.

Moore, R. L., & Faller, D. V. (2013). SIRT1 represses estrogen-signaling, ligand-independent ERalpha-mediated transcription, and cell proliferation in estrogen-responsive breast cells. *The Journal of Endocrinology*, *216*, 273–285.

Murakami, T., Saito, A., Hino, S., Kondo, S., Kanemoto, S., Chihara, K., et al. (2009). Signaling mediated by the endoplasmic reticulum stress transducer OASIS is involved in bone formation. *Nature Cell Biology*, *11*, 1205–1211.

Nadeem, L., Shynlova, O., Matysiak-Zablocki, E., Mesiano, S., Dong, X., & Lye, S. (2016). Molecular evidence of functional progesterone withdrawal in human myometrium. *Nature Communications*, *7*, 11565.

Navathe, R., & Berghella, V. (2016). Progesterone as a tocolytic agent for preterm labor: A systematic review. *Current Opinion in Obstetrics & Gynecology*, *28*, 464–469.

Nishimura, G., Manabe, I., Tsushima, K., Fujiu, K., Oishi, Y., Imai, Y., et al. (2006). DeltaEF1 mediates TGF-beta signaling in vascular smooth muscle cell differentiation. *Developmental Cell*, *11*, 93–104.

Norwitz, E. R., & Caughey, A. B. (2011). Progesterone supplementation and the prevention of preterm birth. *Reviews in Obstetrics & Gynecology*, *4*, 60–72.

Pabona, J. M., Zhang, D., Ginsburg, D. S., Simmen, F. A., & Simmen, R. C. (2015). Prolonged pregnancy in women is associated with attenuated myometrial expression of progesterone receptor co-regulator Kruppel-like Factor 9. *The Journal of Clinical Endocrinology and Metabolism*, *100*, 166–174.

Patel, B., Elguero, S., Thakore, S., Dahoud, W., Bedaiwy, M., & Mesiano, S. (2015). Role of nuclear progesterone receptor isoforms in uterine pathophysiology. *Human Reproduction Update*, *21*, 155–173.

Peters, G. A., Yi, L., Skomorovska-Prokvolit, Y., Patel, B., Amini, P., Tan, H., et al. (2017). Inflammatory stimuli increase progesterone receptor-A stability and transrepressive activity in myometrial cells. *Endocrinology*, *158*, 158–169.

Piekorz, R. P., Gingras, S., Hoffmeyer, A., Ihle, J. N., & Weinstein, Y. (2005). Regulation of progesterone levels during pregnancy and parturition by signal transducer and activator of transcription 5 and 20alpha-hydroxysteroid dehydrogenase. *Molecular Endocrinology*, *19*, 431–440.

Renthal, N. E., Chen, C. C., Williams, K. C., Gerard, R. D., Prange-Kiel, J., & Mendelson, C. R. (2010). miR-200 family and targets, ZEB1 and ZEB2, modulate uterine quiescence and contractility during pregnancy and labor. *Proceedings of the National Academy of Sciences of the United States of America*, *107*, 20828–20833.

Renthal, N. E., Williams, K. C., Montalbano, A. P., Chen, C. C., Gao, L., & Mendelson, C. R. (2015). Molecular regulation of parturition: A myometrial perspective. *Cold Spring Harbor Perspectives in Medicine*, *5*, a023069.

Rothgiesser, K. M., Erener, S., Waibel, S., Luscher, B., & Hottiger, M. O. (2010). SIRT2 regulates NF-kappaB dependent gene expression through deacetylation of p65 Lys310. *Journal of Cell Science*, *123*, 4251–4258.

Rubel, C. A., Wu, S. P., Lin, L., Wang, T., Lanz, R. B., Li, X., et al. (2016). A Gata2-dependent transcription network regulates uterine progesterone responsiveness and endometrial function. *Cell Reports*, *17*, 1414–1425.

Shynlova, O., Kwong, R., & Lye, S. J. (2010). Mechanical stretch regulates hypertrophic phenotype of the myometrium during pregnancy. *Reproduction*, *139*, 247–253.

Shynlova, O., Oldenhof, A., Dorogin, A., Xu, Q., Mu, J., Nashman, N., et al. (2006). Myometrial apoptosis: Activation of the caspase cascade in the pregnant rat myometrium at midgestation. *Biology of Reproduction*, *74*, 839–849.

Stocco, C. O., Zhong, L., Sugimoto, Y., Ichikawa, A., Lau, L. F., & Gibori, G. (2000). Prostaglandin F2alpha-induced expression of 20alpha-hydroxysteroid dehydrogenase involves the transcription factor NUR77. *The Journal of Biological Chemistry*, *275*, 37202–37211.

Stratmann, A., & Haendler, B. (2011). The histone demethylase JARID1A regulates progesterone receptor expression. *The FEBS Journal*, *278*, 1458–1469.

Sugimoto, Y., Yamasaki, A., Segi, E., Tsuboi, K., Aze, Y., Nishimura, T., et al. (1997). Failure of parturition in mice lacking the prostaglandin F receptor. *Science*, *277*, 681–683.

Szwarc, M. M., Kommagani, R., Lessey, B. A., & Lydon, J. P. (2014). The p160/steroid receptor coactivator family: Potent arbiters of uterine physiology and dysfunction. *Biology of Reproduction*, *91*, 122.

Tan, H., Yi, L., Rote, N. S., Hurd, W. W., & Mesiano, S. (2012). Progesterone receptor-A and -B have opposite effects on proinflammatory gene expression in human myometrial cells: Implications for progesterone actions in human pregnancy and parturition. *The Journal of Clinical Endocrinology and Metabolism*, *97*, E719–E730.

Tang, Y., Ji, H., Liu, H., Gu, W., Li, X., & Peng, T. (2015). Identification and functional analysis of microRNA in myometrium tissue from spontaneous preterm labor. *International Journal of Clinical and Experimental Pathology*, *8*, 12811–12819.

Tibbetts, T. A., Conneely, O. M., & O'Malley, B. W. (1999). Progesterone via its receptor antagonizes the pro-inflammatory activity of estrogen in the mouse uterus. *Biology of Reproduction*, *60*, 1158–1165.

Topouzis, S., & Majesky, M. W. (1996). Smooth muscle lineage diversity in the chick embryo. Two types of aortic smooth muscle cell differ in growth and receptor-mediated transcriptional responses to transforming growth factor-beta. *Developmental Biology*, *178*, 430–445.

Udy, G. B., Towers, R. P., Snell, R. G., Wilkins, R. J., Park, S. H., Ram, P. A., et al. (1997). Requirement of STAT5b for sexual dimorphism of body growth rates and liver gene expression. *Proceedings of the National Academy of Sciences of the United States of America*, *94*, 7239–7244.

Vandewalle, C., Van Roy, F., & Berx, G. (2009). The role of the ZEB family of transcription factors in development and disease. *Cellular and Molecular Life Sciences*, *66*, 773–787.

Vegeto, E., Shahbaz, M. M., Wen, D. X., Goldman, M. E., O'Malley, B. W., & McDonnell, D. P. (1993). Human progesterone receptor A form is a cell- and promoter-specific repressor of human progesterone receptor B function. *Molecular Endocrinology*, *7*, 1244–1255.

Walsh, S. W., Stanczyk, F. Z., & Novy, M. J. (1984). Daily hormonal changes in the maternal, fetal, and amniotic fluid compartments before parturition in a primate species. *The Journal of Clinical Endocrinology and Metabolism*, *58*, 629–639.

Williams, K. C., Renthal, N. E., Condon, J. C., Gerard, R. D., & Mendelson, C. R. (2012a). MicroRNA-200a serves a key role in the decline of progesterone receptor function leading to term and preterm labor. *Proceedings of the National Academy of Sciences of the United States of America*, *109*, 7529–7534.

Williams, K. C., Renthal, N. E., Gerard, R. D., & Mendelson, C. R. (2012b). The microRNA (miR)-199a/214 cluster mediates opposing effects of progesterone and estrogen

on uterine contractility during pregnancy and labor. *Molecular Endocrinology*, *26*, 1857–1867.
Yang, H., Zhang, W., Pan, H., Feldser, H. G., Lainez, E., Miller, C., et al. (2012). SIRT1 activators suppress inflammatory responses through promotion of p65 deacetylation and inhibition of NF-kappaB activity. *PloS One*, *7*, e46364.
Yore, M. A., Im, D., Webb, L. K., Zhao, Y., Chadwick, J. G., Jr., Molenda-Figueira, H. A., et al. (2010). Steroid receptor coactivator-2 expression in brain and physical associations with steroid receptors. *Neuroscience*, *169*, 1017–1028.
Yung, H. W., Calabrese, S., Hynx, D., Hemmings, B. A., Cetin, I., Charnock-Jones, D. S., et al. (2008). Evidence of placental translation inhibition and endoplasmic reticulum stress in the etiology of human intrauterine growth restriction. *The American Journal of Pathology*, *173*, 451–462.
Zeng, Z., Velarde, M. C., Simmen, F. A., & Simmen, R. C. (2008). Delayed parturition and altered myometrial progesterone receptor isoform A expression in mice null for Kruppel-like factor 9. *Biology of Reproduction*, *78*, 1029–1037.
Zhang, P., Sun, Y., & Ma, L. (2015). ZEB1: At the crossroads of epithelial-mesenchymal transition, metastasis and therapy resistance. *Cell Cycle*, *14*, 481–487.

CHAPTER SEVEN

Roles of Retinoic Acid in Germ Cell Differentiation

Marius Teletin[*,†,‡,§,¶], **Nadège Vernet**[*,†,‡,§], **Norbert B. Ghyselinck**[*,†,‡,§], **Manuel Mark**[*,†,‡,§,¶,1]

[*]Institut de Génétique et de Biologie Moléculaire et Cellulaire (IGBMC), Illkirch, France
[†]Centre National de la Recherche Scientifique (CNRS), Paris, France
[‡]Institut National de la Santé et de la Recherche Médicale (INSERM), Paris, France
[§]Université de Strasbourg (UNISTRA), Strasbourg, France
[¶]Hôpitaux Universitaires de Strasbourg (HUS), Strasbourg, France
[1]Corresponding author: e-mail address: manuel.mark@igbmc.fr

Contents

Current Topics in Developmental Biology, Volume 125
ISSN 0070-2153
http://dx.doi.org/10.1016/bs.ctdb.2016.11.013

Abstract

The modalities of gametogenesis differ markedly between sexes. Female are born with a definitive reserve of oocytes whose size is crucial to ensure fertility. Male fertility, in contrast, relies on a tightly regulated balance between germ cell self-renewal and differentiation, which operates throughout life, according to recurring spatial and temporal patterns. Genetic and pharmacological studies conducted in the mouse and discussed in this review have revealed that all-*trans* retinoic acid and its nuclear receptors are major players of gametogenesis and are instrumental to fertility in both sexes.

ABBREVIATIONS

A_{al} A aligned spermatogonia
A_{pr} A paired spermatogonia
A_s A single spermatogonia
A_{undiff} undifferentiated spermatogonia
ATRA *all-trans* retinoic acid
BMS-189961 agonistic ligand selective for RARG, from Bristol-Myers Squibb pharmaceuticals
BMS-204493 antagonistic ligand for all RAR isotypes, from Bristol-Myers Squibb pharmaceuticals
CYP26 cytochrome P450 family 26 enzymes
DAPI 4′,6-diamidino-2-phenylindole
ED embryonic day
GC germ cells
IHC immunohistochemistry
Int intermediate spermatogonia
ISH in situ hybridization
MIS meiosis-inducing substance
MPS meiosis-preventing substance
PGC primordial germ cells
PND postnatal day
PNW postnatal week
RAL retinaldehyde
RALDH retinaldehyde dehydrogenase
RAR retinoic acid receptor
RDH retinol dehydrogenase
RXR rexinoid receptor
SE seminiferous epithelium
$Ser^{-/-}$ **(in context of genotypes)** a deletion restricted to Sertoli cells
$Spg^{-/-}$ **(in context of genotypes)** a deletion restricted to spermatogonia and their progeny
vitA vitamin A
WIN-18,446 bis(dichloroacetyl)diamine, inhibitor of retinaldehyde dehydrogenase

1. INTRODUCTION

Germ cells (GC) are unique in their capacity to divide meiotically and to halve their genetic material, generating haploid gametes, oocytes in the female and spermatozoa in the male, that transfer the genetic information from one generation to the next. Due to ethical issues restricting access to human material for research purposes, studies documenting the fate of human GC are sparse and much of what is known about gamete formation has been obtained from the mouse. However, the characteristic cellular events and chronology of the key steps of gametogenesis are similar in mice and humans.

The first GC population established during development, primordial GC (PGC) are segregated from the somatic cell lineage at early gastrulation, around embryonic day (ED) 6.25 in the mouse. PGC proliferate while migrating along the hindgut to reach the gonads during the late embryonic period (i.e., ED10.5–ED11.0) (reviewed in McLaren & Lawson, 2005). Once in the gonad and now known as gonocytes (Clermont & Perey, 1957), GC follow a sexually dimorphic pathway dictated by cues from the somatic cells (Fig. 1). In the fetal ovary, female gonocytes initiate prophase I of meiosis at about ED13.5. As primary oocytes, they pass through the leptotene, zygotene, and pachytene stages before arresting after birth in diplotene, surrounded by a single layer of somatic, follicle cells. Upon completion of puberty, several primary oocytes achieve meiosis I and then divide, yielding haploid secondary oocytes at each menstrual cycle (Fig. 1). Progression through meiosis in oocytes pauses again in metaphase II, shortly before ovulation, to complete the second meiotic division after fertilization in the oviduct. In the fetal testis, male gonocytes, surrounded by somatic cells of the testis cords, undergo a mitotic arrest in G0/G1 (Fig. 1). After birth, they resume proliferation and undergo a transition to spermatogonia. Meiosis in primary spermatocytes begins during the prepubertal period, by the end of the first postnatal week (PNW). The two meiotic divisions proceed without interruption, leading to the formation of haploid secondary spermatocytes and then of round spermatids. The latter undergo morphological changes to finally become spermatozoa (Fig. 1). Spermatogenesis, the process yielding spermatozoa from spermatogonia, is completed for the first time at PNW5 and continues throughout adulthood. It takes place in the seminiferous epithelium (SE) of the testis, under the control of a somatic, supporting cell type called the Sertoli cell (reviewed in Bowles & Koopman, 2007; Griswold, 2016; Kocer, Reichmann, Best, & Adams, 2009).

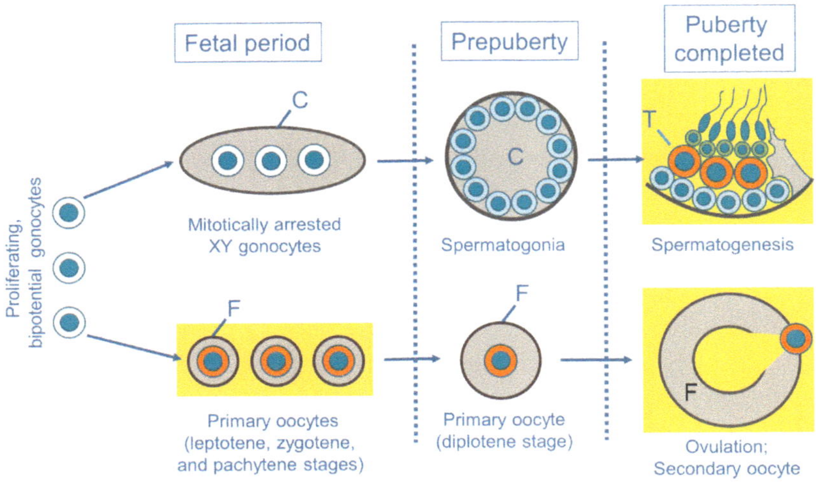

Fig. 1 Key stages of fetal and postnatal GC development. Fetal period: female gonocytes enter meiosis prior to birth, while male gonocytes undergo a mitotic cell cycle arrest. Prepuberty: during the first postnatal week, male gonocytes resume proliferation and migrate from the center of the seminiferous cords to the basement membrane, where they generate spermatogonia. Oogenesis and spermatogenesis are completed at puberty. Meiotic cells (oocytes or spermatocytes) and male postmeiotic cells (spermatids) are in *orange* and *green*, respectively. Supporting somatic cells (follicular or Sertoli cells) are represented in *gray*. Legend: C and T, seminiferous cords and tubules, respectively; F, ovarian follicles. Processes requiring ATRA are drawn on a *yellow* background. See the main text for details.

Spermatogenesis is critically dependent upon vitamin A (vitA), acting through its metabolite, all-*trans* retinoic acid (ATRA). The pioneering study by Wolbach and Howe (1925) showed that spermatogenesis fails in rats fed a vitA-deficient diet. Subsequently it was demonstrated that it can be successfully restored upon systemic administration of ATRA (van Pelt & de Rooij, 1991). Then, several genetic and pharmacological studies have highlighted the importance of ATRA in regulating meiosis.

The canonical pathway for ATRA signaling has been elucidated over the years, mainly though gene-targeting studies in the mouse (Fig. 2). Conversion of vitA to ATRA requires two sequential oxidative steps. The first one is catalyzed by either cytosolic alcohol dehydrogenases or microsomal retinol dehydrogenases (RDH) to generate retinaldehyde (RAL). The second step, generating ATRA, is catalyzed by RAL dehydrogenases (RALDH1, RALDH2, and RALDH3, encoded by the *Aldh1a1*, *Aldh1a2*, and *Aldh1a3* genes, respectively). In cells, ATRA binds to and

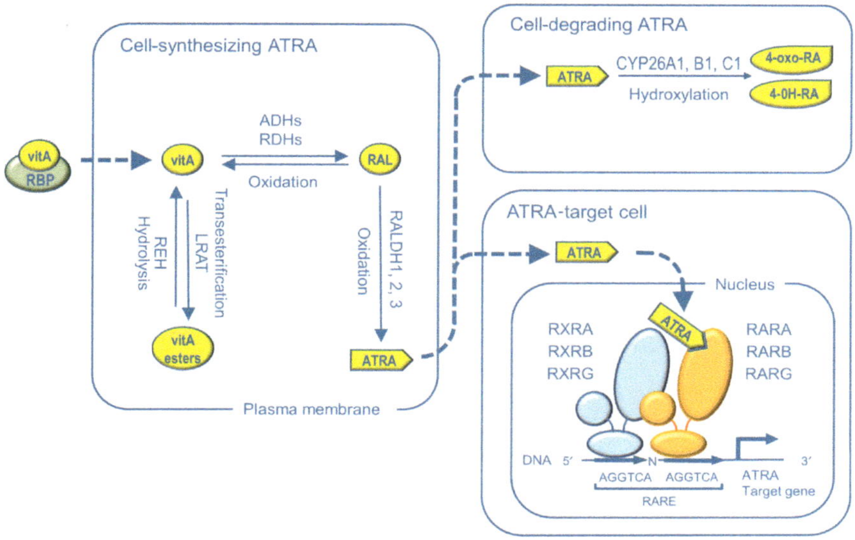

Fig. 2 Regulation of retinoid signaling. Vitamin A (vitA) is provided to cells by the blood, where it circulates bound to the vitA-binding protein (RBP). In cells, vitA is locally stored in the form of vitA esters generated through the transesterification activity of LRAT (lecithin retinol acyltransferase, the most potent vitA-esterifying enzyme). VitA is liberated from vitA-ester stores through the hydrolyzing activity of retinyl ester hydrolases (REH). In ATRA-synthesizing cells, vitA is oxidized into retinaldehyde (RAL) by alcohol dehydrogenases (ADH) and short-chain retinol dehydrogenases (RDH). RAL dehydrogenases (RALDH1, 2, and/or 3) then generate ATRA, which binds within the nucleus to an RAR forming a heterodimer complex with an RXR. RAR/RXR modulate transcription by binding to DNA at RARE motifs located in regulatory regions of ATRA-target genes. In cells expressing CYP26A1, B1, and/or C1 enzymes, ATRA is hydroxylated into polar compounds such as 4-oxo- and 4-hydroxy-retinoic acid (4-oxo-RA and 4-OH-RA, respectively), which are subject to further metabolism and elimination. ATRA often acts in a paracrine manner on neighboring cells, but there also is evidence for autocrine effects in cells that synthesize it (e.g., in Sertoli cells from adults). Examples of cell types expressing both ATRA-synthesizing and -degrading enzymes are rare (e.g., fetal Sertoli cells). *Dashed line* indicates diffusion of retinoids through cell membranes. See the main text for details.

activates nuclear receptors (RARA, RARB, and RARG), which are ligand-dependent transcriptional regulators. They usually function in the form of heterodimers with rexinoid receptors (RXRA, RXRB, and RXRG) to control expression of ATRA-target genes through binding to retinoic acid response elements (RAREs) located in genomic regulatory regions. Aside from synthesis, degradation of ATRA is also an important mechanism that protects cells from inadequate ATRA stimulation. It is catalyzed by three cytochrome P450 hydroxylases (CYP26A1, CYP26B1, and CYP26C1), which hydroxylate ATRA into water-soluble products that are less active and readily excretable (reviewed in Duester, 2008; Samarut & Rochette-Egly, 2012).

In this review, we discuss our current understanding on how ATRA regulates GC differentiation and entry into meiosis. We also provide complementary information and some novel details as to the sources and the functions of ATRA in these processes.

2. ATRA SIGNALING IN THE FETAL GONADS

It has been hypothesized that the decision of gonocytes to enter meiosis is an intrinsic property (i.e., a GC autonomous event, for instance timed by cell divisions), with no reference to somatic cues: they spontaneously enter meiosis unless prevented from doing so by a putative meiosis-preventing substance (MPS). The fact that disaggregated male gonocytes enter meiosis when transferred to lung cell aggregates favors the existence of an MPS synthesized by somatic cells of the fetal testis and acting at a short distance. An alternative hypothesis proposes that gonocytes enter meiosis under the influence of a somatic, diffusible, meiosis-inducing substance (MIS), whose existence in the fetal mouse ovary, but not testis, was deduced from coculture experiments. Such an MIS has to be also present in nongonadal sites because GC that fail to reach the gonad, but instead migrate to ectopic sites such as, for instance, the mesonephros, enter meiosis irrespective of their genetic sex. It is worth noting here that the MPS hypothesis can also account for the fact that male gonocytes enter meiosis in nongonadal sites, where they escape the inhibitory influence of MPS. Whether meiosis entry requires a female MIS only, a male MPS only, or both an MIS and an MPS is still a matter of debate, even though the latter option represents the prevailing model (reviewed in Bowles & Koopman, 2007; Kocer et al., 2009).

2.1 CYP26B1 Acts as an MPS for Male Gonocytes

The possible role for endogenous ATRA in mouse GC differentiation was initially raised by the sexually dimorphic expression of *Cyp26b1*, which is expressed in the embryonic gonads of both sexes at ED11.5, but becomes male specific by ED12.5 with strong expression in the somatic cells of the testis (Bowles et al., 2006; Fig. 3). Pharmacological studies subsequently showed that exposure of embryonic testes in organ culture to exogenous

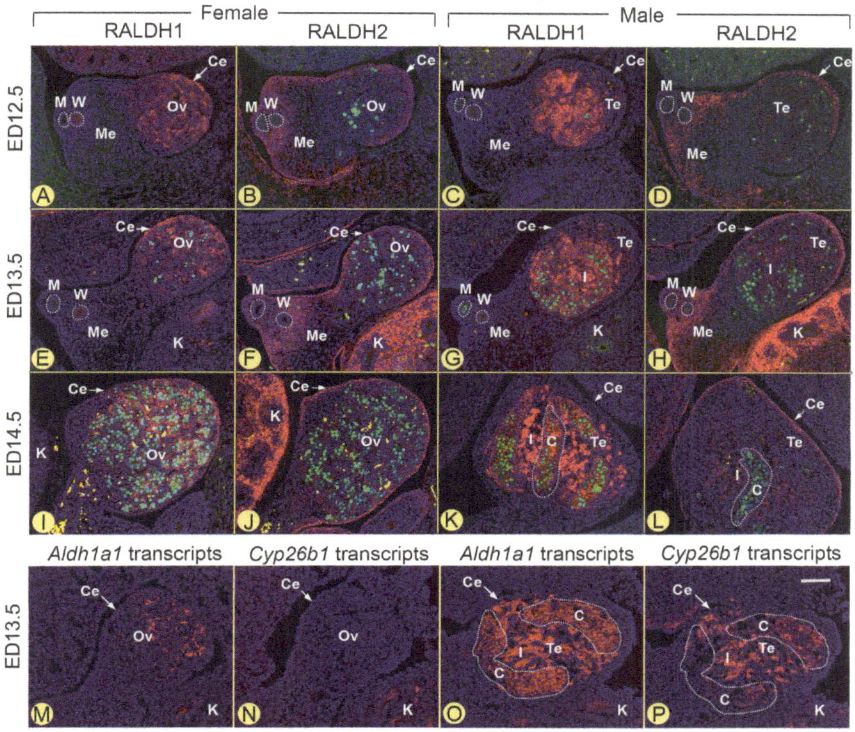

Fig. 3 Distribution patterns of ATRA-synthesizing and -degrading enzymes in the vicinity of gonocytes, around the time of their meiotic entry in the female. (A–L) Transverse histological sections of ED12.5, ED13.5, and ED14.5 fetuses immunolabeled with antibodies to RALDH1 or RALDH2 (from Abcam; *red signals*) and to TRA98 (from Abcam), a nuclear protein specifically expressed in (some) gonocytes (*green signals*). The sections were counterstained with 4′,6-diamidino-2-phenylindole (DAPI; *blue signal*) to label cell nuclei. RALDH1 and RALDH2 are restricted to the somatic cells located in the gonad (ovary, Ov; testis, Te), the coelomic epithelium (Ce), and the mesonephros (Me). Gonocytes, either positive or negative for TRA98, never express any of the two RALDH. (M–P) Adjacent (M, N) or consecutive (O, P) sections hybridized with antisense probes to *Aldh1a1* and *Cyp26b1* as described (Vernet, Dennefeld, Rochette-Egly, et al., 2006). Note that both transcripts are expressed by some of the immature Sertoli cells within the nascent testis cords (C) as well as by some interstitial cells (I) between these cords. The ISH signals were converted to a *red color* and the histological sections were counterstained with DAPI. Abbreviations: *K*, (definitive) kidney; *M*, Müllerian duct; *W*, Wolffian duct. Bar in (P): 160 μm (A–P).

ATRA or to the CYP inhibitor ketoconazole induces (i) aberrant expression of meiotic markers such as *Stra8* (encoding a meiotic gatekeeper, see below), *Dmc1* (encoding a meiotic recombinase) and *Sycp3* (encoding a synaptonemal complex protein) and (ii) the appearance of chromosomal aggregates resembling cytological figures observed during meiotic prophase I (Bowles et al., 2006). As a simultaneous exposure to ketoconazole and to the pan-RAR inverse agonist BMS-204493 prevents these meiotic events, this suggests that ketoconazole, although not CYP26B1 specific, induces untimely meiosis in the testis through an inappropriate induction of the RAR signaling pathway (Koubova et al., 2006). The role of CYP26B1 in preventing meiosis entry in the fetal testis was confirmed in vivo by functional analysis of *Cyp26b1*$^{-/-}$ mutants: meiotic GC expressing *Sycp3* are observed at ED13.5, at the time when GC in ovaries enter meiosis, in sharp contrast to the normal situation where meiosis initiates in the testis by postnatal day (PND) 7 (MacLean, Li, Metzger, Chambon, & Petkovich, 2007). Altogether, these findings are compatible with the possibility that CYP26B1 represents or synthesizes the MPS.

2.2 CYP26B1 Prevents Proliferation As Well As Death of Male Gonocytes

In the mouse testis *Cyp26b1* is notably expressed during the ED12.5–ED14.5 period, when male gonocytes undergo mitotic arrest in G0/G1 (quiescence) (Western, Miles, van den Bergen, Burton, & Sinclair, 2008; Fig. 3). Pharmacological inhibition of CYP26 activity in organ culture of testes prevents the mitotic arrest of male GC and induces their death by apoptosis. These data suggest that impairing CYP26 action in the male fetal gonad not only prevents gonocytes from entering meiosis but also allows the establishment of their quiescent state and survival (Trautmann et al., 2008). Similar conclusions were reached by Li, MacLean, Cameron, Clagett-Dame, and Petkovich (2009) who generated a Sertoli cell-specific knockout of *Cyp26b1* starting at ED15. In these mutants, some gonocytes exit from the G0 stage to reenter the mitotic cycle, express SYCP3, indicating that they undergo inappropriate initiation of meiotic prophase I, and are subsequently lost from the seminiferous cords before birth. It is noteworthy that ablation of *Cyp26b1* in Sertoli cells most probably accounts only very partially for the functions of the enzyme in the developing testis, because the large majority of *Cyp26b1* transcripts are detected outside the fetal seminiferous cords, in the somatic cells located in the interstitial spaces (Abu-Abed et al., 2002; Fig. 3), and are clearly confined to the peritubular myoid cells from birth (PND0) onward (Vernet, Dennefeld, Rochette-Egly, et al., 2006; our unpublished results). It is therefore not surprising that another study shows that ablation of CYP26B1 activity in Sertoli cells has, in fact, little impact on adult spermatogenesis (Hogarth, Evans, et al., 2015).

The combined results of these pharmacological and genetic experiments altogether suggest that increased ATRA levels induce aberrant entry of into meiosis, inappropriate cell proliferation, and massive cell death among male gonocytes.

Thus, if exposure to ATRA is deleterious to male gonocytes, what might be the physiological significance of the strong and steady expression of RALDH1 in the immature Sertoli cells forming the cords of the fetal testis (Bowles et al., 2009; Fig. 3)? In an attempt to answer this question, we analyzed the outcome of ablating all RALDH genes in Sertoli cells, as early as ED15 ($Aldh1a1\text{-}3^{Ser-/-}$ mutants) (Raverdeau et al., 2012). At ED18.5, about 24 h before birth, all gonocytes in testes of wild-type (WT) and $Aldh1a1\text{-}3^{Ser-/-}$ mutants are mitotically arrested (Fig. 4). They express the receptor tyrosine kinase KIT, a general marker of gonocytes (Fig. 4), but never GFRA1 nor ZBTB16, both of which are markers of undifferentiated spermatogonia. Then, WT and $Aldh1a1\text{-}3^{Ser-/-}$ mutant gonocytes express GFRA1 and ZBTB16 at PND1 (Fig. 4), resume proliferation by PND2, and migrate to the basement membrane until PND6 to adopt a spermatogonial phenotype (Kluin, Kramer, & de Rooij, 1982; Nagano et al., 2000). As RALDH activity in Sertoli cells represents the closest and therefore main source of ATRA for male fetal GC, this finding indicates that ATRA is not physiologically required to control the mitotic arrest of gonocytes, their migration from the center of the testicular cords to the basement membrane and their transition to spermatogonia. The reason for the robust expression of RALDH1 in Sertoli cells before birth is therefore still elusive. It is however noteworthy that RALDH1 exhibits a broad substrate specificity and can function as a detoxifying enzyme for xenobiotics (Alnouti & Klaassen, 2008). Whether this is indeed the case during testis development remains to be tested.

2.3 ATRA Is Required for the Production of "Differentiating" Spermatogonia From Male Gonocytes

In the mouse, the transition from gonocytes to spermatogonia occurs during the perinatal period, prior to the onset of prepuberty. Male gonocytes are the precursors of the stem spermatogonia and can also directly differentiate, without passing through the stem cell stage, into differentiating spermatogonia that support the first wave of spermatogenesis (Yoshida et al., 2006). Moreover, it has been suggested that gonocytes that fail to differentiate into spermatogonia may provide the source for the formation of carcinoma in situ, the precursor lesion of most testicular GC cancer in

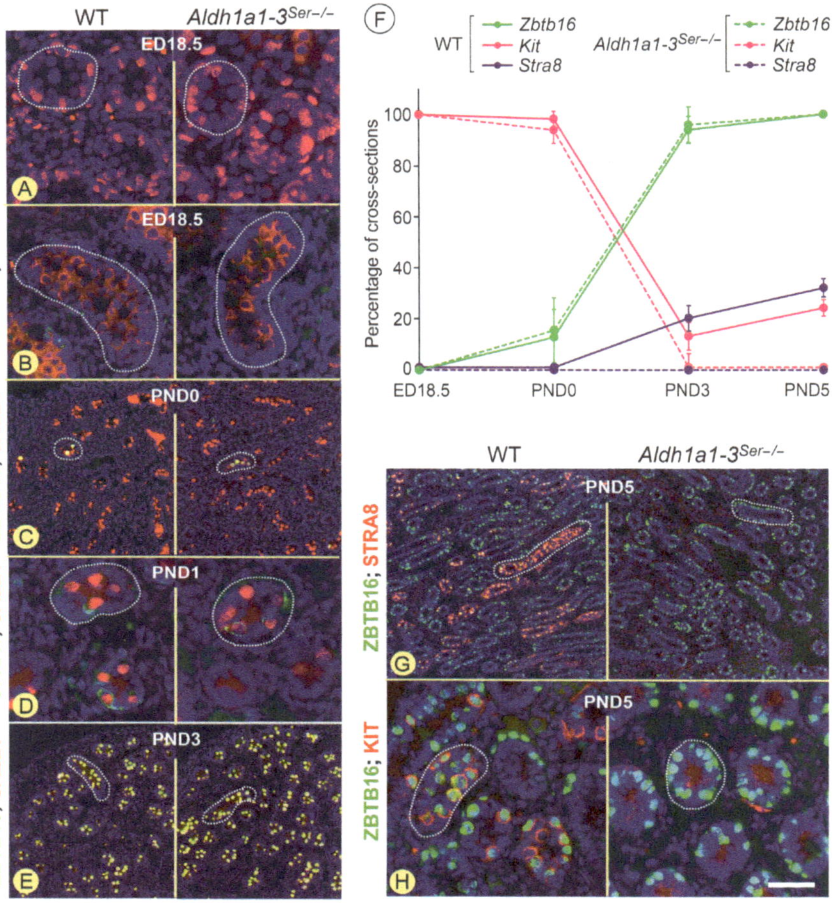

Fig. 4 The transition from gonocytes to differentiating spermatogonia, but not the formation of undifferentiated spermatogonia, is impaired in the absence of ATRA in the testis cords. (A, B) In both *Aldh1a1-3*$^{Ser-/-}$ and WT mice at ED18.5, the proliferation marker MKI67 (*red signal*, A) is strictly confined to the somatic cells, including immature Sertoli cells at the periphery of the seminiferous cords (*dotted white line*), while expression of KIT (B, *red signal*) is identical in GC. (B–E) Between ED18.5 and PND3, expression of ZBTB16 (B, C, and E, *green* and *yellow signals*) and GFRA1 (D, *green signal*) and also is identical in GC of both *Aldh1a1-3*$^{Ser-/-}$ and WT mice. (F) Percentage of cross-sections of seminiferous cords harboring GC expressing: ZBTB16 (*green lines*), KIT (*pink lines*), and STRA8 (*purple lines*) in WT (*solid lines*) and *Aldh1a1-3*$^{Ser-/-}$ (*dotted lines*) mice, at the indicated ages. The bars represent mean ± SEM ($n = 3$ mice, >200 cross-sections per mice). (G, H) STRA8 and KIT (*red signals*) are not expressed in spermatogonia from *Aldh1a1-3*$^{Ser-/-}$ testes, in contrast to the WT situation. Scale bar in (H): 30 μm (A, B, D, H) and 160 μm (C, E, G).

humans (Soone et al., 2009). For these reasons, the transition from gonocytes to spermatogonia is an important step of the differentiation pathway leading to male gametes.

We found that GC from *Aldh1a1-3*$^{Ser-/-}$ mutant definitively lose KIT at PND3 as they do not reinitiate the transcription of the *Kit* gene, as it is normally the case at PND3 (Yoshida et al., 2006; Fig. 4). Moreover, GC from *Aldh1a1-3*$^{Ser-/-}$ never express STRA8 in contrast to WT gonocytes (Fig. 4). This latter observation is in keeping with the previous finding that expression of *Stra8* is delayed in spermatogonia of neonatal *Aldh1a1*$^{-/-}$ mutants (Vernet, Dennefeld, Guillou, et al., 2006). The data presented here thus indicate that ATRA synthesis by Sertoli cells is necessary after birth for the production, from gonocytes, of the spermatogonia that initiate the first wave of spermatogenesis (Yoshida et al., 2006).

2.4 ATRA Possibly Acts as a MIS for Female Gonocytes

It is commonly admitted that ATRA acts as an MIS (Griswold, Hogarth, Bowles, & Koopman, 2012) but the tissue from which originates the ATRA that induces meiosis in fetal ovaries is disputed. Studies of mice carrying an ATRA-dependent reporter transgene (RARE-hsp90-*LacZ*; Rossant, Zirngibl, Cado, Shago, & Giguère, 1991) reveal that the responsiveness to ATRA is weak in the gonads, but robust in the mesonephroi of both sexes (Bowles et al., 2006; Koubova et al., 2006). As, in addition both *Aldh1a2* mRNA (Bowles et al., 2006) and RALDH2 protein (Fig. 3) are detected in mesonephroi of both sexes from ED11.5 to ED13.5, but not in the gonad itself, it was concluded that the source of ATRA detected in the ovary at the time of meiotic initiation is the adjacent mesonephros. A model for the sex-specific timing of meiosis entry was proposed, according to which ATRA synthesized by the mesonephros serves as an extrinsic MIS acting in the adjacent gonad, unless degraded by CYP26B1, the MPS.

It is however noteworthy that the coelomic epithelium, which covers the embryonic gonad, also strongly expresses both *Aldh1a2* transcripts (Niederreither, McCaffery, Dräger, Chambon, & Dollé, 1997) and RALDH2 protein (Fig. 3). Thus, it can also function as a paracrine source of ATRA, closer to GC than the mesonephros. On the other hand, Kumar et al. (2011) have challenged the view that ATRA is required for entry of ovarian gonocytes in meiosis. Actually, by analyzing mice carrying null mutations for RALDH2, singly or in combination with null mutations for RALDH3 (*Aldh1a2*$^{-/-}$ and *Aldh1a2*$^{-/-}$; *Aldh1a3*$^{-/-}$ mutants), they show that the RARE-hsp90-*LacZ* activity typically observed throughout the normal mesonephros and the gonads is totally eliminated, as expected. Yet, both mutant and WT ovaries exhibit similar expressions of meiotic effectors such as *Stra8*, *Sycp3*, and phosphorylated H2AX histone, which marks DNA double-strand breaks occurring during meiotic recombination. The authors conclude that meiosis occurs normally, despite the lack of detectable ATRA in the vicinity of the ovaries. Nevertheless, these observations left open the possibility that RALDH1 might be also involved. This hypothesis has been neglected for a long time likely because (i) *Aldh1a1*$^{-/-}$ mice are viable and fertile (Fan et al., 2003; Matt et al., 2005); (ii) RALDH1 has the lowest affinity for RAL, being about 15-fold less efficient than RALDH2 at producing ATRA (Gagnon, Duester, & Bhat, 2002); and (iii) RALDH1 has been proposed to be marginally involved in ATRA synthesis in vivo (Niederreither, Vermot, Fraulob, Chambon, & Dollé, 2002). The fact remains that both *Aldh1a1* mRNA and RALDH1 protein are detected in somatic cells of the developing gonads that are in direct contact with gonocytes (Fig. 3), and meiosis is indeed delayed in the *Aldh1a1*$^{-/-}$ ovary, as judged by the expression of marker genes (Bowles et al., 2016). Thus, in spite of its low capability of ATRA synthesis, RALDH1 activity appears to participate in meiotic initiation in female GC. By remaining expressed in *Aldh1a2*$^{-/-}$; *Aldh1a3*$^{-/-}$ mutants, it may functionally compensate for the loss of the two other ATRA-synthesizing enzymes. In this context, it is also noteworthy that cultured GC upregulate meiotic genes in response to 1 nM ATRA, which supports the view that ATRA at low concentrations is able to favor meiosis (Bowles et al., 2010; Tedesco, Desimio, Klinger, De Felici, & Farini, 2013).

2.5 *Stra8* as the Meiotic Gatekeeper

A wealth of data support the model whereby ATRA acts as the MIS by inducing the expression of STRA8, a 45-kDa vertebrate-specific protein (Oulad-Abdelghani et al., 1996). STRA8 actually functions as a "meiotic gatekeeper" by controlling the switch between GC division by mitosis and meiosis, as evidenced by gene knockout studies in the mouse (Anderson et al., 2008; Baltus et al., 2006; Mark et al., 2008). The *Stra8* promoter contains an RARE, which can bind to RAR/RXR heterodimers, making this gene a possible direct target of ATRA (Kumar et al., 2011;

Raverdeau et al., 2012). Moreover, the normal expression of *Stra8* in female gonocytes is inhibited in fetal ovaries cultured in the presence of the pan-RAR inverse agonist BMS-204493 (Koubova et al., 2006), and ATRA at supraphysiological levels can stimulate expression of *Stra8* when added to the medium of cultured embryonic testes from mice (Baltus et al., 2006; Bowles et al., 2006; Koubova et al., 2006), humans (Le Bouffant et al., 2010), and amphibians (Piprek et al., 2013; Wallacides, Chesnel, Chardard, Flament, & Dumond, 2009), suggesting that the requirement of ATRA for meiosis entry could be a conserved event.

These data must, however, be interpreted with caution for three reasons. First, it has been recurrently emphasized that developmental alterations generated by added ATRA do not necessarily reflect cellular processes operating in real life (Horton & Maden, 1995; Mark, Ghyselinck, & Chambon, 2006). Second, neither expression of *Stra8* nor upregulation of meiotic markers is synonymous of entry into meiosis. For instance, expression of STRA8 and proteins associated with key meiotic processes occurs in spermatogonia, well before the appearance of premeiotic spermatocytes (Evans, Hogarth, Mitchell, & Griswold, 2014; Mark, Teletin, Vernet, & Ghyselinck, 2015). Third, silencing of *Stra8* expression by the pan-RAR inverse agonist BMS-204493 does not necessarily reflect impairment of its induction by endogenous ATRA-activated RAR because this synthetic retinoid stabilizes interaction between corepressor complexes and RAR, whether naturally bound by ATRA or not. By tethering corepressor-associated histone deacetylase activities, it may silence genes containing RARE but not necessarily activated by endogenous ATRA-bound RAR (Germain et al., 2009). Thus, although STRA8 is indisputably instrumental to proper meiosis, the control of its expression by endogenous ATRA in female GC is more controversial.

2.6 An Alternative Viewpoint of Meiosis Induction in Female Gonocytes

Data on the involvement of ATRA in meiosis are commonly interpreted according to the prevailing idea that both an MPS and an MIS are required to account for the sex-specific timing of meiotic initiation (reviewed in Bowles & Koopman, 2007; Griswold et al., 2012; Kocer et al., 2009). This is mainly due to the pioneering finding that *Cyp26b1*, whose function is supposedly to degrade ATRA, is expressed only in the male, as opposed to the female fetal gonad where meiosis initiates. Since then, Kumar et al. (2011) have proposed that CYP26B1 may instead metabolize a substrate that is distinct from ATRA. It must be emphasized that considering the "MPS-only" hypothesis to interpret the current data would reconcile the apparent discrepancies existing in the literature. According to this scenario, inducing quiescence instead of meiosis in male gonocytes would require a yet unknown substance produced by CYP26B1 (the MPS). On the opposite, entering meiosis in female gonocytes would be an autonomous, intrinsic, property of GC without the requirement of any MIS, including ATRA synthesized by RALDH1 or another RALDH. Interestingly, several authors propose that ATRA in the fetal ovary is involved in GC proliferation rather than meiosis initiation (Jørgensen et al., 2015; Koshimizu, Watanabe, & Nakatsuji, 1995; Morita & Tilly, 1999; Trautmann et al., 2008).

3. ATRA SIGNALING IS INSTRUMENTAL TO SPERMATOGONIA DIFFERENTIATION IN THE PREPUBERTAL AND ADULT TESTIS

Spermatogenesis is initiated during the prepubertal period in the SE and then proceeds throughout adulthood. In the mouse, the initial wave of spermatogenesis, which starts around PND2 and is completed by PND35, involves a subpopulation of gonocytes (Yoshida et al., 2006). The subsequent prepubertal and all adult waves derive only from spermatogonia. These cells have traditionally been classified into "undifferentiated" and "differentiating" based on their ability to self-renew and differentiate (de Rooij & Russell, 2000). Undifferentiated type A (A_{undiff}) spermatogonia are endowed with stem cell functions: they can (i) maintain their own population, thereby ensuring long-term maintenance of the germline and (ii) produce daughter cells that differentiate into spermatozoa. Based on whole-mount analyses of seminiferous tubules, A_{undiff} spermatogonia collectively correspond to A_{single} (A_s, isolated single) spermatogonia which, upon division, give rise to syncytia consisting of 2 A_{paired} (A_{pr}), 4 $A_{aligned\text{-}4}$ ($A_{al\text{-}4}$), 8 ($A_{al\text{-}8}$), 16 ($A_{al\text{-}16}$) cells, and occasionally 32 cells interconnected by cytoplasmic bridges resulting from incomplete cytokinesis (reviewed in de Rooij & Russell, 2000; Fig. 5). Cell lineage tracing experiments indicate that a subpopulation of A_s spermatogonia is responsible for the stem cell function in steady-state spermatogenesis, but its A_{al} progeny retains the potential to revert to As and reconstitute the entire spermatogenesis when the tissue is damaged or when these cells are transplanted into GC-depleted seminiferous tubules (Hara et al., 2014; and references therein;

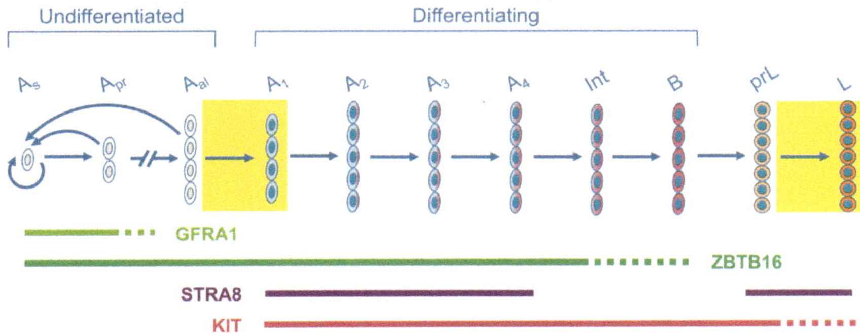

Fig. 5 Spermatogonia differentiation assessed by immunohistochemistry. Because identification of spermatogonia subtypes at the morphological level can only rely on differences in nuclear shape and heterochromatin appearance that are both subtle and variable, immunohistochemical assays represent a valuable alternative. GFRA1 (*light green line*) is strongly expressed at the surface of A_{single} (A_s) and A_{paired} (A_{pr}), and faintly in $A_{aligned\text{-}4}$ ($A_{al\text{-}4}$) spermatogonia (Gely-Pernot et al., 2012). ZBTB16, a nuclear protein (*dark green line*), is frequently considered as a marker A_{undiff} spermatogonia, as it is required for the maintenance of the spermatogonia stem cell pool (Buaas et al., 2004; Costoya et al., 2004). However, it is actually also strongly expressed in A_1–A_4 spermatogonia and, albeit at low levels, also in intermediary (Int) and B spermatogonia (our personal observations). (iii) STRA8 (*purple line*) is expressed in A_1–A_4 spermatogonia, absent in Int and B spermatogonia and reexpressed in preleptotene (prL) and leptotene (L) spermatocytes (Mark et al., 2015). Expression of the tyrosine kinase receptor KIT (*pink line*) is essential to the A_{al}–A_1 transition (Blume-Jensen et al., 2000; Kissel et al., 2000; Schrans-Stassen, van de Kant, de Rooij, & van Pelt, 1999) and is restricted to differentiating (A_1–B) spermatogonia and prL-L spermatocytes (our observations). Note that: (i) the topological arrangement of A_s, A_{pr}, and A_{al} spermatogonia can only be visualized in whole mounts of seminiferous tubules; (ii) there are currently no molecular markers that allow A_1–A_4 spermatogonia to be distinguished from one another. Cellular events requiring ATRA are on a *yellow background*.

Fig. 5). A_{undiff} spermatogonia are evenly distributed along seminiferous tubules. They periodically exit from the stem cell pool and differentiate without a mitotic division into differentiating, type A_1 spermatogonia. This step known as the $A_{aligned}$ to A_1 (A_{al}–A_1) transition is considered as the entrance in the differentiation pathway ultimately yielding spermatozoa. Five mitotic divisions follow A_1 formation, producing sequentially A_2, A_3, A_4, intermediate (Int) and B spermatogonia and accompanying elongation of the syncytial chains. Types A_1 to B spermatogonia referred to as "differentiating spermatogonia" are committed toward the pathway yielding spermatocytes (Fig. 5). They lose their capacity for self-renewal (Shinohara, Orwig, Avarbock, & Brinster, 2000) and differentiate at precise intervals along the seminiferous tubules, each step of differentiation being coupled with a mitotic division (de Rooij, 1998).

KIT and STRA8 are restricted to differentiating spermatogonia and the onset of their synthesis, detected by immunohistochemistry (IHC) and/or by in situ hybridization (ISH) during the progression of spermatogonia differentiation, is a commonly accepted hallmark of the A_{al}−A1 transition (Ghyselinck et al., 2006; Schrans-Stassen et al., 1999). Other proteins that are linked with cell fate or function represent useful tools to characterize the state of spermatogonia differentiation in IHC assays (Fig. 5).

3.1 The Transition From Undifferentiated to Differentiating Spermatogonia Critically Relies on ATRA

The finding that ATRA is a key regulator of the $A_{al}-A_1$ transition was brought about by vitA deficiency studies. Depriving adult rodents of vitA induces a rapid loss of meiotic spermatocytes and spermatids, ultimately yielding seminiferous tubules containing only Sertoli cells, spermatogonia, and, in the case of rats, preleptotene spermatocytes. Based on cell kinetic properties and on the absence of *Kit* expression, the remaining spermatogonia in vitA-deficient animals are arrested just before their differentiation into A_1 spermatogonia. In rats and mice, reintroduction of vitA in the diet or systemic administration of different retinoids induces, within 24 h, the $A_{al}-A_1$ transition and, subsequently, the reinitiation of spermatogenesis in most of the tubules (Gaemers, van Pelt, van der Saag, & de Rooij, 1996; Schrans-Stassen et al., 1999; van Pelt & de Rooij, 1991; van Pelt et al., 1995; and references therein).

The presence of seminiferous tubule segments in which spermatogenesis is not restored several months after vitA replacement in vitA-deficient animals raises the question as to whether retinoids are also required for A_{undiff} spermatogonia self-renewal or survival, in addition to be instrumental to the A_{al}−A1 transition. A recent work using pulse labeling of spermatogonia subtypes in a vitA-deficient mouse model demonstrates that the GFRA1-positive spermatogonia, which comprise the actual stem cell pool, do not depend on ATRA signaling for self-renewal. ATRA is neither required for their differentiation into NGN3-positive A_{undiff} spermatogonia. However, NGN3-positive spermatogonia, a poised cell population that differentiates further into KIT-positive spermatogonia in homeostatic spermatogenesis, die in the absence of vitA (Ikami et al., 2015). Altogether, these data indicate that ATRA is indispensable in vivo to trigger the $A_{al}-A_1$ transition and is independently required for the survival of some A_{undiff}

spermatogonia. In contrast, stem spermatogonia can survive and divide without ATRA. Whether ATRA is also required in differentiating spermatogonia has not been investigated so far.

3.2 The Sources of ATRA Destined for Spermatogenesis Are Intrinsic to the SE

A major source of ATRA in the SE was identified using the *Aldh1a1-3*$^{Ser-/-}$ mutant mouse line as a model. In steady-state conditions, the most advanced GC ever present in *Aldh1a1-3*$^{Ser-/-}$ mutants are ZBTB16-positive, but STRA8- and KIT-negative, A_{undiff} spermatogonia (Raverdeau et al., 2012; Fig.4). Interestingly, almost all these cells strongly express RARG, indicating that they have the capacity to respond to ATRA. However, spermatogonia cannot cell-autonomously trigger the $A_{al}-A_1$ transition because they do not possess the enzymatic machinery to make ATRA (Evans et al., 2014). Within 24 h following an injection of ATRA, a majority of spermatogonia in *Aldh1a1-3*$^{Ser-/-}$ seminiferous tubules switch to an STRA8- and KIT-positive phenotype. It is thus concluded that the only source of ATRA in the SE at the onset of the prepubertal period is RALDH activity in Sertoli cells and that it acts in a paracrine manner to instruct the initial $A_{al}-A_1$ transition.

Quite strikingly, spermatogenesis continues for months in *Aldh1a1-3*$^{Ser-/-}$ mutants after rescue by a single injection of ATRA. This situation cannot be accounted for by a persistent effect of exogenous ATRA because its half-life is less than 1 h in the mouse and direct measurements of ATRA levels in testes indicate that its residual amount 4 days following its injection is null (Hogarth, Arnold, et al., 2015). Thus, once the first wave of spermatogenesis has progressed beyond a certain point, RALDH activity in Sertoli cells is no longer required for spermatogonia differentiation (Raverdeau et al., 2012). In this context, ATRA is necessarily produced either outside or within the SE but not in Sertoli cells. The interstitial spaces between seminiferous tubules contain two potential sources of ATRA for the SE: Leydig cells, which strongly and steadily express RALDH1, and capillaries that convey ATRA derived from the blood (Kane, Folias, Wang, & Napoli, 2008; Vernet, Dennefeld, Rochette-Egly, et al., 2006). ATRA from the interstitial tissue is however unable to trigger spermatogenesis in *Aldh1a1-3*$^{Ser-/-}$ mutants. This is likely due to the activity of peritubular myoid cells that surround the seminiferous tubules. Actually from birth to adulthood, these cells strongly express the CYP26A1 and CYP26B1 enzymes, whereas CYP26 are undetectable by ISH and IHC in the other cell populations of the postnatal

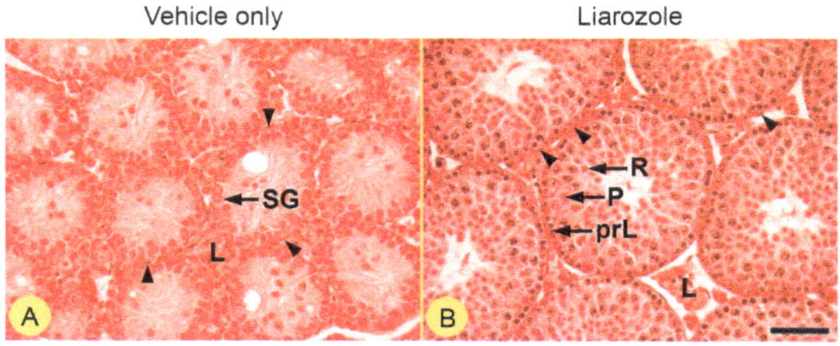

Fig. 6 Pharmacological inhibition of CYP26 enzymes triggers spermatogenesis in a seminiferous epithelium that lacks all ATRA synthetic activities. *Aldh1a1-3*$^{Ser-/-}$ males received intraperitoneal injections of vehicle (a mixture of sunflower oil and ethanol) or liarozole (40 mg/kg) for 3 consecutive days at PNW3 and were sacrificed at PNW6. Testis sections were stained with hematoxylin eosin. (A) After administration of the vehicle only, spermatogenesis remains blocked at the spermatogonia stage. (B) After treatment by liarozole, spermatogenesis progresses synchronously throughout the testis, to round spermatids. Abbreviations: *L*, Leydig cells; *prL* and *P*, preleptotene and pachytene spermatocytes, respectively; *R*, round spermatids; *SG*, spermatogonia. *Arrowheads* point to nuclei of peritubular myoid cells. Bar in (B): 80 μm (A and B).

testis (Vernet, Dennefeld, Rochette-Egly, et al., 2006; Wu, Wang, Guo, & Xu, 2008). Interestingly, administration of the CYP26 inhibitor liarozole to *Aldh1a1-3*$^{Ser-/-}$ mutants can restore a complete spermatogenesis (Fig. 6). This experiment indicates that the catabolic barrier formed by the peritubular myoid cells is efficient in *Aldh1a1-3*$^{Ser-/-}$ mutants and probably also functions in WT mice to degrade any ATRA present in the immediate environment of the seminiferous tubules, thus preventing inopportune occurrences of A_{al}−A_1 transitions. Since ATRA present around the tubules is unable to reach spermatogonia in *Aldh1a1-3*$^{Ser-/-}$ mutants, the activity required to sustain their differentiation is necessarily supported by their progeny: RALDH2 in spermatocytes and spermatids is clearly the only possible source of ATRA for spermatogenesis in these mutants (Sugimoto, Nabeshima, & Yoshida, 2012; Vernet, Dennefeld, Rochette-Egly, et al., 2006), where it acts in a paracrine fashion to induce spermatogonia differentiation from the second round of spermatogonia differentiation onward.

Ablation of *Rdh10* in the SE also causes a block in spermatogenesis at the A_{al}−A_1 transition, which is rescued when RAL or ATRA is administered exogenously. Contrary to *Aldh1a1-3*$^{Ser-/-}$ mice, in which this block is definitive in the absence of retinoid treatment, spermatogenesis is spontaneously restored to normal in *Rdh10*-deficient adults, suggesting that the first step in the synthesis of ATRA (transformation of vitA into RAL) is performed by an enzyme other than RDH10 in the adult testis (Tong, Yang, Davis, & Griswold, 2013). Interestingly, results similar to what is observed in *Aldh1a1-3*$^{Ser-/-}$ and *Rdh10*-deficient mutants are also seen following pharmacological inhibition of RALDH: in prepubertal WT mice treated with WIN-18,446, the only GC type is A_{undiff} spermatogonia (Hogarth et al., 2013).

4. ATRA SIGNALING IS INSTRUMENTAL TO MEIOSIS IN SPERMATOCYTES

It is commonly believed that since expression of *Stra8* in GC is indispensable for their entry into meiosis, then expression of *Stra8* also indicates their irreversible commitment toward meiosis (references in Mark et al., 2015). This may eventually apply to the situation in ovary where *Stra8* expression by fetal gonocytes is temporally tightly correlated with the initiation of meiosis, as it precedes by only a few hours the expression of markers of meiotic prophase I. In contrast, in the developing and adult testis *Stra8* expression in spermatogonia precedes their meiotic entry by almost a week (Drumond, Meistrich, & Chiarini-Garcia, 2011; Kluin et al., 1982; Mark et al., 2015; Zhou et al., 2008). Additionally, synthesis of *Stra8* mRNA and protein is biphasic in spermatogonia: both are detected in differentiating type A spermatogonia, but not in intermediate nor in type B spermatogonia, the latter representing the immediate precursors of preleptotene spermatocytes. Spermatocytes in S (preleptotene) phase reinitiate *Stra8* expression before entering meiosis (Mark et al., 2015; Vernet, Dennefeld, Guillou, et al., 2006; Fig. 5).

In the male mouse, meiotic initiation occurs for the first time during prepuberty at about PND7 and then meiosis recurs repeatedly and continuously throughout adult life (Drumond et al., 2011; Kluin et al., 1982). The involvement of ATRA in this process was investigated using *Aldh1a1-3*$^{Ser-/-}$ mutants in which spermatogenesis was rescued by an injection of ATRA to resume the $A_{al}-A_1$ transition (Raverdeau et al., 2012). Under these conditions, preleptotene spermatocytes expressing *Stra8* and *Rec8* (encoding a meiosis specific recombinase) appear, synchronously, 8 days post-ATRA injection in all seminiferous tubules. Treatment with the pan-RAR inverse agonist BMS-204493 prior to and during meiotic initiation in rescued *Aldh1a1-3*$^{Ser-/-}$ mutants abrogates *Stra8* expression and induces cytological abnormalities, which are identical to those displayed by *Stra8*-null spermatocytes. Actually, in the absence of STRA8 expression, leptotene spermatocytes undergo premature chromosome condensation yielding hypercondensed chromatin figures consisting in 40 univalent chromosomes, which are thus similar to mitotic metaphases (Mark et al., 2008). The potential source of ATRA in *Aldh1a1-3*$^{Ser-/-}$ mutants at the time of meiotic initiation is represented by spermatocytes, which express RALDH2 (Raverdeau et al., 2012; Vernet, Dennefeld, Rochette-Egly, et al., 2006). Keeping in mind the possible artifactual effects of BMS-204493 mentioned earlier, these data may indicate a role for endogenous ATRA in the stable commitment to the meiotic cell cycle in the testis. A recent report further suggests that ATRA may also be required for meiotic recombination (Kent et al., 2016).

5. ATRA METABOLISM WITHIN THE SE CONTROLS THE TIMING AND SPATIAL PATTERNING OF SPERMATOGONIA DIFFERENTIATION

At a given point along a murine seminiferous tubule, the $A_{al}-A_1$ transition proceeds in a periodic manner, every 8.6 days, while its descendants are generated and displaced at a constant pace toward the lumen of the seminiferous tubules. These processes result in the formation of stratified, recurring, cellular associations of fixed composition between precise subtypes of spermatogonia, spermatocytes and spermatids, called stages of the SE cycle, each epithelial stage spanning a short segment along a tubule. Twelve stages (I–XII) are recognizable in the mouse SE, with stage I corresponding to the first step of spermatid maturation (reviewed in de Rooij, 1998; Russell, Ettlin, & Clegg, 1990; Fig. 7). The $A_{al}-A_1$ transition is the point of entry into the SE cycle, and four turns of this cycle are required for a GC to progress from a type A_1 spermatogonium to a spermatozoon that is ready to be released in the tubule lumen. The SE cycle, as well as the repetitive patterning of GC associations along the seminiferous tubules, called wave of the SE, represents the keys for the physiologically asynchronous GC differentiation that allows a constant production of mature spermatids by the testis and a continuous release of spermatozoa in the epididymis. It is therefore paramount to the successful maintenance of fertility throughout the reproductive lifetime of males.

5.1 The SE Cycle and Wave Both Rely on Retinoid Signaling

The idea that retinoids could play a role in controlling the SE cycle came from observations made on vitA-deficient rats in which spermatogenesis

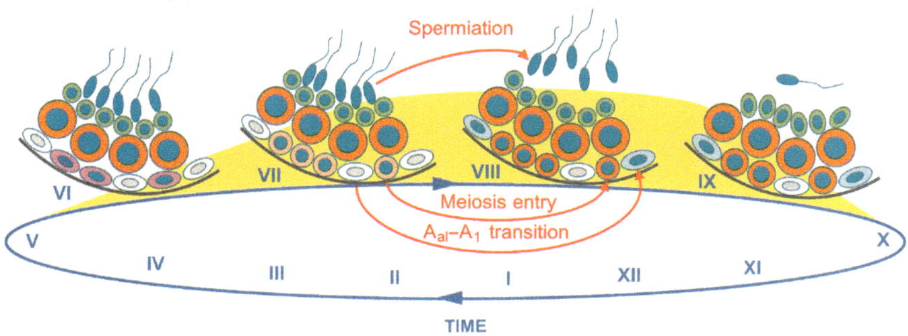

Fig. 7 Diagram depicting the organization of the murine SE and the cellular associations (stages of the SE) present at the time of the ATRA pulse (in *yellow*). GC are organized in a stratified fashion from the periphery to the lumen of the seminiferous epithelium, forming three to four distinct layers. The most peripheral layer, in contact with the basement membrane, contains spermatogonia (*white*, A_{al}; *blue*, A_1; *pink*, B) and preleptotene spermatocytes (*light orange*), while more advanced spermatocytes (*dark orange*) are progressively displaced in the intermediate layer and postmeiotic cells (*green, round*, and *blue, elongating, spermatids*) reside in the layers closest to the lumen of the tubules. Roman numerals designate the 12 stages of the SE cycle. Cellular processes promoted by the rise in ATRA concentrations in the SE are indicated by the *red arrows*. For further details see the main text.

was restored upon systemic administration of vitA and/or ATRA. Actually in retinoid-rescued rat testes, the full representation of the normal 14 stages of the SE is no longer seen. Instead, they contain only three to four stages, which are present at abnormally high frequencies, yielding an apparent "truncation" of the wave. Accordingly, the spermatogenesis of these animals is described as "synchronous," referring to the near-identical timing of GC progression in the entire testis (Morales & Griswold, 1987; Van Beek & Meistrich, 1990; van Pelt & de Rooij, 1990). More recently, it was shown that exposure of vitA-sufficient prepubertal mouse testis to exogenous ATRA drives the entire A_{undiff} population to enter the differentiation pathway in the whole testis, resulting also in synchronous spermatogenesis at adulthood (Davis, Snyder, Hogarth, Small, & Griswold, 2013; Snyder, Davis, Zhou, Evanoff, & Griswold, 2011).

VitA is also indispensable for the maintenance of the epithelial stages. This role was inferred from the study of *Rbp4*$^{-/-}$ mutant mice, which are unable to mobilize their ROL liver stores because they lack the blood carrier retinol-binding protein (Ghyselinck et al., 2006). In these mutants, the timeline of vitA deficiency-induced defects can be drawn, which allows investigation of the target cells and physiological functions of retinoids in the SE. The seminiferous tubules of vitA-deficient *Rbp4*$^{-/-}$ mice display normal segments at any of the 12 stages of the cycle that coexist with segments lacking one or two GC layers (e.g., entire spermatocyte or spermatid layers; Fig. 7). Importantly, these abnormal segments can all be classified as variants of the stages of the SE cycle, and aberrant cellular associations that would not fit in 1 of the canonical 12 stages are not found. Therefore, the tight coordination between spermatogonia differentiation and steps of meiosis and spermiogenesis is maintained in conditions of decreased vitA availability. These observations lead to the conclusion that a major mechanism accounting for vitA deficiency-induced failure of the SE cycle is the normal maturation of spermatocytes and spermatids in a situation where spermatogonia differentiation is temporarily arrested (Ghyselinck et al., 2006).

5.2 Endogenous ATRA Levels in the SE Are Tightly Regulated

Several observations support the notion that endogenous ATRA regulates the SE cycle and wave. First, expression of the ATRA-target genes *Stra8* in spermatocytes, and *Stra6* in Sertoli cells peak at stages VII–VIII of the SE cycle (Vernet, Dennefeld, Guillou, et al., 2006). Second, ATRA is crucial for cellular processes, which occur at stages VII–VIII of the cycle including A_{al}–A_1 transition, entry of spermatocytes into meiosis, and release of mature spermatids (spermiation) into the lumen of the seminiferous tubules (Raverdeau et al., 2012; Fig. 7). Third, measurement of ATRA levels in synchronized mouse testes supports the view that an ATRA pulse occurs at stages VII–IX of the SE cycle (Hogarth, Arnold, et al., 2015; Fig. 7). Altogether, these observations suggest that ATRA is available to the cells only along distinct segments of the seminiferous tubules.

The bioavailability of ATRA is controlled by regulation of the balance between its production and degradation. No variations in the levels of CYP26 enzymes exist across the SE cycle (Hogarth, Evans, et al., 2015; Vernet, Dennefeld, Rochette-Egly, et al., 2006). In contrast, several proteins involved in retinol storage and production of RAL represent potential candidates for contributing to how ATRA levels are regulated in a periodic manner along seminiferous tubules. The distribution pattern of *Lrat* transcripts in spermatids supports the possibility that retinoid storage is promoted during stages I–VI of the SE cycle (Vernet, Dennefeld, Rochette-Egly, et al., 2006). *Rdh10* and dehydrogenase/reductase (SDR family) member 4 (*Dhrs4*) display SE stage-specific expressions, suggesting that conversion of vitA to RAL may occur only at specific sites along seminiferous tubules. However, the fact that 98% of the mutant mice lacking *Rdh10* display a spermatogenesis with a normal frequency of the epithelial stages indicates that RDH10 activity is dispensable to regulate the wave of the SE cycle (Tong et al., 2013).

5.3 RALDH Activity Regulates the Spermatogenic Wave

The expression patterns of *Aldh1a1*–*Aldha3* genes have been extensively investigated: RALDH1 and RALDH2 are the main ATRA-synthesizing enzymes in Sertoli cells and GC, respectively, while RALDH3 is not detected in the SE (Bowles et al., 2009; Kent et al., 2016; Sugimoto et al., 2012; Vernet, Dennefeld, Rochette-Egly, et al., 2006). In Sertoli cells of the adult testis, *Aldh1a1* transcripts are expressed in an SE cycle-dependent manner with the highest levels being found at stages II–VII (Fig. 8). In GC, *Aldh1a2* transcripts are expressed in a periodic manner with a peak in late pachytene and diplotene spermatocytes, at stages VIII–XII (Vernet, Dennefeld, Rochette-Egly, et al., 2006; Fig. 8). These periodic expressions likely reflect a functional role of RALDH. Interestingly, histological analyses show that the spatial distribution of the different SE stages is markedly altered in the testes of *Aldh1a1*$^{-/-}$ males (Fig. 8). Cross-sections of the SE at the same stage are often assembled in defined territories that, according to anatomy (Nakata et al., 2015), correspond to a given seminiferous tubule. This strongly suggests that the segment length occupied by a given stage is increased in *Aldh1a1*$^{-/-}$ mice. In contrast, the topography of the SE stages does not display any defined profile on histological sections from WT testes (Fig. 8). This points to a hitherto ignored function of RALDH1 in regulating the wave of the SE cycle.

RALDH2 in spermatocytes and spermatids is sufficient to maintain an SE cycle in the complete absence of RALDH in Sertoli cells (Raverdeau et al., 2012). This observation suggests a role of ATRA synthesized by GC in regulating the SE cycle or wave. We have proposed that, in adult males, late pachytene spermatocytes present at stage VIII may control the tight coordination between the differentiation of A_{al} spermatogonia and entry into meiosis through cyclically modulating the ATRA environment of the cells (Raverdeau et al., 2012). To further establish the specific role of RALDH2 in the testicular ATRA biosynthesis, a GC-specific ablation of *Aldh1a2* is needed to circumvent the problem of embryonic lethality of the *Aldh1a2* knockout mice.

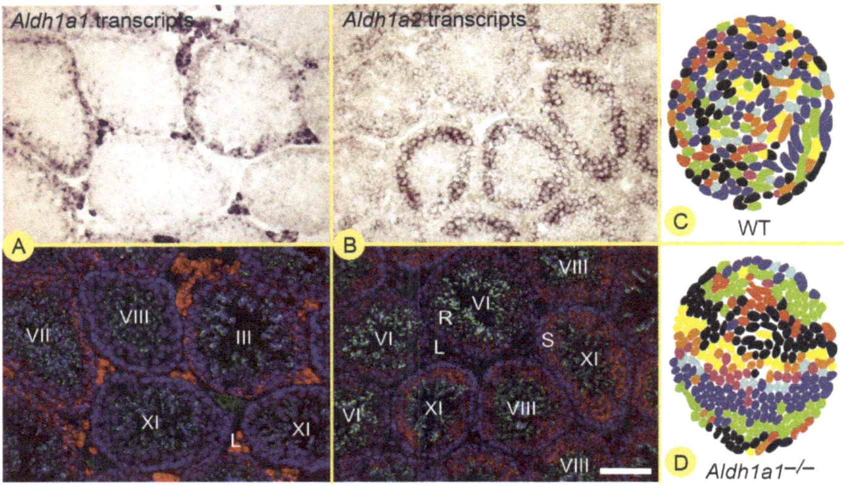

Fig. 8 Roles for ATRA-synthesizing enzymes in regulating the SE cycle and wave. (A) ISH using an antisense probe to *Aldh1a1*: higher mRNA levels (*purple signals*) are detected in Sertoli cells at stages II–VII of the SE cycle, as compared to stages VIII–XII and I. (B) ISH using an antisense probe to *Aldh1a2*: high mRNA levels are detected in spermatocytes at the late pachytene and diplotene stages (present at epithelial stages VIII–XII), as compared to spermatocytes present at stages I–VI; a weak ISH signal is also detected in *round spermatids* at stages II–VI. In the *lower panels* of (A) and (B), the ISH signals were converted to a *red color*, the histological sections were counterstained with 4′,6-diamidino-2-phenylindole (*blue signal*) to label nuclei, and with Alexa Fluor 488-conjugated peanut agglutinin (*green signal*) to label acrosomal systems. Epithelial stages, designated by roman numerals, were defined according to Russell et al. (1990), taking into consideration the shape of acrosomes (*green signal*). Abbreviations: *L*, Leydig cells; *R*, round spermatids; *S*, spermatocytes. (C, D) Transverse histological sections though testes of (C) WT and (D) *Aldh1a1*$^{-/-}$ mutants (see the main text for further details). Each *colored spot* represents a cross-section of a seminiferous tubule. Stages of the SE or group of stages are indicated in different colors: *black*, stages I–III; *red*, stage IV; *green*, stages V and VI; *dark blue*, stages VII and VIII; *light blue*, stage IX; *pink*, stage X; *orange*, stage XI; *yellow*, stage XII.

6. MALE GC ARE BOTH DIRECT AND REMOTE TARGETS OF ATRA ACTION: LESSONS FROM MOUSE MUTANTS LACKING RETINOID RECEPTORS

6.1 RAR in Fetal GC Differentiation

All three RAR isotypes are reported to be expressed in fetal testes and ovaries (Bowles et al., 2006). In mice lacking one RAR isotype, testis differentiation proceeds normally up to the start of puberty (*Rara*$^{-/-}$ and

Rarg$^{-/-}$ mutants) or even beyond its completion (*Rarb*$^{-/-}$ mutants) in males, while females display a normal ovarian morphology and are fertile at adulthood. Compound mutants lacking two RAR isotypes die in utero or at birth and display multiple defects in many tissues, the spectrum of which varies between individuals (Mark et al., 2006). Their gonadal sex can be unambiguously diagnosed based on the arrangement of the somatic cells into compact cords, in male fetuses, as opposed to epithelial clusters with ill-defined borders, in female fetuses. Histological criteria are however far from being sufficient to assess the normality of the developing gonads. This is especially true for young *Rar* compound knockout fetuses, where variations in the timing of GC differentiation, already patent in the WT situation, seem to be exacerbated. It is nevertheless interesting to note that while reanalyzing the serial histological sections obtained more than 20 years ago from the three ED18.5 *Rara*$^{-/-}$;*Rarg*$^{-/-}$ female fetuses ever produced (Mendelsohn et al., 1994), we discovered in preparing this review that one fetus completely lacked meiotic oocytes (Fig. 9). Demonstrating that this finding actually reflects GC autonomous effects of the absence of RAR awaits the generation of a genetic model in which signaling via all 3 RAR is abrogated specifically in GC, as a mean of eliminating any functional redundancy.

6.2 RARG Controls the Capacity of Spermatogonia to Respond to ATRA

A_{undiff} spermatogonia are present throughout the cycle of the SE, including stage VIII when the peak of ATRA concentration supposedly induces the

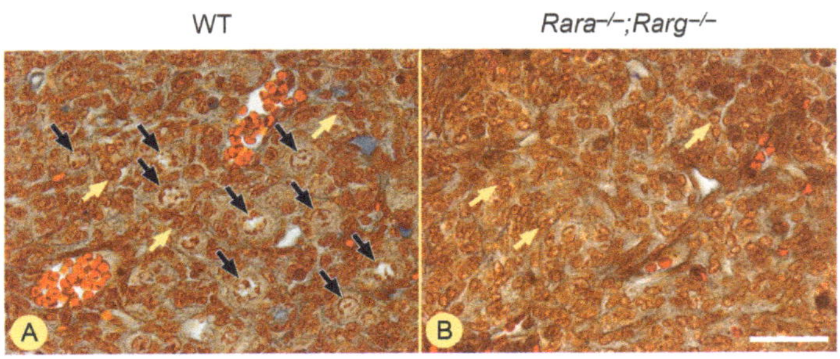

Fig. 9 Meiotic oocytes are lacking in the ovary of some mouse mutant fetuses lacking both RARA and RARG. Representative histological sections of ovaries from ED18.5 WT and *Rara*$^{-/-}$;*Rarg*$^{-/-}$ female mice. See the main text for further details. *Black* and *yellow arrows* point to oocytes and ovarian somatic cells, respectively. Trichrome stain. Bar in (B) = 30 μm (A and B).

A_{al}–A_1 transition (de Rooij & Russell, 2000; Hogarth, Arnold, et al., 2015; Fig. 7). Therefore, a mechanism must exist that ensures the preservation of a stem cell pool while producing, at the same time, differentiating cells upon exposure to ATRA. Recent observations suggest that this mechanism relies on the differential expression of *Rarg* between subpopulations of undifferentiated spermatogonia: as opposed to their A_{al} progeny, A_s spermatogonia do not express *Rarg* and therefore maintain an undifferentiated state and self-renewal capabilities even in the presence of high ATRA levels in the SE (Gely-Pernot et al., 2012; Ikami et al., 2015).

The role of RARG in spermatogonia differentiation was initially inferred from the phenotypic analysis of *Rarg* knockout mice (*Rarg*$^{-/-}$ mutants). The seminiferous tubules of sexually mature, young (2–3-month-old), *Rarg*$^{-/-}$ mutants display segments lacking one or two GC layers around their entire circumference, thereby forming variants of the stages of the SE cycle, identical to those observed in vitA-deficient *Rbp4*$^{-/-}$ mice (Gely-Pernot et al., 2012; Ghyselinck et al., 2006). Since *Rarg*$^{-/-}$ spermatocytes and spermatids do not display features reminiscent of increased apoptosis or delayed differentiation, the presence of these variants most likely reflects the fact that the A_{al}–A_1 transition is temporarily arrested. In some cases the A_{al}–A_1 transition appears to take place in due time, yielding normal stages, while in other instances no A_1 spermatogonia are formed and A_{al} spermatogonia have to wait one (or several) SE cycles to transition to A_1, yielding seminiferous tubule segments with missing generations of spermatocytes and spermatids. Because these variants also occur upon ablation of *Rarg* in spermatogonia only (*Rarg*$^{Spg-/-}$ mutants), it is concluded that RARG functions cell-autonomously in spermatogonia to control the A_{al}–A_1 transition (Gely-Pernot et al., 2012).

With respect to progression with aging of testis defects, *Rarg*$^{-/-}$ mutants are similar, but not fully identical, to vitA-deficient mice. Actually, seminiferous tubules of 1-year-old *Rarg*$^{-/-}$ mutants contain only Sertoli cells and spermatogonia, some of which still express markers of differentiation (e.g., KIT). Thus, in contrast to the vitA-deficient situation, inactivation of *Rarg* is not sufficient to arrest all spermatogonia at the A_{al}–A_1 transition. This discrepancy between the two situations is most readily accounted for by a functional redundancy between RARG and another RAR in spermatogonia, notably RARA that appears to be ubiquitously expressed at low levels not necessarily detected by IHC and ISH (Gely-Pernot et al., 2012). To get rid of any possible functional compensation, mice lacking all RAR in spermatogonia were generated (referred to as *Rara;b;g*$^{Spg-/-}$

mutants) (Gely-Pernot et al., 2015). In these mutants, ablation of *Rar* genes is efficient, as early as PND5, i.e., shortly after the onset of spermatogenesis during prepuberty. In young adults, the phenotypic outcome of ablating all RAR in spermatogonia is undistinguishable from that induced upon *Rarg* knockout, indicating that RARG is indeed the major functional RAR isotype in spermatogonia. In addition, the SE of 1-year-old *Rara;b;g*$^{Spg-/-}$ mutants contains only KIT-negative A_{undiff} spermatogonia (in addition to Sertoli cells), demonstrating that RARA or RARB in spermatogonia can compensate, at least in part, for the loss of RARG.

6.3 The Response of Spermatogonia to ATRA Relies on RXR/RAR Heterodimers

The ATRA signal is generally transduced by RXR/RAR heterodimers, notably during mouse embryonic development (Mark et al., 2006). However, it can also be relayed by RAR independently of RXR, as it is the case in Sertoli cells (Vernet, Dennefeld, Guillou, et al., 2006). To discriminate between these two possibilities in GC, mice lacking all RXR isotypes in spermatogonia, and hence also in their spermatocytes and spermatids descendants, were generated (hereafter called *Rxra;b;g*$^{Spg-/-}$ mutants). Age-matched *Rxra;b;g*$^{Spg-/-}$ and *Rara;b;g*$^{Spg-/-}$ mutants display similar, if not identical, phenotypes, indicating that RXR and RAR exert convergent functions in spermatogonia. As in addition, both anti-RXR and anti-RAR antibodies precipitate, with similar efficiencies, DNA sequences containing the RAR-binding sites of the *Stra8* promoter, it is concluded that RXR/RAR heterodimers are the functional units transducing the ATRA signal in spermatogonia (Gely-Pernot et al., 2015).

To identify the ATRA-controlled genes in spermatogonia, organ cultures of *Aldh1a1-3*$^{Ser-/-}$ testes were treated with an RARG-selective agonist (BMS-189961) to initiate spermatogenesis. Microarray expression profiling identified *Sall4*, which encodes a zinc-finger transcription factor, among the few transcripts that are differentially expressed upon activation of RARG. Interestingly, mice deficient for *Sall4* in spermatogonia display testis defects that resemble those observed in *Rxra;b;g*$^{Spg-/-}$ and *Rara;b;g*$^{Spg-/-}$ mutants, namely lack of differentiating spermatogonia (Hobbs et al., 2012). An RAR-binding region located in the first intron of *Sall4* was identified. This region contains an RARE that binds to RXRA in combination with RARG, but not to RXRA nor to RARG alone, as revealed by electrophoretic mobility shift assays (Gely-Pernot et al., 2015). It is thus proposed that ATRA-activated RXRA/RARG heterodimers increase the

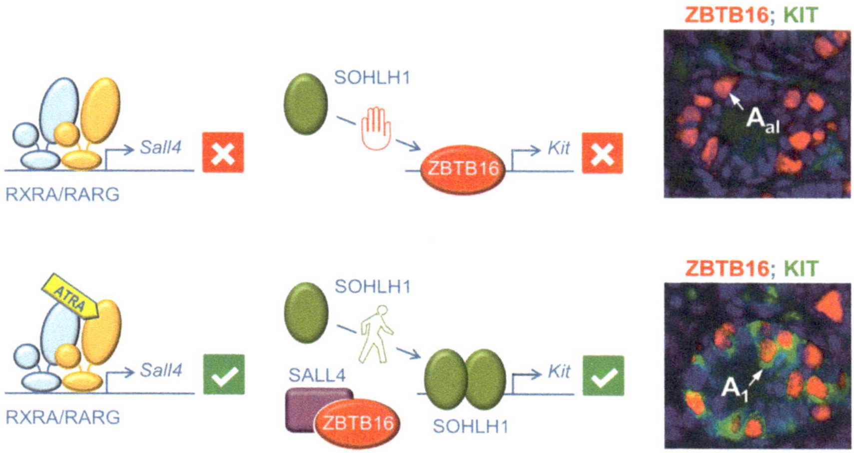

Fig. 10 A model for the regulation of *Kit* expression by ATRA during the transition from A_{al} to A_1 spermatogonia. *Upper panel*: In A_{al} spermatogonia, ATRA is not available to activate the RXRA/RARG heterodimer and thus *Sall4* is transcribed at low levels. Transcription of *Kit* is also low because ZBTB16 is bound to its promoter (Filipponi et al., 2007). *Lower panel*: At the A_{al}–A_1 transition, ATRA activates the RXRA/RARG heterodimer, which increases *Sall4* expression. SALL4 in large amount can then sequester ZBTB16, thereby clearing the *Kit* promoter and relieving the repression normally exerted by ZBTB16 (Hobbs et al., 2012). ATRA is also proposed to increase the level of SOHLH1, which can replace ZBTB16 on regulatory regions to increase *Kit* expression (Barrios et al., 2012).

level of SALL4A protein in spermatogonia, thereby allowing the increase in *Kit* expression (Fig. 10) that is a hallmark of the A_{al}–A_1 transition (Schrans-Stassen et al., 1999). In agreement with this proposal, the amount of SALL4 protein detected in spermatogonia is higher at the A_1 stage than at the A_{al} stage (Hobbs et al., 2012). In addition to SALL4, a recent study shows that ATRA controls spermatogonia differentiation through activating replication-dependent core histone genes (Chen et al., 2016).

6.4 RARA in Sertoli Cells Also Contribute to ATRA Functions in the SE

It is intriguing that spermatogonia differentiation is not stopped straight away in *Rara;b;g*$^{Spg-/-}$ and *Rxra;b;g*$^{Spg-/-}$ mutants, but instead proceeds during several weeks in young adults (Gely-Pernot et al., 2015). This observation suggests the existence of a subpopulation of A_{al} spermatogonia that are committed to become A_1 even though they are unable to cell-autonomously respond to ATRA. Given the pivotal role of ATRA in spermatogonia differentiation, instructing the $A_{al}-A_1$ transition in this subpopulation may nevertheless require an ATRA signal mediated by the Sertoli cell, which expresses RARA and is the only somatic cell type present in the SE. We propose that Sertoli cell responds to ATRA by synthesizing and secreting yet unknown paracrine signals that stimulate a subpopulation of A_{al} spermatogonia to transition to A_1 with no need for functional RXR/RAR heterodimers within these spermatogonia. This proposal is supported by the observation that the only GC ever present in the SE of mice lacking both RARA in Sertoli cells and RARG in spermatogonia (i.e., $Rara^{Ser-/-};Rarg^{-/-}$ mutants) are undifferentiated spermatogonia, as it is the case when ATRA synthesis abolished in the SE (Gely-Pernot et al., 2015).

It is also interesting to note that, in addition to differentiating spermatogonia, spermatocytes are also generated in testes of $Rara;b;g^{Spg-/-}$ and $Rxra;b;g^{Spg-/-}$ mutants. Moreover, meiosis proceeds at a normal pace, despite the absence of either RAR or RXR in all spermatocytes. Therefore, if ATRA is actually a mandatory MIS, then it operates in Sertoli cells rather than in GC (Gely-Pernot et al., 2015).

7. CONCLUDING REMARKS

An active ATRA signaling pathway is instrumental to the onset of oogenesis as well as to the initiation and maintenance of spermatogenesis in the mouse. Interestingly, it is also critically required in this species to disengage spermatozoa from the Sertoli cell cytoplasm during spermiation (Mark et al., 2015; and references therein; Fig. 7) and to promote the expulsion of the secondary oocyte from the mature follicle at ovulation (Kawai, Yanaka, Richards, & Shimada, 2016; Fig. 1). Gametogenesis is similar in mice and humans, notably with respect to the subtypes of GC, the sequence of cellular events involved in the differentiation cascade and the expression of regulatory molecules (Wu et al., 2009). In humans, a direct effect of ATRA on meiosis has been shown only in the ovary (Le Bouffant et al., 2010). Nonetheless, the demonstration that the bis(dichloroacetyl)diamine WIN-18,446 exerts its deleterious effects on human sperm counts and SE histology through its property to inhibit RALDH activity supports the view that ATRA is also required for human spermatogenesis (Hogarth et al., 2013).

In this context, it is worth noting that disturbances of the vitA metabolism caused by exposure to omnipresent environmental pollutants such as dioxin-like compounds or phthalates are strongly suspected to participate in the decline in sperm counts observed in many countries (Novák, Benísek, & Hilscherová, 2008; Sarath Josh et al., 2014). Moreover, possible side effects on human fertility of pharmacological doses of retinoids used in the treatment of acne have not been evaluated so far. It is therefore crucial to decipher the molecular mechanisms by which ATRA controls gametogenesis.

ACKNOWLEDGMENTS

Work from the author's laboratory was supported by grants from CNRS, INSERM, UNISTRA, and Agence Nationale pour la Recherche (ANR; 10-BLAN-1239, 13-BSV6-0003, and 13-BSV2-0017) as well as from EU (FP7-PEOPLE-IEF-2012-331687). These studies were also supported in part by the grant ANR-10-LABX-0030-INRT, a French State fund managed by the ANR under the frame programme Investissements d'Avenir labeled ANR-10-IDEX-0002-02. We thank Betty Féret and Muriel Klopfenstein for their invaluable technical support.

REFERENCES

Abu-Abed, S., MacLean, G., Fraulob, V., Chambon, P., Petkovich, M., & Dollé, P. (2002). Differential expression of the retinoic acid-metabolizing enzymes CYP26A1 and CYP26B1 during murine organogenesis. *Mechanisms of Development, 110*, 173–177.

Alnouti, Y., & Klaassen, C. D. (2008). Tissue distribution, ontogeny, and regulation of aldehyde dehydrogenase (Aldh) enzymes mRNA by prototypical microsomal enzyme inducers in mice. *Toxicological Sciences, 101*, 51–64.

Anderson, E. L., Baltus, A. E., Roepers-Gajadien, H. L., Hassold, T. J., de Rooij, D. G., van Pelt, A. M., et al. (2008). Stra8 and its inducer, retinoic acid, regulate meiotic initiation in both spermatogenesis and oogenesis in mice. *Proceedings of the National Academy of Sciences of the United States of America, 105*, 14976–14980.

Baltus, A. E., Menke, D. B., Hu, Y. C., Goodheart, M. L., Carpenter, A. E., de Rooij, D. G., et al. (2006). In germ cells of mouse embryonic ovaries, the decision to enter meiosis precedes premeiotic DNA replication. *Nature Genetics, 38*, 1430–1434.

Barrios, F., Filipponi, D., Campolo, F., Gori, M., Bramucci, F., Pellegrini, M., et al. (2012). SOHLH1 and SOHLH2 control Kit expression during postnatal male germ cell development. *Journal of Cell Science, 125*, 1455–1464.

Blume-Jensen, P., Jiang, G., Hyman, R., Lee, K. F., O'Gorman, S., & Hunter, T. (2000). Kit/stem cell factor receptor-induced activation of phosphatidylinositol 3′-kinase is essential for male fertility. *Nature Genetics, 24*, 157–162.

Bowles, J., Feng, C. W., Knight, D., Smith, C. A., Roeszler, K. N., Bagheri-Fam, S., et al. (2009). Male-specific expression of Aldh1a1 in mouse and chicken fetal testes: Implications for retinoid balance in gonad development. *Developmental Dynamics, 238*, 2073–2080.

Bowles, J., Feng, C. W., Miles, K., Ineson, J., Spiller, C., & Koopman, P. (2016). ALDH1A1 provides a source of meiosis-inducing retinoic acid in mouse fetal ovaries. *Nature Communications*, 7, 10845.

Bowles, J., Feng, C. W., Spiller, C., Davidson, T. L., Jackson, A., & Koopman, P. (2010). FGF9 suppresses meiosis and promotes male germ cell fate in mice. *Developmental Cell, 19*, 440–449.

Bowles, J., Knight, D., Smith, C., Wilhelm, D., Richman, J., Mamiya, S., et al. (2006). Retinoid signaling determines germ cell fate in mice. *Science*, *312*, 596–600.
Bowles, J., & Koopman, P. (2007). Retinoic acid, meiosis and germ cell fate in mammals. *Development*, *134*, 3401–3411.
Buaas, F. W., Kirsh, A. L., Sharma, M., McLean, D. J., Morris, J. L., Griswold, M. D., et al. (2004). Plzf is required in adult male germ cells for stem cell self-renewal. *Nature Genetics*, *36*, 647–652.
Chen, Y., Ma, L., Hogarth, C., Wei, G., Griswold, M. D., & Tong, M. H. (2016). Retinoid signaling controls spermatogonial differentiation by regulating expression of replication-dependent core histone genes. *Development*, *143*, 1502–1511.
Clermont, Y., & Perey, B. (1957). Quantitative study of the cell population of the seminiferous tubules in immature rats. *The American Journal of Anatomy*, *100*, 241–267.
Costoya, J. A., Hobbs, R. M., Barna, M., Cattoretti, G., Manova, K., Sukhwani, M., et al. (2004). Essential role of Plzf in maintenance of spermatogonial stem cells. *Nature Genetics*, *36*, 653–659.
Davis, J. C., Snyder, E. M., Hogarth, C. A., Small, C., & Griswold, M. D. (2013). Induction of spermatogenic synchrony by retinoic acid in neonatal mice. *Spermatogenesis*, *3*, e23180.
de Rooij, D. G. (1998). Stem cells in the testis. *International Journal of Experimental Pathology*, *79*, 67–80.
de Rooij, D. G., & Russell, L. D. (2000). All you wanted to know about spermatogonia but were afraid to ask. *Journal of Andrology*, *21*, 776–798.
Drumond, A. L., Meistrich, M. L., & Chiarini-Garcia, H. (2011). Spermatogonial morphology and kinetics during testis development in mice: A high-resolution light microscopy approach. *Reproduction*, *142*, 145–155.
Duester, G. (2008). Retinoic acid synthesis and signaling during early organogenesis. *Cell*, *134*, 921–931.
Evans, E., Hogarth, C., Mitchell, D., & Griswold, M. (2014). Riding the spermatogenic wave: Profiling gene expression within neonatal germ and Sertoli cells during a synchronized initial wave of spermatogenesis in mice. *Biology of Reproduction*, *90*, 1–12.
Fan, X., Molotkov, A., Manabe, S., Donmoyer, C. M., Deltour, L., Foglio, M. H., et al. (2003). Targeted disruption of Aldh1a1 (Raldh1) provides evidence for a complex mechanism of retinoic acid synthesis in the developing retina. *Molecular and Cellular Biology*, *23*, 4637–4648.
Filipponi, D., Hobbs, R. M., Ottolenghi, S., Rossi, P., Jannini, E. A., Pandolfi, P. P., et al. (2007). Repression of kit expression by Plzf in germ cells. *Molecular and Cellular Biology*, *27*, 6770–6781.
Gaemers, I. C., van Pelt, A. M., van der Saag, P. T., & de Rooij, D. G. (1996). All-trans-4-oxo-retinoic acid: A potent inducer of in vivo proliferation of growth-arrested A spermatogonia in the vitamin A-deficient mouse testis. *Endocrinology*, *137*, 479–485.
Gagnon, I., Duester, G., & Bhat, P. V. (2002). Kinetic analysis of mouse retinal dehydrogenase type-2 (RALDH2) for retinal substrates. *Biochimica et Biophysica Acta*, *1596*, 156–162.
Gely-Pernot, A., Raverdeau, M., Célébi, C., Dennefeld, C., Feret, B., Klopfenstein, M., et al. (2012). Spermatogonia differentiation requires retinoic acid receptor gamma. *Endocrinology*, *153*, 438–449.
Gely-Pernot, A., Raverdeau, M., Teletin, M., Vernet, N., Féret, B., Klopfenstein, M., et al. (2015). Retinoic acid receptors control spermatogonia cell-fate and induce expression of the SALL4A transcription factor. *PLoS Genetics*, *11*, e1005501.
Germain, P., Gaudon, C., Pogenberg, V., Sanglier, S., Van Dorsselaer, A., Royer, C. A., et al. (2009). Differential action on coregulatory interaction defines inverse retinoid agonists and neutral antagonists. *Chemistry & Biology*, *16*, 479–489.

Ghyselinck, N. B., Vernet, N., Dennefeld, C., Giese, N., Nau, H., Chambon, P., et al. (2006). Retinoids and spermatogenesis: Lessons from mutant mice lacking the plasma retinol binding protein. *Developmental Dynamics*, *235*, 1608–1622.

Griswold, M. D. (2016). Spermatogenesis: The commitment to meiosis. *Physiological Reviews*, *96*, 1–17.

Griswold, M. D., Hogarth, C. A., Bowles, J., & Koopman, P. (2012). Initiating meiosis: The case for retinoic acid. *Biology of Reproduction*, *86*, 35.

Hara, K., Nakagawa, T., Enomoto, H., Suzuki, M., Yamamoto, M., Simons, B. D., et al. (2014). Mouse spermatogenic stem cells continually interconvert between equipotent singly isolated and syncytial states. *Cell Stem Cell*, *14*, 658–672.

Hobbs, R. M., Fagoonee, S., Papa, A., Webster, K., Altruda, F., Nishinakamura, R., et al. (2012). Functional antagonism between Sall4 and Plzf defines germline progenitors. *Cell Stem Cell*, *10*, 284–298.

Hogarth, C. A., Arnold, S., Kent, T., Mitchell, D., Isoherranen, N., & Griswold, M. D. (2015). Processive pulses of retinoic acid propel asynchronous and continuous murine sperm production. *Biology of Reproduction*, *92*, 37.

Hogarth, C. A., Evanoff, R., Mitchell, D., Kent, T., Small, C., Amory, J. K., et al. (2013). Turning a spermatogenic wave into a tsunami: Synchronizing murine spermatogenesis using WIN-18,446. *Biology of Reproduction*, *88*, 40.

Hogarth, C. A., Evans, E., Onken, J., Kent, T., Mitchell, D., Petkovich, M., et al. (2015). CYP26 enzymes are necessary within the postnatal seminiferous epithelium for normal murine spermatogenesis. *Biology of Reproduction*, *93*, 19.

Horton, C., & Maden, M. (1995). Endogenous distribution of retinoids during normal development and teratogenesis in the mouse embryo. *Developmental Dynamics*, *202*, 312–323.

Ikami, K., Tokue, M., Sugimoto, R., Noda, C., Kobayashi, S., Hara, K., et al. (2015). Hierarchical differentiation competence in response to retinoic acid ensures stem cell maintenance during mouse spermatogenesis. *Development*, *142*, 1582–1592.

Jørgensen, A., Nielsen, J. E., Perlman, S., Lundvall, L., Mitchell, R. T., Juul, A., & Rajpert-De Meyts, E. (2015). Ex vivo culture of human fetal gonads: Manipulation of meiosis signalling by retinoic acid treatment disrupts testis development. *Human Reproduction*, *30*, 2351–2363.

Kane, M. A., Folias, A. E., Wang, C., & Napoli, J. L. (2008). Quantitative profiling of endogenous retinoic acid in vivo and in vitro by tandem mass spectrometry. *Analytical Chemistry*, *80*, 1702–1708.

Kawai, T., Yanaka, N., Richards, J. S., & Shimada, M. (2016). De novo-synthesized retinoic acid in ovarian antral follicles enhances FSH-mediated ovarian follicular cell differentiation and female fertility. *Endocrinology*, *157*, 2160–2172.

Kent, T., Arnold, S. L., Fasnacht, R., Rowsey, R., Mitchell, D., Hogarth, C. A., et al. (2016). ALDH enzyme expression is independent of the spermatogenic cycle, and their inhibition causes misregulation of murine spermatogenic processes. *Biology of Reproduction*, *94*, 12.

Kissel, H. I., Timokhina, M. P., Hardy, G., Rothschild, Y., Tajima, V., Soares, M., et al. (2000). Point mutation in kit receptor tyrosine kinase reveals essential roles for kit signaling in spermatogenesis and oogenesis without affecting other kit responses. *The EMBO Journal*, *19*, 1312–1326.

Kluin, P. M., Kramer, M. F., & de Rooij, D. G. (1982). Spermatogenesis in the immature mouse proceeds faster than in the adult. *International Journal of Andrology*, *5*, 282–294.

Kocer, A., Reichmann, J., Best, D., & Adams, I. R. (2009). Germ cell sex determination in mammals. *Molecular Human Reproduction*, *15*, 205–213.

Koshimizu, U., Watanabe, M., & Nakatsuji, N. (1995). Retinoic acid is a potent growth activator of mouse primordial germ cells in vitro. *Developmental Biology*, *168*, 683–685.

Koubova, J., Menke, D. B., Zhou, Q., Capel, B., Griswold, M. D., & Page, D. C. (2006). Retinoic acid regulates sex-specific timing of meiotic initiation in mice. *Proceedings of the National Academy of Sciences of the United States of America, 103*, 2474–2479.

Kumar, S., Chatzi, C., Brade, T., Cunningham, T. J., Zhao, X., & Duester, G. (2011). Sex-specific timing of meiotic initiation is regulated by Cyp26b1 independent of retinoic acid signalling. *Nature Communications, 2*, 151.

Le Bouffant, R., Guerquin, M. J., Duquenne, C., Frydman, N., Coffigny, H., Rouiller-Fabre, V., et al. (2010). Meiosis initiation in the human ovary requires intrinsic retinoic acid synthesis. *Human Reproduction, 25*, 2579–2590.

Li, H., MacLean, G., Cameron, D., Clagett-Dame, M., & Petkovich, M. (2009). Cyp26b1 expression in murine Sertoli cells is required to maintain male germ cells in an undifferentiated state during embryogenesis. *PLoS One, 4,* e7501.

MacLean, G., Li, H., Metzger, D., Chambon, P., & Petkovich, M. (2007). Apoptotic extinction of germ cells in testes of Cyp26b1 knockout mice. *Endocrinology, 148*, 4560–4567.

Mark, M., Ghyselinck, N. B., & Chambon, P. (2006). Function of retinoid nuclear receptors: Lessons from genetic and pharmacological dissections of the retinoic acid signaling pathway during mouse embryogenesis. *Annual Review of Pharmacology and Toxicology, 46*, 451–480.

Mark, M., Jacobs, H., Oulad-Abdelghani, M., Dennefeld, C., Féret, B., Vernet, N., et al. (2008). STRA8-deficient spermatocytes initiate, but fail to complete, meiosis and undergo premature chromosome condensation. *Journal of Cell Science, 121*, 3233–3242.

Mark, M., Teletin, M., Vernet, N., & Ghyselinck, N. B. (2015). Role of retinoic acid receptor (RAR) signaling in post-natal male germ cell differentiation. *Biochimica et Biophysica Acta, 1849*, 84–93.

Matt, N., Dupé, V., Garnier, J. M., Dennefeld, C., Chambon, P., Mark, M., et al. (2005). Retinoic acid-dependent eye morphogenesis is orchestrated by neural crest cells. *Development, 132*, 4789–4800.

McLaren, A., & Lawson, K. A. (2005). How is the mouse germ-cell lineage established? *Differentiation, 73*, 435–437.

Mendelsohn, C., Lohnes, D., Decimo, D., Lufkin, T., LeMeur, M., Chambon, P., & Mark, M. (1994). Function of the retinoic acid receptors (RARs) during development (II). Multiple abnormalities at various stages of organogenesis in RAR double mutants. *Development, 120*, 2749–2771.

Morales, C., & Griswold, M. D. (1987). Retinol-induced stage synchronization in seminiferous tubules of the rat. *Endocrinology, 121*, 432–434.

Morita, Y., & Tilly, J. L. (1999). Segregation of retinoic acid effects on fetal ovarian germ cell mitosis versus apoptosis by requirement for new macromolecular synthesis. *Endocrinology, 140*, 2696–2703.

Nagano, R., Tabata, S., Nakanishi, Y., Ohsako, S., Kurohmaru, M., & Hayashi, Y. (2000). Reproliferation and relocation of mouse male germ cells (gonocytes) during prespermatogenesis. *The Anatomical Record, 258*, 210–220.

Nakata, H., Wakayama, T., Sonomura, T., Honma, S., Hatta, T., & Iseki, S. (2015). Three-dimensional structure of seminiferous tubules in the adult mouse. *Journal of Anatomy, 227*, 686–694.

Niederreither, K., McCaffery, P., Dräger, U. C., Chambon, P., & Dollé, P. (1997). Restricted expression and retinoic acid-induced downregulation of the retinaldehyde dehydrogenase type 2 (RALDH-2) gene during mouse development. *Mechanisms of Development, 62*, 67–78.

Niederreither, K., Vermot, J., Fraulob, V., Chambon, P., & Dollé, P. (2002). Retinaldehyde dehydrogenase 2 (RALDH2)-independent patterns of retinoic acid synthesis in the mouse embryo. *Proceedings of the National Academy of Sciences of the United States of America, 99*, 16111–16116.

Novák, J., Benísek, M., & Hilscherová, K. (2008). Disruption of retinoid transport, metabolism and signaling by environmental pollutants. *Environment International*, *34*, 898–913.

Oulad-Abdelghani, M., Bouillet, P., Decimo, D., Gansmuller, A., Heyberger, S., Dollé, P., et al. (1996). Characterization of a premeiotic germ cell-specific cytoplasmic protein encoded by Stra8, a novel retinoic acid-responsive gene. *The Journal of Cell Biology*, *135*, 469–477.

Piprek, R. P., Pecio, A., Laskowska-Kaszub, K., Kloc, M., Kubiak, J. Z., & Szymura, J. M. (2013). Retinoic acid homeostasis regulates meiotic entry in developing anuran gonads and in Bidder's organ through Raldh2 and Cyp26b1 proteins. *Mechanisms of Development*, *130*, 613–627.

Raverdeau, M., Gely-Pernot, A., Féret, B., Dennefeld, C., Benoit, G., Davidson, I., et al. (2012). Retinoic acid induces Sertoli cell paracrine signals for spermatogonia differentiation but cell autonomously drives spermatocyte meiosis. *Proceedings of the National Academy of Sciences of the United States of America*, *109*, 16582–16587.

Rossant, J., Zirngibl, R., Cado, D., Shago, M., & Giguère, V. (1991). Expression of a retinoic acid response element-hsplacZ transgene defines specific domains of transcriptional activity during mouse embryogenesis. *Genes & Development*, *5*, 1333–1344.

Russell, L. D., Ettlin, A. P. S., & Clegg, R. A. (1990). *Histological and histopathological evaluation of the testis*. Clearwater, FL: Cache River Press.

Samarut, E., & Rochette-Egly, C. (2012). Nuclear retinoic acid receptors: Conductors of the retinoic acid symphony during development. *Molecular and Cellular Endocrinology*, *348*, 348–360.

Sarath Josh, M. K., Pradeep, S., Vijayalekshmi Amma, K. S., Balachandran, S., Abdul Jaleel, U. C., Doble, M., et al. (2014). Phthalates efficiently bind to human peroxisome proliferator activated receptor and retinoid X receptor α, β, γ subtypes: An in silico approach. *Journal of Applied Toxicology*, *34*, 754–765.

Schrans-Stassen, B. H., van de Kant, H. J., de Rooij, D. G., & van Pelt, A. M. (1999). Differential expression of kit in mouse undifferentiated and differentiating type A spermatogonia. *Endocrinology*, *140*, 5894–5900.

Shinohara, T., Orwig, K. E., Avarbock, M. R., & Brinster, R. L. (2000). Spermatogonial stem cell enrichment by multiparameter selection of mouse testis cells. *Proceedings of the National Academy of Sciences of the United States of America*, *97*, 8346–8351.

Snyder, E. M., Davis, J. C., Zhou, Q., Evanoff, R., & Griswold, M. D. (2011). Exposure to retinoic acid in the neonatal but not adult mouse results in synchronous spermatogenesis. *Biology of Reproduction*, *84*, 886–893.

Soone, S. B., Almstrup, K., Dalgaard, M., Juncker, A. S., Edsgard, D., Ruban, L., et al. (2009). Analysis of gene expression profiles of microdissected cell populations indicates that testicular carcinoma in situ is an arrested gonocyte. *Cancer Research*, *69*, 5241–5250.

Sugimoto, R., Nabeshima, Y., & Yoshida, S. (2012). Retinoic acid metabolism links the periodical differentiation of germ cells with the cycle of Sertoli cells in mouse seminiferous epithelium. *Mechanisms of Development*, *128*, 610–624.

Tedesco, M., Desimio, M. G., Klinger, F. G., De Felici, M., & Farini, D. (2013). Minimal concentrations of retinoic acid induce stimulation by retinoic acid 8 and promote entry into meiosis in isolated pregonadal and gonadal mouse primordial germ cells. *Biology of Reproduction*, *88*, 145.

Tong, M. H., Yang, Q. E., Davis, J. C., & Griswold, M. D. (2013). Retinol dehydrogenase 10 is indispensable for spermatogenesis in juvenile males. *Proceedings of the National Academy of Sciences of the United States of America*, *110*, 543–548.

Trautmann, E., Guerquin, M. J., Duquenne, C., Lahaye, J. B., Habert, R., & Livera, G. (2008). Retinoic acid prevents germ cell mitotic arrest in mouse fetal testes. *Cell Cycle*, *7*, 656–664.

Van Beek, M. E., & Meistrich, M. L. (1990). A method for quantifying synchrony in testes of rats treated with vitamin A deprivation and readministration. *Biology of Reproduction*, *42*, 424–431.

van Pelt, A. M., & de Rooij, D. G. (1990). The origin of the synchronization of the seminiferous epithelium in vitamin A-deficient rats after vitamin A replacement. *Biology of Reproduction*, *42*, 677–682.

van Pelt, A. M., & de Rooij, D. G. (1991). Retinoic acid is able to reinitiate spermatogenesis in vitamin A-deficient rats and high replicate doses support the full development of spermatogenic cells. *Endocrinology*, *128*, 697–704.

van Pelt, A. M., van Dissel-Emiliani, F. M., Gaemers, I. C., van der Burg, M. J., Tanke, H. J., & de Rooij, D. G. (1995). Characteristics of A spermatogonia and preleptotene spermatocytes in the vitamin A-deficient rat testis. *Biology of Reproduction*, *53*, 570–578.

Vernet, N., Dennefeld, C., Guillou, F., Chambon, P., Ghyselinck, N. B., & Mark, M. (2006). Prepubertal testis development relies on retinoic acid but not rexinoid receptors in Sertoli cells. *The EMBO Journal*, *25*, 5816–5825.

Vernet, N., Dennefeld, C., Rochette-Egly, C., Oulad-Abdelghani, M., Chambon, P., Ghyselinck, N. B., et al. (2006). Retinoic acid metabolism and signaling pathways in the adult and developing mouse testis. *Endocrinology*, *147*, 96–110.

Wallacides, A., Chesnel, A., Chardard, D., Flament, S., & Dumond, H. (2009). Evidence for a conserved role of retinoic acid in urodele amphibian meiosis onset. *Developmental Dynamics*, *238*, 1389–1398.

Western, P. S., Miles, D. C., van den Bergen, J. A., Burton, M., & Sinclair, A. H. (2008). Dynamic regulation of mitotic arrest in fetal male germ cells. *Stem Cells*, *26*, 339–347.

Wolbach, S. B., & Howe, P. R. (1925). Tissue changes following deprivation of fat-soluble A vitamin. *The Journal of Experimental Medicine*, *42*, 753–777.

Wu, X., Schmidt, J. A., Avarbock, M. R., Tobias, J. W., Carlson, C. A., Kolon, T. F., et al. (2009). Prepubertal human spermatogonia and mouse gonocytes share conserved gene expression of germline stem cell regulatory molecules. *Proceedings of the National Academy of Sciences of the United States of America*, *106*, 21672–21677.

Wu, J. W., Wang, R. Y., Guo, Q. S., & Xu, C. (2008). Expression of the retinoic acid-metabolizing enzymes RALDH2 and CYP26b1 during mouse postnatal testis development. *Asian Journal of Andrology*, *10*, 569–576.

Yoshida, S., Sukeno, M., Nakagawa, T., Ohbo, K., Nagamatsu, G., Suda, T., et al. (2006). The first round of mouse spermatogenesis is a distinctive program that lacks the self-renewing spermatogonia stage. *Development*, *133*, 1495–1505.

Zhou, Q., Li, Y., Nie, R., Friel, P., Mitchell, D., Evanoff, R. M., et al. (2008). Expression of stimulated by retinoic acid gene 8 (Stra8) and maturation of murine gonocytes and spermatogonia induced by retinoic acid in vitro. *Biology of Reproduction*, *78*, 537–545.

CHAPTER EIGHT

Retinoid-Related Orphan Receptor β and Transcriptional Control of Neuronal Differentiation

Hong Liu, Michihiko Aramaki, Yulong Fu, Douglas Forrest[1]
Laboratory of Endocrinology and Receptor Biology, NIDDK, National Institutes of Health, Bethesda, MD, United States
[1]Corresponding author: e-mail address: forrestd@niddk.nih.gov

Contents

Abstract

The ability to generate neuronal diversity is central to the function of the nervous system. Here we discuss the key neurodevelopmental roles of retinoid-related orphan receptor β (RORβ) encoded by the *Rorb* (*Nr1f2*) gene. Recent studies have reported loss of function of the human *RORB* gene in cases of familial epilepsy and intellectual disability. Principal sites of expression of the *Rorb* gene in model species include sensory organs, the spinal cord, and brain regions that process sensory and circadian

Current Topics in Developmental Biology, Volume 125 2017 Published by Elsevier Inc.
ISSN 0070-2153
http://dx.doi.org/10.1016/bs.ctdb.2016.11.009

information. Genetic analyses in mice have indicated functions in circadian behavior, vision, and, at the cellular level, the differentiation of specific neuronal cell types. Studies in the retina and sensory areas of the cerebral cortex suggest that this orphan nuclear receptor acts at decisive steps in transcriptional hierarchies that determine neuronal diversity.

1. INTRODUCTION

A prerequisite for the complex function of the mammalian nervous system is the ability to generate a diverse range of neuronal cell types (Cepko, Austin, Yang, Alexiades, & Ezzeddine, 1996; Greig, Woodworth, Galazo, Padmanabhan, & Macklis, 2013; Jessell, 2000). Elucidation of the events that produce this cellular diversity is central to our understanding of neurodevelopment and how dysfunction of these events can produce neurological disorders. The *Rorb* gene encoding retinoid-related orphan receptor β (RORβ) has been implicated with a specialized role in the nervous system since the observation almost 20 years ago by Becker-André and colleagues that the gene was selectively expressed in the eye, pineal gland, and regions of the brain (Schaeren-Wiemers, Andre, Kapfhammer, & Becker-Andre, 1997). This proposal was supported by early mutagenesis studies in mice that revealed phenotypes in vision, circadian behavior, and locomotion (Andre, Conquet, et al., 1998).

Renewed interest in this orphan receptor has arisen from the finding that at the cellular level it controls key steps in transcriptional pathways that direct the differentiation of neuronal cell types. Studies in mice indicate that the *Rorb* gene is a potent determinant of neuronal cell fates in the retina (Jia et al., 2009; Liu et al., 2013) and sensory areas of the cerebral cortex (Jabaudon, Shnider, Tischfield, Galazo, & Macklis, 2012; Oishi, Aramaki, & Nakajima, 2016). The expression of *Rorb* in immature neuronal cells suggests that it is suitably poised to influence the cell fate outcome in several differentiation pathways (Chow, Levine, & Reh, 1998; Fu et al., 2014; Nakagawa & O'Leary, 2003). This orphan receptor has also attracted attention following recent reports that the human *RORB* gene is disrupted in cases of epilepsy and intellectual disability (Boudry-Labis et al., 2013; Rudolf et al., 2016), suggesting that there are conserved functions for the gene in the human nervous system. Another point of interest for orphan receptors like RORβ that lack known natural ligands in vivo (Giguere, 1999; Mullican, Dispirito, & Lazar, 2013) is the scope they offer as targets for synthetic ligands that might find future uses in therapeutic applications

(Kojetin & Burris, 2014). The growing evidence of the role of RORβ in neuronal differentiation, and neurological disease may add impetus to the search for RORβ-selective agents.

Here, we review the role of RORβ in neuronal differentiation, based mainly on genetic studies in mice, and the emerging evidence of its involvement in human disease. We also discuss a transcriptional model by which RORβ contributes to neuronal differentiation.

2. THE *RORB* GENE AND RORβ PROTEIN ISOFORMS

The *Rorb* gene differentially expresses two isoforms, RORβ1 and RORβ2, that differ only in their short N-terminal domains (2 and 13 amino acid residues, respectively) (Andre, Gawlas, & Becker-Andre, 1998; Liu et al., 2013) (Fig. 1). These isoforms are identical in the DNA-binding domain and the C-terminal domain that is partly homologous to the ligand-binding domain of some other nuclear receptors including retinoid receptors (Carlberg, Hooft van Huijsduijnen, Staple, DeLamarter, & Becker-Andre, 1994). The N-termini of RORβ1 and RORβ2 are among the shortest in the superfamily of nuclear receptors and lack the longer N-terminal activation domain found in some other receptors (Evans, 1988; Green & Chambon, 1988). Both isoforms can activate similar reporter genes in transfected cells (Srinivas, Ng, Liu, Jia, & Forrest, 2006) and in retinal explants (Fu et al., 2014), suggesting that regardless of their divergent N-termini, they may control similar genes in tissues where they are coexpressed.

In this chapter, we use the terms RORβ1 and RORβ2 where the evidence concerns a specific isoform but otherwise use the generic term "RORβ" if a property is common to both isoforms or has not been assigned to a specific isoform.

RORβ1 is expressed relatively widely including in the brain, retina, cochlea, spinal cord, pineal gland, and pituitary gland in the rat and mouse (Liu et al., 2013; Schaeren-Wiemers et al., 1997). In contrast, RORβ2 is restricted, being detected primarily in the pineal gland (Andre, Gawlas, et al., 1998) and retinal photoreceptors (Fu et al., 2014). The differential expression of RORβ1 and RORβ2 is directed by individual promoters in the *Rorb* gene.

The identification of mutations and microdeletions that disrupt the human *RORB* gene in epilepsy and other neurological disorders (Fig. 1B and C; discussed in a later section) (Rudolf et al., 2016) supports a conserved

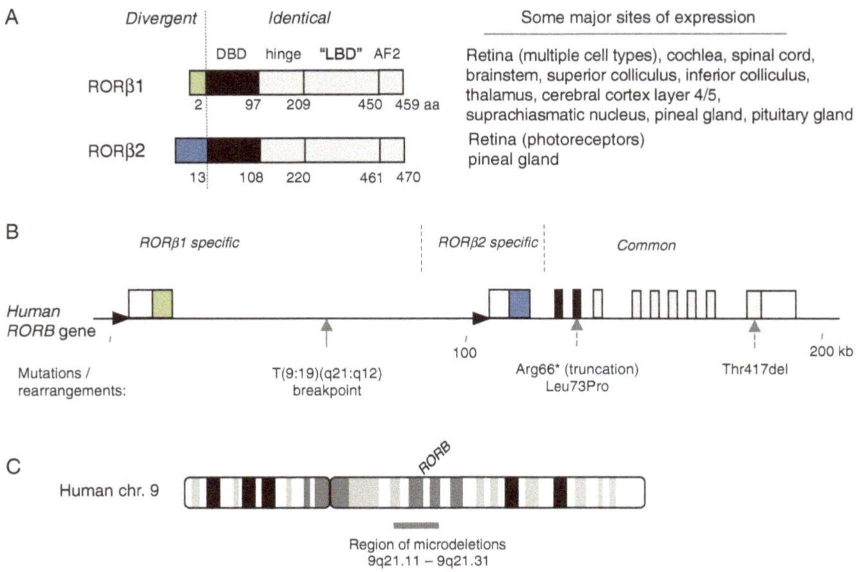

Fig. 1 RORβ protein isoforms and the *RORB* gene. (A) RORβ1 and RORβ2 isoforms are identical in the DNA-binding domain (DBD), ligand-binding domain ("LBD," although no ligand is known in vivo), and AF2 activation domain. RORβ1 and RORβ2 are differentially expressed in tissues in mice (*right*). RORβ1 is the most widely detected isoform. Lengths of domains are indicated below the proteins (in amino acid residues, aa). (B) The human *RORB* gene structure is representative of the mammalian gene, which is highly conserved. *Black triangles*, promoters for RORβ1- and RORβ2-specific exons; *white boxes*, untranslated sequence. Mutations (*dashed vertical arrows*) and a balanced chromosomal translocation (*solid vertical arrow*) found in human epilepsy and neurological disorders are indicated below the gene. Not drawn strictly to scale. (C) In some human patients, microdeletions on chromosome 9 between 9q21.11 and 9q21.31 encompass the *RORB* gene or part of the *RORB* gene.

neurodevelopmental function for the gene in different species. The gene structure is largely conserved in mammals, birds, reptiles, amphibians, and fish. Insects also possess an ortholog of the retinoid-related orphan receptor genes, *Hr46* (*Dhr3* or *Nr1f4*). In *Drosophila*, *Hr46* is required for embryogenesis (Lam, Hall, Bender, & Thummel, 1999) and formation of the peripheral nervous system (Carney, Wade, Sapra, Goldstein, & Bender, 1997). *Hr46* promotes condensation of the ventral nerve cord (Ruaud, Lam, & Thummel, 2010), consistent with a neurodevelopmental role for this class of orphan receptor in diverse animal species.

In vertebrate species, RORβ1 is conserved in the N-terminus, DNA-binding domain, and C-terminal domain (Fig. 2A). RORβ2 has also been

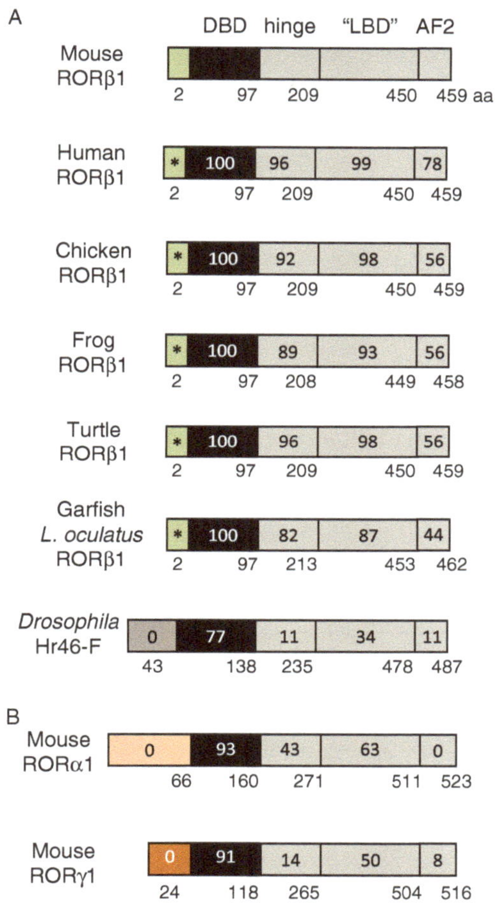

Fig. 2 Conservation of RORβ proteins in representative species. (A) Identity relative to mouse RORβ1 is denoted by *percentage numbers* within protein domains. The short N-terminus of RORβ1 (2 amino acids, Met-Arg) is 100% conserved in the vertebrate species shown (denoted *). Lengths of domains are indicated below proteins (amino acid residues, aa). Accession numbers (NCBI): mouse, *Mus musculus*, NP_001036819; human, *Homo sapiens*, NP_008845; chicken, *Gallus gallus*, NP_990424; frog, *Xenopus tropicalis*, XP_002940077; turtle, *Chelonia mydas*, XP_007065615; fish, *Lepisosteus oculatus*, XP_015192150; *Drosophila*, *D. melanogaster*, NP_788303. RORβ2 displays a similar level of identity across mammalian species and is therefore not shown. Representative RORβ2 accession numbers (sometimes called variant *x*1 or *x*2): mouse, NM_146095.4; human, ENST00000396204.2, KY037794; cow, XM_010807895; bat, *Myotis brandtii*, XM_014541277.1; monkey, *Mandrillus leucophaeus*, XM_011968250. (B) Homology of RORα1 and RORγ1 proteins relative to RORβ1 in the mouse. Accession numbers: RORα1, NM_013646; RORγ1, NM_011281.

identified in a growing number of species including mouse and human (Fig. 2 legend). The previously poor representation, or absence, of RORβ2 in databases for some species probably reflects its tissue-restricted expression rather than a species difference. We have cloned human RORβ2 and RORβ1 cDNAs from Weri retinoblastoma cells (accession numbers in Fig. 2 legend), which are thought to arise from a cone-like photoreceptor precursor cell (Xu et al., 2009).

RORβ is related to RORα and RORγ proteins encoded by the *Rora* (*Nr1f1*) and *Rorc* (*Nr1f3*) genes, respectively, in vertebrate species (Jetten, 2009) (Fig. 2B). For RORβ and RORα isoforms, the conservation, especially in the DNA-binding and C-terminal domains, suggests that they may serve overlapping functions in several regions of the nervous system where both are expressed. *Rora* mediates actions in the nervous system as well as in the immune and metabolic systems (Jetten, 2009; Takeda et al., 2014). *Rorc* mediates major actions in the immune system (Ivanov et al., 2006).

3. RESPONSE ELEMENTS FOR RORβ PROTEINS

Analysis of artificial elements (Andre, Conquet, et al., 1998; Gawlas & Stunnenberg, 2000) and several biologically important target genes (Fig. 3)

Fig. 3 Binding sites for RORβ proteins in natural target genes. Response elements in the indicated genes possess a consensus motif for nuclear receptor binding (AGGTCA) with an AT-rich extension on the 5′-side. The *Opn1sw* and *Brn2* genes each possess two elements. The consensus noted below was derived from the above binding sites using MEME Motif Discovery.

indicates that RORβ proteins can bind as monomers to DNA elements that contain a single motif for nuclear receptor binding (5′-AGGTCA-3′) with an AT-rich extension on the 5′-side. The few known target genes for RORβ include genes encoding lineage-determining transcription factors (*Nrl*, *Prdm1*, *Ptf1a*, *Brn2*) as well as photoreceptor differentiation markers (*Opn1sw*, short-wavelength-sensitive, S opsin). More extensive investigation of target genes and their response elements is warranted given the functional importance of RORβ in vivo.

The consensus binding site for RORα1 determined in vitro (Giguere et al., 1994) is similar to that for RORβ, supporting the proposal that they may coregulate some common genes in vivo. The expression profile of *Rora* partly overlaps with that of *Rorb*, for example, in cone photoreceptors (Fujieda, Bremner, Mears, & Sasaki, 2009; Srinivas et al., 2006), the thalamus, and cerebral cortex (Nakagawa & O'Leary, 2003). In cone photoreceptors, it has been proposed that RORβ initiates activation of the *Opn1sw* gene (Srinivas et al., 2006) and that at later stages, RORα1 augments this role (Fujieda et al., 2009).

4. MOUSE STRAINS WITH *RORB* MUTATIONS

The following discussion of the functions of the *Rorb* gene is largely based on mouse genetic models, as listed in Table 1. Published strains carry targeted mutations in the *Rorb* gene that inactivate all known products (Andre, Conquet, et al., 1998) or selectively inactivate RORβ1 (Liu et al., 2013) or RORβ2 (Fu et al., 2014). Another spontaneously arising strain, highstepper, carries a 300-kb duplication that includes the RORβ1-specific exon region of the gene (Fairfield et al., 2015). Some other unpublished alleles are listed at Mouse Genome Informatics (http://www.informatics.jax.org).

5. EXPRESSION PATTERNS OF THE *RORB* GENE IN THE NERVOUS SYSTEM

The *Rorb* gene offers the opportunity for investigation of sensory pathways at multiple levels based on its intriguing pattern of expression at ascending anatomical stages in the visual, auditory, and somatosensory systems, first observed in the rat (Schaeren-Wiemers et al., 1997). Thus, *Rorb* is expressed in the peripheral sensory organ (retina, cochlea) or site for integration of somatosensory input (dorsal horn of the spinal cord) as well as in brain

Table 1 Published Alleles of the *Rorb* Gene in Mice

Rorb Allele	Description	Main Phenotypes Reported	References
tm1Mba (*Rorb*−)	Targeted knockout all known isoforms (RORβ1, RORβ2); LacZ insertion in exon encoding second zinc finger of DBD	• Locomotor defect in hind limbs • Transiently impaired fertility in males • Abnormal circadian behavior • Loss of visual function • Severe loss of rods/ gain of cone-like cells	Andre, Conquet, et al. (1998) and Jia et al. (2009)
tm1Df (*Rorb* b1gfp)	Targeted knockout RORβ1; GFP reporter insertion in RORβ1-specific exon	• Locomotor defect in hind limbs • Transiently impaired fertility in males • Loss of horizontal and amacrine cells • Modest loss of rods/ gain of cones • Loss of photoreceptor function	Liu et al. (2013)
tm2Df (*Rorb* b2lacz)	Targeted knockout RORβ2; LacZ insertion in RORβ2-specific exon	• Modest loss of rods/ gain of cones	Fu et al. (2014)
hstp highstepper	Spontaneous allele; duplication of 300-kb region including RORβ1-specific exon	• Locomotor defect in hind limbs	Fairfield et al. (2015)

Other unpublished alleles (targeted and chemically induced) are listed at http://www.informatics.jax.org at The Jackson Laboratory.

regions that process sensory information. Ultimately, different forms of sensory information are relayed to the thalamus and then to sensory areas of the cerebral cortex. *Rorb* marks certain thalamic nuclei and neurons of layer IV in the cortex that receive input from thalamocortical projections (Nakagawa & O'Leary, 2003; Schaeren-Wiemers et al., 1997). Fig. 4 shows

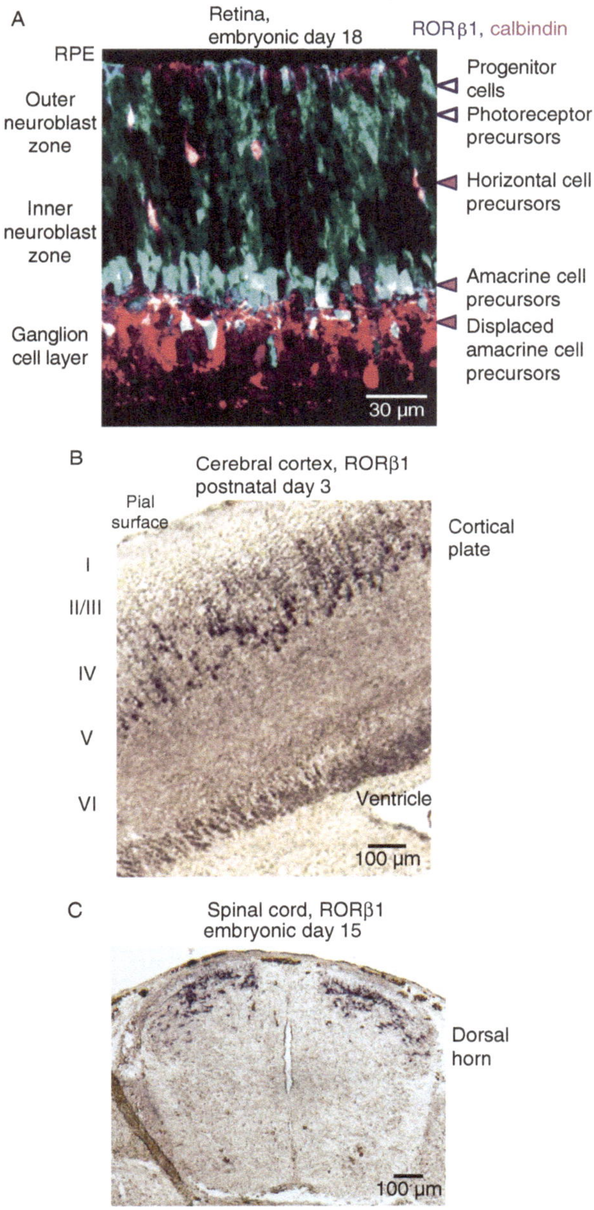

Fig. 4 Expression patterns of RORβ1 in the nervous system in mice. (A) In the immature retina, RORβ1 (*turquoise*) is expressed in a dynamically shifting pattern in undifferentiated progenitor cells and in precursors of photoreceptor (*open arrowheads*) and horizontal and amacrine interneuron lineages (*filled arrowheads*). Expression levels decrease at older, postnatal ages. Horizontal cells and a subpopulation of amacrine cells *(Continued)*

examples of RORβ1 expression patterns in the retina, cerebral cortex, and spinal cord in immature mice.

5.1 Sensory Systems

Briefly, in the visual pathway, *Rorb* is expressed in the retina, the dorsal lateral geniculate nucleus of the thalamus, which receives input from retinal ganglion cells, the superior colliculus, and the primary visual cortex. In the auditory pathway, *Rorb* is expressed in the cochlear duct, dorsal cochlear nucleus in the brainstem, inferior colliculus, medial geniculate nucleus in the thalamus, and the primary auditory cortex. In the somatosensory pathway, sites of expression include the second and third neuronal layers of the dorsal horn of the spinal cord, the ventroposterior medial, ventroposterior lateral and centromedial nuclei in the thalamus, and the primary somatosensory cortex (Nakagawa & O'Leary, 2003; Schaeren-Wiemers et al., 1997).

At the cellular level, *Rorb* has been investigated in detail in only a few of these regions, mainly in the retina and cortex. However, RORβ1 recently proved useful for defining subtypes of neurons in the superficial region of the superior colliculus, which integrates input from retinal ganglion cells and the cortex to influence eye and head movement (Byun et al., 2016). Another recent study took advantage of RORβ and other markers to define subtypes of sensory interneurons in the dorsal spinal cord in mice (Del Barrio et al., 2013).

5.2 Circadian System

The *Rorb* gene is also expressed at key sites for circadian activity, including the suprachiasmatic nucleus (SCN) in the hypothalamus and the pineal gland (Schaeren-Wiemers et al., 1997). The SCN acts as a central oscillator that influences rhythmic activity in many physiological systems in the body in accord with the day–night cycle (Zhao et al., 2014). The SCN receives input

Fig. 4—Cont'd are identified with calbindin staining (*purple* or *whitish double stain*). Many cells have not yet migrated to their final location in the retina. *RPE*, retinal pigmented epithelium. (B) RORβ1 (*dark stain*) expression in layer IV/V in the cortical plate; sagittal section at P3. Most of these neurons have migrated near to their final location but remain immature. Expression persists into adulthood. Cortical layers are numbered on the *left*. (C) RORβ1 (*dark stain*) marks a select population of neurons in the dorsal horn of the spinal cord; transverse section at E15. RORβ1 expression persists in layer II/III of the dorsal horn at more mature ages after weaning. All images obtained in the authors' lab using *Rorb* +/1gfp mice carrying a GFP reporter knocked-in at the RORβ1-specific exon (Liu et al., 2013).

from the retina through the retinohypothalamic tract and in turn influences other neuronal and neuroendocrine responses (Kalsbeek & Buijs, 2002). Expression of *Rorb* itself oscillates in the SCN with a peak in the light period (Agez, Laurent, Pevet, Masson-Pevet, & Gauer, 2007; Sumi et al., 2002) and in the pineal gland, with a peak during the dark period (Andre, Gawlas, et al., 1998; Baler, Coon, & Klein, 1996). Adrenergic agonists and cyclic AMP can stimulate *Rorb* expression in the rat pineal gland, suggesting that the *Rorb* gene may mediate some responses of the gland (Baler et al., 1996).

6. THE *RORB* GENE AND NEURONAL DIFFERENTIATION

The functions of *Rorb* in differentiation are discussed in the next sections with a focus on the cerebral cortex and retina, the two most studied systems to date. The cortex illustrates a cell-restricted role of *Rorb*, whereas the retina illustrates a more developmentally dynamic situation in which *Rorb* acts in multiple cell types (Fig. 4).

7. *RORB* AND DIFFERENTIATION IN THE CEREBRAL CORTEX

The formation of distinct neuronal cell types in organized layers is critical for the function of the cerebral cortex in perception, cognition, and other processes. The layered structure of the cortex is formed according to an inside-out process of neuronal migration and differentiation (Greig et al., 2013). In the mouse embryo after mid-gestation, progenitor cells in the ventricular zone and subventricular zone generate postmitotic cells that migrate toward the pia to form the rudimentary neuronal layer V of the cortex (Fig. 5A). Subsequent waves of migrating cells progressively form the more superficial layer IV and then the upper cortical layer II/III. The *Rorb* gene is a selective marker of layers IV and V in different species, including the rat, mouse, and ferret (Rowell, Mallik, Dugas-Ford, & Ragsdale, 2010; Schaeren-Wiemers et al., 1997; Tasic et al., 2016), which has prompted questions about its role in cortical differentiation or function.

Rorb marks many layer IV and some layer V neurons at immature stages when neuronal identities are being determined. Analyses in mice suggest an interplay between *Rorb* and *Brn2*, a marker of layer II/III neurons in determining neuronal identity in the cortical plate (Oishi, Aramaki, & Nakajima, 2016). Ectopic expression of RORβ1 in embryonic migrating cells resulted in loss of some markers of layer II/III neurons and gain of markers of layer

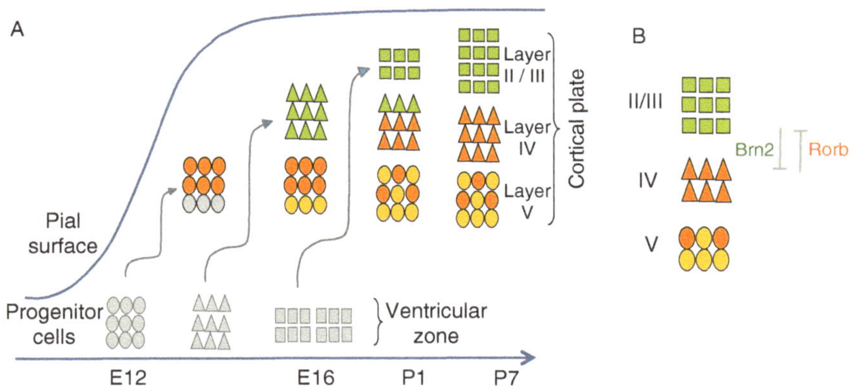

Fig. 5 *Rorb* and neuronal identity in the cerebral cortex. (A) Progenitor cells in the ventricular and subventricular zones generate postmitotic, immature neurons that migrate to the cortical plate in an inside-out fashion. Migrating cells first form layer V (*ovals*), then layer IV (*triangles*), and then layer II/III (*squares*). Layer IV is initially marked by expression of *Brn2* (*green*) and then progressively by *Rorb* (*orange*) as these neurons differentiate. Layer II/III is marked by *Brn2* (*green*). Some layer V neurons may be positive for *Rorb* (*orange*) and others negative (*yellow*). Layers are established over the developmental stages indicated in mice. (B) Proposed mutual repression by *Brn2* and *Rorb* genes to determine neuronal identity in layers II/III and IV.

IV/V neurons when analyzed in the postnatal period. Conversely, ectopic expression of BRN2 led to loss of some markers of layer IV/V neurons and a gain in markers of layer II/III neurons. The study also indicated that RORβ1 represses the endogenous *Brn2* gene and conversely, and that BRN2 represses the *Rorb* gene. The findings suggest that the identity of layer II/III and layer IV/V neurons and their type of projections are determined in part as a result of mutual repression between the *Rorb* and *Brn2* genes. The identification of binding sites for RORβ1 in the *Brn2* gene suggests a direct form of repression (Fig. 5B).

The *Rorb* gene also marks the barrel cortex, consisting of clusters of neurons in somatosensory areas of the cortex (Jabaudon et al., 2012; Nakagawa & O'Leary, 2003). The barrel cortex in rodents receives tactile input from the whiskers that are especially important in sensing environmental surroundings in the confined or dark habitats of some species. The barrels are arranged somatotopically in the cortex in a similar array as the whiskers on the face (Petersen, 2007). Ectopic expression of RORβ1 in progenitor cells in the embryonic cortex was reported to produce barrel-like structures and also to stimulate the attraction of thalamocortical projections to these structures (Jabaudon et al., 2012).

The function of RORβ1 in determining neuronal identity in the cortical layers is probably coordinated with other signals that contribute to cell fate decisions (Greig et al., 2013). These signals may include other transcription factors, or positional cues that coordinate the differentiation of a migrating cell with an appropriate location in the cortical layers, or the arrival of thalamocortical projections during development (Oishi, Nakagawa, et al., 2016).

The *Rorb* gene is expressed in layer IV/V along most of the rostrocaudal length of the cortex, with some gradients of expression evident during development (Nakagawa & O'Leary, 2003; Schaeren-Wiemers et al., 1997; Sumi et al., 2002). Thus, *Rorb* might potentially mediate distinct functions in somatosensory, auditory, visual, motor, or other areas of the cortex. It is also not excluded that *Rorb* promotes the function of mature cortical neurons because the gene continues to be expressed in this layer in adulthood.

8. *RORB* AND DIFFERENTIATION IN THE RETINA

In the retina, progenitor cells produce postmitotic progeny that differentiate as diverse cell types, including six major classes of neurons as well as Müller glial cells (Fig. 6A). During development, progenitor cells progressively change their competence for generating different cell lineages (Agathocleous & Harris, 2009; Brzezinski & Reh, 2015; Carter-Dawson & LaVail, 1979; Livesey & Cepko, 2001). Transcription factors play a major role in specifying these cell lineages and do so by acting in progenitors and postmitotic precursor cells to determine the differentiation fate as the cell acquires its final form and function (Swaroop, Kim, & Forrest, 2010; Xiang, 2013). RORβ1 is expressed in subpopulations of progenitor cells and in precursor cells for amacrine, horizontal, and photoreceptor lineages (Liu et al., 2013) and accordingly can influence decisive steps in several differentiation pathways.

A salient feature of some genes downstream of *Rorb* in the retina, including *Ptf1a*, *Prdm1*, and *Nrl*, is that they are critical for cell lineage determination. Indeed, some of these effector genes exert dual functions in directing a given cell fate while suppressing alternative cell fates (Fig. 6B). For example, the *Nrl* gene, which stimulates rod differentiation, induces rod genes but also suppresses cone genes (Mears et al., 2001). Thus, *Rorb* serves a particularly key role in the control of other control genes in a transcriptional hierarchy that underlies retinal cell differentiation.

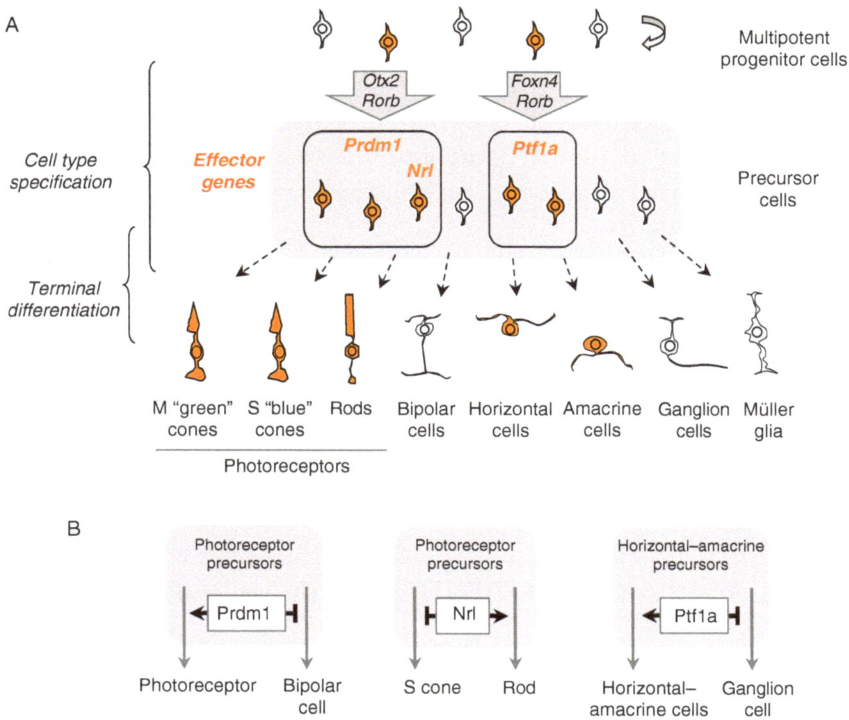

Fig. 6 *Rorb* and neuronal identity in the retina. (A) This scheme highlights the proposed role of *Rorb* in specific neuronal lineages. Proliferative progenitor cells generate postmitotic precursors for diverse cell lineages. *Rorb* pairs with other lineage-determining factors (*Otx2*, *Foxn4*) in subpopulations of progenitors and newly born precursors (*orange* cells) to induce effector genes that direct a given cell fate. In combination with *Otx2*, *Rorb* induces *Prdm1* and *Nrl* effectors. *Prdm1* promotes photoreceptor fates and *Nrl* specifically promotes a rod fate. In combination with *Foxn4*, *Rorb* induces *Ptf1a* to promote horizontal and amacrine cell fates. (B) Dual functions of the effector genes that act downstream of *Rorb*. *Prdm1* promotes a photoreceptor fate and suppresses a bipolar cell fate. *Nrl* promotes a rod fate and suppresses an S cone fate. *Ptf1a* promotes horizontal and amacrine cell fates but suppresses a ganglion cell fate.

8.1 Horizontal and Amacrine Interneurons

Horizontal and amacrine cells integrate visual information as it is relayed from the photoreceptors to the ganglion cells that form the optic nerve (Masland, 2001). Deletion of RORβ1 in mice results in loss of almost all horizontal and amacrine cells, reflecting a role in the initial specification of these cell lineages (Liu et al., 2013) (Fig. 6A). Diverse subtypes of amacrine cells are lost including GABAergic, glycinergic, and cholinergic cell types, in accord with RORβ1 acting early in differentiation before divergence of these sublineages (Balasubramanian & Gan, 2014; Boije, Shirazi Fard, Edqvist, & Hallbook, 2016). Furthermore, overexpression of RORβ1 in neonatal retina induces the formation of amacrine cells (Liu et al., 2013).

This phenotype of RORβ1-deficient mice can be explained by a failure to induce the *Ptf1a* gene (Liu et al., 2013) encoding PTF1A (pancreatic transcription factor 1a), a basic-helix–loop–helix transcription factor that is essential for the differentiation of horizontal and amacrine cell lineages (Fujitani et al., 2006; Nakhai et al., 2007). Moreover, mice deficient for RORβ1 or PTF1A not only lack horizontal and amacrine cells but also display an excess of ganglion cells at early developmental stages (Fujitani et al., 2006; Liu et al., 2013; Nakhai et al., 2007), suggesting a partial switch in cell fate. Thus, *Ptf1a* directs the differentiation of horizontal and amacrine cells and also suppresses a tendency for precursors to form ganglion cells. The *Ptf1a* gene is induced by RORβ1 and by FOXN4, a forkhead transcription factor that is expressed in progenitor cells and is also essential for specification of horizontal and amacrine cell lineages (Li et al., 2004).

8.2 Rod and Cone Photoreceptors

Vision is facilitated by rod photoreceptors that function in dim light and cones that function in bright light. Cones also mediate color vision, which in most mammals is determined by subpopulations that express opsin photopigments for response to medium-long (M, "green") or -short (S, "blue") wavelengths of light (Nathans, 1999) (Fig. 6A). There is much interest in understanding the events that generate these different types of photoreceptors and their relevance to causes of photoreceptor loss and retinal degeneration (Swaroop et al., 2010; Tran & Chen, 2014; Wright, Chakarova, Abd El-Aziz, & Bhattacharya, 2010).

Rorb−/− mice display a nearly complete absence of rods and an excess of cone-like photoreceptors, which is explained by a failure to induce the effector gene *Nrl* (Jia et al., 2009), encoding the rod-determining transcription factor NRL (neural retina leucine zipper factor) (Mears et al., 2001). *Nrl−/−* mice resemble *Rorb−/−* mice, as they lack rods and have an excess of S cone-like photoreceptors, suggestive of a cell fate switch. Reexpression of NRL in *Rorb−/−* mice partly rescues rod photoreceptors, consistent with *Nrl* acting downstream of *Rorb* in rod differentiation (Jia et al., 2009). An enhancer in the *Nrl* gene binds RORβ and the retinal homeodomain factors OTX2 or CRX (Kautzmann, Kim, Felder-Schmittbuhl, & Swaroop, 2011;

Montana et al., 2011). OTX2 is essential for the generation of photoreceptors (Nishida et al., 2003) and is induced during terminal mitosis as progenitor cells produce postmitotic progeny cells (Emerson & Cepko, 2011; Muranishi et al., 2011). CRX promotes terminal differentiation of photoreceptors and is induced by OTX2 in photoreceptor precursors (Chen et al., 1997; Nishida et al., 2003).

Rorb has also been proposed to regulate *Prdm1* (*Blimp1*), another gene required for photoreceptor differentiation (Wang, Sengel, Emerson, & Cepko, 2014). Deletion of PRDM1 (PR domain-containing 1 protein) results in a reduced number of photoreceptors and excess bipolar cells (Brzezinski, Lamba, & Reh, 2010; Katoh et al., 2010), suggesting that this effector also suppresses bipolar cell differentiation. An enhancer in the *Prdm1* gene binds RORβ1 and OTX2 (Wang et al., 2014). Thus, RORβ1 may serve at least two roles in photoreceptor precursors: first, via the *Prdm1* gene, in determining a photoreceptor rather than bipolar cell fate, and second, via *Nrl*, in determining that a population of precursors differentiate as rods rather than cone-like cells.

The only known function for RORβ2 is in rod differentiation, which is thought to involve a positive feedback loop between the *Rorb* and *Nrl* genes. RORβ1 initially induces *Nrl* at early stages and NRL protein in turn induces the RORβ2-specific promoter of the *Rorb* gene (Fu et al., 2014). Subsequently, both RORβ1 and RORβ2 may induce *Nrl* in the expanding, postnatal population of rods. Positive feedback may therefore promote the terminal differentiation of the growing rod population. In accord with this proposed shared function, deletion of both RORβ isoforms produces a severe loss of rods and a large excess of cone-like cells, whereas deletion of RORβ1 or RORβ2 individually produces a limited loss of rods and modest increase in cones (Fu et al., 2014; Jia et al., 2009).

A common feature of *Rorb* and some other genes that initiate cell fate decisions is that their expression decreases during terminal differentiation. This decline suggests the existence of transient periods of transcriptional plasticity during which cell fates are determined. For example, RORβ1 and FOXN4 decline in horizontal and amacrine cell lineages, whereas RORβ1 and OTX2 decline in photoreceptor lineages. In some cases, a related factor may later take over functions in the terminal differentiation or maintenance of the cell. In photoreceptors, OTX2 is succeeded by CRX in intermediary phases of differentiation (Koike et al., 2007; Roger et al., 2014) and RORβ1 is succeeded by RORβ2 (Fu et al., 2014). The transition from an immature state to a fully differentiated state is likely to involve terminal transcription programs and epigenetic events as a neuron acquires its function at mature stages. For example, in the development of the mouse retina (Wilken et al., 2015) or photoreceptors (Kim et al., 2016; Mo et al., 2016) changes have been observed in DNA methylation, DNase I hypersensitive sites, and histone modifications.

9. PARTNERS IN PLASTICITY: A COMBINATORIAL MODEL FOR DIFFERENTIATION

A fascinating property of RORβ is its ability to direct diverse neuronal fates. How can this orphan receptor direct such different outcomes as a rod photoreceptor and amacrine interneuron? Here, using the retina as the best-studied system, we propose a transcriptional model in which RORβ selectively pairs with other early-acting factors, namely OTX2 and FOXN4, to induce specific cell fates (Fig. 7). Partnership with stronger activators may be a necessity given that in transfection assays, RORβ is a relatively weak activator. Cooperation in some cases is known to result in synergistic activation of target gene enhancers at a much higher level than that induced by either RORβ or its partner alone (Liu et al., 2013; Srinivas et al., 2006).

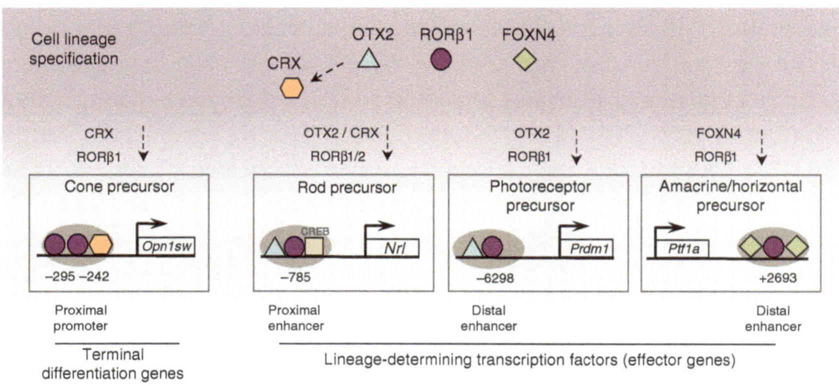

Fig. 7 RORβ and transcriptional control of neuronal identity in the retina. In this model, RORβ1 selectively cooperates with partner factors to bind to enhancer elements in target genes. RORβ1 and FOXN4 induce *Ptf1a* to promote horizontal and amacrine cell fates. RORβ1 and OTX2 induce *Prdm1* to promote photoreceptor differentiation and also induce *Nrl* to promote rod differentiation. In photoreceptors, RORβ2 may take over some functions as RORβ1 declines. An initial action of OTX2 is to induce CRX in photoreceptor precursors. RORβ proteins and CRX together induce the *Opn1sw* differentiation marker (S opsin). Other factors may contribute to enhancer-binding complexes, such as CREB on the *Nrl* gene. RORβ-binding enhancers may reside in proximal or distal locations relative to the promoter. *Numbers* below genes denote enhancer locations relative to translation start site of the gene.

Progenitor and precursor cells are presumably in a state of transcriptional flux and express varying combinations and varying levels of transcription factors (Brzezinski & Reh, 2015; Swaroop et al., 2010; Xiang, 2013). Thus, in a given subpopulation of progenitor cells, if RORβ1 and FOXN4 (Li et al., 2004) attain a combined threshold of expression, this synergistically activates an enhancer in the *Ptf1a* gene, a key gene for the differentiation of horizontal and amacrine cells (Liu et al., 2013). RORβ1 may provide highly specific recognition of the target gene via its extended DNA-binding site of >10 bp, whereas recruitment of FOXN4, a more promiscuous factor, provides strong activation functions.

In other subpopulations of cells, if RORβ1 and OTX2 attain a combined threshold of expression, this pair of factors may activate an enhancer in the *Prdm1* gene (Wang et al., 2014), a key gene for the differentiation of photoreceptor lineages. These same two factors also bind an enhancer in the *Nrl* gene (Kautzmann et al., 2011; Montana et al., 2011), a key gene for rod differentiation. Thus, RORβ1 and OTX2 may form a potent partnership for the control of effector genes in photoreceptor lineages.

RORβ1 (or RORβ2) also initiates expression of the *Opn1sw* gene in cones in combination with CRX (Srinivas et al., 2006). These factors synergistically activate the *Opn1sw* promoter, and again, RORβ may confer specific recognition of the gene, whereas CRX, a more promiscuous binding factor that can recruit histone-acetylating complexes, provides strong activation (Corbo et al., 2010; Peng & Chen, 2007). During terminal differentiation, RORβ and CRX may coregulate other photoreceptor genes that are required for the biosynthesis of outer segments since *Crx*−/− and *Rorb*−/− mouse strains both lack these structures that contain opsins and phototransduction proteins (Furukawa, Morrow, & Cepko, 1997; Jia et al., 2009).

If synergy is mainly governed by independent binding of RORβ and its partner to enhancer elements, this may explain how RORβ cooperates functionally with diverse classes of activator proteins. Protein–protein interactions are not excluded, but evidence suggests that these are weak and may be secondary to DNA binding (Liu et al., 2013; Srinivas et al., 2006). At present, little is known of other factors and chromatin-modifying complexes that interact with RORβ to form enhancer-binding complexes on target genes. However, neuronal-specific cofactors influence activity, as RORβ1 activates transcription in vitro in the presence of nuclear extracts from Neuro2A neuronal cells but not HeLa nonneuronal cells (Gawlas & Stunnenberg, 2001). A cofactor for RORβ1, NIX1 (Nrip2), has also been isolated from brain (Greiner et al., 2000).

For most known target genes, RORβ proteins are activators, but little is known of the extent to which they repress other genes. The *Brn2* gene is one example of a gene that may be suppressed by RORβ1 in the cortical plate (Oishi, Aramaki, & Nakajima, 2016) (Fig. 3). Conceivably, repression of negative response genes by RORβ requires the recruitment of repressive cofactors, analogous to the recruitment of coactivators in the induction of positive response genes.

A caveat of this model is that it views RORβ and its partner factors in the context of a simplified scheme of retinal cell differentiation, although many other transcription factors are involved and may prime a cell for adopting a particular fate (Haider et al., 2000; Hatakeyama & Kageyama, 2004; Sapkota et al., 2014; Xiang, 2013). Thus, it should be borne in mind that RORβ would exert its key actions as part of an integrated network of signals, including cell-to-cell signals such as notch, that influence how and when differentiation occurs (Agathocleous & Harris, 2009; Livesey & Cepko, 2001; Luo et al., 2012; Riesenberg, Liu, Kopan, & Brown, 2009).

10. THE *RORB* GENE AND HUMAN DISEASE

Studies of familial epilepsy and intellectual disability have identified varied forms of disruption of the *RORB* gene (Fig. 1B and C). Microdeletions ranging from 52 kb to 12 Mb in length have been mapped in the region of chromosome 9q21 that encompasses the *RORB* gene, or part of the *RORB* gene (Baglietto et al., 2014; Bartnik et al., 2012; Boudry-Labis et al., 2013; Lal et al., 2015; Rudolf et al., 2016). A comprehensive report also identified mutations in coding exons of the *RORB* gene including a nonsense variant (Arg66*), which would truncate both RORβ1 and RORβ2 in the DNA-binding domain, a missense variant (Leu73Pro) in the DNA-binding domain, and a deletion in the C-terminal coding exon (Thr417del) (Rudolf et al., 2016). The consequences of these coding changes on protein function remain to be demonstrated biochemically but are presumed to inactivate RORβ.

These genetic disruptions are thought to inactivate one allele of the *RORB* gene, such that symptoms arise from haploinsufficiency, or partial loss of function (Rudolf et al., 2016). The mode of inheritance in familial cases is autosomal dominant. There is heterogeneity of symptoms but frequently observed features include generalized epilepsy, abnormal cortical responses to photic stimulation, and intellectual disability. In one patient with autism and an abnormal sleep cycle but not epilepsy, a balanced chromosomal translocation [t(9;19)(q21;q12)dn] split the *RORB* gene between the RORβ1- and RORβ2-specific exons.

The C-terminal coding change (Thr417del) may interfere with the ability of RORβ to interact with transcriptional coactivators (Rudolf et al., 2016). If so, the symptoms in this case may not simply reflect loss of function but possibly also gain of dominant negative activity by the protein, analogous to the consequences of similar mutations found in thyroid hormone receptors in the syndrome of resistance to thyroid hormone (Dumitrescu & Refetoff, 2013; Schoenmakers et al., 2013). C-terminal coding changes in thyroid hormone receptors typically impair the ability of the receptor to bind ligand, alter its interactions with cofactors, and result in inhibitory activity over normal receptors in transactivation assays.

Based on evidence in mice, defective cellular differentiation in the cerebral cortex, thalamus, or other brain regions may cause neuronal dysfunction that contributes to the epilepsy and intellectual disability in human patients with loss of function of the *RORB* gene. Overt ocular symptoms have not been noted in these patients, but further examination of retinal function is of interest given the roles of the gene in mice. Potentially, in humans as in mice, homozygous loss of function of *RORB* is necessary before retinal defects become obvious. A possible role for the *RORB* gene in the human retina is supported by studies of in vitro differentiation of H9 human embryonic stem cells (Kaewkhaw et al., 2015). Following three-dimensional retinal differentiation in culture, photoreceptors expressed *OTX2*, *PRDM1*, and *NRL* consistent with the presence of a similar network of transcription factors in photoreceptors as in mice. *RORB* mRNA was detected in cone- and rod-like precursor cells.

The *RORB* gene has also been associated with bipolar disorder based on analysis of single-nucleotide polymorphisms (McGrath et al., 2009). The authors suggested that genes like *RORB* with a possible role in circadian rhythms may be involved in bipolar disorder.

11. CIRCADIAN RHYTHMS, LOCOMOTION, AND OTHER FUNCTIONS

11.1 Circadian Rhythms

Although this chapter focuses on neurodevelopment, we mention here other functions for the *Rorb* gene observed in mice. A potentially important role of *Rorb* is in setting circadian rhythms, based on its expression in the retina, pineal gland, and the SCN, the central oscillator that controls the activity of different physiological systems. *Rorb−/−* mice display an abnormal activity (free-running period) in response to photic entrainment (Andre, Conquet, et al., 1998). When the light:dark cycle is changed to continuous darkness, the circadian period of *Rorb−/−* mice becomes longer in contrast to wild-type mice in which the period shortens (Andre, Conquet, et al., 1998; Masana, Sumaya, Becker-Andre, & Dubocovich, 2007). In a regular light:dark cycle, *Rorb−/−* mice do not display obviously abnormal circadian behavior possibly because of substitution of function by the *Rora* gene, which also mediates circadian activity (Akashi & Takumi, 2005; Sato et al., 2004). Both *Rorb* and *Rora* are expressed in the SCN (VanDunk, Hunter, & Gray, 2011). Combined knockouts of *Rorb* and *Rora* genes or tissue-specific knockouts may help to define the cellular basis of these functions, whether in the SCN or elsewhere. It is also important to exclude confounding defects caused by visual and retinal phenotypes in *Rorb−/−* mice.

At present, we understand little of the role of *Rorb* in circadian control and further investigation is necessary. For example, is *Rorb* involved in the differentiation of cell types in the SCN and pineal gland or does it modulate rhythmic responses in these tissues, as suggested by the cyclic expression of the gene in both locations (Baler et al., 1996; Sumi et al., 2002)? Does *Rorb* have a role in the rhythmic activity of other neuronal or peripheral tissues? Retinoid-related orphan receptors bind to similar DNA response elements as Reverb orphan receptors and it has been proposed that these two classes of proteins oppose each other by activating and repressing, respectively, the clock gene networks that set circadian rhythms (Zhao et al., 2014). However, little is known of the role of RORβ in these mechanisms.

11.2 Abnormal Gait and Other Functions

Another overt phenotype of mice lacking all RORβ isoforms (Andre, Conquet, et al., 1998) or RORβ1 (Liu et al., 2013) is an exaggerated, clasping gait for the hind limbs (Table 1). The highstepper mouse with a duplication of part of the *Rorb* gene displays a similar abnormal gait (Fairfield et al., 2015). The cellular basis is unclear but might involve defects in the spinal cord or motor control pathways. Interestingly, the *Rora* gene, like *Rorb*, is also expressed in the dorsal horn of the spinal cord (Del Barrio et al., 2013; Schaeren-Wiemers et al., 1997). Inactivation of *Rora* in interneurons in this location results in foot slippage, suggesting a loss of corrective movement and fine motor control (Bourane et al., 2015). Some excitatory neurons in the dorsal horn coexpress both *Rorb* and *Rora* in mice, suggesting possible cooperative functions (Del Barrio et al., 2013).

Expression of *Rorb* has also been reported in the epididymis and vas deferens in mice (Andre, Conquet, et al., 1998). *Rorb*-deficient male mice (Andre, Conquet, et al., 1998; Liu et al., 2013) have transiently impaired fertility as young adults, but the reason is unknown. Expression of *Rorb* has also been noted in osteoblast cultures in which it may limit osteogenic differentiation (Roforth, Khosla, & Monroe, 2013; Roforth, Liu, Khosla, & Monroe, 2012). Future analysis may reveal other, previously unrecognized sites of expression of *Rorb* in novel cell types.

12. POTENTIAL LIGANDS FOR RORβ PROTEINS

A natural, high-affinity ligand for RORβ in vivo is currently unknown. Moreover, in transactivation assays in cultured cells (Gawlas & Stunnenberg, 2001; Liu et al., 2013; Srinivas et al., 2006) or in explanted retina (Fu et al., 2014), RORβ proteins are constitutively active without added ligand. Nonetheless, the ligand-binding pocket in the C-terminus of RORβ accommodates small hydrophobic molecules such as stearic acid (Stehlin et al., 2001) and all-*trans* retinoic acid (Stehlin-Gaon et al., 2003) in vitro, suggesting that small molecules may exist in vivo that bind to the protein. Binding of ligand to a nuclear receptor is thought to stabilize helix 12 in the AF2 domain in a conformation that allows recruitment of coactivators (Rastinejad, Huang, Chandra, & Khorasanizadeh, 2013). However, it is debatable whether small molecules that bind to RORβ act like specific, low abundance ligands that regulate the activity of classical nuclear receptors such as thyroid hormone receptors. An alternative possibility is that lipophilic molecules that are available in certain host cell types may bind RORβ as a means of stabilizing the protein to allow recruitment of coactivators (Rastinejad et al., 2013).

As with other orphan receptors, the search for synthetic ligands for RORβ offers the hope that such compounds could find therapeutic applications in modifying cellular responses in disease states. The search for RORβ-selective synthetic ligands has lagged behind the identification of ligands for RORα and RORγ (Kojetin & Burris, 2014). However, a compound in a class of *N*-(5-(arylcarbonyl)thiazol-2-yl)amides has been described as an RORβ-selective inverse agonist that might be tested in a biological model system (Gege, Schluter, & Hoffmann, 2014). The evidence of involvement of the *RORB* gene in human disease should encourage further investigation of RORβ-selective compounds and potential therapeutic applications.

13. CONCLUDING REMARKS

The *Rorb* gene promotes neuronal differentiation at least in part by regulating key effector genes at decisive steps in the determination of cell fates. This is best understood in the retina, but it is plausible that the gene is equally critical in the brain and spinal cord. We propose that *Rorb* acts at critical early steps in transcriptional hierarchies that promote neuronal diversity. Intriguingly, there may be conserved functions for *Hr46* (*Dhr3*), an insect ortholog of the *Rorb* gene, which similarly induces a downstream effector, βFTZf1, in a transcriptional cascade for embryogenesis and neurodevelopment in *Drosophila* (Ruaud et al., 2010).

We suggest that the ability of RORβ to direct diverse neuronal outcomes is facilitated by selective partnership with other fate-determining transcription factors. Such cooperation would occur during periods of transcriptional plasticity when cell fates remain in an unconsolidated or pliable state. However, much remains to be learned of the target gene networks and the cofactors that associate with RORβ to regulate enhancers in effector genes during differentiation. It is possible that the retina is a special case concerning multilineage control by RORβ. In other locations, such as the cerebral cortex or the dorsal horn of the spinal cord, RORβ displays more lineage-restricted expression and may operate through other forms of transcriptional networks.

It is an open possibility that the *Rorb* gene also promotes the function and maintenance of some types of neurons during adulthood and aging. Another area that merits more thorough investigation is the role of *Rorb* in circadian activity.

The expression pattern of the *Rorb* gene suggests that it serves a coordinating role in sensory or circadian pathways at multiple anatomical levels. Thus, the gene promotes the ability both to acquire information at the level of the receptor organ and to process the information at higher levels in the brain. Impairment at any of these levels might contribute to the susceptibility to seizures and intellectual disability observed in human cases where *RORB* gene function is lost. Further study of the human *RORB* gene will be of great interest in revealing the extent of its involvement in intellectual disability, epilepsy, or other neurological disorders.

ACKNOWLEDGMENT

This work was supported by the intramural research program at NIDDK at the National Institutes of Health.

REFERENCES

Agathocleous, M., & Harris, W. A. (2009). From progenitors to differentiated cells in the vertebrate retina. *Annual Review of Cell and Developmental Biology*, *25*, 45–69.

Agez, L., Laurent, V., Pevet, P., Masson-Pevet, M., & Gauer, F. (2007). Melatonin affects nuclear orphan receptors mRNA in the rat suprachiasmatic nuclei. *Neuroscience*, *144*(2), 522–530.

Akashi, M., & Takumi, T. (2005). The orphan nuclear receptor RORα regulates circadian transcription of the mammalian core-clock Bmal1. *Nature Structural & Molecular Biology*, *12*(5), 441–448.

Andre, E., Conquet, F., Steinmayr, M., Stratton, S. C., Porciatti, V., & Becker-Andre, M. (1998). Disruption of retinoid-related orphan receptor β changes circadian behavior, causes retinal degeneration and leads to vacillans phenotype in mice. *The EMBO Journal*, *17*(14), 3867–3877.

Andre, E., Gawlas, K., & Becker-Andre, M. (1998). A novel isoform of the orphan nuclear receptor RORβ is specifically expressed in pineal gland and retina. *Gene*, *216*(2), 277–283.

Baglietto, M. G., Caridi, G., Gimelli, G., Mancardi, M., Prato, G., Ronchetto, P., et al. (2014). RORB gene and 9q21.13 microdeletion: Report on a patient with epilepsy and mild intellectual disability. *European Journal of Medical Genetics*, *57*(1), 44–46.

Balasubramanian, R., & Gan, L. (2014). Development of retinal amacrine cells and their dendritic stratification. *Current Ophthalmology Reports*, *2*(3), 100–106.

Baler, R., Coon, S., & Klein, D. C. (1996). Orphan nuclear receptor RZRβ: Cyclic AMP regulates expression in the pineal gland. *Biochemical and Biophysical Research Communications*, *220*(3), 975–978.

Bartnik, M., Szczepanik, E., Derwinska, K., Wisniowiecka-Kowalnik, B., Gambin, T., Sykulski, M., et al. (2012). Application of array comparative genomic hybridization in 102 patients with epilepsy and additional neurodevelopmental disorders. *American Journal of Medical Genetics. Part B, Neuropsychiatric Genetics*, *159B*(7), 760–771.

Boije, H., Shirazi Fard, S., Edqvist, P. H., & Hallbook, F. (2016). Horizontal cells, the odd ones out in the retina, give insights into development and disease. *Frontiers in Neuroanatomy*, *10*, 77.

Boudry-Labis, E., Demeer, B., Le Caignec, C., Isidor, B., Mathieu-Dramard, M., Plessis, G., et al. (2013). A novel microdeletion syndrome at 9q21.13 characterised by mental retardation, speech delay, epilepsy and characteristic facial features. *European Journal of Medical Genetics*, *56*(3), 163–170.

Bourane, S., Grossmann, K. S., Britz, O., Dalet, A., Del Barrio, M. G., Stam, F. J., et al. (2015). Identification of a spinal circuit for light touch and fine motor control. *Cell*, *160*(3), 503–515.

Brzezinski, J. A., Lamba, D. A., & Reh, T. A. (2010). Blimp1 controls photoreceptor versus bipolar cell fate choice during retinal development. *Development*, *137*(4), 619–629.

Brzezinski, J. A., & Reh, T. A. (2015). Photoreceptor cell fate specification in vertebrates. *Development*, *142*(19), 3263–3273.

Byun, H., Kwon, S., Ahn, H. J., Liu, H., Forrest, D., Demb, J. B., et al. (2016). Molecular features distinguish ten neuronal types in the mouse superficial superior colliculus. *The Journal of Comparative Neurology*, *524*(11), 2300–2321.

Carlberg, C., Hooft van Huijsduijnen, R., Staple, J. K., DeLamarter, J. F., & Becker-Andre, M. (1994). RZRs, a new family of retinoid-related orphan receptors that function as both monomers and homodimers. *Molecular Endocrinology*, *8*(6), 757–770.
Carney, G. E., Wade, A. A., Sapra, R., Goldstein, E. S., & Bender, M. (1997). DHR3, an ecdysone-inducible early-late gene encoding a Drosophila nuclear receptor, is required for embryogenesis. *Proceedings of the National Academy of Sciences of the United States of America*, *94*(22), 12024–12029.
Carter-Dawson, L. D., & LaVail, M. M. (1979). Rods and cones in the mouse retina. II. Autoradiographic analysis of cell generation using tritiated thymidine. *The Journal of Comparative Neurology*, *188*(2), 263–272.
Cepko, C. L., Austin, C. P., Yang, X., Alexiades, M., & Ezzeddine, D. (1996). Cell fate determination in the vertebrate retina. *Proceedings of the National Academy of Sciences of the United States of America*, *93*(2), 589–595.
Chen, S., Wang, Q. L., Nie, Z., Sun, H., Lennon, G., Copeland, N. G., et al. (1997). Crx, a novel Otx-like paired-homeodomain protein, binds to and transactivates photoreceptor cell-specific genes. *Neuron*, *19*(5), 1017–1030.
Chow, L., Levine, E. M., & Reh, T. A. (1998). The nuclear receptor transcription factor, retinoid-related orphan receptor β, regulates retinal progenitor proliferation. *Mechanisms of Development*, *77*(2), 149–164.
Corbo, J. C., Lawrence, K. A., Karlstetter, M., Myers, C. A., Abdelaziz, M., Dirkes, W., et al. (2010). CRX ChIP-seq reveals the cis-regulatory architecture of mouse photoreceptors. *Genome Research*, *20*(11), 1512–1525.
Del Barrio, M. G., Bourane, S., Grossmann, K., Schule, R., Britsch, S., O'Leary, D. D., et al. (2013). A transcription factor code defines nine sensory interneuron subtypes in the mechanosensory area of the spinal cord. *PLoS One*, *8*(11), e77928.
Dumitrescu, A. M., & Refetoff, S. (2013). The syndromes of reduced sensitivity to thyroid hormone. *Biochimica et Biophysica Acta*, *1830*(7), 3987–4003.
Emerson, M. M., & Cepko, C. L. (2011). Identification of a retina-specific Otx2 enhancer element active in immature developing photoreceptors. *Developmental Biology*, *360*(1), 241–255.
Evans, R. M. (1988). The steroid and thyroid hormone receptor superfamily. *Science*, *240*, 889–895.
Fairfield, H., Srivastava, A., Ananda, G., Liu, R., Kircher, M., Lakshminarayana, A., et al. (2015). Exome sequencing reveals pathogenic mutations in 91 strains of mice with Mendelian disorders. *Genome Research*, *25*(7), 948–957.
Fu, Y., Liu, H., Ng, L., Kim, J. W., Hao, H., Swaroop, A., et al. (2014). Feedback induction of a photoreceptor-specific isoform of retinoid-related orphan nuclear receptor β by the rod transcription factor NRL. *The Journal of Biological Chemistry*, *289*(47), 32469–32480.
Fujieda, H., Bremner, R., Mears, A. J., & Sasaki, H. (2009). Retinoic acid receptor-related orphan receptor α regulates a subset of cone genes during mouse retinal development. *Journal of Neurochemistry*, *108*, 91–101.
Fujitani, Y., Fujitani, S., Luo, H., Qiu, F., Burlison, J., Long, Q., et al. (2006). Ptf1a determines horizontal and amacrine cell fates during mouse retinal development. *Development*, *133*(22), 4439–4450.
Furukawa, T., Morrow, E. M., & Cepko, C. L. (1997). Crx, a novel otx-like homeobox gene, shows photoreceptor-specific expression and regulates photoreceptor differentiation. *Cell*, *91*(4), 531–541.
Gawlas, K., & Stunnenberg, H. G. (2000). Differential binding and transcriptional behaviour of two highly related orphan receptors, RORα(4) and RORβ(1). *Biochimica et Biophysica Acta*, *1494*(3), 236–241.

Gawlas, K., & Stunnenberg, H. G. (2001). Differential transcription of the orphan receptor RORβ in nuclear extracts derived from Neuro2A and HeLa cells. *Nucleic Acids Research*, *29*(16), 3424–3432.

Gege, C., Schluter, T., & Hoffmann, T. (2014). Identification of the first inverse agonist of retinoid-related orphan receptor (ROR) with dual selectivity for RORβ and RORγt. *Bioorganic & Medicinal Chemistry Letters*, *24*(22), 5265–5267.

Giguere, V. (1999). Orphan nuclear receptors: From gene to function. *Endocrine Reviews*, *20*(5), 689–725.

Giguere, V., Tini, M., Flock, G., Ong, E., Evans, R. M., & Otulakowski, G. (1994). Isoform-specific amino-terminal domains dictate DNA-binding properties of RORα, a novel family of orphan hormone nuclear receptors. *Genes & Development*, *8*(5), 538–553.

Green, S., & Chambon, P. (1988). Nuclear receptors enhance our understanding of transcription regulation. *Trends in Genetics*, *4*, 309–314.

Greig, L. C., Woodworth, M. B., Galazo, M. J., Padmanabhan, H., & Macklis, J. D. (2013). Molecular logic of neocortical projection neuron specification, development and diversity. *Nature Reviews. Neuroscience*, *14*(11), 755–769.

Greiner, E. F., Kirfel, J., Greschik, H., Huang, D., Becker, P., Kapfhammer, J. P., et al. (2000). Differential ligand-dependent protein-protein interactions between nuclear receptors and a neuronal-specific cofactor. *Proceedings of the National Academy of Sciences of the United States of America*, *97*(13), 7160–7165.

Haider, N. B., Jacobson, S. G., Cideciyan, A. V., Swiderski, R., Streb, L. M., Searby, C., et al. (2000). Mutation of a nuclear receptor gene, NR2E3, causes enhanced S cone syndrome, a disorder of retinal cell fate. *Nature Genetics*, *24*(2), 127–131.

Hatakeyama, J., & Kageyama, R. (2004). Retinal cell fate determination and bHLH factors. *Seminars in Cell & Developmental Biology*, *15*(1), 83–89.

Ivanov, I. I., McKenzie, B. S., Zhou, L., Tadokoro, C. E., Lepelley, A., Lafaille, J. J., et al. (2006). The orphan nuclear receptor RORγt directs the differentiation program of proinflammatory IL-17+ T helper cells. *Cell*, *126*(6), 1121–1133.

Jabaudon, D., Shnider, S. J., Tischfield, D. J., Galazo, M. J., & Macklis, J. D. (2012). RORβ induces barrel-like neuronal clusters in the developing neocortex. *Cerebral Cortex*, *22*(5), 996–1006.

Jessell, T. M. (2000). Neuronal specification in the spinal cord: Inductive signals and transcriptional codes. *Nature Reviews. Genetics*, *1*(1), 20–29.

Jetten, A. M. (2009). Retinoid-related orphan receptors (RORs): Critical roles in development, immunity, circadian rhythm, and cellular metabolism. *Nuclear Receptor Signaling*, 7, e003.

Jia, L., Oh, E. C., Ng, L., Srinivas, M., Brooks, M., Swaroop, A., et al. (2009). Retinoid-related orphan nuclear receptor RORβ is an early-acting factor in rod photoreceptor development. *Proceedings of the National Academy of Sciences of the United States of America*, *106*(41), 17534–17539.

Kaewkhaw, R., Kaya, K. D., Brooks, M., Homma, K., Zou, J., Chaitankar, V., et al. (2015). Transcriptome dynamics of developing photoreceptors in three-dimensional retina cultures recapitulates temporal sequence of human cone and rod differentiation revealing cell surface markers and gene networks. *Stem Cells*, *33*(12), 3504–3518.

Kalsbeek, A., & Buijs, R. M. (2002). Output pathways of the mammalian suprachiasmatic nucleus: Coding circadian time by transmitter selection and specific targeting. *Cell and Tissue Research*, *309*, 109–118.

Katoh, K., Omori, Y., Onishi, A., Sato, S., Kondo, M., & Furukawa, T. (2010). Blimp1 suppresses Chx10 expression in differentiating retinal photoreceptor precursors to ensure proper photoreceptor development. *The Journal of Neuroscience*, *30*(19), 6515–6526.

Kautzmann, M. A., Kim, D. S., Felder-Schmittbuhl, M. P., & Swaroop, A. (2011). Combinatorial regulation of photoreceptor differentiation factor, neural retina leucine zipper gene NRL, revealed by in vivo promoter analysis. *The Journal of Biological Chemistry*, *286*(32), 28247–28255.

Kim, J.-W., Yang, H.-J., Brooks, M. J., Zelinger, L., Karakülah, G., Gotoh, N., et al. (2016). NRL-regulated transcriptome dynamics of developing rod photoreceptors. *Cell Reports*, *17*, 1–14.
Koike, C., Nishida, A., Ueno, S., Saito, H., Sanuki, R., Sato, S., et al. (2007). Functional roles of Otx2 transcription factor in postnatal mouse retinal development. *Molecular and Cellular Biology*, *27*(23), 8318–8329.
Kojetin, D. J., & Burris, T. P. (2014). REV-ERB and ROR nuclear receptors as drug targets. *Nature Reviews. Drug Discovery*, *13*(3), 197–216.
Lal, D., Ruppert, A. K., Trucks, H., Schulz, H., de Kovel, C. G., Kasteleijn-Nolst Trenite, D., et al. (2015). Burden analysis of rare microdeletions suggests a strong impact of neurodevelopmental genes in genetic generalised epilepsies. *PLoS Genetics*, *11*(5) e1005226.
Lam, G., Hall, B. L., Bender, M., & Thummel, C. S. (1999). DHR3 is required for the prepupal-pupal transition and differentiation of adult structures during Drosophila metamorphosis. *Developmental Biology*, *212*(1), 204–216.
Li, S., Mo, Z., Yang, X., Price, S. M., Shen, M. M., & Xiang, M. (2004). Foxn4 controls the genesis of amacrine and horizontal cells by retinal progenitors. *Neuron*, *43*(6), 795–807.
Liu, H., Kim, S. Y., Fu, Y., Wu, X., Ng, L., Swaroop, A., et al. (2013). An isoform of retinoid-related orphan receptor β directs differentiation of retinal amacrine and horizontal interneurons. *Nature Communications*, *4*, 1813.
Livesey, F. J., & Cepko, C. L. (2001). Vertebrate neural cell-fate determination: Lessons from the retina. *Nature Reviews. Neuroscience*, *2*(2), 109–118.
Luo, H., Jin, K., Xie, Z., Qiu, F., Li, S., Zou, M., et al. (2012). Forkhead box N4 (Foxn4) activates Dll4-Notch signaling to suppress photoreceptor cell fates of early retinal progenitors. *Proceedings of the National Academy of Sciences of the United States of America*, *109*(9), E553–E562.
Masana, M. I., Sumaya, I. C., Becker-Andre, M., & Dubocovich, M. L. (2007). Behavioral characterization and modulation of circadian rhythms by light and melatonin in C3H/HeN mice homozygous for the RORβ knockout. *American Journal of Physiology. Regulatory. Integrative and Comparative Physiology*, *292*(6), R2357–R2367.
Masland, R. H. (2001). The fundamental plan of the retina. *Nature Neuroscience*, *4*(9), 877–886.
McGrath, C. L., Glatt, S. J., Sklar, P., Le-Niculescu, H., Kuczenski, R., Doyle, A. E., et al. (2009). Evidence for genetic association of RORB with bipolar disorder. *BMC Psychiatry*, *9*, 70.
Mears, A. J., Kondo, M., Swain, P. K., Takada, Y., Bush, R. A., Saunders, T. L., et al. (2001). Nrl is required for rod photoreceptor development. *Nature Genetics*, *29*(4), 447–452.
Mo, A., Luo, C., Davis, F. P., Mukamel, E. A., Henry, G. L., Nery, J. R., et al. (2016). Epigenomic landscapes of retinal rods and cones. *eLife*, *5*, e11613.
Montana, C. L., Lawrence, K. A., Williams, N. L., Tran, N. M., Peng, G. H., Chen, S., et al. (2011). Transcriptional regulation of neural retina leucine zipper (Nrl), a photoreceptor cell fate determinant. *The Journal of Biological Chemistry*, *286*(42), 36921–36931.
Mullican, S. E., Dispirito, J. R., & Lazar, M. A. (2013). The orphan nuclear receptors at their 25-year reunion. *Journal of Molecular Endocrinology*, *51*(3), T115–T140.
Muranishi, Y., Terada, K., Inoue, T., Katoh, K., Tsujii, T., Sanuki, R., et al. (2011). An essential role for RAX homeoprotein and NOTCH-HES signaling in Otx2 expression in embryonic retinal photoreceptor cell fate determination. *The Journal of Neuroscience*, *31*(46), 16792–16807.
Nakagawa, Y., & O'Leary, D. D. (2003). Dynamic patterned expression of orphan nuclear receptor genes RORα and RORβ in developing mouse forebrain. *Developmental Neuroscience*, *25*(2–4), 234–244.

Nakhai, H., Sel, S., Favor, J., Mendoza-Torres, L., Paulsen, F., Duncker, G. I., et al. (2007). Ptf1a is essential for the differentiation of GABAergic and glycinergic amacrine cells and horizontal cells in the mouse retina. *Development*, *134*(6), 1151–1160.
Nathans, J. (1999). The evolution and physiology of human color vision: Insights from molecular genetic studies of visual pigments. *Neuron*, *24*(2), 299–312.
Nishida, A., Furukawa, A., Koike, C., Tano, Y., Aizawa, S., Matsuo, I., et al. (2003). Otx2 homeobox gene controls retinal photoreceptor cell fate and pineal gland development. *Nature Neuroscience*, *6*(12), 1255–1263.
Oishi, K., Aramaki, M., & Nakajima, K. (2016). Mutually repressive interaction between Brn1/2 and Rorb contributes to the establishment of neocortical layer 2/3 and layer 4. *Proceedings of the National Academy of Sciences of the United States of America*, *113*(12), 3371–3376.
Oishi, K., Nakagawa, N., Tachikawa, K., Sasaki, S., Aramaki, M., Hirano, S., et al. (2016). Identity of neocortical layer 4 neurons is specified through correct positioning into the cortex. *eLife*, *5*, e10907.
Peng, G. H., & Chen, S. (2007). Crx activates opsin transcription by recruiting HAT-containing co-activators and promoting histone acetylation. *Human Molecular Genetics*, *16*(20), 2433–2452.
Petersen, C. C. (2007). The functional organization of the barrel cortex. *Neuron*, *56*(2), 339–355.
Rastinejad, F., Huang, P., Chandra, V., & Khorasanizadeh, S. (2013). Understanding nuclear receptor form and function using structural biology. *Journal of Molecular Endocrinology*, *51*(3), T1–T21.
Riesenberg, A. N., Liu, Z., Kopan, R., & Brown, N. L. (2009). Rbpj cell autonomous regulation of retinal ganglion cell and cone photoreceptor fates in the mouse retina. *The Journal of Neuroscience*, *29*(41), 12865–12877.
Roforth, M. M., Khosla, S., & Monroe, D. G. (2013). Identification of RORβ targets in cultured osteoblasts and in human bone. *Biochemical and Biophysical Research Communications*, *440*(4), 768–773.
Roforth, M. M., Liu, G., Khosla, S., & Monroe, D. G. (2012). Examination of nuclear receptor expression in osteoblasts reveals *Rorb* as an important regulator of osteogenesis. *Journal of Bone and Mineral Research*, *27*(4), 891–901.
Roger, J. E., Hiriyanna, A., Gotoh, N., Hao, H., Cheng, D. F., Ratnapriya, R., et al. (2014). OTX2 loss causes rod differentiation defect in CRX-associated congenital blindness. *The Journal of Clinical Investigation*, *124*(2), 631–643.
Rowell, J. J., Mallik, A. K., Dugas-Ford, J., & Ragsdale, C. W. (2010). Molecular analysis of neocortical layer structure in the ferret. *The Journal of Comparative Neurology*, *518*(16), 3272–3289.
Ruaud, A. F., Lam, G., & Thummel, C. S. (2010). The Drosophila nuclear receptors DHR3 and betaFTZ-F1 control overlapping developmental responses in late embryos. *Development*, *137*(1), 123–131.
Rudolf, G., Lesca, G., Mehrjouy, M. M., Labalme, A., Salmi, M., Bache, I., et al. (2016). Loss of function of the retinoid-related nuclear receptor (RORB) gene and epilepsy. *European Journal of Human Genetics*, *24*, 1761–1770. http://dx.doi.org/10.1038/ejhg.2016.80.
Sapkota, D., Chintala, H., Wu, F., Fliesler, S. J., Hu, Z., & Mu, X. (2014). Onecut1 and Onecut2 redundantly regulate early retinal cell fates during development. *Proceedings of the National Academy of Sciences of the United States of America*, *111*(39), E4086–E4095.
Sato, T. K., Panda, S., Miraglia, L. J., Reyes, T. M., Rudic, R. D., McNamara, P., et al. (2004). A functional genomics strategy reveals Rora as a component of the mammalian circadian clock. *Neuron*, *43*(4), 527–537.

Schaeren-Wiemers, N., Andre, E., Kapfhammer, J. P., & Becker-Andre, M. (1997). The expression pattern of the orphan nuclear receptor RORβ in the developing and adult rat nervous system suggests a role in the processing of sensory information and in circadian rhythm. *The European Journal of Neuroscience*, *9*(12), 2687–2701.
Schoenmakers, N., Moran, C., Peeters, R. P., Visser, T., Gurnell, M., & Chatterjee, K. (2013). Resistance to thyroid hormone mediated by defective thyroid hormone receptor α. *Biochimica et Biophysica Acta*, *1830*(7), 4004–4008.
Srinivas, M., Ng, L., Liu, H., Jia, L., & Forrest, D. (2006). Activation of the blue opsin gene in cone photoreceptor development by retinoid-related orphan receptor β. *Molecular Endocrinology*, *20*(8), 1728–1741.
Stehlin, C., Wurtz, J. M., Steinmetz, A., Greiner, E., Schule, R., Moras, D., et al. (2001). X-ray structure of the orphan nuclear receptor RORβ ligand-binding domain in the active conformation. *The EMBO Journal*, *20*(21), 5822–5831.
Stehlin-Gaon, C., Willmann, D., Zeyer, D., Sanglier, S., Van Dorsselaer, A., Renaud, J. P., et al. (2003). All-trans retinoic acid is a ligand for the orphan nuclear receptor RORβ. *Nature Structural Biology*, *10*(10), 820–825.
Sumi, Y., Yagita, K., Yamaguchi, S., Ishida, Y., Kuroda, Y., & Okamura, H. (2002). Rhythmic expression of RORβ mRNA in the mice suprachiasmatic nucleus. *Neuroscience Letters*, *320*(1–2), 13–16.
Swaroop, A., Kim, D., & Forrest, D. (2010). Transcriptional regulation of photoreceptor development and homeostasis in the mammalian retina. *Nature Reviews. Neuroscience*, *11*(8), 563–576.
Takeda, Y., Kang, H. S., Freudenberg, J., DeGraff, L. M., Jothi, R., & Jetten, A. M. (2014). Retinoic acid-related orphan receptor gamma (RORγ): A novel participant in the diurnal regulation of hepatic gluconeogenesis and insulin sensitivity. *PLoS Genetics*, *10*(5), e1004331.
Tasic, B., Menon, V., Nguyen, T. N., Kim, T. K., Jarsky, T., Yao, Z., et al. (2016). Adult mouse cortical cell taxonomy revealed by single cell transcriptomics. *Nature Neuroscience*, *19*(2), 335–346.
Tran, N. M., & Chen, S. (2014). Mechanisms of blindness: Animal models provide insight into distinct CRX-associated retinopathies. *Developmental Dynamics*, *243*(10), 1153–1166.
VanDunk, C., Hunter, L. A., & Gray, P. A. (2011). Development, maturation, and necessity of transcription factors in the mouse suprachiasmatic nucleus. *The Journal of Neuroscience*, *31*(17), 6457–6467.
Wang, S., Sengel, C., Emerson, M. M., & Cepko, C. L. (2014). A gene regulatory network controls the binary fate decision of rod and bipolar cells in the vertebrate retina. *Developmental Cell*, *30*(5), 513–527.
Wilken, M. S., Brzezinski, J. A., LaTorre, A., Siebenthall, K., Thurman, R., Sabo, P., et al. (2015). DNase I hypersensitivity analysis of the muse brain and retina identifies region-specific regulatory elements. *Epigenetics & Chromatin*, *8*, 8.
Wright, A. F., Chakarova, C. F., Abd El-Aziz, M. M., & Bhattacharya, S. S. (2010). Photoreceptor degeneration: Genetic and mechanistic dissection of a complex trait. *Nature Reviews. Genetics*, *11*(4), 273–284.
Xiang, M. (2013). Intrinsic control of mammalian retinogenesis. *Cellular and Molecular Life Sciences*, *70*, 2519–2532. http://dx.doi.org/10.1007/s00018-012-1183-2.
Xu, X. L., Fang, Y., Lee, T. C., Forrest, D., Gregory-Evans, C., Almeida, D., et al. (2009). Retinoblastoma has properties of a cone precursor tumor and depends upon cone-specific MDM2 signaling. *Cell*, *137*(6), 1018–1031.
Zhao, X., Cho, H., Yu, R. T., Atkins, A. R., Downes, M., & Evans, R. M. (2014). Nuclear receptors rock around the clock. *EMBO Reports*, *15*(5), 518–528.

CHAPTER NINE

Nuclear Receptor TLX in Development and Diseases

Guoqiang Sun*, Qi Cui*,†, Yanhong Shi*,†,1
*Beckman Research Institute of City of Hope, Duarte, CA, United States
†Irell & Manella Graduate School of Biological Sciences, Beckman Research Institute of City of Hope, Duarte, CA, United States
[1]Corresponding author: e-mail address: yshi@coh.org

Contents

Abstract

The nuclear receptor TLX (NR2E1) is a transcription factor that is critical for neural development and adult neurogenesis through its actions in regulating neural stem cell proliferation, self-renewal, and fate determination. These roles are primarily executed by regulating TLX downstream target genes involved in myriad pathways such as cell cycle progression, RNA processing, angiogenesis, and senescence. Recent studies suggest that dysregulation of TLX pathways plays an important role in the pathogenesis of human neurological disorders and brain tumors. Here, we will highlight recent progress in the roles of TLX in brain development and adult neurogenesis, and the relevance of TLX to neurological diseases and brain tumors. We will also discuss the potential of TLX as a therapeutic target for these disorders.

1. INTRODUCTION

Nuclear receptors are a superfamily of ligand-dependent transcription factors that govern many aspects of development and diseases. TLX is a member of the nuclear receptor superfamily and is designated as nuclear

Current Topics in Developmental Biology, Volume 125
ISSN 0070-2153
http://dx.doi.org/10.1016/bs.ctdb.2016.12.003

receptor subfamily 2 group E member 1 (NR2E1) (Mangelsdorf et al., 1995). TLX is evolutionarily conserved among vertebrates and is highly homologous to the *Drosophila* tailless (Pignoni et al., 1990). The mammalian TLX is expressed predominantly in the brain (Monaghan, Grau, Bock, & Schutz, 1995; Yu, McKeown, Evans, & Umesono, 1994). TLX has been shown to play important roles in maintaining neural stem cell (NSC) self-renewal in both the developing and adult brain (Li et al., 2008; Shi et al., 2004). Dysregulation of TLX contributes to tumorigenesis and neurological diseases such as schizophrenia (Cui et al., 2016; Liu et al., 2010; Murai et al., 2016; Park et al., 2010; Zhu et al., 2014; Zou et al., 2012).

Our understanding of the molecular regulation of TLX and its downstream target genes (see later) and the solving of TLX structure (Zhi et al., 2015) shed light on the contribution of TLX in multiple biological processes and disease conditions (Fig. 1). In this review, we will discuss various regulatory roles of TLX in brain development and adult neurogenesis. Moreover, we will review the potential roles of TLX in brain aging and

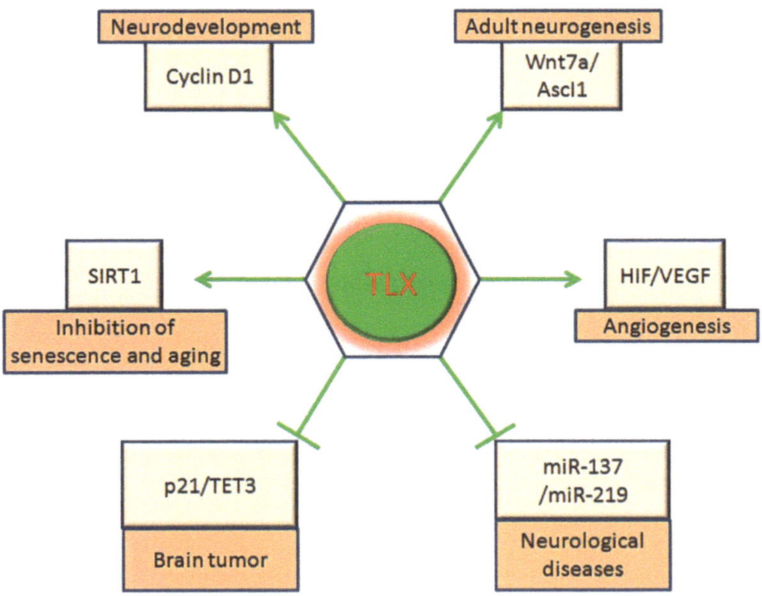

Fig. 1 TLX regulates downstream target genes that are involved in multiple biological processes and disease conditions. Examples of TLX target genes are listed in *yellow boxes* and the biological processes regulated by TLX are listed in *orange boxes* underneath. *Arrow lines* indicate activation of gene expression by TLX while *cut lines* indicate repression of gene expression by TLX.

highlight how dysregulation of TLX is associated with neurological disorders and brain tumorigenesis.

2. TLX IN DEVELOPMENT AND DISEASES

2.1 Molecular Regulation of TLX and Its Target Genes

The human *TLX* gene was originally isolated from a 24-kb region on chromosomal band 6q21 (Jackson, Panayiotidis, & Foroni, 1998). This gene includes 9 exons and encodes a protein of 385 amino acids. Like its homologues in *Drosophila*, zebrafish, and mouse, the human TLX protein contains a highly conserved DNA-binding domain (DBD), a hinge region, and a moderately conserved C-terminal ligand-binding domain (LBD). TLX lacks an N-terminal activation function domain that usually contains ligand-independent activation or repression functions. Recently, three compounds were found to bind the recombinant TLX LBD, however these compounds failed to activate TLX (Benod et al., 2014).

TLX has been shown to function primarily through transcriptional repression of downstream target genes by forming a complex with transcriptional corepressors, such as the lysine-specific histone demethylase 1 (LSD1) (Sun et al., 2010, 2011; Yokoyama, Takezawa, Schule, Kitagawa, & Kato, 2008), histone deacetylases (HDACs) (Sun, Yu, Evans, & Shi, 2007), and nuclear corepressor atrophin (Davis, Thomas, Nomie, Huang, & Dierick, 2014; Wang, Rajan, Pitman, McKeown, & Tsai, 2006; Zhang, Zou, Yu, Gage, & Evans, 2006; Zhi et al., 2015). In addition, the screen by Estruch et al. identified B-cell lymphoma/leukemia 11A/CTIP1 (BCL11A), an oncoprotein and transcription factor, as a TLX coregulator (Estruch et al., 2012). This selective interaction with TLX relies on two copies of the signature motif F/YSXXLXXL/Y in BCL11A (Chan et al., 2013).

TLX regulates a variety of cellular activities by regulating the expression of its downstream target genes. Various TLX target genes have been identified, including Pax2, GFAP, p21 (Cdkn1a), Pten, miR-137, miR-9, BMP4, Wnt7a, Ascl1, HIF-2α, MMP-2, Oct4, SIRT1, and retina-specific homeobox gene 2 (rx2) (Chavali, Saini, Matsumoto, Agren, & Funa, 2011; Chavali et al., 2014; Elmi et al., 2010; Hu et al., 2014; Iwahara, Hisahara, Hayashi, & Horio, 2009; Qin, Niu, Iqbal, Smith, & Zhang, 2014; Qu et al., 2010; Reinhardt et al., 2015; Shi et al., 2004; Sun & Shi, 2010; Sun et al., 2011, 2007; Wu et al., 2015; Yu et al., 1994; Zhang, Zou, He, Gage, & Evans, 2008; Zeng et al., 2012; Zhao, Sun, Li, & Shi,

2009). TLX represses the expression of p21 and Pten, but activates Wnt signaling, to promote NSC proliferation and self-renewal (Qu et al., 2010; Sun et al., 2010, 2011, 2007; Zhang et al., 2008). TLX crosstalks with microRNAs (miRNAs), such as let-7b, let-7d, miR-9, and miR-378, to regulate NSC fate determination (Hu et al., 2014; Huang, Liu, & Wang, 2015; Ni et al., 2014; Zhao et al., 2010, 2009; Zhao, Sun, Ye, Li, & Shi, 2013). Moreover, TLX represses GFAP and BMP signaling to inhibit gliogenesis (Qin et al., 2014; Shi et al., 2004), but activates Ascl1 to promote neuronal induction (Elmi et al., 2010).

Specifically, by analyzing RNAs isolated from adult brains of wild type and *Tlx*-knockdown mice, Qu et al. identified Wnt7a as a downstream target of Tlx (Qu et al., 2010). Wnt proteins bind to cell membrane Frizzled receptors to activate Wnt signaling, which leads to translocation of the cytoplasmic β-catenin to the nucleus, where β-catenin binds to T cell factor family (TCF) transcription factors to activate target gene expression. TLX activates Wnt7a to promote the proliferation and self-renewal of NSCs (Qu et al., 2010). TLX also regulates NSC proliferation through modulating the expression of cyclin-dependent kinase inhibitors, such as p21 (Sun et al., 2010, 2011, 2007), p57 (Cdkn1c, Kip2), and several genes downstream of p53 (Zhang et al., 2008), all of which are also involved in brain aging and tumorigenesis.

SRY-box-containing gene 2 (Sox2) was demonstrated to be an upstream regulator of TLX. TLX and Sox2 form a molecular network in adult NSCs and retinal stem cells (Islam et al., 2015; Reinhardt et al., 2015; Shimozaki et al., 2012). Both Sox2 and Tlx proteins bind to the *Tlx* gene promoter. Sox2 positively regulates *Tlx* gene expression, whereas the binding of Tlx to its own promoter suppresses its transcription (Shimozaki et al., 2012). The Tlx-mediated suppression can be alleviated by overexpressing Sox2 (Shimozaki et al., 2012).

Recently, TLX was found to be involved in posttranscriptional regulation of miRNA processing (Murai et al., 2016). In the *Tlx*-knockout mouse brains, the precursor form of miR219 (pre-miR219) increased significantly, compared to that in the wild-type brains. However, no marked alteration was found in the level of primary miR-219 (pri-miR-219) expression. Similar change in the expression of miR-219 species was observed in mouse NSCs with TLX knockdown. The upregulation of pre-miR-219 and mature miR-219 by *Tlx* knockdown was not affected by actinomycin D treatment, indicating that the regulation is at posttranscriptional level.

Instead, TLX inhibits miR-219 processing by interacting with the RNA helicase p68- and p68-associated Drosha and DGCR8, the two main components of miRNA processing machinery. These interactions in turn prevent the miRNA processing machinery from binding to miR-219 primary form. This study uncovers a new role of TLX beyond transcriptional regulation (Murai et al., 2016). TLX can also physically bind to the von Hippel–Lindau protein (VHL) to stabilize HIF-2α and activate the production of angiogenic factors, such as VEGF, which may contribute to the angiogenic activity of TLX (Zeng et al., 2012).

2.2 TLX in Brain Development and Adult Neurogenesis

TLX has been shown to be highly expressed in the ventricular and subventricular zones (SVZs) of embryonic mouse brains (Li et al., 2008). Knockout of *Tlx* in mice leads to significant microcephaly and severe retinopathies (Chiang & Evans, 1997; Corso-Diaz & Simpson, 2015; Monaghan et al., 1997; Young et al., 2002; Yu et al., 2000). These mice exhibit cortical hypoplasia, limbic system abnormalities, cognitive impairment, and abnormal social behavior, such as aggressive violence (Chiang & Evans, 1997; Monaghan et al., 1997; Murai et al., 2014; Shi et al., 2004; Wong et al., 2010; Young et al., 2002; Zhang et al., 2008).

In the developing mouse brain, the primary role of Tlx is to maintain NSCs in an undifferentiated state (Li et al., 2008). The undifferentiated neural precursor cells in the embryonic brains help to establish a functional central nervous system. Deletion of the *Tlx* gene at embryonic stages results in significant thinning of the neocortex, reduction of cell proliferation, and enlargement of the ventricle. This is consistent with the considerable decrease in neural progenitor cell numbers and reduced proliferative capability of neural progenitor cells in the germinal region of embryonic *Tlx*-knockout brains (Li et al., 2008). The reduced proliferation in embryonic neural progenitor cells could be explained by prolonged cell cycles, increased cell cycle exit, and precocious differentiation of embryonic neural progenitor cells as demonstrated by transient knockdown of *Tlx* via in utero electroporation in embryonic mouse brain cortex (Li et al., 2008). These findings support a critical role of Tlx in controlling cell cycle progression of neural progenitor cells in the developing brains.

In adult mouse brains, *Tlx* is expressed in at least two discrete regions: the subgranular zone (SGZ) of the dentate gyrus (DG) in the hippocampus and the SVZ of the lateral ventricles (Shi et al., 2004). These two regions represent the adult neurogenic niche, where neural stem/progenitor cells generate new neurons in adult brains, a fundamental process known as adult neurogenesis (Gage, 2002).

The Tlx-positive cells in the adult SVZ are relatively quiescent stem cells and knockout of *Tlx* in these cells leads to loss of neurogenesis in the SVZ (Li, Sun, Murai, Ye, & Shi, 2012; Liu et al., 2008, 2010; Shi et al., 2004). *Tlx* is also expressed in activated NSCs and transit-amplifying progenitors in SVZ (Li et al., 2012; Niu, Zou, Shen, & Zhang, 2011; Obernier et al., 2011). Neuroblasts produced from the SVZ migrate along the rostral migratory stream to reach the olfactory bulb, where they become mature neurons.

The Tlx-positive NSCs in the DG produce neurons associated with hippocampal-dependent learning and memory (Murai et al., 2014; Zhang et al., 2008). The Tlx-expressing cells in the adult SGZ include the relatively inactive type 1 NSCs that have radial glia-like processes, and the transiently amplifying type 2 cells that are derived from a subset of type 1 NSCs. The type 2 cells have short processes and rapidly proliferate to generate type 3 cells. The type 3 cells represent immature neuroblasts that can mature into granule neurons that exhibit the capacity to functionally integrate into the existing neural networks (Islam & Zhang, 2015). Loss of *Tlx* expression in the DG results in deficiency of transiently amplifying type 2 cells and neuroblasts (Shi et al., 2004; Zhang et al., 2008).

Conditional knockout of *Tlx* in adult mice induced impaired spatial learning and memory (Zhang et al., 2008). On the other hand, overexpression of *Tlx* using a *Tlx* transgene under the promoter of nestin, a neural progenitor marker, led to increased number of proliferating neural progenitor cells, and newborn neurons in the hippocampal DG (Murai et al., 2014). Accordingly, the *Tlx*-transgenic mice exhibited enhanced spatial learning and memory capacity (Murai et al., 2014). It is worth noting that no tumor was detected in the TLX transgenic mice throughout the course of the study (Murai et al., 2014). Thus, TLX expression in neural progenitor cells is important for adult hippocampal neurogenesis and memory.

2.3 TLX in Senescence and Aging of the Brain

Senescence is the gradual deterioration of functions characteristic of most complex life forms. Studies have linked TLX to cellular senescence and aging (Iwahara et al., 2009; O'Loghlen et al., 2015). TLX has been shown to bind to the promoter of the CBX7 gene to regulate the expression of CBX7, a component of polycomb repressive complex (PRC) that regulates cell senescence (O'Loghlen et al., 2015). CBX7 also represses TLX as part of a regulatory loop. Ectopic TLX expression inhibits cellular senescence, extending cellular lifespan in fibroblasts via CBX7-mediated regulation of p16INK4a, and direct repression of p21CIP1 by TLX. In addition, the expression of TLX also counteracts oncogene-induced senescence (O'Loghlen et al., 2015). Modulation of polycomb function and control of cellular senescence may represent two additional mechanisms by which TLX regulates NSC self-renewal and tumorigenesis (O'Loghlen et al., 2015).

TLX could regulate aging through activating the expression of SIRT1 (Iwahara et al., 2009), a HDAC that has been implicated in aging (Herranz & Serrano, 2010). SIRT1 is highly expressed in the neural progenitor cells and participates in neurogenesis (Hisahara et al., 2008). *Tlx* knockdown reduces SIRT1 expression in neural progenitor cells (Iwahara et al., 2009). SIRT1-mediated deacetylation of genes, such as PGC-1α, FOXO3, p53, and NF-κB, has profound effect on aging, mitochondrial function, apoptosis, and inflammation (Mazucanti et al., 2015). These biological processes and functions are critical in aging and longevity. The association of TLX with the inhibition of cellular senescence and activation of SIRT1 expression suggests that TLX may be a potential target for antiaging and antiage-related neurological diseases.

2.4 TLX in Neurological Diseases

TLX is a key regulator of embryonic and adult neurogenesis and has been genetically linked to bipolar disorder by a genetic association study (Kumar et al., 2008). The nestin promoter-driven *Tlx* transgene expression in *Tlx*-transgenic mice led to improved hippocampus neurogenesis and enhanced learning and memory (Murai et al., 2014), whereas adult-stage conditional knockout of *Tlx* in mice resulted in reduced learning and memory (Zhang et al., 2008). Human TLX shares 99% identity with mouse Tlx in amino acid sequence. The introduction of a human TLX transgene under its own promoter into "Fierce" mice that have germ line *Tlx* deletion was able to correct the pathological defect caused by *Tlx* gene deletion in these mice (Abrahams et al., 2005).

TLX crosstalks with multiple miRNAs, including miR137 and miR219, both of which have been shown to be associated with neurological disorders. miR-137 exhibits the highest association with schizophrenia and bipolar in GWAS study (Ripke et al., 2011). Overexpression of miR-137 in mouse DG by lentiviral transduction resulted in changes in synaptic vesicle pool distribution, impaired induction of mossy fiber long-term potentiation, and deficits in hippocampus-dependent learning and memory (Siegert et al., 2015). This study provides a model in which a psychiatric risk-associated miRNA can regulate multiple targets in a pathway to modulate synaptic function (Siegert et al., 2015).

Elevated expression of miR-219 has been observed in brains of schizophrenia patients (Beveridge, Gardiner, Carroll, Tooney, & Cairns, 2010; Beveridge et al., 2008; Smalheiser et al., 2014), suggesting a potential role of miR-219 in schizophrenia pathogenesis. A 4-bp deletion in the coding region of the disrupted in schizophrenia 1 (*DISC1*) gene has been shown to cosegregate with major psychiatric disorders such as schizophrenia (Sachs et al., 2005). This deletion causes a frameshift and premature termination of translation of DISC1 (Wen et al., 2014). We differentiated induced pluripotent stem cells (iPSCs) derived from schizophrenia patients with the *DISC1* mutation and their isogenic controls with the wild-type *DISC1* gene (Wen et al., 2014) into NSCs (Murai et al., 2016). NSCs derived from schizophrenia patient iPSCs exhibited increased expression of miR-219 and reduced expression of TLX, a negative regulator of miR-219 expression (Murai et al., 2016). Consistent with the essential role of TLX in NSC proliferation and self-renewal (Shi et al., 2004), we observed reduced cell proliferation in schizophrenia NSCs. Overexpression of TLX rescued the proliferative defect of schizophrenia NSCs. Moreover, inhibition of miR219 in schizophrenia NSCs could also rescue the proliferative defect of these cells (Murai et al., 2016) (Fig. 2). This study provides a direct link between dysregulated expression of TLX/miR-219 and DISC1 function as well as schizophrenia and presents a mechanism underlying defective NSC proliferation in schizophrenia. Moreover, this study suggests that TLX and miRNAs could serve as potential therapeutic targets for neuropsychiatric disorders, including schizophrenia.

2.5 TLX in Glioblastoma Tumorigenesis

Glioblastoma multiforme, a grade IV astrocytoma, is the most common and lethal type of brain tumor in adults (Louis et al., 2007). Despite efforts

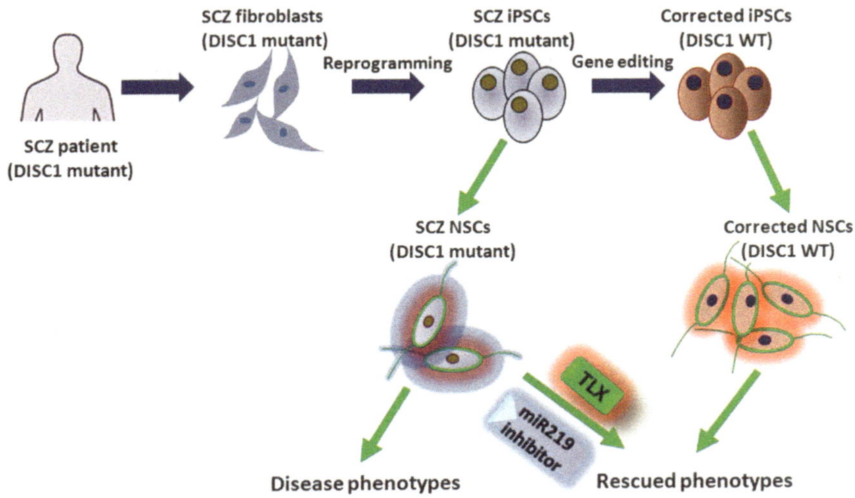

Fig. 2 Disease modeling of schizophrenia (SCZ) using patient iPSC-derived NSCs. Fibroblast cells are isolated from SCZ patients with DISC1 mutations and reprogrammed into SCZ iPSCs. Neural stem cells (NSCs) are derived from DISC1 mutant SCZ iPSCs. The SCZ iPSC-derived NSCs exhibit defects in cell proliferation. Through gene editing, the mutant DISC1 gene can be corrected into wild-type (WT) DISC1. NSCs derived from the gene-corrected iPSCs show normal cell proliferation. The SCZ NSCs can be rescued to assume normal cell proliferation by either overexpression of TLX or treatment with a miR219 inhibitor.

for several decades, the clinical outcome of GBM has not been significantly improved regardless of multimodal treatments including surgery, radiation therapy, and chemotherapy (Sathornsumetee & Rich, 2008). The tumor recurrence is inevitable due to the invasive nature of glioma and the resistance of tumor cells to current therapies (Giese, Bjerkvig, Berens, & Westphal, 2003). Specifically, it is believed that current therapies remain palliative because the treatment is not able to target the glioblastoma stem cell (GSC) subpopulation. The cancer stem cell hypothesis proposes that this GSC represents a population at the top hierarchy of brain tumor, which is more resistant to current therapy than bulk tumor cells, and is able to repopulate the tumor bulk (Bao et al., 2006; Eramo et al., 2006; Galli et al., 2004; Liu et al., 2006; Singh et al., 2004).

TLX has been demonstrated to play a role in brain tumor development. Overexpression of TLX has been observed in subsets of brain tumors (Modena et al., 2006; Taylor et al., 2005). Several studies identified TLX as a critical regulator for gliomagenesis. Liu et al. found that overexpression of TLX along with loss of p53 expression induced glioma in a mouse model (Liu et al., 2010). In addition, TLX is required for the development of brain tumors that are derived from NSCs with deletions of known tumor suppressors such as p21, p53, and Pten (Zou et al., 2012). TLX has also been shown to regulate GSCs. The expression of TLX is elevated in GSCs compared to normal cells and overexpression of TLX in mouse astrocytes promotes the genesis of cells that resemble GSC (Park et al., 2010). A lineage tracing study demonstrated that TLX regulates the self-renewal of GSC in mouse models (Zhu et al., 2014). The TLX-expressing tumor cells are quiescent and can generate TLX-negative tumor cells, indicating that the TLX-expressing tumor cells represent a population of GSC.

Our recent study demonstrated that TLX is critical for GSC self-renewal and tumorigenesis (Cui et al., 2016). We showed that primary GSCs express TLX and nestin. Knockdown of TLX through small hairpin interference RNA (shRNA) dramatically reduced the growth and self-renewal of GSCs. Excitingly, knockdown of TLX dramatically reduced the growth of tumors derived from GSCs and increased the survival of GSC-grafted mice. More significantly, knockdown of TLX in vivo by viral delivery of TLX shRNA suppressed the progression of GSC-initiated tumor in a xenograft tumor model. To be more clinically applicable, TLX siRNA was delivered by a dendrimer-based nanoparticle. A poly(amidoamine) dendermer with a triethanolamine core and branching units growing five generations from the core can form stable nanoparticle complex with TLX siRNA. Treatment of GSC-grafted mice with TLX siRNA-dendrimer nanoparticles resulted in significant suppression of GSC-derived tumor development and extended survival of GSC-grafted mice. This study suggests that targeting TLX in GSC could serve as a potential therapeutic strategy to improve the outcome and survival of glioblastoma patients.

We recently identified TET3 as a novel target of TLX in GSC (Cui et al., 2016). TET3 is a member of TET family proteins that control DNA demethylation by converting 5-methylcytosine (5mC) into 5-hydroxymethylcytosine (5hmC) (Kriaucionis & Heintz, 2009; Tahiliani et al., 2009). The level of 5hmC is decreased in various types of cancers, including malignant gliomas (Jin et al., 2011; Orr, Haffner, Nelson, Yegnasubramanian, & Eberhart, 2012; Xu et al., 2011). We showed that TET3 works as a tumor suppressor downstream of TLX to regulate GSC growth and self-renewal. Knockdown of TLX in GSCs induces TET3 expression. TET3 suppresses the self-renewal and tumorigenesis of GSCs. We further showed that TET3 controls the 5hmC levels at the promoter regions of its downstream targets, such as BTG2 and PPP2R1B, and regulates the expressions of these tumor suppressor genes in GSCs. Future studies to elucidate the function of the TLX–TETs regulatory cascade in GSCs may open new avenues for glioblastoma therapy.

Recently, a three-dimensional organoid culture system was developed to model human glioblastomas. This system can recapitulate the original tumor microenvironment and tumor organization (Hubert et al., 2016). The tumor organoids can grow for months and exhibit regional heterogeneity containing rapidly dividing cells. Located at the outer layer are TLX-expressing cells, surrounding the core. The tumor organoids can be self-organized, similar to primary tumor, with stem cells densely located near organoids periphery where they are exposed to nutrients, oxygen, and growth factors, while inner core mimicking resource-poor hypoxia condition (Hubert et al., 2016). These tumor organoids can potentially be used to identify molecular targets and screen drugs for glioblastoma.

3. PERSPECTIVES

Adult neurogenesis in elderly and neurodegenerative disease patients is an attractive subject because boosting endogenous neurogenesis in the elderly could have a positive impact on the cognition and health of these individuals. In the near future, we may see lot of studies in the converged field of neurodegenerative diseases and adult neurogenesis, aiming to identify factors that can prevent brain deterioration in neurodegenerative diseases. Modulation of TLX expression or activity through miRNAs, upstream regulators, or ligands may provide potential therapeutic strategy for the treatment of aging and age-related neurodegenerative diseases.

Like neurodegenerative diseases, diabetes is another emerging social problem in many countries. Interestingly, the pancreas shares many features and signaling pathways with the brain. The recent findings on TLX in the pancreas and in the diabetes (Shi et al., 2016; Sun et al., 2015) offer a new perspective for the role of TLX in organs besides the brain.

Although biological functions of TLX have been extensively studied in mice, we know little about the function of TLX in humans during brain development. However, the ability of human TLX gene to rescue the brain abnormality and aggressive behavior in *Tlx*-null mice suggests that human TLX has a function in brain development that is similar to Tlx in mice.

Recent breakthroughs in iPSC technology have enabled a new in vitro system for functional modeling of human brain development. These advances, in combination with improved NSC differentiation methods, have established the brain organoid approach (Lancaster & Knoblich, 2014; Lancaster et al., 2013). This is a fundamentally new approach that may allow for in vitro recapitulation of aspects of human brain development. If combined with genome editing technology, such as CRISPR/Cas9 (Cong et al., 2013; Jinek et al., 2012), brain organoids may provide an excellent model to study the functions of TLX in human brain development and diseases.

ACKNOWLEDGMENTS

We apologize to colleagues whose work could not be cited due to space limitations. We would like to thank Jeremy Klein and Jesse Rivas for their critical reading of the manuscript. G.S. was a Herbert Horvitz Fellow. This work was supported by the Sidell Kagan Foundation and California Institute for Regenerative Medicine TR2-01832, RB4-06277, and TRAN1-08525.

Conflict of interest: The authors declare no conflict of interest.

REFERENCES

Abrahams, B. S., Kwok, M. C., Trinh, E., Budaghzadeh, S., Hossain, S. M., & Simpson, E. M. (2005). Pathological aggression in "fierce" mice corrected by human nuclear receptor 2E1. *Journal of Neuroscience*, *25*(27), 6263–6270.

Bao, S., Wu, Q., McLendon, R. E., Hao, Y., Shi, Q., Hjelmeland, A. B., et al. (2006). Glioma stem cells promote radioresistance by preferential activation of the DNA damage response. *Nature*, *444*(7120), 756–760.

Benod, C., Villagomez, R., Filgueira, C. S., Hwang, P. K., Leonard, P. G., Poncet-Montange, G., et al. (2014). The human orphan nuclear receptor tailless (TLX, NR2E1) is druggable. *PLoS One*, *9*(6), e99440.

Beveridge, N. J., Gardiner, E., Carroll, A. P., Tooney, P. A., & Cairns, M. J. (2010). Schizophrenia is associated with an increase in cortical microRNA biogenesis. *Molecular Psychiatry*, *15*(12), 1176–1189.

Beveridge, N. J., Tooney, P. A., Carroll, A. P., Gardiner, E., Bowden, N., Scott, R. J., et al. (2008). Dysregulation of miRNA 181b in the temporal cortex in schizophrenia. *Human Molecular Genetics*, *17*(8), 1156–1168.

Chan, C. M., Fulton, J., Montiel-Duarte, C., Collins, H. M., Bharti, N., Wadelin, F. R., et al. (2013). A signature motif mediating selective interactions of BCL11A with the NR2E/F subfamily of orphan nuclear receptors. *Nucleic Acids Research*, *41*(21), 9663–9679.

Chavali, P. L., Saini, R. K., Matsumoto, Y., Agren, H., & Funa, K. (2011). Nuclear orphan receptor TLX induces Oct-3/4 for the survival and maintenance of adult hippocampal progenitors upon hypoxia. *Journal of Biological Chemistry*, *286*(11), 9393–9404.

Chavali, P. L., Saini, R. K., Zhai, Q., Vizlin-Hodzic, D., Venkatabalasubramanian, S., Hayashi, A., et al. (2014). TLX activates MMP-2, promotes self-renewal of tumor spheres in neuroblastoma and correlates with poor patient survival. *Cell Death & Disease*, *5*, e1502.

Chiang, M. Y., & Evans, R. M. (1997). *Reverse genetic analysis of nuclear receptors, RXRg, RARb, and TLX in mice*. Dissertation San Diego, La Jolla, CA: University of California.

Cong, L., Ran, F. A., Cox, D., Lin, S., Barretto, R., Habib, N., et al. (2013). Multiplex genome engineering using CRISPR/Cas systems. *Science*, *339*(6121), 819–823.

Corso-Diaz, X., & Simpson, E. M. (2015). Nr2e1 regulates retinal lamination and the development of Muller glia, S-cones, and glycineric amacrine cells during retinogenesis. *Molecular Brain*, *8*, 37.

Cui, Q., Yang, S., Ye, P., Tian, E., Sun, G., Zhou, J., et al. (2016). Downregulation of TLX induces TET3 expression and inhibits glioblastoma stem cell self-renewal and tumorigenesis. *Nature Communications*, 7, 10637.

Davis, S. M., Thomas, A. L., Nomie, K. J., Huang, L., & Dierick, H. A. (2014). Tailless and Atrophin control Drosophila aggression by regulating neuropeptide signalling in the pars intercerebralis. *Nature Communications*, *5*, 3177.

Elmi, M., Matsumoto, Y., Zeng, Z. J., Lakshminarasimhan, P., Yang, W., Uemura, A., et al. (2010). TLX activates MASH1 for induction of neuronal lineage commitment of adult hippocampal neuroprogenitors. *Molecular and Cellular Neuroscience*, *45*(2), 121–131.

Eramo, A., Ricci-Vitiani, L., Zeuner, A., Pallini, R., Lotti, F., Sette, G., et al. (2006). Chemotherapy resistance of glioblastoma stem cells. *Cell Death and Differentiation*, *13*(7), 1238–1241.

Estruch, S. B., Buzon, V., Carbo, L. R., Schorova, L., Luders, J., & Estebanez-Perpina, E. (2012). The oncoprotein BCL11A binds to orphan nuclear receptor TLX and potentiates its transrepressive function. *PLoS One*, 7(6), e37963.

Gage, F. H. (2002). Neurogenesis in the adult brain. *Journal of Neuroscience*, *22*(3), 612–613.

Galli, R., Binda, E., Orfanelli, U., Cipelletti, B., Gritti, A., De Vitis, S., et al. (2004). Isolation and characterization of tumorigenic, stem-like neural precursors from human glioblastoma. *Cancer Research*, *64*(19), 7011–7021.

Giese, A., Bjerkvig, R., Berens, M. E., & Westphal, M. (2003). Cost of migration: Invasion of malignant gliomas and implications for treatment. *Journal of Clinical Oncology*, *21*(8), 1624–1636.

Herranz, D., & Serrano, M. (2010). SIRT1: Recent lessons from mouse models. *Nature Reviews Cancer*, *10*(12), 819–823.

Hisahara, S., Chiba, S., Matsumoto, H., Tanno, M., Yagi, H., Shimohama, S., et al. (2008). Histone deacetylase SIRT1 modulates neuronal differentiation by its nuclear translocation. *Proceedings of the National Academy of Sciences of the United States of America*, *105*(40), 15599–15604.

Hu, Y., Luo, M., Ni, N., Den, Y., Xia, J., Chen, J., et al. (2014). Reciprocal actions of microRNA-9 and TLX in the proliferation and differentiation of retinal progenitor cells. *Stem Cells and Development*, *23*(22), 2771–2781.

Huang, Y., Liu, X., & Wang, Y. (2015). MicroRNA-378 regulates neural stem cell proliferation and differentiation in vitro by modulating Tailless expression. *Biochemical and Biophysical Research Communications*, *466*(2), 214–220.

Hubert, C. G., Rivera, M., Spangler, L. C., Wu, Q., Mack, S. C., Prager, B. C., et al. (2016). A three-dimensional organoid culture system derived from human glioblastomas recapitulates the hypoxic gradients and cancer stem cell heterogeneity of tumors found in vivo. *Cancer Research*, *76*(8), 2465–2477.

Islam, M. M., Smith, D. K., Niu, W., Fang, S., Iqbal, N., Sun, G., et al. (2015). Enhancer analysis unveils genetic interactions between TLX and SOX2 in neural stem cells and in vivo reprogramming. *Stem Cell Reports*, *5*(5), 805–815.

Islam, M. M., & Zhang, C. L. (2015). TLX: A master regulator for neural stem cell maintenance and neurogenesis. *Biochimica et Biophysica Acta*, *1849*(2), 210–216.

Iwahara, N., Hisahara, S., Hayashi, T., & Horio, Y. (2009). Transcriptional activation of NAD+-dependent protein deacetylase SIRT1 by nuclear receptor TLX. *Biochemical and Biophysical Research Communications*, *386*(4), 671–675.

Jackson, A., Panayiotidis, P., & Foroni, L. (1998). The human homologue of the Drosophila tailless gene (TLX): Characterization and mapping to a region of common deletion in human lymphoid leukemia on chromosome 6q21. *Genomics*, *50*(1), 34–43.

Jin, S. G., Jiang, Y., Qiu, R., Rauch, T. A., Wang, Y., Schackert, G., et al. (2011). 5-Hydroxymethylcytosine is strongly depleted in human cancers but its levels do not correlate with IDH1 mutations. *Cancer Research*, *71*(24), 7360–7365.

Jinek, M., Chylinski, K., Fonfara, I., Hauer, M., Doudna, J. A., & Charpentier, E. (2012). A programmable dual-RNA-guided DNA endonuclease in adaptive bacterial immunity. *Science*, *337*(6096), 816–821.

Kriaucionis, S., & Heintz, N. (2009). The nuclear DNA base 5-hydroxymethylcytosine is present in Purkinje neurons and the brain. *Science*, *324*(5929), 929–930.

Kumar, R. A., McGhee, K. A., Leach, S., Bonaguro, R., Maclean, A., Aguirre-Hernandez, R., et al. (2008). Initial association of NR2E1 with bipolar disorder and identification of candidate mutations in bipolar disorder, schizophrenia, and aggression through resequencing. *American Journal of Medical Genetics. Part B, Neuropsychiatric Genetics: The Official Publication of the International Society of Psychiatric Genetics*, *147B*(6), 880–889.

Lancaster, M. A., & Knoblich, J. A. (2014). Organogenesis in a dish: Modeling development and disease using organoid technologies. *Science*, *345*(6194), 1247125.

Lancaster, M. A., Renner, M., Martin, C. A., Wenzel, D., Bicknell, L. S., Hurles, M. E., et al. (2013). Cerebral organoids model human brain development and microcephaly. *Nature*, *501*(7467), 373–379.

Li, S., Sun, G., Murai, K., Ye, P., & Shi, Y. (2012). Characterization of TLX expression in neural stem cells and progenitor cells in adult brains. *PLoS One*, 7(8), e43324.

Li, W., Sun, G., Yang, S., Qu, Q., Nakashima, K., & Shi, Y. (2008). Nuclear receptor TLX regulates cell cycle progression in neural stem cells of the developing brain. *Molecular Endocrinology*, *22*(1), 56–64.

Liu, H. K., Belz, T., Bock, D., Takacs, A., Wu, H., Lichter, P., et al. (2008). The nuclear receptor tailless is required for neurogenesis in the adult subventricular zone. *Genes and Development*, *22*(18), 2473–2478.

Liu, H. K., Wang, Y., Belz, T., Bock, D., Takacs, A., Radlwimmer, B., et al. (2010). The nuclear receptor tailless induces long-term neural stem cell expansion and brain tumor initiation. *Genes and Development*, *24*(7), 683–695.

Liu, G., Yuan, X., Zeng, Z., Tunici, P., Ng, H., Abdulkadir, I. R., et al. (2006). Analysis of gene expression and chemoresistance of CD133+ cancer stem cells in glioblastoma. *Molecular Cancer*, *5*, 67.

Louis, D. N., Ohgaki, H., Wiestler, O. D., Cavenee, W. K., Burger, P. C., Jouvet, A., et al. (2007). The 2007 WHO classification of tumours of the central nervous system. *Acta Neuropathologica*, *114*(2), 97–109.

Mangelsdorf, D. J., Thummel, C., Beato, M., Herrlich, P., Schutz, G., Umesono, K., et al. (1995). The nuclear receptor superfamily: The second decade. *Cell*, *83*(6), 835–839.

Mazucanti, C. H., Cabral-Costa, J. V., Vasconcelos, A. R., Andreotti, D. Z., Scavone, C., & Kawamoto, E. M. (2015). Longevity pathways (mTOR, SIRT, insulin/IGF-1) as key modulatory targets on aging and neurodegeneration. *Current Topics in Medicinal Chemistry*, *15*(21), 2116–2138.

Modena, P., Lualdi, E., Facchinetti, F., Veltman, J., Reid, J. F., Minardi, S., et al. (2006). Identification of tumor-specific molecular signatures in intracranial ependymoma and association with clinical characteristics. *Journal of Clinical Oncology*, *24*(33), 5223–5233.

Monaghan, A. P., Bock, D., Gass, P., Schwager, A., Wolfer, D. P., Lipp, H. P., et al. (1997). Defective limbic system in mice lacking the tailless gene. *Nature*, *390*(6659), 515–517.

Monaghan, A. P., Grau, E., Bock, D., & Schutz, G. (1995). The mouse homolog of the orphan nuclear receptor tailless is expressed in the developing forebrain. *Development*, *121*(3), 839–853.

Murai, K., Qu, Q., Sun, G., Ye, P., Li, W., Asuelime, G., et al. (2014). Nuclear receptor TLX stimulates hippocampal neurogenesis and enhances learning and memory in a transgenic mouse model. *Proceedings of the National Academy of Sciences of the United States of America*, *111*(25), 9115–9120.

Murai, K., Sun, G., Ye, P., Tian, E., Yang, S., Cui, Q., et al. (2016). The TLX-miR-219 cascade regulates neural stem cell proliferation in neurodevelopment and schizophrenia iPSC model. *Nature Communications*, *7*, 10965.

Ni, N., Zhang, D., Xie, Q., Chen, J., Wang, Z., Deng, Y., et al. (2014). Effects of let-7b and TLX on the proliferation and differentiation of retinal progenitor cells in vitro. *Scientific Reports*, *4*, 6671.

Niu, W., Zou, Y., Shen, C., & Zhang, C. L. (2011). Activation of postnatal neural stem cells requires nuclear receptor TLX. *Journal of Neuroscience*, *31*(39), 13816–13828.

Obernier, K., Simeonova, I., Fila, T., Mandl, C., Holzl-Wenig, G., Monaghan-Nichols, P., et al. (2011). Expression of Tlx in both stem cells and transit amplifying progenitors regulates stem cell activation and differentiation in the neonatal lateral subependymal zone. *Stem Cells*, *29*(9), 1415–1426.

O'Loghlen, A., Martin, N., Krusche, B., Pemberton, H., Alonso, M. M., Chandler, H., et al. (2015). The nuclear receptor NR2E1/TLX controls senescence. *Oncogene*, *34*(31), 4069–4077.

Orr, B. A., Haffner, M. C., Nelson, W. G., Yegnasubramanian, S., & Eberhart, C. G. (2012). Decreased 5-hydroxymethylcytosine is associated with neural progenitor phenotype in normal brain and shorter survival in malignant glioma. *PLoS One*, *7*(7), e41036.

Park, H. J., Kim, J. K., Jeon, H. M., Oh, S. Y., Kim, S. H., Park, M. J., et al. (2010). The neural stem cell fate determinant TLX promotes tumorigenesis and genesis of cells resembling glioma stem cells. *Molecules and Cells*, *30*(5), 403–408.

Pignoni, F., Baldarelli, R. M., Steingrimsson, E., Diaz, R. J., Patapoutian, A., Merriam, J. R., et al. (1990). The Drosophila gene tailless is expressed at the embryonic termini and is a member of the steroid receptor superfamily. *Cell*, *62*(1), 151–163.

Qin, S., Niu, W., Iqbal, N., Smith, D. K., & Zhang, C. L. (2014). Orphan nuclear receptor TLX regulates astrogenesis by modulating BMP signaling. *Frontiers in Neuroscience*, *8*, 74.

Qu, Q., Sun, G., Li, W., Yang, S., Ye, P., Zhao, C., et al. (2010). Orphan nuclear receptor TLX activates Wnt/beta-catenin signalling to stimulate neural stem cell proliferation and self-renewal. *Nature Cell Biology*, *12*(1), 31–40. Suppl. pp. 31–39.

Reinhardt, R., Centanin, L., Tavhelidse, T., Inoue, D., Wittbrodt, B., Concordet, J. P., et al. (2015). Sox2, Tlx, Gli3, and Her9 converge on Rx2 to define retinal stem cells in vivo. *The EMBO Journal*, *34*(11), 1572–1588.

Ripke, S., Sanders, A. R., Kendler, K. S., Levinson, D. F., Sklar, P., Holmans, P. A., et al. (2011). Genome-wide association study identifies five new schizophrenia loci. *Nature Genetics*, *43*(10), 969–976.

Sachs, N. A., Sawa, A., Holmes, S. E., Ross, C. A., DeLisi, L. E., & Margolis, R. L. (2005). A frameshift mutation in disrupted in schizophrenia 1 in an American family with schizophrenia and schizoaffective disorder. *Molecular Psychiatry*, *10*(8), 758–764.

Sathornsumetee, S., & Rich, J. N. (2008). Designer therapies for glioblastoma multiforme. *The Annals of the New York Academy of Sciences*, *1142*, 108–132.

Shi, X., Deng, H., Dai, Z., Xu, Y., Xiong, X., Ma, P., et al. (2016). Nr2e1 deficiency augments palmitate-induced oxidative stress in beta cells. *Oxidative Medicine and Cellular Longevity*, *2016*, 9648769.

Shi, Y., Lie, C. D., Taupin, P., Nakashima, K., Ray, J., Yu, R. T., et al. (2004). Expression and function of orphan nuclear receptor TLX in adult neural stem cells. *Nature*, *427*(6969), 78–83.

Shimozaki, K., Zhang, C. L., Suh, H., Denli, A. M., Evans, R. M., & Gage, F. H. (2012). SRY-box-containing gene 2 regulation of nuclear receptor tailless (Tlx) transcription in adult neural stem cells. *Journal of Biological Chemistry*, *287*(8), 5969–5978.

Siegert, S., Seo, J., Kwon, E. J., Rudenko, A., Cho, S., Wang, W., et al. (2015). The schizophrenia risk gene product miR-137 alters presynaptic plasticity. *Nature Neuroscience*, *18*(7), 1008–1016.

Singh, S. K., Hawkins, C., Clarke, I. D., Squire, J. A., Bayani, J., Hide, T., et al. (2004). Identification of human brain tumour initiating cells. *Nature*, *432*(7015), 396–401.

Smalheiser, N. R., Lugli, G., Zhang, H., Rizavi, H., Cook, E. H., & Dwivedi, Y. (2014). Expression of microRNAs and other small RNAs in prefrontal cortex in schizophrenia, bipolar disorder and depressed subjects. *PLoS One*, *9*(1), e86469.

Sun, G., Alzayady, K., Stewart, R., Ye, P., Yang, S., Li, W., et al. (2010). Histone demethylase LSD1 regulates neural stem cell proliferation. *Molecular and Cellular Biology*, *30*(8), 1997–2005.

Sun, L., Deng, H., He, L., Hu, X., Huang, Q., Xue, J., et al. (2015). The relationship between NR2E1 and subclinical inflammation in newly diagnosed type 2 diabetic patients. *Journal of Diabetes and Its Complications*, *29*(4), 589–594.

Sun, G., & Shi, Y. (2010). Nuclear receptors in stem cells and their therapeutic potential. *Advanced Drug Delivery Reviews*, *62*(13), 1299–1306.

Sun, G., Ye, P., Murai, K., Lang, M. F., Li, S., Zhang, H., et al. (2011). miR-137 forms a regulatory loop with nuclear receptor TLX and LSD1 in neural stem cells. *Nature Communications*, *2*, 529.

Sun, G., Yu, R. T., Evans, R. M., & Shi, Y. (2007). Orphan nuclear receptor TLX recruits histone deacetylases to repress transcription and regulate neural stem cell proliferation. *Proceedings of the National Academy of Sciences of the United States of America*, *104*(39), 15282–15287.

Tahiliani, M., Koh, K. P., Shen, Y., Pastor, W. A., Bandukwala, H., Brudno, Y., et al. (2009). Conversion of 5-methylcytosine to 5-hydroxymethylcytosine in mammalian DNA by MLL partner TET1. *Science*, *324*(5929), 930–935.

Taylor, M. D., Poppleton, H., Fuller, C., Su, X., Liu, Y., Jensen, P., et al. (2005). Radial glia cells are candidate stem cells of ependymoma. *Cancer Cell*, *8*(4), 323–335.

Wang, L., Rajan, H., Pitman, J. L., McKeown, M., & Tsai, C. C. (2006). Histone deacetylase-associating atrophin proteins are nuclear receptor corepressors. *Genes and Development*, *20*(5), 525–530.

Wen, Z., Nguyen, H. N., Guo, Z., Lalli, M. A., Wang, X., Su, Y., et al. (2014). Synaptic dysregulation in a human iPS cell model of mental disorders. *Nature*, *515*, 414–418.

Wong, B. K., Hossain, S. M., Trinh, E., Ottmann, G. A., Budaghzadeh, S., Zheng, Q. Y., et al. (2010). Hyperactivity, startle reactivity and cell-proliferation deficits are resistant to chronic lithium treatment in adult Nr2e1(frc/frc) mice. *Genes, Brain and Behavior*, *9*(7), 681–694.

Wu, D., Yu, S., Jia, L., Zou, C., Xu, Z., Xiao, L., et al. (2015). Orphan nuclear receptor TLX functions as a potent suppressor of oncogene-induced senescence in prostate cancer via its transcriptional co-regulation of the CDKN1A (p21(WAF1) (/) (CIP1)) and SIRT1 genes. *Journal of Pathology*, *236*(1), 103–115.

Xu, W., Yang, H., Liu, Y., Yang, Y., Wang, P., Kim, S. H., et al. (2011). Oncometabolite 2-hydroxyglutarate is a competitive inhibitor of alpha-ketoglutarate-dependent dioxygenases. *Cancer Cell*, *19*(1), 17–30.

Yokoyama, A., Takezawa, S., Schule, R., Kitagawa, H., & Kato, S. (2008). Transrepressive function of TLX requires the histone demethylase LSD1. *Molecular and Cellular Biology*, *28*(12), 3995–4003.

Young, K. A., Berry, M. L., Mahaffey, C. L., Saionz, J. R., Hawes, N. L., Chang, B., et al. (2002). Fierce: A new mouse deletion of Nr2e1; violent behaviour and ocular abnormalities are background-dependent. *Behavioural Brain Research, 132*(2), 145–158.
Yu, R. T., Chiang, M. Y., Tanabe, T., Kobayashi, M., Yasuda, K., Evans, R. M., et al. (2000). The orphan nuclear receptor Tlx regulates Pax2 and is essential for vision. *Proceedings of the National Academy of Sciences of the United States of America, 97*(6), 2621–2625.
Yu, R. T., McKeown, M., Evans, R. M., & Umesono, K. (1994). Relationship between Drosophila gap gene tailless and a vertebrate nuclear receptor Tlx. *Nature, 370*(6488), 375–379.
Zeng, Z. J., Johansson, E., Hayashi, A., Chavali, P. L., Akrap, N., Yoshida, T., et al. (2012). TLX controls angiogenesis through interaction with the von Hippel–Lindau protein. *BiologyOpen, 1*(6), 527–535.
Zhang, C. L., Zou, Y., He, W., Gage, F. H., & Evans, R. M. (2008). A role for adult TLX-positive neural stem cells in learning and behaviour. *Nature, 451*(7181), 1004–1007.
Zhang, C. L., Zou, Y., Yu, R. T., Gage, F. H., & Evans, R. M. (2006). Nuclear receptor TLX prevents retinal dystrophy and recruits the corepressor atrophin1. *Genes and Development, 20*(10), 1308–1320.
Zhao, C., Sun, G., Li, S., Lang, M., Yang, S., Li, W., et al. (2010). microRNA let-7b regulates neural stem cell proliferation and differentiation by targeting nuclear receptor TLX signaling. *Proceedings of the National Academy of Sciences of the United States of America, 107*, 1876–1881.
Zhao, C., Sun, G., Li, S., & Shi, Y. (2009). A feedback regulatory loop involving microRNA-9 and nuclear receptor TLX in neural stem cell fate determination. *Nature Structural and Molecular Biology, 16*(4), 365–371.
Zhao, C., Sun, G., Ye, P., Li, S., & Shi, Y. (2013). MicroRNA let-7d regulates the TLX/microRNA-9 cascade to control neural cell fate and neurogenesis. *Scientific Reports, 3*, 1329.
Zhi, X., Zhou, X. E., He, Y., Searose-Xu, K., Zhang, C. L., Tsai, C. C., et al. (2015). Structural basis for corepressor assembly by the orphan nuclear receptor TLX. *Genes and Development, 29*(4), 440–450.
Zhu, Z., Khan, M. A., Weiler, M., Blaes, J., Jestaedt, L., Geibert, M., et al. (2014). Targeting self-renewal in high-grade brain tumors leads to loss of brain tumor stem cells and prolonged survival. *Cell Stem Cell, 15*, 185–198.
Zou, Y., Niu, W., Qin, S., Downes, M., Burns, D. K., & Zhang, C. L. (2012). The nuclear receptor TLX is required for gliomagenesis within the adult neurogenic niche. *Molecular and Cellular Biology, 32*(23), 4811–4820.

CHAPTER TEN

COUP-TF Genes, Human Diseases, and the Development of the Central Nervous System in Murine Models

Xiong Yang, Su Feng[1], Ke Tang[2]
Institute of Life Science, Nanchang University, Nanchang, Jiangxi, China
[2]Corresponding author: e-mail address: ktang.sc@gmail.com

Contents

Abstract

COUP-TFI and -TFII are members of the steroid/thyroid nuclear receptor superfamily. Recent clinical studies reveal that *COUP-TFI* gene mutations are associated with Bosch–Boonstra–Schaaf optic atrophy syndrome displaying symptoms of optic atrophy,

[1] Present address: Institute of Biochemistry and Cell Biology, Shanghai Institutes for Biological Sciences, Chinese Academy of Sciences, Shanghai 200031, China.

Current Topics in Developmental Biology, Volume 125
ISSN 0070-2153
http://dx.doi.org/10.1016/bs.ctdb.2016.12.002

intellectual disability, hypotonia, seizure, autism spectrum disorders, oromotor dysfunction, thin corpus callosum, or hearing defects, and *COUP-TFII* gene mutations lead to congenital heart defects and/or congenital diaphragmatic hernia with developmental delay and mental defects. In this review, we first describe the functions of *COUP-TF* genes in the morphogenesis of mouse forebrain including cerebral cortex, hippocampus, amygdala complex, hypothalamus, and cortical interneuron. Then, we address their roles in the development of cerebellum, glial cells, neural crest cells, and adult neuronal stem cells. Clearly, the investigations on the functions of *COUP-TF* genes in the developing mouse central nervous system will benefit not only the understanding of neurodevelopment, but also the etiology of human mental diseases.

ABBREVIATIONS

BBSOAS Bosch–Boonstra–Schaaf optic atrophy syndrome
BMA basomedial amygdala nucleus
CA cornu ammonis field
CDH congenital diaphragmatic hernia
CGE caudal ganglionic eminence
CHD congenital heart defects
ChIP chromatin immunoprecipitation
CNS central nervous system
COUP-TFs chick ovalbumin upstream promoter-transcription factors
CR calretinin
DBD DNA-binding domain
DG dentate gyrus
ESCs embryonic stem cells
hNCCs human neural crest cells
LBD ligand-binding domain
LGE lateral ganglionic eminence
MC4R melanocortin-4 receptor
MGE medial ganglionic eminence
NCCs neural crest cells
NPCs neuronal progenitor cells
NSCs neural stem cells
OB olfactory bulb
POA preoptic area
PV parvalbumin
RMS rostral migratory stream
SST somatostatin
SVZ subventricular zone
VIP vasoactive intestinal peptide
VMH ventromedial hypothalamic nucleus
VZ ventricular zone

1. INTRODUCTION

Chick ovalbumin upstream promoter-transcription factors (COUP-TFs) belong to the steroid/thyroid nuclear receptor superfamily. In human, there are two COUP-TFs: COUP-TFI (also known as NR2F1, according to the Nuclear Receptors Nomenclature Committee 1999) and COUP-TFII (also known as NR2F2). COUP-TFs contain a DNA-binding domain (DBD) and a putative ligand-binding domain (LBD). Their DBDs and LBDs are almost identical among different species. The N-terminal domains of COUP-TFs are diverse with only 45% identity. Both COUP-TFI and -TFII can either positively or negatively modulate the expression of their downstream genes through different mechanisms (Lin, Qin, Tang, Tsai, & Tsai, 2011; Tang, Tsai, & Tsai, 2015; Tsai & Tsai, 1997).

In the central nervous system (CNS), *COUP-TFI* and *-TFII* genes are expressed in broad areas including brain, spinal cord, and eye (Armentano et al., 2007; Armentano, Filosa, Andolfi, & Studer, 2006; Faedo et al., 2008; Lodato et al., 2011; Lutz et al., 1994; Qiu et al., 1994; Tang, Rubenstein, Tsai, & Tsai, 2012; Tang et al., 2010). The roles of *COUP-TF* genes in the eye and the ocular disease have recently been reviewed (Tang et al., 2015). In this chapter, we are mainly focusing on the overview of *COUP-TF* genes in human diseases and the development of the CNS, especially the forebrain.

2. *COUP-TF* GENES AND HUMAN DISEASES

The human *COUP-TFI* gene is mapped to chromosome 5 at 5q14. The previous clinical findings revealed that 5q14.3-15 deletions, including *COUP-TFI* gene, are associated with optic atrophy, dysmorphism, global developmental delay, mental retardation, and epilepsy (Al-Kateb et al., 2013; Brown et al., 2009; Cardoso et al., 2009). The latest clinical studies advance the understanding of *COUP-TFI* gene in ocular defects with intellectual disability. Except for the deletion mutations, various point mutations at the coding or the noncoding region of *COUP-TFI* gene are the reason for Bosch–Boonstra–Schaaf optic atrophy syndrome (BBSOAS). The majority of point mutations are identified in the DBD of COUP-TFI protein (Fig. 1A and B) (Bosch et al., 2016, 2014; Chen et al., 2016). BBSOAS patients also display

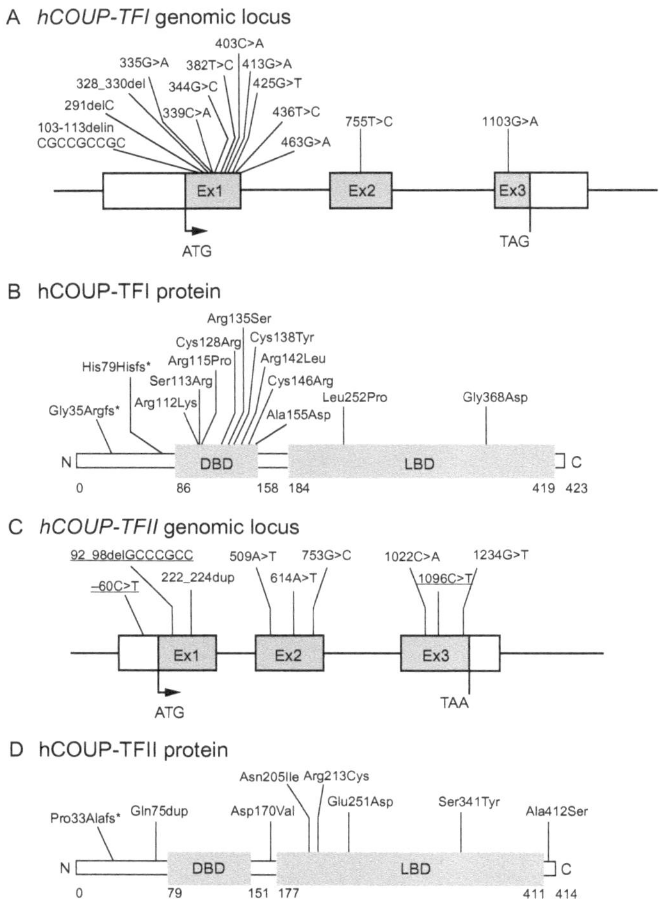

Fig. 1 The summary of the point mutations of *COUP-TFI* and *-TFII* genes associated with human diseases. (A) The point mutations of the *COUP-TFI* gene identified in Bosch–Boonstra–Schaaf optic atrophy syndrome. (B) The coding changes caused by the point mutations in COUP-TFI protein. (C) The point mutations of the *COUP-TFII* gene identified in congenital heart defects (CHD) and congenital diaphragmatic hernia (CDH). The mutations underlined are from CDH patients. (D) The coding changes caused by the point mutations in COUP-TFII protein. *C*, C-terminus; *DBD*, DNA-binding domain; *Ex*, exon; *LBD*, ligand-binding domain; *N*, N-terminus.

symptoms of hypotonia, seizure, autism spectrum disorders, oromotor dysfunction, thin corpus callosum, or hearing defects (Chen et al., 2016).

The human *COUP-TFII* gene is mapped to chromosome 15q at 15q26. Accumulating clinical evidence reveals that 15q26 deletions, in which resides the *COUP-TFII* gene, are highly associated with congenital diaphragmatic hernia (CDH) (Arrington et al., 2012; Klaassens et al., 2005; Slavotinek et al., 2006). Especially, coding variants in *COUP-TFII* gene are possible causes in some CDH patients (High et al., 2016; Longoni et al., 2015). Moreover, patients with various 15q26 deletions generate heart defects accompanying with or without CDH symptom (Arrington et al., 2012; Klaassens et al., 2005; Nakamura et al., 2011; Poot et al., 2007; Rump et al., 2008; Slavotinek et al., 2006). Recent studies show that either missense mutations or in-frame duplication mutations at *COUP-TFII* gene coding region lead to congenital heart defects (CHD) or CDH (Al Turki et al., 2014; High et al., 2016; Longoni et al., 2015). However, none of the point mutations identified in CDH or CHD patients is located in the DBD of COUP-TFII protein (Fig. 1C and D). Consistently, heart defects and CDH phenotype are observed in different *COUP-TFII* gene conditional knockout mouse models, respectively (Lin et al., 2012; Pereira, Qiu, Zhou, Tsai, & Tsai, 1999; You et al., 2005). Intriguingly, many patients with the 15q26 deletions also generate intrauterine growth retardation, developmental delay, and mental defects.

Nonetheless, so far, how *COUP-TFI* gene mutations lead to hypotonia, seizure, autism spectrum disorders, oromotor dysfunction, whether and how *COUP-TFII* gene mutations contribute to mental defects, have not been clearly elucidated. The studies on the functions of *COUP-TF* genes in the developing mouse CNS will benefit our understanding of the etiology of the associated human mental defects.

3. BRIEF OVERVIEW OF THE EARLY CNS DEVELOPMENT

CNS development is one of the most complicated and hierarchical events in the mammalian development. In murine models, the CNS originates from dorsal ectoderm as indicated by the presence of neuroepithelial cells at the neural plate. Neuroepithelial cells in the neural plate proliferate quickly. The neural plate then generates the neural fold, and subsequently closes to form the neural tube. During the early patterning of the CNS, forebrain, midbrain, hindbrain, and spinal cord are specified along the anterior–posterior axis of the neural tube. Following the patterning of the CNS, other critical events such as regionalization, neurogenesis, and neural circuit formation contribute to the formation of a mature and functional CNS (Hebert & Fishell, 2008; Miller & Gauthier, 2007; Nord, Pattabiraman, Visel, & Rubenstein, 2015; Rallu, Corbin, & Fishell, 2002; Temple, 2001).

The forebrain, a rostral part of the CNS, is comprised of telencephalon and diencephalon. The development of the telencephalon is one of the best characterized. In the early mouse embryo, the telencephalon is divided into pallium and subpallium along the dorsal–ventral axis. There are four major proliferating zones in the pallium including dorsal pallium, medial pallium, lateral pallium, and ventral pallium. Following the development, dorsal pallium mainly gives rise to cerebral cortex, medial pallium to hippocampus, lateral pallium to olfactory cortex and amygdala complex, and ventral pallium to claustrum and amygdala complex. There are three major proliferating zones in the subpallium including lateral ganglionic eminence (LGE), medial ganglionic eminence (MGE), and caudal ganglionic eminence (CGE). LGE gives rise to caudate putamen and interneurons in olfactory bulb (OB), MGE to globus pallidus and interneurons, and CGE to amygdala complex and interneurons (Guillemot, 2005; Hebert & Fishell, 2008; Nord et al., 2015). So far, the intrinsic molecular mechanisms mediating the telencephalon development have been fully elucidated.

4. *COUP-TF* GENES AND THE DEVELOPMENT OF DORSAL FOREBRAIN

4.1 *COUP-TFI* Gene and the Regionalization of Cerebral Cortex

The cerebral cortex originates from the dorsal pallium of the telencephalon and is a key region in the CNS to process different kinds of sensory, motor, and higher order information (Sur & Rubenstein, 2005). Tangentially, the cerebral cortex is organized around four main primary areas including sensory cortex, visual cortex, auditory cortex, and motor cortex. These primary areas form reciprocal area-specific connections with thalamic nuclei, such as sensory cortex with ventroposterior nucleus, visual cortex with dorsal lateral geniculate, auditory cortex with medial geniculate, and motor cortex with ventrolateral and ventromedial thalamic nuclei (Rakic, 1988). Accumulating evidence in mouse studies reveals that intrinsic regulatory networks initiated by secreted morphogens participate in the regulation of the regionalization of the cerebral cortex (Hebert & Fishell, 2008; Hirabayashi & Gotoh, 2010; Hoerder-Suabedissen & Molnar, 2015).

At E11.5, at the onset of the cerebral corticogenesis in mouse, COUP-TFI expression is higher at lateral and caudal cortex (Zhou, Tsai, & Tsai, 2001). The high caudolateral expression gradient of COUP-TFI in the cortex is maintained after birth. The expression profiles of COUP-TFI in the developing cortex suggest that *COUP-TFI* gene may play crucial functions during the regionalization of the cerebral cortex. Indeed, in *COUP-TFI* null mutant, the expression patterns of region-specific marker genes such as *Id2*, *RORβ*, and *Cadherin 8* are altered, accompanied with compromised connections between the cortex and the thalamus (Zhou et al., 2001). Moreover, evidence from *COUP-TFI* gene conditional knockout mice with *Emx1Cre* reveals that compared with the control (Fig. 2A), the primary motor cortex is expanded in the mutant at the expense of the primary somatosensory cortex, primary visual cortex, and primary auditory cortex (Fig. 2B) (Armentano et al., 2007). *COUP-TFI* gene may also function in postmitotic neurons to specify sensory cortex during the corticogenesis (Alfano, Magrinelli, Harb, Hevner, & Studer, 2014). Pax6 and Emx2 have been identified as two essential intrinsic factors to ensure the regionalization of the cerebral cortex (Bishop, Goudreau, & O'Leary, 2000; Mallamaci, Muzio, Chan, Parnavelas, & Boncinelli, 2000). Interestingly, their expression is not altered in *COUP-TFI* null mutant (Zhou et al., 2001). Thus, COUP-TFI may cooperate with other key factors such as, Pax6, Emx2, and Sp8, to program the regionalization

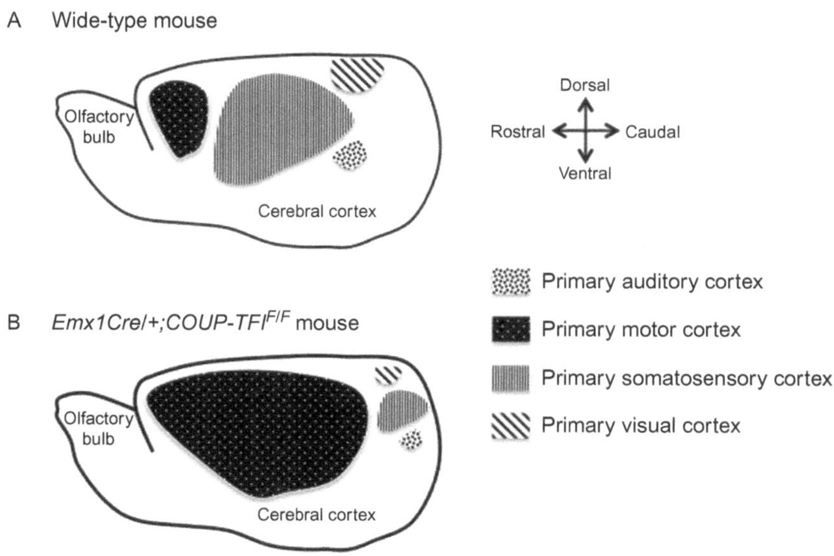

Fig. 2 *COUP-TFI* gene is essential for the regionalization of the cerebral cortex. Compared with the control (A), the primary motor cortex is expanded at the expense of the primary somatosensory cortex, primary visual cortex, and primary auditory cortex in *Emx1Cre/+;COUP-TFI$^{F/F}$* mutant mouse (B).

of the cerebral cortex (Greig, Woodworth, Galazo, Padmanabhan, & Macklis, 2013).

4.2 *COUP-TFI* Gene and Neurogenesis in Cerebral Cortex

There are six layers in the mammalian cortex, and each layer contains various sorts of excitatory projection neurons with distinct morphologies and connectivity (Arlotta et al., 2005; Joshi et al., 2008; Lai et al., 2008; Leone, Srinivasan, Chen, Alcamo, & McConnell, 2008; Molyneaux, Arlotta, Hirata, Hibi, & Macklis, 2005; Molyneaux, Arlotta, Menezes, & Macklis, 2007). Cortical projection neurons are derived from neuroepithelial cells localized at the dorsal pallium. Following the development of the early mouse embryo, neuroepithelial cells proliferate quickly to generate radial glial cells with the characteristics of neural stem cells (NSCs). Radial glial cells give rise to a series of neuronal progenitor cells (NPCs), and then differentiate to the mature cortical projection neurons. In the dorsal telencephalon, the earliest differentiated cortical projection neurons migrate into the preplate to form the cortical plate, which separates the preplate into an upper marginal zone and a lower subplate layer. Layer VI projection neurons settle down in the cortical plate first. Later-forming cortical neurons migrate past the previous layers and form superficial layers of layer V, IV, III, II projection neurons in an "inside out" pattern (Angevine & Sidman, 1961; Berry & Rogers, 1965; Luskin & Shatz, 1985; McConnell, 1995; Rakic, 1974). Multiple transcription factors such as Fezf2, Ctip2, Bhlhb5, Lom4, and Cux2 play crucial roles in the specification of distinct neuron identities in the cerebral cortex (Greig et al., 2013; Hoerder-Suabedissen & Molnar, 2015).

COUP-TFI is highly expressed in the NPCs of dorsal telencephalon and the mature neurons in the adult cortex. In *COUP-TFI* gene null mutant mice, the differentiation of subplate neurons is compromised. The defects of subplate neurons lead to the failed formation of the connection between layer IV neurons and thalamic nuclei, which results in the abnormal apoptosis of layer IV neurons in *COUP-TFI* null mouse (Zhou et al., 1999). There are two groups of neurons in layer V, which express Fez1 or ER81, respectively (Molnar & Cheung, 2006). There are more Fez1+ neurons at the expense of ER81+ neurons in *COUP-TFI* null mutant (Faedo et al., 2008). Furthermore, the presumptive corticothalamic neurons of layer IV gain the characteristics of corticospinal motor neurons of layer V and VI in the sensory cortex of *Emx1Cre/+;COUP-TFI*$^{F/F}$ mutant mouse

(Tomassy et al., 2010). The naïve corticospinal motor neurons usually send axons to cerebral peduncle; however, the mutant corticospinal motor neurons establish abnormal connections between the prospective sensory cortex and cervical, thoracic, lumbar spinal cord segments, which could be the cause for the impaired skilled motor behavior of the mutant mice. The *COUP-TFI* gene may program the temporal neurogenesis of corticospinal motor neurons in sensory cortex to ensure the fine motor skills of the adult mouse (Tomassy et al., 2010). Clearly, *COUP-TFI* gene is required for neurogenesis of some cortical projection neurons in layer IV, V, and VI.

4.3 *COUP-TFI* and the Development of Dorsal Hippocampus

The hippocampus, which originates from the medial pallium of the dorsal telencephalon, plays imperative roles in learning and memory. The mammalian hippocampus mainly contains cornu ammonis field (CA) and dentate gyrus (DG). There are CA1, CA2, and CA3 in the rodent hippocampus, and an additional CA4 in the primate and the human hippocampus (Strange, Witter, Lein, & Moser, 2014). Several recent studies reveal that dorsal hippocampus and ventral hippocampus are not only molecularly distinguished, but also functionally segregated. The dorsal hippocampus is associated with the spatial learning and memory, and the ventral hippocampus is involved in the emotional behavior (Dong, Swanson, Chen, Fanselow, & Toga, 2009; Fanselow & Dong, 2010; Thompson et al., 2008).

Following the development of the mouse hippocampus, COUP-TFI is constitutively expressed in the early medial pallium, developing Ammon's horns and DG, and mature CA1, CA2, CA3, and DG (Flore et al., 2016). COUP-TFII is highly expressed in the ventral hippocampus of the adult mouse. In *COUP-TFI* conditional knockout mouse with *Emx1Cre*, the development of the dorsal hippocampus is severely compromised with normal field specification and neuronal distribution. Furthermore, the connection between the dorsal hippocampus and the dorsal entorhinal cortex is impaired in the mutant. *COUP-TFI* mutant mice display behavior defects in spatial memory, but not in nonspatial learning and anxiety tasks (Flore et al., 2016). Most likely, the function of *COUP-TFI* gene in the hippocampus is associated with some symptoms reported in the BBSOAS patients such as intellectual disability (Bosch et al., 2014; Chen et al., 2016).

5. *COUP-TF* GENES AND THE DEVELOPMENT OF VENTRAL FOREBRAIN

5.1 *COUP-TF* genes and the Differentiation of Cortical Interneurons

The cerebral cortex contains two types of neurons, glutamatergic excitatory projection neurons and GABAergic inhibitory interneurons. Different from the local origin of the projection neurons, the cortical interneurons are mainly derived from MGE and CGE in the subpallium. The NPCs in MGE mainly generate parvalbumin (PV)- and somatostatin (SST)-expressing interneurons, which are mediated by transcription factors such as Nkx2.1 and Lhx6. The NPCs in CGE give rise to calretinin (CR)- and vasoactive intestinal peptide (VIP)-expressing interneurons (Fishell & Rudy, 2011; Kepecs & Fishell, 2014; Wonders & Anderson, 2006). So far, the key functional regulators, specifically programming the differentiation of CGE-originated interneurons, have not been identified.

In the developing mouse embryo, COUP-TFI is expressed in the subpallium, including LGE, MGE, CGE, and preoptic area (POA) (Armentano et al., 2007; Faedo et al., 2008; Lodato et al., 2011). In the cortex of the postnatal mouse, COUP-TFI expression is observed in many SST-, NPY-, VIP-, and CR-interneurons, and a few PV-interneurons (Lodato et al., 2011). COUP-TFII is preferentially expressed in CGE, and the boundary between LGE and MGE at E12.5. In the adult mouse cortex, COUP-TFII is mainly expressed in CGE-originated CR-/VIP-interneurons, but not in MGE-originated PV-/SST-interneurons (Tang et al., 2012). The dynamic expression profiles of *COUP-TFI* and *-TFII* genes indicate that they may play essential roles in the differentiation of cortical interneurons.

COUP-TFI is highly expressed in both the ventricular zone (VZ) and the subventricular zone (SVZ) in the subpallium. Dlx5/6Cre recombinase excises the *COUP-TFI-floxed* allele in the SVZ efficiently, but not in the VZ (Lodato et al., 2011). In the cortex of *Dlx5/6Cre/+; COUP-TFI*$^{F/F}$ mutant mice, the number of MGE-derived PV-interneurons is increased significantly; in contrast, the number of CGE-derived CR- and VIP-interneurons is reduced dramatically (Lodato et al., 2011). *RXCre* mouse is used to generate a ventral forebrain-specific *COUP-TFII* gene knockout (*RXCre/+;COUP-TFII*$^{F/F}$) mouse. However, neither CGE-originated CR-/VIP-interneurons, nor MGE-originated PV-/SST-interneurons display obvious developmental defects in the adult mutant cortex (Tang et al., 2012). Both *COUP-TFI* and *-TFII* genes are expressed in the embryonic CGE; therefore, the presence of *COUP-TFI* gene expression in the CGE of *COUP-TFII* mutant mouse may compensate for the loss-of-function of the *COUP-TFII* gene to maintain the appropriate differentiation of CGE-derived interneurons.

Except for LGE, MGE, and CGE, POA at rostral diencephalon also gives rise to interneurons in the developing mouse brain (Gelman et al., 2011, 2009; Hirata et al., 2009). One recent study reveals that one group of POA-originated COUP-TFII-expressing GABAergic cells initially generate a bundle-like migration stream in anterior peduncular area on the route to ventral CGE. After those cells reach to CGE, some of them are differentiated in CGE and ventrocaudal cortex, and others migrate and mature in dorsal cortex. The expression of COUP-TFII and Nrp2 is reduced in the POA-originated migratory interneuron to the cortex. Thus, the expression level of COUP-TFII and Nrp2 is essential to determine the migration route and the destination of these POA-derived interneurons to the cortex or the medial part of amygdala complex (Kanatani et al., 2015).

5.2 *COUP-TFII* Gene and the Development of Amygdala Complex

The amygdala complex, located in the caudoventral forebrain, participates in the regulation of emotion and stress-related behaviors (LeDoux, 2007; Swanson & Petrovich, 1998). The mouse amygdala complex is comprised of about a dozen nuclei, such as the center amygdala nucleus, lateral amygdala nucleus, basolateral amygdala nucleus, basomedial amygdala nucleus (BMA), medial amygdala nucleus, and layer 2 nucleus of lateral olfactory tract. Various neurons in different amygdala nuclei have diverse origins. Neurons in lateral amygdala nucleus, basolateral amygdala nucleus, and layer 2 nucleus of lateral olfactory tract are derived from the NPCs at the pallium (Medina et al., 2004; Remedios et al., 2007). Neurons of center amygdala nucleus, inhibitory neurons of medial amygdala nucleus, and intercalated cells are originated from the NPCs at the subpallium (Hirata et al., 2009; Waclaw, Ehrman, Pierani, & Campbell, 2010; Wang, Lufkin, & Rubenstein, 2011). Interestingly, the previous evidence on the origin of BMA neurons is controversial. Gene expression profiling supports the dorsal origin of the BMA nucleus (Medina et al., 2004); in contrast, lineage-tracing analysis in mouse in vivo demonstrates that neurons in the BMA nucleus may be derived from the NPCs in the CGE (Nery, Fishell, & Corbin, 2002).

COUP-TFII is expressed in the developing and the mature BMA neurons and is also preferentially expressed in the CGE NPCs (Kanatani, Yozu, Tabata, & Nakajima, 2008; Tang et al., 2012; Willi-Monnerat et al., 2008). The agenesis of the BMA nucleus is observed in the ventral forebrain-specific *RXCre/+;COUP-TFII*$^{F/F}$ mouse, indicating a ventral origin of the BMA neurons (Tang et al., 2012). Furthermore, it is the excitatory projection neurons but not the inhibitory interneurons in the BMA nucleus that are affected in the mutant. In the subpallium, Pax6+ migrating cells are originated from either dorsal LGE or CGE (Stoykova, Treichel, Hallonet, & Gruss, 2000; Tang et al., 2012; Toresson, Potter, & Campbell, 2000; Yun, Potter, & Rubenstein, 2001). The migration of the CGE-derived Pax6+ neurons, which coexpress COUP-TFII, Nrp1, or Nrp2, is specifically compromised in the mutant. Most likely, the reduced expression of *Nrp1* and *Nrp2* genes is main cause for the failure of Pax6+ migrating neurons settling down to the BMA locus (Tang et al., 2012). Thus, *COUP-TFII* gene directs the CGE NPCs to generate the BMA nucleus of amygdala complex.

5.3 The Function of *COUP-TFII* Gene in Hypothalamus

The hypothalamus, which originates from the diencephalon, contains a number of nuclei, such as arcuate nucleus of the hypothalamus, paraventricular nucleus of the hypothalamus, lateral hypothalamic area, dorsomedial hypothalamic nucleus, and the ventromedial hypothalamic nucleus (VMH). Several types of neurons in different mouse hypothalamic nuclei play essential roles in the modulation of food intake and energy expenditure, two key components of energy balance system (Gautron, Elmquist, & Williams, 2015; Morton, Cummings, Baskin, Barsh, & Schwartz, 2006; Morton, Meek, & Schwartz, 2014). Those neurons are POMC neurons and AgRP neurons in arcuate nucleus (Graham, Shutter, Sarmiento, Sarosi, & Stark, 1997; Ollmann et al., 1997; Yaswen, Diehl, Brennan, & Hochgeschwender, 1999), melanocortin-4 receptor (MC4R) neurons and oxytocin neurons in paraventricular nucleus of the hypothalamus (Atasoy, Betley, Su, & Sternson, 2012; Balthasar et al., 2005; Fan, Boston, Kesterson, Hruby, & Cone, 1997; Huszar et al., 1997), as well as MCH neurons and Orexin neurons in lateral hypothalamic area (Hara et al., 2001; Sakurai et al., 1998; Shimada, Tritos, Lowell, Flier, & Maratos-Flier, 1998). Sf1 neurons in VMH nucleus, which is located in the medial hypothalamus adjacent to the third ventricle, are involved in the regulation of glucose homeostasis and energy balance (Borg et al., 1994; Borg, Sherwin, Borg, Tamborlane, & Shulman, 1997; Majdic et al., 2002; Tong et al., 2007).

Microarray assay reveals that *COUP-TFII* mRNA is highly enriched in the VMH nucleus of the neonatal mouse (Kurrasch et al., 2007). Immunostaining data show that COUP-TFII is preferentially expressed in the adult VMH nucleus, but not other hypothalamic nuclei. COUP-TFII protein is not only detected in 65% of Sf1+ neurons, but also colocalized with MC4R protein in some VMH neurons (Sabra-Makke et al., 2010). Interestingly, the hypothalamic expression of *COUP-TFII* gene is induced by both fed state and MC3/4-R activation in the adult mice, indicating that *COUP-TFII* gene could play a crucial role in the VMH nucleus (Sabra-Makke et al., 2010).

Sf1Cre mouse is used to generate a VMH-specific *COUP-TFII* gene mutant mouse model. Unfortunately, only 2% of *Sf1Cre/+;COUP-TFII*$^{F/F}$ mutant mice are observed at birth, indicating the embryonic lethality of the conditional homozygous mutants (Sabra-Makke et al., 2013). *Sf1Cre/+;COUP-TFII*$^{F/+}$ heterozygous mutant mice survive to the adult stage with expected Mendelian distribution. Western blot assay shows that the expression of COUP-TFII protein in the heterozygous mutant hypothalamus is reduced by 45%, compared with that of the control. The body weight is similar between the heterozygous mutant and the control at 3 months after birth; however, the mutant mice are slimmer with less fat mass than the control mice at 6 months. The body weight loss of the mutant is caused by the increase of physical activity associated with the upregulation of UCP3 expression in the muscle, but not basal and resting energy expenditure, nor food intake (Sabra-Makke et al., 2013). Moreover, *Sf1Cre/+; COUP-TFII*$^{F/+}$ heterozygous mutant mice display hypothalamic hypersensitivity to insulin, and become more hypoglycemic in comparison to the control mice. The appropriate expression of *COUP-TFII* gene in the VMH neurons is essential to prevent the onset of hypoglycemia-associated autonomic failure (Sabra-Makke et al., 2013).

COUP-TFII gene heterozygous mutant mice are hypersensitive to insulin and slim, mainly because of the defects in adipose tissue (Li et al., 2009). Furthermore, *COUP-TFII* gene is also crucial for pancreatic glucose-dependent insulin secretion (Bardoux et al., 2005; Perilhou, Tourrel-Cuzin, Kharroubi, et al., 2008; Perilhou, Tourrel-Cuzin, Zhang, et al., 2008). The evidence from several different *COUP-TFII* gene mutant mouse models suggests that *COUP-TFII* gene is a key regulator of glucose homeostasis through its function in multiple organs, including pancreas, adipose tissue, muscle, and hypothalamus, especially VMH nucleus.

6. *COUP-TFII* GENE AND THE DEVELOPMENT OF CEREBELLUM

The cerebellum, a part of hindbrain, contributes not only to motor coordination, but also to cognitive and affective functions (Ramnani, 2006). Postmortem studies reveal loss of Purkinje cells in autism spectrum disorder patients (Amaral, Schumann, & Nordahl, 2008; Bauman & Kemper, 2005), indicating that the cerebellum may be associated with autism. There are at least five types of neurons in the cerebellum, such as Purkinje cells, granule cells, Bergmann glia cells, Golgi cells, and Basket/Stellate cells (Wang & Zoghbi, 2001).

In the mouse, the expression of COUP-TFII is detected in the middle region of cerebellar anlage at E18.5, the anterior region of Purkinje cell layer at P0, and Purkinje cells at P7 and P21. NSE-Cre recombinase can excise *COUP-TFII* gene in the mouse cerebellum before birth (Kim, Takamoto, Yan, Tsai, & Tsai, 2009). *NSECre/+;COUP-TFII*$^{F/F}$ mutant mouse displays reduced cerebellum and abnormal foliation pattern. Both the decreased proliferation and the enhanced apoptosis of granule cell precursors are possible causes for the reduction of cerebellum size. Compared with the control, the dendrite branching of Purkinje cells is decreased in the mutant. The expression of insulin-like growth factor-1 (IGF-1), which is generated in Purkinje cells, is downregulated significantly in the *COUP-TFII* mutant. The presence of IGF-1 in cultured cerebellar slice rescues the defects of granule cell precursors and Purkinje cells in the mutant. Chromatin immunoprecipitation (*ChIP*) assays reveal that COUP-TFII may activate the expression of *IGF-1* gene through a Sp1-binding site in the *IGF-1* promoter. Thus, *COUP-TFII* gene is important to ensure the appropriate development of mouse cerebellum by regulating IGF-1 signaling (Kim et al., 2009).

7. *COUP-TF* GENES AND GLIOGENESIS

The CNS mainly contains neuronal cells and glial cells, which include astrocytes and oligodendrocytes. Both neuronal cells and glial cells originate from NSCs. Neurogenesis occurs earlier than gliogenesis during the CNS development. In the past decades, many efforts have been made to investigate the molecular and cellular mechanisms involved in neurogenesis (Freeman & Rowitch, 2013; Gotz & Huttner, 2005; Guillemot, 2005; Rowitch, 2004). The study of oligogenesis has become one of the hottest topics in the neuroscience field with the newly discovered functions of glial cells in the CNS. Both extrinsic and intrinsic factors participate in the regulation of gliogenesis. Extrinsic signals such as BMPs, Notch, and Il-6 family of cytokines, can activate astrocyte genes such as *Gfap* (Grandbarbe et al., 2003; Gross et al., 1996; Koblar et al., 1998; Nakashima et al., 2001, 1999). Epigenetic factors including Brg1 may promote gliogenesis through DNA methylation and chromatin remodeling of astrocytic genes (Matsumoto et al., 2006; Song & Ghosh, 2004; Takizawa et al., 2001). Neurogenic genes such as *Ngn* genes enhance neurogenesis, but prevent gliogenesis (Sun et al., 2001).

7.1 *COUP-TF* Genes Control Temporal Gliogenesis In Vitro and In Vivo

COUP-TFI/-TFII gene double knockdown enhances the neuronal differentiation of embryonic stem cells (ESCs) at the expense of the glial cell development (Naka, Nakamura, Shimazaki, & Okano, 2008). A Stat3-binding site in the *Gfap* promoter is epigenetically silenced during the early neurogenesis (Song & Ghosh, 2004; Takizawa et al., 2001). *ChIP* data at the Stat3-binding site reveal that the levels of silencing dimethylated H3K9 are significant higher, while the levels of activating acetylated histone H3 and dimethylated H3K4 are lower in *COUP-TFI/-TFII* knockdown cells. The higher CpG methylation in the *Gfap* promoter and the failed response to gliogenic cytokines such as LIF and BMP2 further support that *COUP-TF* genes are necessary for the temporal acquisition of gliogenic competency in ESCs in vitro (Naka et al., 2008). Coexpression of COUP-TFI and -TFII is detected in the VZ of mouse brain at E12.5, and their expression is reduced before the specification of astrocyte, indicating that they may also regulate gliogenesis in the telencephalon in vivo. As expected, *COUP-TFI/-TFII* double knockdown in the early embryonic mouse telencephalon phenocopies the observation in vitro that cells infected with *COUP-TFI/-TFII* knockdown lentivirus are prone to acquire neuronal fate in vivo. Thus, as transcriptional factors, COUP-TFs participate in the intrinsic regulation of the glial cell specification (Naka et al., 2008).

7.2 *COUP-TFI* Gene and the Differentiation of Oligodendrocyte

COUP-TFI is expressed in progenitor oligodendrocytes, oligodendrocytes, and type-2 astrocytes of the mouse optic nerve (Yamaguchi et al., 2004). The appropriate differentiation of oligodendrocytes is required for axon myelination. Unmyelinated or hypomyelinated optic nerve is observed in the postnatal *COUP-TFI* null mutant mice, because the initiation of myelination is delayed in the mutant. Moreover, hypomyelination also occurs in the white matter and the striatum of the *COUP-TFI* mutant at P21, indicating that *COUP-TFI* gene plays a general role in the differentiation of oligodendrocytes. The expression of *Oct6* gene, encoding a POU domain transcription factor, is reduced in the brain and the optic nerve of *COUP-TFI* mutant mouse. Thus, the *COUP-TFI* gene may guarantee the appropriate differentiation of oligodendrocytes through regulating the expression of *Oct6* gene (Yamaguchi et al., 2004).

8. *COUP-TF* GENES AND NEURAL CREST CELLS

Neural crest cells (NCCs), originated from ectoderm, are a group of transient embryonic cells in the vertebrate. Once detached from the dorsal neural tube, NCCs migrate throughout the body to generate the peripheral nervous system, craniofacial bones, and cartilages, as well as pigment cells (Gammill & Bronner-Fraser, 2003; Sauka-Spengler & Bronner-Fraser, 2008). In mouse, COUP-TFI is expressed in both premigratory and migratory NCCs. In *COUP-TFI* null mutant embryos, the morphogenesis of the ninth cranial ganglion is compromised with the reduced number of neurons, as well as aberrant projection and arborization. The *COUP-TFI* gene is necessary for the survival of NCCs differentiating to the ninth cranial ganglion (Qiu et al., 1997). The function of *COUP-TFI* gene in the development of the NCCs is conserved in different organisms, since craniofacial dysmorphisms are also observed in *Xenopus laevis*, when *COUP-TFI* expression is repressed by morpholino oligonucleotides (Rada-Iglesias et al., 2012).

The expression of *COUP-TFI* and *-TFII* transcripts is dramatically enhanced during the differentiation of human ESCs to human neural crest cells (hNCCs) (Rada-Iglesias et al., 2012). Interestingly, knockdown of *COUP-TFI* expression by shRNA results in the decreased expression of NCC-specific genes. Epigenomically annotated hNCC enhancers are identified by *ChIP-seq* assays with antibodies recognizing p300, H3K4me1, H3K27ac, H3K4me3, and H3K27me3. *ChIP-seq* data with COUP-TFI and -TFII antibodies reveal that COUP-TFI/-TFII-binding motifs are not only located at the crucial hNCC enhancers, but also required for the activity of these enhancers. COUP-TFI/-TFII may activate the expression of NCC genes through cooperating with TFAP2A, a master NCC regulator, at the crucial hNCC enhancers (Rada-Iglesias et al., 2012).

9. *COUP-TF* GENES AND ADULT NEURONAL STEM CELLS

Neurogenesis occurs not only in the developing embryonic brain, but also in the adult mammalian brain. Adult neuronal stem cells have been detected in two main niches in the mature mouse brain, the SVZ of lateral ventricles and the subgranular zone of DG. Most likely, the appropriate development of adult NSCs is necessary to maintain certain functions of OB and hippocampus (Ming & Song, 2005; Zhao, Deng, & Gage, 2008).

In the adult mouse brain, the expression of COUP-TFI is detected in the NSCs and the NPCs in the SVZ of lateral ventricle, neuroblasts of rostral migratory stream (RMS), and mature OB interneuron. The expression of COUP-TFII in SVZ–RMS–OB pathway is mainly detected in the olfactory ensheathing cells of olfactory nerve layer. Interestingly, COUP-TFII expression is significantly enhanced in the SVZ and the RMS of COUP-TFI conditional knockout with *hGFAPCre* (Zhou et al., 2015). The accumulation of DCX+, NeuN+, and CR+ neurons is observed in *COUP-TFI* conditional knockout mutant mouse with either *hGFAPCre* or *Emx1Cre*. Meanwhile, the most obvious defect in the double knockout mouse with *hGFAPCre* is the increase of DCX+ and DCX+/Pax6+ neurons in the caudal SVZ. *COUP-TFI* and *-TFII* genes are associated with proliferation, migration, and survival of cells in the SVZ–RMS–OB pathway. Thus, *COUP-TFI* and *-TFII* genes may cooperate with each other to regulate the adult neurogenesis in the SVZ of lateral ventricle (Zhou et al., 2015).

10. CONCLUSION AND PERSPECTIVES

The developmental defects in brain are the causes of neurodevelopmental diseases such as intellectual disability, autism spectrum disorders, schizophrenia, and attention deficit hyperactivity disorder. BBSOAS patients associated with the *COUP-TFI* gene mutations display various symptoms including optic atrophy, mental retardation, epilepsy, and autism spectrum disorders. Except for CHD and CDH, mental defects and developmental delay are also observed in the patients with different mutations of *COUP-TFII* gene. All these clinical findings suggest that the functions of *COUP-TF* genes are imperative to ensure the appropriate development of the nervous system, and then to prevent neurodevelopmental diseases.

The study of the function of *COUP-TFII* gene in the CNS is still at its infant stage. Especially, at present, mouse studies are mainly focusing on the roles of *COUP-TF* genes in the developing forebrain. Nevertheless, COUP-TFI and -TFII generate much broader expression profiles in forebrain, midbrain, and hindbrain in the early mouse embryo at E9.5 (Fig. 3; and K. Tang's unpublished data). What are their roles in midbrain and hindbrain? Next, COUP-TFI and -TFII are also expressed in various regions in the adult mouse brain. Do they participate in the maintenance of the appropriate functions of mature neurons and if so, how? Furthermore, double conditional knockout mice always display more obvious defects than single

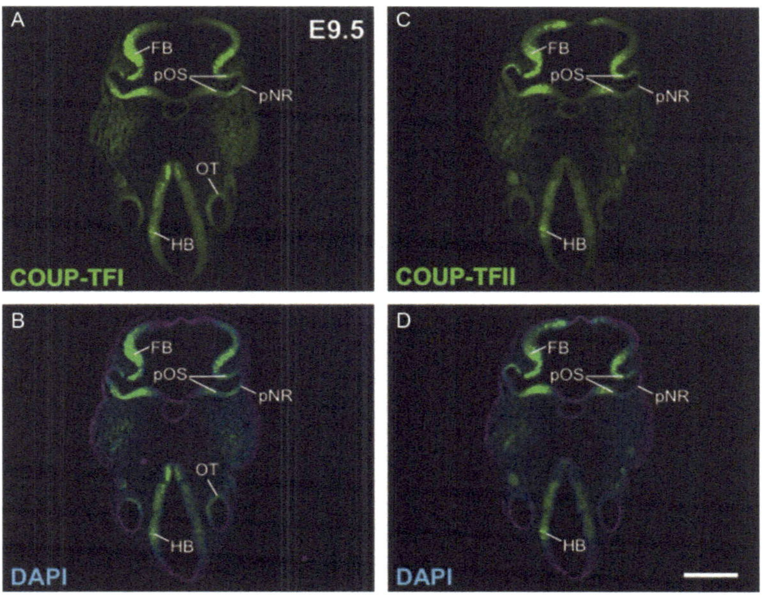

Fig. 3 The expression of COUP-TFI and -TFII in both forebrain and hindbrain of mouse embryo at E9.5. (A, B) COUP-TFI is expressed in the forebrain, hindbrain, presumptive neural retina, presumptive optic stalk, and otic vesicle of the mouse embryo at E9.5. (C, D) COUP-TFII is expressed in the forebrain, hindbrain, and presumptive optic stalk of the mouse embryo at E9.5. *FB*, forebrain; *HB*, hindbrain; *OT*, otic vesicle; *pNR*, presumptive neural retina; *pOS*, presumptive optic stalk. Scale bar, 400 μm.

gene mutant mice. How do *COUP-TFI* and *-TFII* genes cooperate with each other to program various neurodevelopmental events? Lastly, the patients with the point mutations generate more severe symptoms than the patients harboring the deletion mutations. Why? Compared with the naïve COUP-TF proteins, do the point mutations in COUP-TFI/-TFII cause loss-of-function or gain-of-function? The efforts to pursue the answers of these questions will benefit not only the understanding of neurodevelopment, but also the understanding of the etiology of human diseases.

ACKNOWLEDGMENTS

We thank Ms. Yu Deng for assistance with the manuscript. This work was supported by National Natural Science Foundation of China (81360124, 31671508 to K.T.).

REFERENCES

Al Turki, S., Manickaraj, A. K., Mercer, C. L., Gerety, S. S., Hitz, M. P., Lindsay, S., … Hurles, M. E. (2014). Rare variants in NR2F2 cause congenital heart defects in humans. *The American Journal of Human Genetics*, *94*(4), 574–585. http://dx.doi.org/10.1016/j.ajhg.2014.03.007.

Alfano, C., Magrinelli, E., Harb, K., Hevner, R. F., & Studer, M. (2014). Postmitotic control of sensory area specification during neocortical development. *Nature Communications*, *5*, 5632. http://dx.doi.org/10.1038/ncomms6632.

Al-Kateb, H., Shimony, J. S., Vineyard, M., Manwaring, L., Kulkarni, S., & Shinawi, M. (2013). NR2F1 haploinsufficiency is associated with optic atrophy, dysmorphism and global developmental delay. *American Journal of Medical Genetics. Part A*, *161A*(2), 377–381. http://dx.doi.org/10.1002/ajmg.a.35650.

Amaral, D. G., Schumann, C. M., & Nordahl, C. W. (2008). Neuroanatomy of autism. *Trends in Neurosciences*, *31*(3), 137–145. http://dx.doi.org/10.1016/j.tins.2007.12.005.

Angevine, J. B., Jr., & Sidman, R. L. (1961). Autoradiographic study of cell migration during histogenesis of cerebral cortex in the mouse. *Nature*, *192*, 766–768.

Arlotta, P., Molyneaux, B. J., Chen, J., Inoue, J., Kominami, R., & Macklis, J. D. (2005). Neuronal subtype-specific genes that control corticospinal motor neuron development in vivo. *Neuron*, *45*(2), 207–221. http://dx.doi.org/10.1016/j.neuron.2004.12.036.

Armentano, M., Chou, S. J., Tomassy, G. S., Leingartner, A., O'Leary, D. D., & Studer, M. (2007). COUP-TFI regulates the balance of cortical patterning between frontal/motor and sensory areas. *Nature Neuroscience*, *10*(10), 1277–1286. http://dx.doi.org/10.1038/nn1958.

Armentano, M., Filosa, A., Andolfi, G., & Studer, M. (2006). COUP-TFI is required for the formation of commissural projections in the forebrain by regulating axonal growth. *Development*, *133*(21), 4151–4162. http://dx.doi.org/10.1242/dev.02600.

Arrington, C. B., Bleyl, S. B., Matsunami, N., Bowles, N. E., Leppert, T. I., Demarest, B. L., … Brunelli, L. (2012). A family-based paradigm to identify candidate chromosomal regions for isolated congenital diaphragmatic hernia. *American Journal of Medical Genetics. Part A*, *158A*(12), 3137–3147. http://dx.doi.org/10.1002/ajmg.a.35664.

Atasoy, D., Betley, J. N., Su, H. H., & Sternson, S. M. (2012). Deconstruction of a neural circuit for hunger. *Nature*, *488*(7410), 172–177. http://dx.doi.org/10.1038/nature11270.

Balthasar, N., Dalgaard, L. T., Lee, C. E., Yu, J., Funahashi, H., Williams, T., ... Lowell, B. B. (2005). Divergence of melanocortin pathways in the control of food intake and energy expenditure. *Cell*, *123*(3), 493–505. http://dx.doi.org/10.1016/j.cell.2005.08.035.

Bardoux, P., Zhang, P., Flamez, D., Perilhou, A., Lavin, T. A., Tanti, J. F., ... Vasseur-Cognet, M. (2005). Essential role of chicken ovalbumin upstream promoter-transcription factor II in insulin secretion and insulin sensitivity revealed by conditional gene knockout. *Diabetes*, *54*(5), 1357–1363.

Bauman, M. L., & Kemper, T. L. (2005). Neuroanatomic observations of the brain in autism: A review and future directions. *International Journal of Developmental Neuroscience*, *23*(2–3), 183–187. http://dx.doi.org/10.1016/j.ijdevneu.2004.09.006.

Berry, M., & Rogers, A. W. (1965). The migration of neuroblasts in the developing cerebral cortex. *Journal of Anatomy*, *99*(Pt. 4), 691–709.

Bishop, K. M., Goudreau, G., & O'Leary, D. D. (2000). Regulation of area identity in the mammalian neocortex by Emx2 and Pax6. *Science*, *288*(5464), 344–349.

Borg, W. P., During, M. J., Sherwin, R. S., Borg, M. A., Brines, M. L., & Shulman, G. I. (1994). Ventromedial hypothalamic lesions in rats suppress counterregulatory responses to hypoglycemia. *Journal of Clinical Investigation*, *93*(4), 1677–1682. http://dx.doi.org/10.1172/JCI117150.

Borg, M. A., Sherwin, R. S., Borg, W. P., Tamborlane, W. V., & Shulman, G. I. (1997). Local ventromedial hypothalamus glucose perfusion blocks counterregulation during systemic hypoglycemia in awake rats. *Journal of Clinical Investigation*, *99*(2), 361–365. http://dx.doi.org/10.1172/JCI119165.

Bosch, D. G., Boonstra, F. N., de Leeuw, N., Pfundt, R., Nillesen, W. M., de Ligt, J., ... de Vries, B. B. (2016). Novel genetic causes for cerebral visual impairment. *European Journal of Human Genetics*, *24*(5), 660–665. http://dx.doi.org/10.1038/ejhg.2015.186.

Bosch, D. G., Boonstra, F. N., Gonzaga-Jauregui, C., Xu, M., de Ligt, J., Jhangiani, S., ... Schaaf, C. P. (2014). NR2F1 mutations cause optic atrophy with intellectual disability. *The American Journal of Human Genetics*, *94*(2), 303–309. http://dx.doi.org/10.1016/j.ajhg.2014.01.002.

Brown, K. K., Alkuraya, F. S., Matos, M., Robertson, R. L., Kimonis, V. E., & Morton, C. C. (2009). NR2F1 deletion in a patient with a de novo paracentric inversion, inv(5)(q15q33.2), and syndromic deafness. *American Journal of Medical Genetics. Part A*, *149A*(5), 931–938. http://dx.doi.org/10.1002/ajmg.a.32764.

Cardoso, C., Boys, A., Parrini, E., Mignon-Ravix, C., McMahon, J. M., Khantane, S., ... Guerrini, R. (2009). Periventricular heterotopia, mental retardation, and epilepsy associated with 5q14.3-q15 deletion. *Neurology*, *72*(9), 784–792. http://dx.doi.org/10.1212/01.wnl.0000336339.08878.2d.

Chen, C. A., Bosch, D. G., Cho, M. T., Rosenfeld, J. A., Shinawi, M., Lewis, R. A., ... Schaaf, C. (2016). The expanding clinical phenotype of Bosch-Boonstra-Schaaf optic atrophy syndrome: 20 new cases and possible genotype-phenotype correlations. *Genetics in Medicine*, *18*, 1143–1150. http://dx.doi.org/10.1038/gim.2016.18.

Dong, H. W., Swanson, L. W., Chen, L., Fanselow, M. S., & Toga, A. W. (2009). Genomic-anatomic evidence for distinct functional domains in hippocampal field CA1. *Proceedings of the National Academy of Sciences of the United States of America*, *106*(28), 11794–11799. http://dx.doi.org/10.1073/pnas.0812608106.

Faedo, A., Tomassy, G. S., Ruan, Y., Teichmann, H., Krauss, S., Pleasure, S. J., ... Rubenstein, J. L. (2008). COUP-TFI coordinates cortical patterning, neurogenesis, and laminar fate and modulates MAPK/ERK, AKT, and beta-catenin signaling. *Cerebral Cortex*, *18*(9), 2117–2131. http://dx.doi.org/10.1093/cercor/bhm238.

Fan, W., Boston, B. A., Kesterson, R. A., Hruby, V. J., & Cone, R. D. (1997). Role of melanocortinergic neurons in feeding and the agouti obesity syndrome. *Nature, 385*(6612), 165–168. http://dx.doi.org/10.1038/385165a0.

Fanselow, M. S., & Dong, H. W. (2010). Are the dorsal and ventral hippocampus functionally distinct structures? *Neuron, 65*(1), 7–19. http://dx.doi.org/10.1016/j.neuron.2009.11.031.

Fishell, G., & Rudy, B. (2011). Mechanisms of inhibition within the telencephalon: "Where the wild things are". *Annual Review of Neuroscience, 34*, 535–567. http://dx.doi.org/10.1146/annurev-neuro-061010-113717.

Flore, G., Di Ruberto, G., Parisot, J., Sannino, S., Russo, F., Illingworth, E. A., ... De Leonibus, E. (2016). Gradient COUP-TFI expression is required for functional organization of the hippocampal septo-temporal longitudinal axis. *Cerebral Cortex*. http://dx.doi.org/10.1093/cercor/bhv336.

Freeman, M. R., & Rowitch, D. H. (2013). Evolving concepts of gliogenesis: A look way back and ahead to the next 25 years. *Neuron, 80*(3), 613–623. http://dx.doi.org/10.1016/j.neuron.2013.10.034.

Gammill, L. S., & Bronner-Fraser, M. (2003). Neural crest specification: Migrating into genomics. *Nature Reviews. Neuroscience, 4*(10), 795–805. http://dx.doi.org/10.1038/nrn1219.

Gautron, L., Elmquist, J. K., & Williams, K. W. (2015). Neural control of energy balance: Translating circuits to therapies. *Cell, 161*(1), 133–145. http://dx.doi.org/10.1016/j.cell.2015.02.023.

Gelman, D., Griveau, A., Dehorter, N., Teissier, A., Varela, C., Pla, R., ... Marin, O. (2011). A wide diversity of cortical GABAergic interneurons derives from the embryonic preoptic area. *Journal of Neuroscience, 31*(46), 16570–16580. http://dx.doi.org/10.1523/JNEUROSCI.4068-11.2011.

Gelman, D. M., Martini, F. J., Nobrega-Pereira, S., Pierani, A., Kessaris, N., & Marin, O. (2009). The embryonic preoptic area is a novel source of cortical GABAergic interneurons. *Journal of Neuroscience, 29*(29), 9380–9389. http://dx.doi.org/10.1523/JNEUROSCI.0604-09.2009.

Gotz, M., & Huttner, W. B. (2005). The cell biology of neurogenesis. *Nature Reviews. Molecular Cell Biology, 6*(10), 777–788. http://dx.doi.org/10.1038/nrm1739.

Graham, M., Shutter, J. R., Sarmiento, U., Sarosi, I., & Stark, K. L. (1997). Overexpression of Agrt leads to obesity in transgenic mice. *Nature Genetics, 17*(3), 273–274. http://dx.doi.org/10.1038/ng1197-273.

Grandbarbe, L., Bouissac, J., Rand, M., Hrabe de Angelis, M., Artavanis-Tsakonas, S., & Mohier, E. (2003). Delta-Notch signaling controls the generation of neurons/glia from neural stem cells in a stepwise process. *Development, 130*(7), 1391–1402.

Greig, L. C., Woodworth, M. B., Galazo, M. J., Padmanabhan, H., & Macklis, J. D. (2013). Molecular logic of neocortical projection neuron specification, development and diversity. *Nature Reviews. Neuroscience, 14*(11), 755–769. http://dx.doi.org/10.1038/nrn3586.

Gross, R. E., Mehler, M. F., Mabie, P. C., Zang, Z., Santschi, L., & Kessler, J. A. (1996). Bone morphogenetic proteins promote astroglial lineage commitment by mammalian subventricular zone progenitor cells. *Neuron, 17*(4), 595–606.

Guillemot, F. (2005). Cellular and molecular control of neurogenesis in the mammalian telencephalon. *Current Opinion in Cell Biology, 17*(6), 639–647. http://dx.doi.org/10.1016/j.ceb.2005.09.006.

Hara, J., Beuckmann, C. T., Nambu, T., Willie, J. T., Chemelli, R. M., Sinton, C. M., ... Sakurai, T. (2001). Genetic ablation of orexin neurons in mice results in narcolepsy, hypophagia, and obesity. *Neuron, 30*(2), 345–354.

Hebert, J. M., & Fishell, G. (2008). The genetics of early telencephalon patterning: Some assembly required. *Nature Reviews. Neuroscience*, *9*(9), 678–685. http://dx.doi.org/10.1038/nrn2463.

High, F. A., Bhayani, P., Wilson, J. M., Bult, C. J., Donahoe, P. K., & Longoni, M. (2016). De novo frameshift mutation in COUP-TFII (NR2F2) in human congenital diaphragmatic hernia. *American Journal of Medical Genetics. Part A*, *170*(9), 2457–2461. http://dx.doi.org/10.1002/ajmg.a.37830.

Hirabayashi, Y., & Gotoh, Y. (2010). Epigenetic control of neural precursor cell fate during development. *Nature Reviews. Neuroscience*, *11*(6), 377–388. http://dx.doi.org/10.1038/nrn2810.

Hirata, T., Li, P., Lanuza, G. M., Cocas, L. A., Huntsman, M. M., & Corbin, J. G. (2009). Identification of distinct telencephalic progenitor pools for neuronal diversity in the amygdala. *Nature Neuroscience*, *12*(2), 141–149. http://dx.doi.org/10.1038/nn.2241.

Hoerder-Suabedissen, A., & Molnar, Z. (2015). Development, evolution and pathology of neocortical subplate neurons. *Nature Reviews. Neuroscience*, *16*(3), 133–146. http://dx.doi.org/10.1038/nrn3915.

Huszar, D., Lynch, C. A., Fairchild-Huntress, V., Dunmore, J. H., Fang, Q., Berkemeier, L. R., … Lee, F. (1997). Targeted disruption of the melanocortin-4 receptor results in obesity in mice. *Cell*, *88*(1), 131–141.

Joshi, P. S., Molyneaux, B. J., Feng, L., Xie, X., Macklis, J. D., & Gan, L. (2008). Bhlhb5 regulates the postmitotic acquisition of area identities in layers II-V of the developing neocortex. *Neuron*, *60*(2), 258–272. http://dx.doi.org/10.1016/j.neuron.2008.08.006.

Kanatani, S., Honda, T., Aramaki, M., Hayashi, K., Kubo, K., Ishida, M., … Nakajima, K. (2015). The COUP-TFII/Neuropilin-2 is a molecular switch steering diencephalon-derived GABAergic neurons in the developing mouse brain. *Proceedings of the National Academy of Sciences of the United States of America*, *112*(36), E4985–E4994. http://dx.doi.org/10.1073/pnas.1420701112.

Kanatani, S., Yozu, M., Tabata, H., & Nakajima, K. (2008). COUP-TFII is preferentially expressed in the caudal ganglionic eminence and is involved in the caudal migratory stream. *Journal of Neuroscience*, *28*(50), 13582–13591. http://dx.doi.org/10.1523/JNEUROSCI.2132-08.2008.

Kepecs, A., & Fishell, G. (2014). Interneuron cell types are fit to function. *Nature*, *505*(7483), 318–326. http://dx.doi.org/10.1038/nature12983.

Kim, B. J., Takamoto, N., Yan, J., Tsai, S. Y., & Tsai, M. J. (2009). Chicken ovalbumin upstream promoter-transcription factor II (COUP-TFII) regulates growth and patterning of the postnatal mouse cerebellum. *Developmental Biology*, *326*(2), 378–391. http://dx.doi.org/10.1016/j.ydbio.2008.11.001.

Klaassens, M., van Dooren, M., Eussen, H. J., Douben, H., den Dekker, A. T., Lee, C., … de Klein, A. (2005). Congenital diaphragmatic hernia and chromosome 15q26: Determination of a candidate region by use of fluorescent in situ hybridization and array-based comparative genomic hybridization. *The American Journal of Human Genetics*, *76*(5), 877–882. http://dx.doi.org/10.1086/429842.

Koblar, S. A., Turnley, A. M., Classon, B. J., Reid, K. L., Ware, C. B., Cheema, S. S., … Bartlett, P. F. (1998). Neural precursor differentiation into astrocytes requires signaling through the leukemia inhibitory factor receptor. *Proceedings of the National Academy of Sciences of the United States of America*, *95*(6), 3178–3181.

Kurrasch, D. M., Cheung, C. C., Lee, F. Y., Tran, P. V., Hata, K., & Ingraham, H. A. (2007). The neonatal ventromedial hypothalamus transcriptome reveals novel markers with spatially distinct patterning. *Journal of Neuroscience*, *27*(50), 13624–13634. http://dx.doi.org/10.1523/JNEUROSCI.2858-07.2007.

Lai, T., Jabaudon, D., Molyneaux, B. J., Azim, E., Arlotta, P., Menezes, J. R., & Macklis, J. D. (2008). SOX5 controls the sequential generation of distinct corticofugal neuron subtypes. *Neuron*, *57*(2), 232–247. http://dx.doi.org/10.1016/j.neuron.2007.12.023.

LeDoux, J. (2007). The amygdala. *Current Biology*, *17*(20), R868–R874. http://dx.doi.org/10.1016/j.cub.2007.08.005.

Leone, D. P., Srinivasan, K., Chen, B., Alcamo, E., & McConnell, S. K. (2008). The determination of projection neuron identity in the developing cerebral cortex. *Current Opinion in Neurobiology*, *18*(1), 28–35. http://dx.doi.org/10.1016/j.conb.2008.05.006.

Li, L., Xie, X., Qin, J., Jeha, G. S., Saha, P. K., Yan, J., … Tsai, M. J. (2009). The nuclear orphan receptor COUP-TFII plays an essential role in adipogenesis, glucose homeostasis, and energy metabolism. *Cell Metabolism*, *9*(1), 77–87. http://dx.doi.org/10.1016/j.cmet.2008.12.002.

Lin, F. J., Qin, J., Tang, K., Tsai, S. Y., & Tsai, M. J. (2011). Coup d'Etat: An orphan takes control. *Endocrine Reviews*, *32*(3), 404–421. http://dx.doi.org/10.1210/er.2010-0021.

Lin, F. J., You, L. R., Yu, C. T., Hsu, W. H., Tsai, M. J., & Tsai, S. Y. (2012). Endocardial cushion morphogenesis and coronary vessel development require chicken ovalbumin upstream promoter-transcription factor II. *Arteriosclerosis, Thrombosis, and Vascular Biology*, *32*(11), e135–e146. http://dx.doi.org/10.1161/ATVBAHA.112.300255.

Lodato, S., Tomassy, G. S., De Leonibus, E., Uzcategui, Y. G., Andolfi, G., Armentano, M., … Studer, M. (2011). Loss of COUP-TFI alters the balance between caudal ganglionic eminence- and medial ganglionic eminence-derived cortical interneurons and results in resistance to epilepsy. *Journal of Neuroscience*, *31*(12), 4650–4662. http://dx.doi.org/10.1523/JNEUROSCI.6580-10.2011.

Longoni, M., Russell, M. K., High, F. A., Darvishi, K., Maalouf, F. I., Kashani, A., … Donahoe, P. K. (2015). Prevalence and penetrance of ZFPM2 mutations and deletions causing congenital diaphragmatic hernia. *Clinical Genetics*, *87*(4), 362–367. http://dx.doi.org/10.1111/cge.12395.

Luskin, M. B., & Shatz, C. J. (1985). Neurogenesis of the cat's primary visual cortex. *Journal of Comparative Neurology*, *242*(4), 611–631. http://dx.doi.org/10.1002/cne.902420409.

Lutz, B., Kuratani, S., Cooney, A. J., Wawersik, S., Tsai, S. Y., Eichele, G., & Tsai, M. J. (1994). Developmental regulation of the orphan receptor COUP-TF II gene in spinal motor neurons. *Development*, *120*(1), 25–36.

Majdic, G., Young, M., Gomez-Sanchez, E., Anderson, P., Szczepaniak, L. S., Dobbins, R. L., … Parker, K. L. (2002). Knockout mice lacking steroidogenic factor 1 are a novel genetic model of hypothalamic obesity. *Endocrinology*, *143*(2), 607–614. http://dx.doi.org/10.1210/endo.143.2.8652.

Mallamaci, A., Muzio, L., Chan, C. H., Parnavelas, J., & Boncinelli, E. (2000). Area identity shifts in the early cerebral cortex of Emx2-/- mutant mice. *Nature Neuroscience*, *3*(7), 679–686. http://dx.doi.org/10.1038/76630.

Matsumoto, S., Banine, F., Struve, J., Xing, R., Adams, C., Liu, Y., … Sherman, L. S. (2006). Brg1 is required for murine neural stem cell maintenance and gliogenesis. *Developmental Biology*, *289*(2), 372–383. http://dx.doi.org/10.1016/j.ydbio.2005.10.044.

McConnell, S. K. (1995). Constructing the cerebral cortex: Neurogenesis and fate determination. *Neuron*, *15*(4), 761–768.

Medina, L., Legaz, I., Gonzalez, G., De Castro, F., Rubenstein, J. L., & Puelles, L. (2004). Expression of Dbx1, Neurogenin 2, Semaphorin 5A, Cadherin 8, and Emx1 distinguish ventral and lateral pallial histogenetic divisions in the developing mouse claustroamygdaloid complex. *Journal of Comparative Neurology*, *474*(4), 504–523. http://dx.doi.org/10.1002/cne.20141.

Miller, F. D., & Gauthier, A. S. (2007). Timing is everything: Making neurons versus glia in the developing cortex. *Neuron*, *54*(3), 357–369. http://dx.doi.org/10.1016/j.neuron.2007.04.019.

Ming, G. L., & Song, H. (2005). Adult neurogenesis in the mammalian central nervous system. *Annual Review of Neuroscience*, *28*, 223–250. http://dx.doi.org/10.1146/annurev.neuro.28.051804.101459.

Molnar, Z., & Cheung, A. F. (2006). Towards the classification of subpopulations of layer V pyramidal projection neurons. *Neuroscience Research*, *55*(2), 105–115. http://dx.doi.org/10.1016/j.neures.2006.02.008.

Molyneaux, B. J., Arlotta, P., Hirata, T., Hibi, M., & Macklis, J. D. (2005). Fezl is required for the birth and specification of corticospinal motor neurons. *Neuron*, *47*(6), 817–831. http://dx.doi.org/10.1016/j.neuron.2005.08.030.

Molyneaux, B. J., Arlotta, P., Menezes, J. R., & Macklis, J. D. (2007). Neuronal subtype specification in the cerebral cortex. *Nature Reviews. Neuroscience*, *8*(6), 427–437. http://dx.doi.org/10.1038/nrn2151.

Morton, G. J., Cummings, D. E., Baskin, D. G., Barsh, G. S., & Schwartz, M. W. (2006). Central nervous system control of food intake and body weight. *Nature*, *443*(7109), 289–295. http://dx.doi.org/10.1038/nature05026.

Morton, G. J., Meek, T. H., & Schwartz, M. W. (2014). Neurobiology of food intake in health and disease. *Nature Reviews. Neuroscience*, *15*(6), 367–378. http://dx.doi.org/10.1038/nrn3745.

Naka, H., Nakamura, S., Shimazaki, T., & Okano, H. (2008). Requirement for COUP-TFI and II in the temporal specification of neural stem cells in CNS development. *Nature Neuroscience*, *11*(9), 1014–1023. http://dx.doi.org/10.1038/nn.2168.

Nakamura, E., Makita, Y., Okamoto, T., Nagaya, K., Hayashi, T., Sugimoto, M., … Fujieda, K. (2011). 5.78 Mb terminal deletion of chromosome 15q in a girl, evaluation of NR2F2 as candidate gene for congenital heart defects. *European Journal of Medical Genetics*, *54*(3), 354–356. http://dx.doi.org/10.1016/j.ejmg.2010.12.004.

Nakashima, K., Takizawa, T., Ochiai, W., Yanagisawa, M., Hisatsune, T., Nakafuku, M., … Taga, T. (2001). BMP2-mediated alteration in the developmental pathway of fetal mouse brain cells from neurogenesis to astrocytogenesis. *Proceedings of the National Academy of Sciences of the United States of America*, *98*(10), 5868–5873. http://dx.doi.org/10.1073/pnas.101109698.

Nakashima, K., Wiese, S., Yanagisawa, M., Arakawa, H., Kimura, N., Hisatsune, T., … Taga, T. (1999). Developmental requirement of gp130 signaling in neuronal survival and astrocyte differentiation. *Journal of Neuroscience*, *19*(13), 5429–5434.

Nery, S., Fishell, G., & Corbin, J. G. (2002). The caudal ganglionic eminence is a source of distinct cortical and subcortical cell populations. *Nature Neuroscience*, *5*(12), 1279–1287. http://dx.doi.org/10.1038/nn971.

Nord, A. S., Pattabiraman, K., Visel, A., & Rubenstein, J. L. (2015). Genomic perspectives of transcriptional regulation in forebrain development. *Neuron*, *85*(1), 27–47. http://dx.doi.org/10.1016/j.neuron.2014.11.011.

Ollmann, M. M., Wilson, B. D., Yang, Y. K., Kerns, J. A., Chen, Y., Gantz, I., & Barsh, G. S. (1997). Antagonism of central melanocortin receptors in vitro and in vivo by agouti-related protein. *Science*, *278*(5335), 135–138.

Pereira, F. A., Qiu, Y., Zhou, G., Tsai, M. J., & Tsai, S. Y. (1999). The orphan nuclear receptor COUP-TFII is required for angiogenesis and heart development. *Genes and Development*, *13*(8), 1037–1049.

Perilhou, A., Tourrel-Cuzin, C., Kharroubi, I., Henique, C., Fauveau, V., Kitamura, T., … Vasseur-Cognet, M. (2008a). The transcription factor COUP-TFII is negatively regulated by insulin and glucose via Foxo1- and ChREBP-controlled pathways. *Molecular and Cellular Biology*, *28*(21), 6568–6579. http://dx.doi.org/10.1128/MCB.02211-07.

Perilhou, A., Tourrel-Cuzin, C., Zhang, P., Kharroubi, I., Wang, H., Fauveau, V., … Vasseur-Cognet, M. (2008b). The MODY1 gene for hepatocyte nuclear factor 4alpha and a feedback loop control COUP-TFII expression in pancreatic beta cells. *Molecular and Cellular Biology*, *28*(14), 4588–4597. http://dx.doi.org/10.1128/MCB.01191-07.

Poot, M., Eleveld, M. J., van't Slot, R., van Genderen, M. M., Verrijn Stuart, A. A., Hochstenbach, R., & Beemer, F. A. (2007). Proportional growth failure and oculocutaneous albinism in a girl with a 6.87 Mb deletion of region 15q26.2– > qter. *European Journal of Medical Genetics*, *50*(6), 432–440. http://dx.doi.org/10.1016/j.ejmg.2007.08.003.

Qiu, Y., Cooney, A. J., Kuratani, S., DeMayo, F. J., Tsai, S. Y., & Tsai, M. J. (1994). Spatiotemporal expression patterns of chicken ovalbumin upstream promoter-transcription factors in the developing mouse central nervous system: Evidence for a role in segmental patterning of the diencephalon. *Proceedings of the National Academy of Sciences of the United States of America*, *91*(10), 4451–4455.

Qiu, Y., Pereira, F. A., DeMayo, F. J., Lydon, J. P., Tsai, S. Y., & Tsai, M. J. (1997). Null mutation of mCOUP-TFI results in defects in morphogenesis of the glossopharyngeal ganglion, axonal projection, and arborization. *Genes and Development*, *11*(15), 1925–1937.

Rada-Iglesias, A., Bajpai, R., Prescott, S., Brugmann, S. A., Swigut, T., & Wysocka, J. (2012). Epigenomic annotation of enhancers predicts transcriptional regulators of human neural crest. *Cell Stem Cell*, *11*(5), 633–648. http://dx.doi.org/10.1016/j.stem.2012.07.006.

Rakic, P. (1974). Neurons in rhesus monkey visual cortex: Systematic relation between time of origin and eventual disposition. *Science*, *183*(4123), 425–427.

Rakic, P. (1988). Specification of cerebral cortical areas. *Science*, *241*(4862), 170–176.

Rallu, M., Corbin, J. G., & Fishell, G. (2002). Parsing the prosencephalon. *Nature Reviews. Neuroscience*, *3*(12), 943–951. http://dx.doi.org/10.1038/nrn989.

Ramnani, N. (2006). The primate cortico-cerebellar system: Anatomy and function. *Nature Reviews. Neuroscience*, *7*(7), 511–522. http://dx.doi.org/10.1038/nrn1953.

Remedios, R., Huilgol, D., Saha, B., Hari, P., Bhatnagar, L., Kowalczyk, T., … Tole, S. (2007). A stream of cells migrating from the caudal telencephalon reveals a link between the amygdala and neocortex. *Nature Neuroscience*, *10*(9), 1141–1150. http://dx.doi.org/10.1038/nn1955.

Rowitch, D. H. (2004). Glial specification in the vertebrate neural tube. *Nature Reviews. Neuroscience*, *5*(5), 409–419. http://dx.doi.org/10.1038/nrn1389.

Rump, P., Dijkhuizen, T., Sikkema-Raddatz, B., Lemmink, H. H., Vos, Y. J., Verheij, J. B., & van Ravenswaaij, C. M. (2008). Drayer's syndrome of mental retardation, microcephaly, short stature and absent phalanges is caused by a recurrent deletion of chromosome 15(q26.2– > qter). *Clinical Genetics*, *74*(5), 455–462. http://dx.doi.org/10.1111/j.1399-0004.2008.01064.x.

Sabra-Makke, L., Maritan, M., Planchais, J., Boutant, M., Pegorier, J. P., Even, P. C., … Bossard, P. (2013). Hypothalamic ventromedial COUP-TFII protects against hypoglycemia-associated autonomic failure. *Proceedings of the National Academy of Sciences of the United States of America*, *110*(11), 4333–4338. http://dx.doi.org/10.1073/pnas.1219262110.

Sabra-Makke, L., Tourrel-Cuzin, C., Denis, R. G., Moldes, M., Pegorier, J. P., Luquet, S., … Bossard, P. (2010). The nutritional induction of COUP-TFII gene expression in ventromedial hypothalamic neurons is mediated by the melanocortin pathway. *Plos One*, *5*(10), e13464. http://dx.doi.org/10.1371/journal.pone.0013464.

Sakurai, T., Amemiya, A., Ishii, M., Matsuzaki, I., Chemelli, R. M., Tanaka, H., … Yanagisawa, M. (1998). Orexins and orexin receptors: A family of hypothalamic neuropeptides and G protein-coupled receptors that regulate feeding behavior. *Cell*, *92*(5), 573–585. 1 page following 696.

Sauka-Spengler, T., & Bronner-Fraser, M. (2008). A gene regulatory network orchestrates neural crest formation. *Nature Reviews. Molecular Cell Biology*, *9*(7), 557–568. http://dx.doi.org/10.1038/nrm2428.

Shimada, M., Tritos, N. A., Lowell, B. B., Flier, J. S., & Maratos-Flier, E. (1998). Mice lacking melanin-concentrating hormone are hypophagic and lean. *Nature*, *396*(6712), 670–674. http://dx.doi.org/10.1038/25341.

Slavotinek, A. M., Moshrefi, A., Davis, R., Leeth, E., Schaeffer, G. B., Burchard, G. E., … Pennacchio, L. A. (2006). Array comparative genomic hybridization in patients with congenital diaphragmatic hernia: Mapping of four CDH-critical regions and sequencing of candidate genes at 15q26.1-15q26.2. *European Journal of Human Genetics*, *14*(9), 999–1008. http://dx.doi.org/10.1038/sj.ejhg.5201652.

Song, M. R., & Ghosh, A. (2004). FGF2-induced chromatin remodeling regulates CNTF-mediated gene expression and astrocyte differentiation. *Nature Neuroscience*, 7(3), 229–235. http://dx.doi.org/10.1038/nn1192.

Stoykova, A., Treichel, D., Hallonet, M., & Gruss, P. (2000). Pax6 modulates the dorsoventral patterning of the mammalian telencephalon. *Journal of Neuroscience*, *20*(21), 8042–8050.

Strange, B. A., Witter, M. P., Lein, E. S., & Moser, E. I. (2014). Functional organization of the hippocampal longitudinal axis. *Nature Reviews. Neuroscience*, *15*(10), 655–669. http://dx.doi.org/10.1038/nrn3785.

Sun, Y., Nadal-Vicens, M., Misono, S., Lin, M. Z., Zubiaga, A., Hua, X., … Greenberg, M. E. (2001). Neurogenin promotes neurogenesis and inhibits glial differentiation by independent mechanisms. *Cell*, *104*(3), 365–376.

Sur, M., & Rubenstein, J. L. (2005). Patterning and plasticity of the cerebral cortex. *Science*, *310*(5749), 805–810. http://dx.doi.org/10.1126/science.1112070.

Swanson, L. W., & Petrovich, G. D. (1998). What is the amygdala? *Trends in Neurosciences*, *21*(8), 323–331.

Takizawa, T., Nakashima, K., Namihira, M., Ochiai, W., Uemura, A., Yanagisawa, M., … Taga, T. (2001). DNA methylation is a critical cell-intrinsic determinant of astrocyte differentiation in the fetal brain. *Developmental Cell*, *1*(6), 749–758.

Tang, K., Rubenstein, J. L., Tsai, S. Y., & Tsai, M. J. (2012). COUP-TFII controls amygdala patterning by regulating neuropilin expression. *Development*, *139*(9), 1630–1639. http://dx.doi.org/10.1242/dev.075564.

Tang, K., Tsai, S. Y., & Tsai, M. J. (2015). COUP-TFs and eye development. *Biochimica et Biophysica Acta*, *1849*(2), 201–209. http://dx.doi.org/10.1016/j.bbagrm.2014.05.022.

Tang, K., Xie, X., Park, J. I., Jamrich, M., Tsai, S., & Tsai, M. J. (2010). COUP-TFs regulate eye development by controlling factors essential for optic vesicle morphogenesis. *Development*, *137*(5), 725–734. http://dx.doi.org/10.1242/dev.040568.

Temple, S. (2001). The development of neural stem cells. *Nature*, *414*(6859), 112–117. http://dx.doi.org/10.1038/35102174.

Thompson, C. L., Pathak, S. D., Jeromin, A., Ng, L. L., MacPherson, C. R., Mortrud, M. T., … Lein, E. S. (2008). Genomic anatomy of the hippocampus. *Neuron*, *60*(6), 1010–1021. http://dx.doi.org/10.1016/j.neuron.2008.12.008.

Tomassy, G. S., De Leonibus, E., Jabaudon, D., Lodato, S., Alfano, C., Mele, A., … Studer, M. (2010). Area-specific temporal control of corticospinal motor neuron differentiation by COUP-TFI. *Proceedings of the National Academy of Sciences of the United States of America*, *107*(8), 3576–3581. http://dx.doi.org/10.1073/pnas.0911792107.

Tong, Q., Ye, C., McCrimmon, R. J., Dhillon, H., Choi, B., Kramer, M. D., … Lowell, B. B. (2007). Synaptic glutamate release by ventromedial hypothalamic neurons is part of the neurocircuitry that prevents hypoglycemia. *Cell Metabolism*, *5*(5), 383–393. http://dx.doi.org/10.1016/j.cmet.2007.04.001.

Toresson, H., Potter, S. S., & Campbell, K. (2000). Genetic control of dorsal-ventral identity in the telencephalon: Opposing roles for Pax6 and Gsh2. *Development*, *127*(20), 4361–4371.

Tsai, S. Y., & Tsai, M. J. (1997). Chick ovalbumin upstream promoter-transcription factors (COUP-TFs): Coming of age. *Endocrine Reviews*, *18*(2), 229–240. http://dx.doi.org/10.1210/edrv.18.2.0294.

Waclaw, R. R., Ehrman, L. A., Pierani, A., & Campbell, K. (2010). Developmental origin of the neuronal subtypes that comprise the amygdalar fear circuit in the mouse. *Journal of Neuroscience*, *30*(20), 6944–6953. http://dx.doi.org/10.1523/JNEUROSCI.5772-09.2010.

Wang, B., Lufkin, T., & Rubenstein, J. L. (2011). Dlx6 regulates molecular properties of the striatum and central nucleus of the amygdala. *Journal of Comparative Neurology*, *519*(12), 2320–2334. http://dx.doi.org/10.1002/cne.22618.

Wang, V. Y., & Zoghbi, H. Y. (2001). Genetic regulation of cerebellar development. *Nature Reviews. Neuroscience*, *2*(7), 484–491. http://dx.doi.org/10.1038/35081558.

Willi-Monnerat, S., Migliavacca, E., Surdez, D., Delorenzi, M., Luthi-Carter, R., & Terskikh, A. V. (2008). Comprehensive spatiotemporal transcriptomic analyses of the ganglionic eminences demonstrate the uniqueness of its caudal subdivision. *Molecular and Cellular Neurosciences*, *37*(4), 845–856. http://dx.doi.org/10.1016/j.mcn.2008.01.009.

Wonders, C. P., & Anderson, S. A. (2006). The origin and specification of cortical interneurons. *Nature Reviews. Neuroscience*, 7(9), 687–696. http://dx.doi.org/10.1038/nrn1954.

Yamaguchi, H., Zhou, C., Lin, S. C., Durand, B., Tsai, S. Y., & Tsai, M. J. (2004). The nuclear orphan receptor COUP-TFI is important for differentiation of oligodendrocytes. *Developmental Biology*, *266*(2), 238–251.

Yaswen, L., Diehl, N., Brennan, M. B., & Hochgeschwender, U. (1999). Obesity in the mouse model of pro-opiomelanocortin deficiency responds to peripheral melanocortin. *Nature Medicine*, *5*(9), 1066–1070. http://dx.doi.org/10.1038/12506.

You, L. R., Takamoto, N., Yu, C. T., Tanaka, T., Kodama, T., Demayo, F. J., ... Tsai, M. J. (2005). Mouse lacking COUP-TFII as an animal model of Bochdalek-type congenital diaphragmatic hernia. *Proceedings of the National Academy of Sciences of the United States of America*, *102*(45), 16351–16356. http://dx.doi.org/10.1073/pnas.0507832102.

Yun, K., Potter, S., & Rubenstein, J. L. (2001). Gsh2 and Pax6 play complementary roles in dorsoventral patterning of the mammalian telencephalon. *Development*, *128*(2), 193–205.

Zhao, C., Deng, W., & Gage, F. H. (2008). Mechanisms and functional implications of adult neurogenesis. *Cell*, *132*(4), 645–660. http://dx.doi.org/10.1016/j.cell.2008.01.033.

Zhou, X., Liu, F., Tian, M., Xu, Z., Liang, Q., Wang, C., ... Yang, Z. (2015). Transcription factors COUP-TFI and COUP-TFII are required for the production of granule cells in the mouse olfactory bulb. *Development*, *142*(9), 1593–1605. http://dx.doi.org/10.1242/dev.115279.

Zhou, C., Qiu, Y., Pereira, F. A., Crair, M. C., Tsai, S. Y., & Tsai, M. J. (1999). The nuclear orphan receptor COUP-TFI is required for differentiation of subplate neurons and guidance of thalamocortical axons. *Neuron*, *24*(4), 847–859.

Zhou, C., Tsai, S. Y., & Tsai, M. J. (2001). COUP-TFI: An intrinsic factor for early regionalization of the neocortex. *Genes and Development*, *15*(16), 2054–2059. http://dx.doi.org/10.1101/gad.913601.

CHAPTER ELEVEN

Genetic Investigation of Thyroid Hormone Receptor Function in the Developing and Adult Brain

Frédéric Flamant[1], Karine Gauthier, Sabine Richard
Institut de Génomique Fonctionnelle de Lyon, Université de Lyon, Université Lyon 1, CNRS UMR 5242, INRA USC 1370, Ecole Normale Supérieure de Lyon, Lyon cedex, France
[1]Corresponding author: e-mail address: frederic.flamant@ens-lyon.fr

Contents

Abstract

Thyroid hormones exert a broad influence on brain development and function, which has been extensively studied over the years. Mouse genetics has brought an important contribution, allowing precise analysis of the interplay between TRα1 and TRβ1 nuclear receptors in neural cells. However, the exact contribution of each receptor, the possible intervention of nongenomic signaling, and the nature of the genetic program that is controlled by the receptors remain poorly understood.

1. INTRODUCTION

Thyroid hormones (TH, including 3,3′,5-triiodo-L-thyronine (T3) which is the most active form, and thyroxine (T4), its poorly active precursor) are required for proper neurodevelopment and exert important

Current Topics in Developmental Biology, Volume 125
ISSN 0070-2153
http://dx.doi.org/10.1016/bs.ctdb.2017.01.001

functions in the adult brain throughout life. Early TH deficiency, also called congenital hypothyroidism, is often caused by iodine deficiency and maternal hypothyroidism. It has irreversible and severe consequences on cognitive functions if not handled properly soon after birth. By contrast, adult hypothyroidism alters memory and motor coordination and increases the risk of depression, but these symptoms can be reversed by replacement therapy.

An abundant literature has described the multiple consequences of T3 deficiency in the brain, using mainly rats and mice as experimental models (Berbel, Navarro, & Roman, 2014; Bernal, 2007). It seems that all aspects of brain development and physiology are influenced, making a complete understanding of T3 action in the brain a daunting task. However, T3 directly influences gene expression, by binding to nuclear receptors, which is a key advantage to tackle these questions. The nuclear receptors of T3 (TRα1, TRβ1, and TRβ2, collectively called TRs) are transcription factors, which act mainly as heterodimers with RXRs, another type of nuclear receptor. TR/RXR heterodimers bind DNA at specific locations, called T3 response elements, and regulate transcription initiation of proximal genes, called target genes (Fig. 1). TRs are encoded by two genes called *Thra* (Fig. 2) and *Thrb* (Fig. 3) in mice, or *THRA* and *THRB* in humans.

This review focuses specifically on the progress brought by mouse genetic studies in the understanding of the functions of TRs in the brain. Such studies can serve two purposes. They provide an access to the genetic program of normal brain development, in which TR target genes must play an important role. It is also a way to unravel the multiple consequences of alterations of TH signaling, due to either TH deficiency or to genetic mutations. Notably, human germline mutations of *THRA* and *THRB* lead to two different genetic diseases, now called RTHα and RTHβ (RTH standing for resistance to thyroid hormone) (Refetoff et al., 2014). RTHα was discovered only in 2012. Up to now, it has been reported for only 25 patients, with a phenotype reminiscent of congenital hypothyroidism, but with only marginally subnormal levels of circulating TH. The absence of a specific biochemical marker makes the diagnosis of RTHα very difficult, suggesting that this rare genetic disease often remains undiagnosed. By contrast, RTHβ is marked by a combination of increased levels of circulating T4 and T3, due to altered feedback regulation of the hypothalamus–pituitary–thyroid gland axis, and decreased sensitivity to T3 of cell types where TRβ1 or TRβ2 is the main receptor. RTHβ has been known for a long time and is present in hundreds of families.

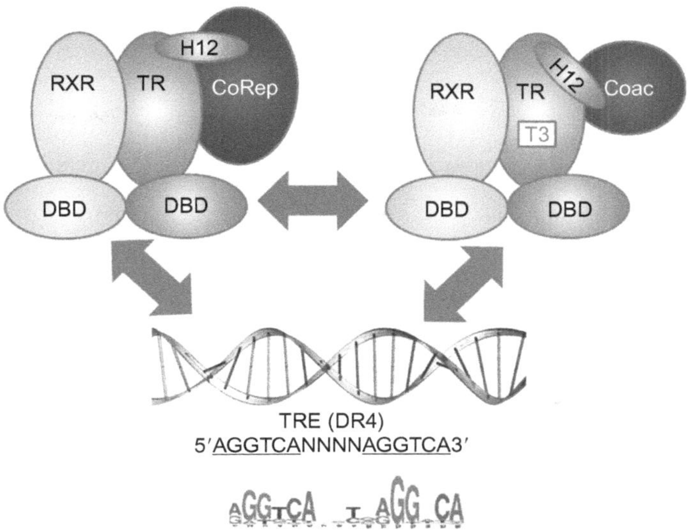

Fig. 1 Simplified model for TR-mediated transactivation. TRs form heterodimers with RXR nuclear receptors, which can bind DNA at T3 response elements (TREs) through their DNA-binding domains (DBD). The consensus DR4 sequences have a tandem duplication of the 5′AGGTCA3′ half-site separated by a 4-nucleotide spacer. This motif has been identified by de novo search in a TR ChipSeq dataset obtained in neural cells (Chatonnet, Guyot, Benoit, & Flamant, 2013). Unliganded TRs form stable complexes with transcription repressors, which possess histone deacetylase activity. T3 entry in the ligand-binding pocket provokes a configuration change, repositioning helix 12 (H12). This favors the interaction between liganded TRs and coactivators, among which are histone acetyl-transferases, at the expense of corepressors. While the associations between heterodimers and DNA may be somewhat labile (Grontved et al., 2015), coactivator- and corepressor-containing complexes leave long-lasting marks on histone tails and define the recruitment of RNA polymerase II at proximal transcription start sites.

2. TRs IN THE BRAIN

THRA and *THRB* are broadly expressed, in many organs including the brain. TRα1 mRNA is ubiquitous in the rodent brain, while *TRβ1* mRNA only appears in the brain at a late developmental stage and has a more restricted distribution (Bradley, Young, & Weinberger, 1989; Mellstrom, Naranjo, Santos, Gonzalez, & Bernal, 1991). *TRβ2* mRNA is found in few cell types, predominantly in the hypothalamus, pituitary, retina, and cochlea (Jones, Srinivas, Ng, & Forrest, 2003). Accordingly, most functional studies conclude that there is a predominant role of TRα1 in

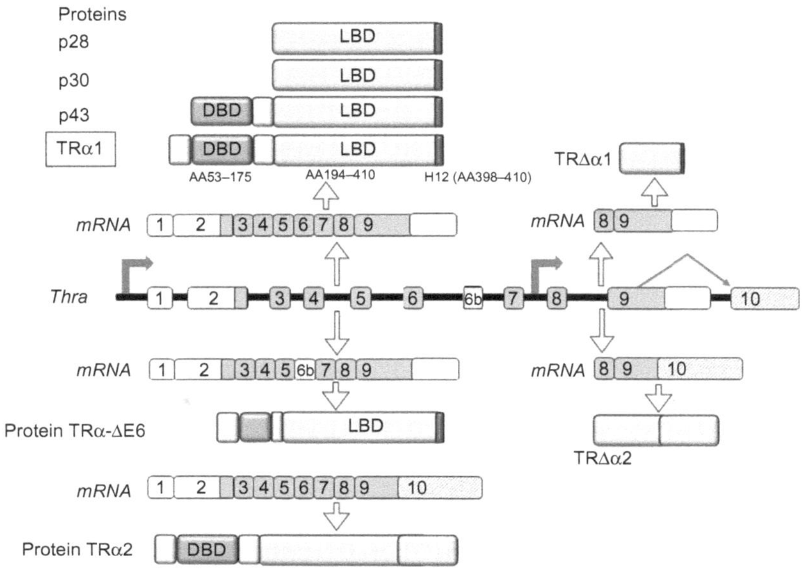

Fig. 2 Structure and expression of the *Thra* gene. All the alternate mRNA and protein products are represented, except for TRα3, which is very similar to TRα2, except for the alternate splicing acceptor site that is located downstream on the same exon. As a result, TRα3 has a shorter C-terminal extension (83 AA instead of 122 AA for TRα2). TRα1 is the only genuine nuclear receptor, present in the nucleus, able to bind DNA and T3, with a C-terminal helix 12 for coactivator recruitment.

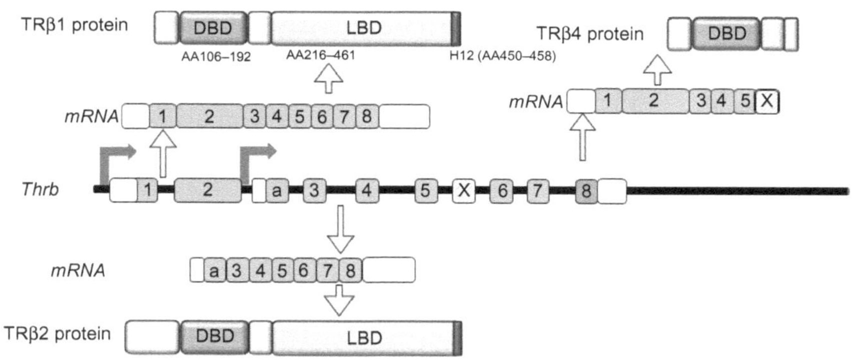

Fig. 3 Structure and expression of the *Thrb* gene. TRβ1 and TRβ2 are the main protein products present in all mammals examined. The functions of other *Thrb* products are less documented. Alternate promoter usage produces TRβ2, which also acts as a nuclear receptor of T3, in few cell types, mainly in retina, pituitary, hypothalamus, and inner ear. TRβ3 seems to be a rat-specific protein (Williams, 2000). Alternate splicing produces TRΔβ3, also rat specific and TRβ4, both of which do not have transactivation or T3-binding capacity.

brain development, and a selective influence of TRβ1/2 in the development of sensory functions and the maturation of some specific neuronal populations (Nunez, Celi, Ng, & Forrest, 2008). However, as most available antibodies are of insufficient specificity, the relative abundance of TRα1 and TRβ1 proteins in the different brain areas and cell types is unclear. Whether rodent and human expression patterns are similar is also unknown.

3. ANIMAL MODELS WITH TR MUTATIONS

A number of mouse models with *Thra* or *Thrb* mutations have been produced. The proliferation of recombinant alleles (Tables 1 and 2; Figs. 4 and 5) can be a source of confusion, which we clarify here. For knock-out mutations of *Thrb*, efforts from several groups generated seemingly equivalent models, which received identical names. However, the constructs differed by details that might have subtle consequences. Alternate promoter usage in *Thrb* produces TRβ2, which also acts as a nuclear receptor of T3, in few cell types, mainly in the retina, pituitary, hypothalamus, and inner ear, and there is genetic evidence that this protein has an important function, at least for mouse retina cone differentiation (Ng et al., 2001) and TRH feedback regulation (Abel, Ahima, Boers, Elmquist, & Wondisford, 2001). Moreover, the recent identification of TRβ4 isoform in human cells (Moriyama et al., 2016; Tagami et al., 2010) following the discovery of the rat specific TRβ3 and TRβΔ3 (Williams, 2000) suggests that the *Thrb* locus may still hold surprises.

The situation for *Thra* is different, as initial analysis of the first knock-outs revealed that the *Thra* locus has a more complex structure than the *Thrb* locus. We now know that *Thra* not only encodes the nuclear receptor TRα1, but also a number of proteins whose functions are still unclear: p43 and p30 are alternate translation products of the TRα1 mRNA which can bind T3 (Bigler, Hokanson, & Eisenman, 1992) but are not nuclear proteins (Casas et al., 1999; Kalyanaraman et al., 2014). Alternate splicing also gives rise to TRα2 (Lazar, Hodin, & Chin, 1989), TRα3 (Mitsuhashi, Tennyson, & Nikodem, 1988), and TRαΔE6 (Casas et al., 2006), which do not bind T3. An intronic transcription promoter also produces TRΔα1 and TRΔα2, which lack the DNA-binding domain, the hinge region and part of the ligand-binding domain, and lose the capacity to bind T3 (Chassande et al., 1997). These discoveries stimulated the generation of new models to selectively eliminate some of these protein products. These isoforms will not be discussed further, as their

Table 1 *Thra* Mouse Mutations

Allele Name	Designation	Mutation Type	TRα1 Mutation	Proteins Maintained				References
				TRα1	TRα2	TRΔα1	TRΔα2	
$TR\alpha^{0}$	$Thra^{tm2Jas}$	KO						Gauthier et al. (2001)
$TR\alpha1^{-}$	$Thra^{tm1Ven}$	KO			√		√	Wikstrom et al. (1998)
$TR\alpha2^{-}$	$Thra^{tm2Ven}$	KO		√		√		Salto et al. (2001)
$TR\alpha^{-}$	$Thra^{tm1Jas}$	KO				√	√	Fraichard et al. (1997)
$TR\alpha^{7}$	$Thra^{tm3Jas}$	KO				√	√	Plateroti et al. (2001)
P43-	$Thra^{tm1.1Fcas}$	KO	M37L	√	√	√	√	Blanchet et al. (2012)
$Thra^{f/f}$		KO (floxed)				√?	√?	Kalyanaraman et al. (2014)
	$Thra^{tm1b\ EUCOMM)Wtsi/J}$	KO (floxed)				√	√	Unpublished[a]
$TR\alpha1^{GFP}$	$Thra^{tm4.1Ven}$	KI	Fusion with Gfp	√	√	√	√	Wallis et al. (2010)
$TR\alpha1^{P398H}$	$Thra^{tm1Brnt}$	KI	P398H[b]	√	√	√	√	Liu, Schultz, and Brent (2003)
$TR\alpha1^{PV}$	$Thra^{tm1Syc}$	KI	C-term.[c]	√		√		Kaneshige et al. (2001)
$TR\alpha1^{R384C}$	$Thra^{tm3Ven}$	KI	R384C	√		√		Tinnikov et al. (2002)
$TR\alpha^{AMI}$	$Thra^{tm1Ffla}$	KI (floxed)	L400R	√				Quignodon, Vincent, Winter, Samarut, and Flamant (2007)

[a]Available at the Jackson laboratories.
[b]Due to poor fertility, this mouse strain has been lost.
[c]The C-terminal end (AA392–410) is replaced by CPHRTLPPFVLGSVRVLD, as in a human TRβ1 mutation.
Allele designation is according to Mouse Genome Informatics database.

Table 2 Mouse Target Mutations in *Thrb*

Allele Name	Designation	Mutation Type	Mutation in Thrb Reading Frame	References
TRβ^{-}	Thrb^{tm1Df}	KO		Forrest et al. (1996)
TRβ^{-}	Thrb^{tm1Olc}	KO		Gauthier et al. (1999)
TRβ^{-}	Thrb^{tm4Few}	KO		Shibusawa et al. (2003)
Thrb^{b1}	Thrb^{tm3Df}	KO	Selective TRβ1 elimination	Ng et al. (2015)
TRβ2^{-}	Thrb^{tm1Few}	KO	Selective TRβ2 elimination	Abel et al. (1999)
TRβ2^{-}	Thrb^{tm2Df}	KO	Selective TRβ2 elimination	Ng et al. (2001)
TRβ^{lox}	Thrb^{tm1Mkni}	KO (floxed)		Winter et al. (2009)
THRB$^{f/f}$		KO (floxed)		Kalyanaraman et al. (2014)
TRβ^{GS}	Thrb^{tm3Few}	KI	E125G+G126S	Shibusawa et al. (2003)
TRβ^{Y147F}	Thrb^{tm1Ehs}	KI	Y147F	Martin et al. (2014)
TRβ$^{\Delta 337T}$	Thrb^{tm2Few}	KI	Δ337T	Hashimoto et al. (2001)
TRβ^{R429Q}	Thrb$^{tm6.1Few}$	KI	R429Q	Machado et al. (2009)
TRβ^{E457A}	Thrb^{tm5Few}	KI	E457A	Ortiga-Carvalho et al. (2005)
TRβ^{PV}	Thrb^{tm1Syc}	KI	C-term.[a]	Kaneshige et al. (2000)

[a]The C-terminal end (AA443–461) is replaced by CPHRTLPPFVLGSVRVLD as in a human mutation.
Allele designation is according to Mouse Genome Informatics database.

function in the central nervous system has not been specifically investigated. However, genetic evidence suggests that some of them have developmental functions (reviewed in Flamant & Samarut, 2003).

TH signaling has been conserved during vertebrate evolution, due to strong selective pressure. The addition of other nonmammalian animal models would thus be highly profitable, notably in species where an extensive genetic toolbox is available, like in zebrafish. In this species the *Thra* gene is duplicated, in *Thraa* and *Thrab*. Systematic mutagenesis programs have generated point mutations (2 for *Thraa,* 1 for *Thrab,* 1 for *Thrb)* and

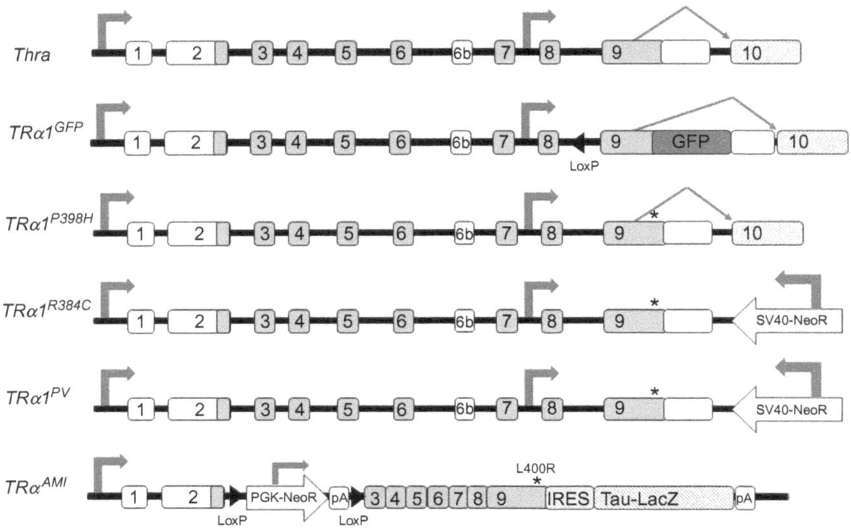

Fig. 4 Design of *Thra* knock-in alleles. Except for $TR\alpha1^{P398H}$, which was produced using a hit-and-run method, all *Thra* knock-in alleles contain superfluous sequences used for drug selection in embryonic stem cells, or carry reporter genes (GFP and TauLacZ). $TR\alpha1^{gfp}$ was designed to make *Thra*-expressing cells fluorescent. However, the C-terminal extension is expected to reduce the repressing activity of the receptor. The alternate splicing that produces TRα2, TRα3, and TRΔα2 is eliminated in $TR\alpha1^{PV}$ and $TR\alpha1^{R384C}$. $TR\alpha^{AMI}$ (for *A*ctivation function 2 *M*utation, *I*nducible), is a null allele, due to the presence of a polyadenylation signal (pA) which impairs transcription elongation. Elimination of the PGK-NeoR-pA cassette by Cre/loxP recombination triggers the expression of the $TR\alpha1^{L400R}$ reading frame. An internal ribosome entry site (IRES) allows internal translation initiation for the production of the TaulacZ reporter enzyme, which has β-galactosidase activity. For each knock-in allele, the *asterisk* (*) indicates the position of the codon substitution.

insertion mutations (referenced in the Zfin.org database) but phenotypes resulting from these mutations have not yet been investigated or reported in the literature. Morpholino knock-down has been recently explored as a rapid and versatile alternative to classical mutagenesis to study *Thraa* (Marelli et al., 2016) and *Thrb* function (Suzuki et al., 2013).

4. INTERPRETATION OF PHENOTYPES RESULTING FROM KNOCK-IN AND KNOCK-OUT MUTATIONS

Both knock-out mutations, which eliminate at least one protein, and knock-in mutations, which change the amino acid sequence of one

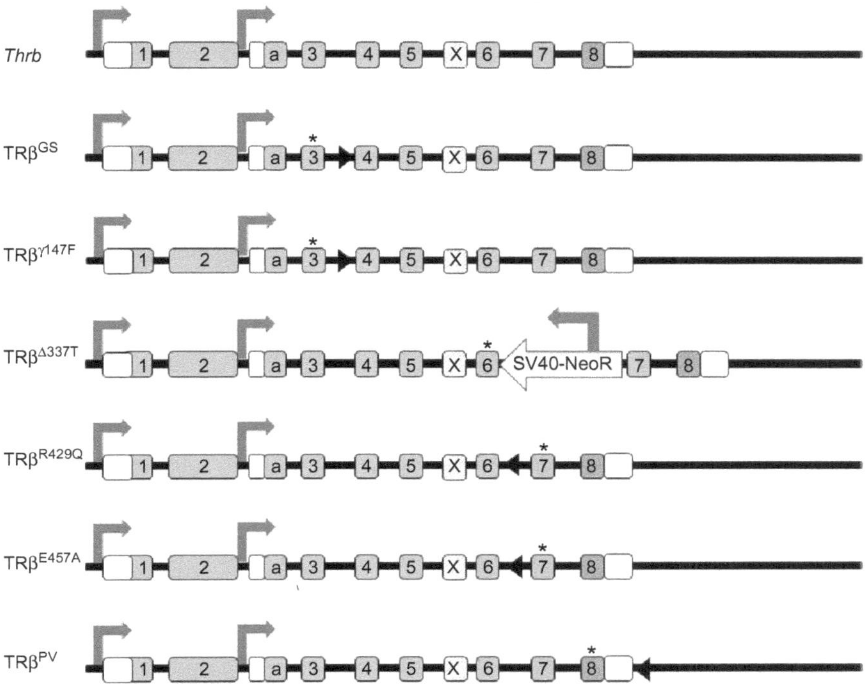

Fig. 5 Design of *Thrb* knock-in alleles. The situation is simpler than for *Thra*, as the construction design is similar in all cases, except for TRβ$^{\Delta 337T}$ in which a residual selection marker is present. For each knock-in allele, the *asterisk* (*) indicates the position of the codon substitution. Alternate use of exon X results in early translation termination, generating the TRβ4 variant.

or several proteins, were produced in mice. However, the generation of a knock-in allele often necessitates using a selection gene, to provide drug resistance to embryonic stem cells. In the case of *Thra* (Fig. 4), this constraint led to design of alleles where changes in the amino acid sequence of one isoform were combined with the elimination of other splice variant proteins encoded by the locus. Although homozygosity usually aggravates the phenotype (Portella et al., 2010; Tinnikov et al., 2002), it is prudent to study the consequences of knock-in mutations in heterozygous mice, as the intact gene copy ensures the maintenance of all encoded proteins.

There are striking differences between the outcome of knock-out and knock-in mutations. *Thrb* knock-out mainly alters retina and inner ear maturation (Jones et al., 2003), while *Thra* knock-out only has minimal neurodevelopmental consequences. More specifically, the only reported defect in *Thra* knock-out is a reduction in GABAergic synaptic density in hippocampus (Guadano-Ferraz et al., 2003). By contrast, knock-in mutations often cause severe histological defects in cerebellum and cortex and entail major behavior abnormalities. It is also clear that the lethality induced by hypothyroidism reflects the fact that unliganded TRs have a strong capacity to repress transcription, as they efficiently recruit corepressors with histone deacetylase activity on chromatin. The main evidence supporting this interpretation is that some developmental defects caused by congenital hypothyroidism, notably in the brain, can be at least partially rescued by knocking-out TRα1 (Flamant et al., 2002; Morte, Manzano, Scanlan, Vennstrom, & Bernal, 2002). Furthermore, the reduction of corepressor activity, either by pharmacological treatment, or by gene knock-out, improves the condition of mice with *Thra* or *Thrb* knock-in mutations (Fozzatti et al., 2013, 2011; Kim, Park, Willingham, & Cheng, 2014). Therefore, the key difference between knock-out and knock-in models is that knock-outs abrogate both the capacity of the receptor to transactivate target genes upon ligand binding and the possibility for the receptor to recruit transcription corepressors.

Another parameter, which may aggravate the phenotype resulting from knock-in mutations as compared to knock-outs, is the ability of a mutated receptor to antagonize both TRα1 and TRβ1/2 functions. Like in human patients, the mutated TR exerts a dominant-negative influence over the intact TR, making the mutation dominant. The underlying mechanism of such dominance is unclear, but an explanation could stem from the lability of the association between TR and chromatin (Grontved et al., 2015). A likely hypothesis, awaiting definitive experimental evidence, is that coactivator recruitment destabilizes chromatin association, as suggested for other nuclear receptors (Chen, Lin, Xie, Wilpitz, & Evans, 1999). Therefore, inactive TRs may have longer residence times in chromatin and impose negative regulation marks on histone tails, which would last longer than positive regulation marks. In vitro, mutated TRs have been shown to exert this dominant-negative effect on all TRs in the cell, but with various efficacies. For example, TRα1^{R384C} suppresses transactivation by TRα1 and TRβ2, but is less efficient against TRβ1 (Tinnikov et al., 2002). A similar differential dominance has been observed for TRα1^{L400R}. Accordingly, TRα1^{L400R} induces major defects in mouse tissues that have previously been identified as sensitive to *Thra* knock-out, while it preserves TRβ1 known functions. However, an alternative explanation of these in vivo observations would be that the maintenance of TRβ1 functions in TRα1^{L400R} expressing mice only reflects the higher abundance of TRβ1, relative to TRα1, in the cell types where its function is best characterized (Quignodon, Vincent, et al., 2007).

While the phenotypes of knock-in mice usually suggest that the functions of TRα1 and TRβ1 are clearly distinct, knock-out studies rather indicate a redundancy between the functions of these two receptors. The main observation supporting this view is that the combination of *Thra* and *Thrb* knock-out mutations induces a synergetic effect. For example, TSH deregulation and reduction in body temperature were found to be maximal when both knock-outs were combined (Gauthier et al., 1999; Gothe et al., 1999). Thus, several studies based on knock-out phenotyping have concluded that the functions of TRα1 and TRβ1 are interchangeable, and that their respective importance in a given cell type is mainly governed by their respective abundance. However, there is limited published evidence supporting that the combination of *Thra* or *Thrb* knock-out aggravates neurodevelopmental defects. A rare example is that combined knockout mice have exacerbated cochlear abnormalities (Rusch et al., 2001). More generally, the relatively mild phenotype of mice devoid of all TRs (Gauthier et al., 2001; Gothe et al., 1999) came as a surprise, considering that a complete TH deficiency is lethal in mice, within about 3 weeks after birth (Mansouri, Chowdhury, & Gruss, 1998). Furthermore, it has been shown for other nuclear receptors that the compensation taking place after a single knock-out is to some extent an artifact, the absence of one receptor broadening the repertoire of target genes regulated by the remaining receptor (Taneja et al., 1996). For example, in a pituitary thyrotrophic Tα1T cell line, direct biochemical evidence indicates that chromatin access to the *TSHb* gene promoter becomes possible for TRα1 only when TRβ1 is eliminated (Chiamolera et al., 2012). Therefore, knock-out analysis does not provide definitive evidence that TRα1 and TRβ1 are interchangeable.

Knock-in models offer an advantage over knock-outs to study the function of TRs in the brain, as they may produce more obvious neurodevelopmental defects. Knock-in mutations are also better models of the human genetic diseases RTHα and RTHβ. Like mouse knock-in mutations, these diseases are generally dominant caused by missense or frameshift mutations, which are carried and transmitted by heterozygous patients (Fig. 6).

Wild-type C-terminus of TRα1 (AA370-410)	
	KVTDLRMIGACHASRFLHMKVECPTELFPPLFLEVFEDQEV
	Helix 12
Human *THRA* missense mutations	
R384H	KVTDLRMIGACHAS**H**FLHMKVECPTELFPPLFLEVFEDQEV
R384C	KVTDLRMIGACHAS**C**FLHMKVECPTELFPPLFLEVFEDQEV
P398R	KVTDLRMIGACHASRFLHMKVECPTELF**R**PLFLEVFEDQEV
Human *THRA* frameshift mutations	
C380fs387X	KVTDLRMIGA**TPAASST**
A382Pfs393X	KVTDLRMIGACH**PAASST**
C392X	KVTDLRMIGACHASRFLHMKVEC
F397fs406X	KVTDLRMIGACHASRFLHMKVECPTELFPP**TLPRGL**
E403X	KVTDLRMIGACHASRFLHMKVECPTELFPPLFL**E**
E403K	KVTDLRMIGACHASRFLHMKVECPTELFPPLFL**K**
Mouse *Thra* knock-in mutations	
R384C	KVTDLRMIGACHAS**C**FLHMKVECPTELFPPLFLEVFEDQEV
P398H	KVTDLRMIGACHASRFLHMKVECPTELF**H**PLFLEVFEDQEV
L400R	KVTDLRMIGACHASRFLHMKVECPTELFPP**R**FLEVFEDQEV
PV	KVTDLRMIGACHASRFLHMKVECP**HRTLPPFVLGSVRGLD**

Fig. 6 C-terminal mutations in TRα1. Human mutations located after AA370 in the C-terminal AF2 domain are all represented. This leaves aside four missense mutations located upstream (D211G, A263S, A263V, and N359Y), which alter T3 binding. *Bold characters* are for changed amino acids. Note that the R384C mutation, first produced in mice, was then discovered in a human RTHα patient, while the P398R missense mutation was introduced on a residue later found to be mutated in a patient. The PV mutation, an artificial construct copied from a human *THRB* mutation, eliminates the C-terminal helix, which is required for coactivator recruitment, like several human *THRA* mutations. Finally L400R, which is lethal in mice, does not have an exact equivalent among the 14 known human *THRA* mutations and hundreds of *THRB* mutations.

5. RESPECTIVE FUNCTIONS OF TRs IN NEURAL CELL DIFFERENTIATION

The influence of T3, TRα1 and TRβ1 on cell differentiation has been studied in detail for only a few cell types. We focus here on a glial cell type, oligodendrocytes, and a neuronal cell type, Purkinje cells. Both examples illustrate how generating somatic mutations in chosen cell types by Cre/loxP recombination enables studies to unravel the complex cellular interactions, which underlie neurodevelopment, and grasp the time course of in vivo TR action.

Oligodendrocytes produce myelin, which wraps around axons at a late stage of central nervous system development. Two populations of oligodendrocyte progenitor cells (OPCs) can be distinguished. Early OPCs divide rapidly and ensure the initial myelination during brain maturation. A different population of slowly proliferating OPCs ensures the renewal of the oligodendrocyte population, and the repair of myelin, in the adult brain (Zhu et al., 2011). The process of OPCs differentiation has been extensively analyzed in primary cell culture, where it appears to be strictly dependent on T3 (Barres, Lazar, & Raff, 1994; Dugas, Ibrahim, & Barres, 2012). However, these in vitro observations seem to only reflect the behavior of adult OPCs. The initial myelination, and the early OPCs postnatal differentiation, is only slightly delayed in *Thra*, but not in *Thrb*, knockout (Billon, Jolicoeur, Tokumoto, Vennstrom, & Raff, 2002). By contrast, the population of slow cycling adult OPCs is amplified in the optic nerve of mice devoid of all receptors, suggesting a significant impairment of cell-cycle exit (Baas, Legrand, Samarut, & Flamant, 2002). Cre/loxP manipulation of *Thra* has confirmed a different behavior of these two populations of OPCs. The late OPCs display cell-autonomous response to T3 (Picou, Fauquier, Chatonnet, & Flamant, 2012). By contrast, the differentiation of early OPCs and axon initial myelination are only indirectly influenced by T3. This indirect influence of T3 on early OPCs is a consequence of neurotrophins secretion by other cell types, a process which is itself dependent on T3 (Barres et al., 1994; Bouslama-Oueghlani et al., 2012; Dugas et al., 2012; Picou et al., 2012).

Purkinje cells of the postnatal cerebellum cortex (Sotelo & Dusart, 2009) offer a favorable model to study the influence of T3 on neuronal terminal differentiation (Fig. 7). These cells first express TRα1 and then TRβ1, the latter becoming predominant over time, probably around the second postnatal week (Bradley et al., 1989; Mellstrom et al., 1991; Wallis et al., 2010). The stepwise process of Purkinje cell maturation, during which the most visible outcome is the generation of a large dendritic tree, can also be observed in isolated cells, in a primary cell culture (Boukhtouche et al., 2010, 2006). That isolated Purkinje cells are sensitive to T3 is an indication that their in vivo response is cell-autonomous, i.e., independent of a possible influence of T3 on their microenvironment. However, Purkinje cell maturation is also sensitive to the microenvironment of the cerebellum. Their differentiation is notably accelerated by neurotrophin 3, a protein that is secreted in the cerebellum by neighboring granule cells, this secretion being itself stimulated by T3. However, the stimulation of production of

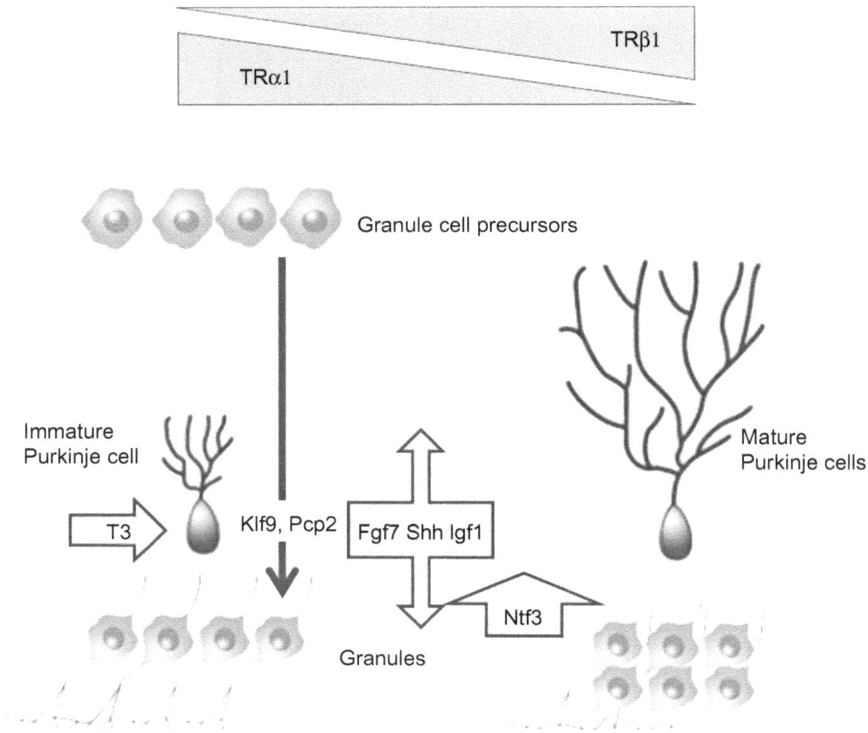

Fig. 7 Thyroid hormone influence on the interaction between neuronal cell populations during rodent cerebellum maturation. This simplified model depicts the postnatal evolution of Purkinje cells and granule cells. Additional cell types (not represented) also participate in the local network of cellular interactions. Notably, Bergmann glia cells guide the inward migration of granule cells that follows cell-cycle exit and favor synapse formation. Purkinje cells express TRα1 and then TRβ1, the latter becoming predominant around the second postnatal week. Granule cells express TRα1, but they display little cell-autonomous response to T3. Purkinje cells secrete a number of growth factors, which favor the cell-cycle exit of granule cell precursors and their differentiation. Granule cells, whose numbers quickly increase after birth, also secrete growth factors, notably neurotrophin 3, which facilitates Purkinje cell maturation. The cross-talk between Purkinje cells and granule cells supports a self-regulatory mechanism: Purkinje cells promote the expansion of the granule cell population and thus favor the secretion of the neurotrophins by these neurons. These neurotrophins are necessary for the Purkinje cell maturation. This transient-positive feedback loop greatly amplifies the initial response of Purkinje cells to T3.

neurotrophin 3 by granule cells is also not a direct effect of T3. It is probably the indirect consequence of the stimulation of granule cells by growth factors secreted by Purkinje cells themselves. Therefore, T3 response in the integrated context of the developing cerebellum (Fig. 7) is probably amplified by a positive feedback loop involving the secretion of neurotrophin 3 and of other proteins secreted by neighboring cells (Fauquier et al., 2014; Lindholm et al., 1993; Neveu & Arenas, 1996; Picou et al., 2012). The in vitro response of Purkinje cells to T3 is lost after *Thra,* but not *Thrb* knock-out, showing a predominant intervention of TRα1 in Purkinje cell maturation (Heuer & Mason, 2003). However, both *Thra* (Fauquier et al., 2011) and *Thrb* (Portella et al., 2010) knock-in mutations alter this maturation process, reducing dendritic arborization. The comparison between these in vitro and in vivo data suggests that TRβ1 action is rather on the extracellular feedback loop that involves neurotrophin 3. There is, however, strong genetic evidence that both receptors exert their influence in a cell-autonomous manner in Purkinje cells. On the one hand, expressing $TR\beta^{G345R}$ only in Purkinje cells efficiently prevents their maturation in mice (Yu et al., 2015). On the other hand, the use of the *Ptf1aCre* transgene to restrict the expression $TR\alpha1^{L400R}$ to Purkinje cells has detrimental consequences on their differentiation (Fauquier et al., 2014). Interestingly, using a *Cre* transgene driven by a *L7/Pcp2* transcription promoter, which also activates $TR\alpha1^{L400R}$ expression selectively in Purkinje cells, but at a later stage of postnatal development, does not alter their maturation in the same way. This suggests that the role of TRα1 in Purkinje cell maturation is predominant during the first postnatal week, when there is little TRβ1 expression. Taken together, these data lead to the conclusion that the relative stoichiometry of TRα1 and TRβ1 is a key parameter. It defines the respective contribution of each receptor to the response of Purkinje cells to T3, which changes over time. Stoichiometry also determines the capacity of mutated receptors to hamper dendritic arborization.

The sequential functions of TRα1 and TRβ1 in Purkinje cells suggest that at some stage, both receptors are present in the cells. This raises the question of whether they fulfill different or redundant functions when expressed at similar levels in the same cells. This question has not yet been investigated in any neuronal cell type. A favorable situation has been found in sensory systems; however, in the inner ear, where outer hair cells express both TRα1 and TRβ1 and need T3 for terminal maturation. This maturation is marked by the appearance of the Kcnq4 potassium channel, and the polarized redistribution of prestin, a transmembrane protein required for outer hair cell electromotility. Hair cells provide an interesting model to evaluate the capacity of mutant receptors to extend their dominant-negative effect to all TRs. In this cell type, $TR\alpha1^{L400R}$ down-regulates the Kcnq4 channel (Quignodon, Vincent, et al., 2007), leaving prestin protein distribution unchanged, while *Thrb* knock-out selectively alters prestin distribution without changing Kcnq4 level (Winter et al., 2006). Therefore, some T3 response persists in cells with one TR mutation, indicating that TRα1 and TRβ1 can serve distinct functions within the same cell. The data also indicate that receptor selective functions are not always sensitive to the dominant-negative influence exerted by the other mutated receptor.

Although phenotype interpretation should carefully consider the limits of both knock-out and knock-in models, these examples illustrate the capacity of available mouse models to clarify the respective functions of the different TRs in an intact organism. Another way to analyze the respective contributions of TRα1 and TRβ1/2 in neurodevelopment is to use artificial ligands designed to bind only one type of receptor. However, this is difficult because available ligands display only limited specificity and may not be devoid of side effects (Grijota-Martinez, Samarut, Scanlan, Morte, & Bernal, 2011; Manzano, Morte, Scanlan, & Bernal, 2003; Nguyen et al., 2002).

6. NONGENOMIC SIGNALING IN THE BRAIN

The focus on *Thra* and *Thrb* functions is based on the assumption that all T3 action is mediated by the TRα1, TRβ1, and TRβ2 nuclear receptors. However, a controversial literature reports that T3 stimulates the generation of second messengers such as Ca2+, NO, inositol trisphosphate, and cAMP in various cells in culture. This mode of action takes place within minutes, a speed which is hardly compatible with a transcriptional response (Cheng, Leonard, & Davis, 2010). These rapid effects are thus called "nongenomic," although, in the long-term, any change in cellular physiology should translate into some changes in gene expression.

A first possibility for a nongenomic response would be that T3, or some of its metabolites, which are not considered here (Goglia, 2005; Piehl, Hoefig, Scanlan, & Kohrle, 2011), can act in a TR-independent manner. It has been suggested that T4 can initiate a cellular response at the plasma membrane, using integrin αvβ3 as a receptor (Cheng et al., 2010). Another intriguing possibility would be that T3 simply acts as an allosteric regulator of a metabolic enzyme. T3 has been known to bind with low affinity to μ-crystallin, also called NADP-regulated thyroid hormone-binding protein (THBP), Cytosolic 3,5,3′-T3-binding protein (CTBP) or Crym. Crym is a cytoplasmic protein, which has been initially considered as a component of a storage compartment for T3 (Mori et al., 2002; Suzuki et al., 2007). However, Crym has recently been identified as the enzyme that reduces the sulfur-containing cyclic ketimines, which are potential neurotransmitters. T3 appears to be a strong inhibitor of this enzymatic activity (Hallen, Cooper, Jamie, & Karuso, 2015). When overexpressed, Crym exerts a neuroprotective effect in a model of Huntington's disease, suggesting that T3 could modulate this function (Francelle et al., 2015).

The second possible type of nongenomic response relies on alternate functions of TRs. A recent unbiased genome-wide analysis has indicated that gene expression of cultured cortical neurons lacking both *Thra* and *Thrb* is insensitive to T3 (Gil-Ibanez, Garcia-Garcia, Dopazo, Bernal, & Morte, 2015). This observation implies that, if nongenomic signaling takes place in these cell types, it must involve proteins encoded by *Thra* and/or *Thrb*. Several possibilities received experimental support, which all ultimately converged on the rapid activation of the phosphatidylinositol-4,5-bisphosphate 3-kinase(PI3K)/Protein kinase B (Akt) pathway by T3. One possibility is based on the *Thra*-encoded p30 protein, whose translation is initiated at a downstream AUG codon of the TRα1 mRNA and thus lacks a DNA-binding domain, but still binds T3 (Fig. 2). This protein has been localized at the plasma membrane, where it stimulates calcium entry upon T3 binding by an unknown mechanism, indirectly stimulating nitric oxide synthase and PI3K/Akt pathway (Kalyanaraman et al., 2014). Although this pathway has been only illustrated in chondrocytes, none of the components is specific from this cell type, and it should operate in any *Thra*-expressing cell. A previous study proposed that the full-length TRα1 receptor was also able to trigger the same type of cellular response, this time by directly interacting with the p85α subunit of PI3K (Hiroi et al., 2006). Other studies concluded that neither TRα1 nor p30, but TRβ1 is responding to T3 at the plasma membrane (Martin et al., 2014; Storey et al., 2006). Unlike TRα1, which lacks a critical tyrosine residue in its DNA-binding domain, TRβ1 can serve as an intermediate between tyrosine kinase receptors and PI3K. It can also interact with the p85α subunit of PI3K. While it was observed that T3 stabilizes the TRα1/p85α interaction (Hiroi et al., 2006), the opposite conclusion was reached for the TRβ1/p85α interaction (Martin et al., 2014). Another indirect, Akt independent, consequence of the activation of PI3K is the activation of the Kv11.1 potassium channel (Gentile et al., 2006, 2008). Interestingly, this channel is encoded by the *Kcna1* gene, whose expression is regulated by the genomic pathway, raising the possibility that nongenomic and genomic signaling can cross-talk at this level (Chatonnet et al., 2013; Chatonnet, Guyot, Picou, Bondesson, & Flamant, 2012; Dong et al., 2005).

Given the importance of PI3K/Akt and ion channels in the developing and mature brain, the clarification of the relevance of all these conflicting data is of importance. One step in this direction has been made by introducing a missense mutation in *Thrb*. The Y147F substitution introduced in the DNA-binding domain of mouse TRβ1, which at least at first glance does not alter the genomic response, abrogates the capacity of TRβ1 to activate PI3K (Martin et al., 2014). Electrophysiological recordings on hippocampus slices prepared from $TR\beta1^{Y147F/+}$ mice revealed a decreased synaptic strength and decreased long-term potentiation in response to high-frequency stimulation. According to the proposed mechanism, these deficiencies could be direct consequences of altered PI3K pathway response. However, the possibility that these are indirect consequences of neurodevelopmental defects has not been addressed. This preliminary in vivo investigation may be an indication that nongenomic signaling contributes to the influence of TH in memory and learning (Beydoun et al., 2013; Rovet, 2014; Shrestha et al., 2016).

7. THE ORIGIN OF PHENOTYPE VARIABILITY

Major differences in phenotype have been noticed between the various mouse models. These differences are especially noteworthy between the four published mouse strains with *Thra* knock-in mutations (Mittag, Wallis, & Vennstrom, 2010). Three of these mutations (L400R, P398H, PV) alter the C-terminal helix 12 required for coactivator recruitment. The fourth mutation (R384C) is also located in the ligand-binding domain but its effect is different. It does not directly influence protein/protein interactions but reduces the affinity of TRα1 for T3. Consequently, the detrimental effects caused by $TR\alpha1^{R384C}$ can be attenuated by an excess of TH. Understanding the cause underlying this phenotype variability would be of interest, expecting that the variability of the phenotype in mice parallels the clinical variability observed in humans, where 14 mutations have been reported to date. Like the mouse mutations, the human TRα1 mutations alter the receptor function either by preventing coactivator recruitment by helix 12 or by reducing ligand affinity. Although disease severity can vary within a family among carriers of the same mutation, it mainly depends on the residual activity of the mutated receptor and its capacity to recruit corepressors. For example, IQ deficit is generally moderate when the capacity of TRα1 to release corepressors and recruit coactivators is not fully impaired. However, some traits, reported for one or few patients, could reflect specific properties of some mutant receptors. For example, among 25 known patients with RTHα, one has been identified in a cohort of patients with autism (Yuen et al., 2015), while another one suffered from epilepsy (Moran et al., 2013).

Thra knock-in models have not been phenotyped side-by-side and comparisons should be considered with caution, as some differences may reflect differences between genetic backgrounds, breeding conditions, or phenotyping methods. Furthermore, different strategies have been used to introduce the mutations, and the genetic architecture of the modified locus may influence differently the expression levels of the mutated TRα1 receptor and of the other *Thra*-encoded proteins. Small changes in the balance between *Thra*-encoded proteins may also have an influence on the phenotype of the mice. However, these uncontrolled factors can hardly account for at least two striking differences: the first one is spontaneous obesity, which was observed only in mice expressing $TR\alpha1^{P398H}$. Unfortunately, the $TR\alpha1^{P398H/+}$ mouse line has been lost, due to very low fertility (G Brent personal communication). It is thus impossible to verify that obesity and *Thra* mutation cosegregated over the generations among these mice. Such a genetic test is required to rule out the existence of an additional and unidentified genetic mutation, possibly on the same chromosome, brought by the stem cells that were used to create the mutation. The second discrepancy is lethality, which was observed only in mice expressing $TR\alpha1^{L400R}$ throughout development. $TR\alpha1^{L400R}$ is slightly overexpressed, compared to TRα1, as alternate splicing has been eliminated in this mutant allele, and all *Thra* mutant allele transcription is dedicated to $TR\alpha1^{L400R}$ (Fig. 4). However, a similar overexpression is also expected for the *Thra* mutant alleles that were designed to express $TR\alpha1^{PV}$ and $TR\alpha1^{R384C}$. Therefore, the relatively high level of expression of $TR\alpha1^{L400R}$ in itself is unlikely to account for mouse lethality and the peculiar structure of the $TR\alpha1^{L400R}$ receptor probably contributes to lethality: the L400R mutation was chosen based on structural data, because the equivalent mutation in TRβ was among the few which fully eliminated the interaction with NCoA-2 (also called Grip-1 or SRC-2), a member of the p160 coactivators family which is required for T3-mediated transactivation (Darimont et al., 1998). Epileptic seizures were observed in juvenile mice expressing $TR\alpha1^{L400R}$, and these may be sufficient to explain lethality (Faingold, Randall, & Tupal, 2010).

8. DISTINCTION BETWEEN DEVELOPMENTAL AND ADULT FUNCTIONS OF TRs IN THE BRAIN

Adult onset of hypothyroidism in humans is responsible for mood disorders, which are independent of the neurodevelopment disorders caused by congenital hypothyroidism. These two T3 functions may, however, display some analogies. For example, part of the influence of T3 in the adult brain is to regulate adult neural stem cell differentiation (Remaud, Gothie, Morvan-Dubois, & Demeneix, 2014), which is, in essence, a developmental program. Furthermore, developmental studies pinpoint inhibitory GABAergic neurons as central targets of T3/TRα1 action (Fauquier et al., 2014; Wallis et al., 2008), while a recent study showed that the memory deficit in $TR\alpha1^{R384C/+}$ mice could be rescued at an adult stage by pharmacological intervention, using low dose of pentylenetetrazol, an antagonist of the $GABA_A$ receptor. The mechanism underlying this memory rescue remains, however, unclear and paradoxical. First the use of an inhibitor is not expected to restore GABA inhibitory function, which is deficient in these mice. Second the observed effect lasts much longer than the drug treatment. Gene expression profiling confirms that pentylenetetrazol treatment has long lasting effect and resets the expression of genes involved in the balance between synaptic inhibition and excitation (Wang et al., 2016).

Telling apart developmental from adult functions of T3 in the brain is not an easy matter. If only minor developmental defects are present in TR knock-out mice, as unpublished data suggest, it would be tempting to interpret any behavioral phenotype seen at adult stages in these mice as a manifestation of the direct influence of T3 in adult brain cells. This conclusion should, however, be taken with caution, because systemic disorders can influence brain function. Notably, *Thrb* knock-out results in increased TH levels in the serum, which may well explain the observed increased proliferation of hippocampal progenitor cell proliferation (Kapoor, Ghosh, Nordstrom, Vennstrom, & Vaidya, 2011). Knock-out studies indicate that *Thra* is the main effector of T3 in learning, memory, and anxiety (Guadano-Ferraz et al., 2003; Wilcoxon, Nadolski, Samarut, Chassande, & Redei, 2007). One interesting attempt to uncouple the developmental and adult functions of TRα1 relied on the residual sensitivity of $TR\alpha 1^{R384C/+}$ mice to T3, as the mutation only reduces T3 affinity. Treating juvenile $TR\alpha 1^{R384C/+}$ mice with TH, or using $TR\alpha 1^{R384C/+}$ mice born from mothers with an excess of circulating TH, is a way to counteract the negative effect of the mutation on neurodevelopment and to analyze the residual influence of the mutation in the adult brain. Reciprocally, TH treatment has been done at an adult stage, assuming that it would allow identification of deficiencies that are not indirect consequences of neurodevelopmental defects. The depressive-like and anxious phenotype of $TR\alpha 1^{R384C/+}$ mice proved to be reversed by TH treatment at an adult stage (Pilhatsch et al., 2010), while only juvenile treatment could restore locomotor activity. Therefore, only the defect in adult locomotion must have a neurodevelopmental origin (Venero et al., 2005).

9. TR TARGET GENES DEFINITION

Once the influence of T3 signaling on neural cell differentiation has been established, the main challenge is to identify the repertoire of TR target genes in the relevant brain cell types. This task is far from completed due to a number of technical difficulties. The current knowledge of TR target genes in brain cells is very limited and mainly based on transcriptome analysis (Chatonnet, Flamant, & Morte, 2015). However, genes should be defined as TR target genes only if they fulfill the following parameters:

(1) T3 replacement in hypothyroid animals should induce a clear change in mRNA level. Genome-wide scale analysis of transcriptome, currently performed by RNAseq, has identified hundreds of such genes. This is, however, far from sufficient to identify TR target genes as some of the observed changes are indirect consequences of T3 stimulation, notably of the secretion of neurotrophins. Neurotrophins exert a broad influence on neurodevelopment and indirectly affect gene expression in many cell types few hours after T3 stimulation.

(2) The mRNA level increase should be a cell-autonomous consequence of T3 stimulation, and not secondary to the activation of growth factor secretion in the cell microenvironment. This can be tested either in primary cell cultures, which is feasible for only few neural cell types, or in mice, using the Cre/loxP technology.

(3) The mRNA level increase should be a direct consequence of T3 action, and not secondary to the induction by genuine T3 target genes encoding transcription factors. The possibility of an intermediate transcription factor is not easy to assess in vivo. It is usually tested in primary culture cells or cell lines, by using a pharmacological inhibitor of mRNA translation, like cycloheximide: only the T3 response of genuine TR target genes is not sensitive to this treatment. A recent study which combined RNAseq, primary cultures of cortical neurons, and translation inhibition with cycloheximide, suggests that only a small fraction of the T3-regulated genes (106/1045) are *bona fide* TR target genes (Gil-Ibanez et al., 2015). This study also concludes that expression of a small subset of these likely target genes, 17 out of 106, is negatively regulated by liganded TR, by some unknown mechanism, while the remaining others are probably regulated by the canonical pathway depicted in Fig. 1.

(4) TRs should be associated to chromatin in the enhancer or promoter regions, which control the gene expression level. ChipSeq analysis permits definition of chromatin occupancy at genome-wide scale. However, for the moment, this type of analysis has been done in vivo only for the liver (Grontved et al., 2015; Ramadoss et al., 2014) and in vitro for only one mouse neural cell line (Chatonnet et al., 2013). An additional difficulty is that the definition of regulatory sequences, enhancers, or promoters, is a complex matter, and the capacity of enhancers to act at long distance is a subject of intense investigations (Buisine et al., 2015; Dekker & Mirny, 2016). Up to now, it has been arbitrarily considered that TR binding can influence gene transcription only if it takes place within a reasonable distance (20–30 kb) from the transcription start site. It is, however, known that chromatin looping permits regulation at much longer distance. In the future, the general influence of T3 on genome function will have to be defined more precisely by other means, such as identification of nascent transcripts (Danko et al., 2015), coactivator recruitment (Zwart et al., 2011), histone tail modifications, and changes in chromosome conformation (Dekker, Marti-Renom, & Mirny, 2013).

(5) Elimination of the RXR/TR heterodimer-binding site(s) suspected to impact the transcription of a candidate TR target gene should eliminate its T3 response. Transient expression assays have been used for such demonstration, but their relevance is questionable. Transfected cells contain hundreds of copies of exogenous genes, which are, for a large part, not chromatinized. Experience has proved that these are not regulated like the single copy endogenous gene. In the future, in vivo genetic evidence and mutation of chromosomal DR4 elements will be required. The CRISPR/Cas9 methodology has already made it feasible in cell lines (Kyono et al., 2016). While the DR4 consensus binding site is very frequent within the thousands of genomic-binding sites identified by TR ChIPseq, this is not reciprocal: DR4 elements are very common in the genome (>100,000). Although these would be expected to bind heterodimers in vitro, most of them are not occupied by TRs in the chromatin context. Although the choice of several statistical thresholds and other technical details influence the calculation, the ratio of occupied DR4 elements can be estimated between 0.7% and 7% from available ChipSeq datasets. The presence of a proximal DR4 consensus sequence, which has often been used in the past to infer that a gene is a TR target, is thus a very weak indication.

10. T3 TARGET GENE FUNCTIONS

An important issue is whether T3 action during neurodevelopment is different from its action in adult brain cells. Another unsolved question is whether different cell types, which all express at least one TR, will share the same set of target genes or not. Neural cells can be glial cells (microglial cells, astrocytes, oligodendrocytes, ependymal cells, radial glial cells) or neurons. Neuronal cell types are extremely diverse and defined using the following criteria: cell position, cell morphology, synaptic connections, cellular content, and electrophysiological properties. If each cell type possesses a different set of TR target genes, the inventory of these may be hardly feasible. However, at least two genuine TR target genes, *Hr* and *Klf9*, are upregulated by T3 in many cell types. These regulations take place both in the brain and other organs. A favorable possibility would thus be that neural cell types share at least a common subset of TR target genes. Under this hypothesis, the cellular context would mainly define the basal expression level of TR target genes, and the secondary consequences of the direct T3 response, while TRs would access chromatin essentially in the same way in all neural cell types. The comparison of ChipSeq data between liver and neural cells support this view, providing indications that, unlike other nuclear receptors, TRs can occupy the same genomic-binding site in very different cellular contexts. As a result, a significant fraction of TR target genes may be shared even between unrelated cell types (Chatonnet et al., 2015).

Experience proves, however, that the capacity to respond to T3 stimulation widely varies between different neural cell types, brain areas, and developmental stages for some unknown reason, which is not related to variations in *Thra* or *Thrb* expression levels. Cerebellum granule cells, which have been estimated to represent more than 70% of the whole brain neurons, and can be easily maintained in culture, display very limited response to T3

(Chatonnet et al., 2012; Quignodon, Grijot-Martinez, et al., 2007). The number of T3-sensitive genes, and the induction rate, is comparatively much higher in the striatum, where many neuronal cell types are inhibitory GABAergic neurons (Diez et al., 2008; Gil-Ibanez, Morte, & Bernal, 2013).

The fact that several important neural cell types display very limited autonomous response to T3 represents a major obstacle for the identification of TR target genes, as transcriptome analyses in the brain can be informative only if performed at a favorable developmental stage, in a suitable brain area, and preferably on purified or enriched cell populations. For this reason, the most convincing data were drawn from transcriptome studies performed on primary cultures of cortical neurons (Gil-Ibanez et al., 2015) or glial cells (Dugas, Tai, Speed, Ngai, & Barres, 2006).

From this preliminary survey of TR target genes, it is fair to recognize that, although they can provide a number of interesting hypotheses for later investigations, transcriptome datasets did not yet bring much novelty to our understanding of the physiological functions of T3. However, combining all these datasets, a small set of genes emerges, which are reproducibly detected as TH-responsive in several neural models and in which proximal regulatory sequences are occupied by TRs in the C17.2 neural cells (Chatonnet et al., 2015).

Several of these genes encode transcription factors *(Klf9, Dbp, Stat5a, Fos)*, cofactors *(Hr)*, or global regulators of gene expression (*Dnmt3a*) (Kyono et al., 2016), which probably mediate the secondary, cell-autonomous response to T3. The consequences of their regulation by TH have been studied only for *Klf9*, which is probably a TR target gene in all vertebrate species (Bagamasbad et al., 2015; Hoopfer, Huang, & Denver, 2002). *Klf9* encodes a Krüppel-like transcription factor with pleiotropic functions and may relay T3 action in several cell types. It upregulates levels of intracellular reactive oxygen species in cultured fibroblasts (Zucker et al., 2014) and favors differentiation of stem cells at the expense of self-renewal (Cvoro et al., 2015; Ying et al., 2014). *Klf9* induction by T3 is also necessary to promote OPCs differentiation in vitro (Dugas et al., 2012), and the terminal maturation of Purkinje neurons, a process during which these cells lose the capacity to regenerate an axon (Avci et al., 2012; Lebrun et al., 2013). Few of the other TR target genes provide immediate hypotheses on the mechanisms underlying T3 neurodevelopmental function, or its pathological deregulation, in either hypothyroid, RTHα or RTHβ patients. A noticeable exception is *Kcna1*, encoding the Kv1.1 potassium channel, as *Kcna1* knock-out mice spontaneously display epileptic seizures (Gautier & Glasscock, 2015). The recent technical progress in genetics and gene expression analysis described earlier will certainly provide important new information in the coming years, which should greatly improve our understanding of the influence of T3 on brain development and function.

ACKNOWLEDGMENT

Work in our group is supported by Agence Nationale de la Recherche (Thyromut2 program; ANR-15-CE14-0011-01).

REFERENCES

Abel, E. D., Ahima, R. S., Boers, M. E., Elmquist, J. K., & Wondisford, F. E. (2001). Critical role for thyroid hormone receptor beta2 in the regulation of paraventricular thyrotropin-releasing hormone neurons. *The Journal of Clinical Investigation, 107*, 1017–1023.

Abel, E. D., Boers, M. E., Pazos-Moura, C., Moura, E., Kaulbach, H., Zakaria, M., et al. (1999). Divergent roles for thyroid hormone receptor beta isoforms in the endocrine axis and auditory system. *The Journal of Clinical Investigation, 104*, 291–300.

Avci, H. X., Lebrun, C., Wehrle, R., Doulazmi, M., Chatonnet, F., Morel, M. P., et al. (2012). Thyroid hormone triggers the developmental loss of axonal regenerative capacity via thyroid hormone receptor alpha1 and kruppel-like factor 9 in Purkinje cells. *Proceedings of the National Academy of Sciences of the United States of America, 109*, 14206–14211.

Baas, D., Legrand, C., Samarut, J., & Flamant, F. (2002). Persistence of oligodendrocyte precursor cells and altered myelination in optic nerve associated to retina degeneration in mice devoid of all thyroid hormone receptors. *Proceedings of the National Academy of Sciences of the United States of America, 99*, 2907–2911.

Bagamasbad, P. D., Bonett, R. M., Sachs, L., Buisine, N., Raj, S., Knoedler, J. R., et al. (2015). Deciphering the regulatory logic of an ancient, ultraconserved nuclear receptor enhancer module. *Molecular Endocrinology, 29*, 856–872.

Barres, B. A., Lazar, M. A., & Raff, M. C. (1994). A novel role for thyroid hormone, glucocorticoids and retinoic acid in timing oligodendrocyte development. *Development, 120*, 1097–1108.

Berbel, P., Navarro, D., & Roman, G. C. (2014). An evo-devo approach to thyroid hormones in cerebral and cerebellar cortical development: Etiological implications for autism. *Frontiers in Endocrinology, 5*, 146.

Bernal, J. (2007). Thyroid hormone receptors in brain development and function. *Nature Clinical Practice. Endocrinology & Metabolism, 3*, 249–259.

Beydoun, M. A., Beydoun, H. A., Kitner-Triolo, M. H., Kaufman, J. S., Evans, M. K., & Zonderman, A. B. (2013). Thyroid hormones are associated with cognitive function: Moderation by sex, race and depressive symptoms. *The Journal of Clinical Endocrinology and Metabolism, 98*, 3470–3481.

Bigler, J., Hokanson, W., & Eisenman, R. N. (1992). Thyroid hormone receptor transcriptional activity is potentially autoregulated by truncated forms of the receptor. *Molecular and Cellular Biology, 12*, 2406–2417.

Billon, N., Jolicoeur, C., Tokumoto, Y., Vennstrom, B., & Raff, M. (2002). Normal timing of oligodendrocyte development depends on thyroid hormone receptor alpha 1 (TRalpha1). *The EMBO Journal, 21*, 6452–6460.

Blanchet, E., Bertrand, C., Annicotte, J. S., Schlernitzauer, A., Pessemesse, L., Levin, J., et al. (2012). Mitochondrial T3 receptor p43 regulates insulin secretion and glucose homeostasis. *FASEB Journal, 26*, 40–50.

Boukhtouche, F., Brugg, B., Wehrle, R., Bois-Joyeux, B., Danan, J. L., Dusart, I., et al. (2010). Induction of early Purkinje cell dendritic differentiation by thyroid hormone requires RORalpha. *Neural Development*, *5*, 18.

Boukhtouche, F., Janmaat, S., Vodjdani, G., Gautheron, V., Mallet, J., Dusart, I., et al. (2006). Retinoid-related orphan receptor alpha controls the early steps of Purkinje cell dendritic differentiation. *The Journal of Neuroscience*, *26*, 1531–1538.

Bouslama-Oueghlani, L., Wehrle, R., Doulazmi, M., Chen, X. R., Jaudon, F., Lemaigre-Dubreuil, Y., et al. (2012). Purkinje cell maturation participates in the control of oligodendrocyte differentiation: Role of sonic hedgehog and vitronectin. *PloS One*, *7*, e49015

Bradley, D. J., Young, W. S., 3rd, & Weinberger, C. (1989). Differential expression of alpha and beta thyroid hormone receptor genes in rat brain and pituitary. *Proceedings of the National Academy of Sciences of the United States of America*, *86*, 7250–7254.

Buisine, N., Ruan, X., Bilesimo, P., Grimaldi, A., Alfama, G., Ariyaratne, P., et al. (2015). Xenopus tropicalis genome re-scaffolding and re-annotation reach the resolution required for in vivo ChIA-PET analysis. *PloS One*, *10*, e0137526

Casas, F., Busson, M., Grandemange, S., Seyer, P., Carazo, A., Pessemesse, L., et al. (2006). Characterization of a novel thyroid hormone receptor alpha variant involved in the regulation of myoblast differentiation. *Molecular Endocrinology*, *20*, 749–763.

Casas, F., Rochard, P., Rodier, A., Cassar-Malek, I., Marchal-Victorion, S., Wiesner, R. J., et al. (1999). A variant form of the nuclear triiodothyronine receptor c-ErbAalpha1 plays a direct role in regulation of mitochondrial RNA synthesis. *Molecular and Cellular Biology*, *19*, 7913–7924.

Chassande, O., Fraichard, A., Gauthier, K., Flamant, F., Legrand, C., Savatier, P., et al. (1997). Identification of transcripts initiated from an internal promoter in the c-erbA alpha locus that encode inhibitors of retinoic acid receptor-alpha and triiodothyronine receptor activities. *Molecular Endocrinology*, *11*, 1278–1290.

Chatonnet, F., Flamant, F., & Morte, B. (2015). A temporary compendium of thyroid hormone target genes in brain. *Biochimica et Biophysica Acta*, *1849*, 122–129.

Chatonnet, F., Guyot, R., Benoit, G., & Flamant, F. (2013). Genome-wide analysis of thyroid hormone receptors shared and specific functions in neural cells. *Proceedings of the National Academy of Sciences of the United States of America*, *110*, E766–E775.

Chatonnet, F., Guyot, R., Picou, F., Bondesson, M., & Flamant, F. (2012). Genome-wide search reveals the existence of a limited number of thyroid hormone receptor alpha target genes in cerebellar neurons. *PloS One*, *7*, e30703

Chen, H., Lin, R. J., Xie, W., Wilpitz, D., & Evans, R. M. (1999). Regulation of hormone-induced histone hyperacetylation and gene activation via acetylation of an acetylase. *Cell*, *98*, 675–686.

Cheng, S. Y., Leonard, J. L., & Davis, P. J. (2010). Molecular aspects of thyroid hormone actions. *Endocrine Reviews*, *31*, 139–170.

Chiamolera, M. I., Sidhaye, A. R., Matsumoto, S., He, Q., Hashimoto, K., Ortiga-Carvalho, T. M., et al. (2012). Fundamentally distinct roles of thyroid hormone receptor isoforms in a thyrotroph cell line are due to differential DNA binding. *Molecular Endocrinology*, *26*, 926–939.

Cvoro, A., Devito, L., Milton, F. A., Noli, L., Zhang, A., Filippi, C., et al. (2015). A thyroid hormone receptor/KLF9 axis in human hepatocytes and pluripotent stem cells. *Stem Cells*, *33*, 416–428.

Danko, C. G., Hyland, S. L., Core, L. J., Martins, A. L., Waters, C. T., Lee, H. W., et al. (2015). Identification of active transcriptional regulatory elements from GRO-seq data. *Nature Methods*, *12*, 433–438.

Darimont, B. D., Wagner, R. L., Apriletti, J. W., Stallcup, M. R., Kushner, P. J., Baxter, J. D., et al. (1998). Structure and specificity of nuclear receptor-coactivator interactions. *Genes & Development, 12*, 3343–3356.

Dekker, J., Marti-Renom, M. A., & Mirny, L. A. (2013). Exploring the three-dimensional organization of genomes: Interpreting chromatin interaction data. *Nature Reviews. Genetics, 14*, 390–403.

Dekker, J., & Mirny, L. (2016). The 3D Genome as Moderator of Chromosomal Communication. *Cell, 164*, 1110–1121.

Diez, D., Grijota-Martinez, C., Agretti, P., De Marco, G., Tonacchera, M., Pinchera, A., et al. (2008). Thyroid hormone action in the adult brain: Gene expression profiling of the effects of single and multiple doses of triiodo-L-thyronine in the rat striatum. *Endocrinology, 149*, 3989–4000.

Dong, H., Wade, M., Williams, A., Lee, A., Douglas, G. R., & Yauk, C. (2005). Molecular insight into the effects of hypothyroidism on the developing cerebellum. *Biochemical and Biophysical Research Communications, 330*, 1182–1193.

Dugas, J. C., Ibrahim, A., & Barres, B. A. (2012). The T3-induced gene KLF9 regulates oligodendrocyte differentiation and myelin regeneration. *Molecular and Cellular Neurosciences, 50*, 45–57.

Dugas, J., Tai, Y. C., Speed, T. P., Ngai, J., & Barres, B. A. (2006). Functional genomic analysis of oligodendrocyte differentiation. *The Journal of Neuroscience, 26*, 10967–10983.

Faingold, C. L., Randall, M., & Tupal, S. (2010). DBA/1 mice exhibit chronic susceptibility to audiogenic seizures followed by sudden death associated with respiratory arrest. *Epilepsy & Behavior, 17*, 436–440.

Fauquier, T., Chatonnet, F., Picou, F., Richard, S., Fossat, N., Aguilera, N., et al. (2014). Purkinje cells and Bergmann glia are primary targets of the TRalpha1 thyroid hormone receptor during mouse cerebellum postnatal development. *Development, 141*, 166–175.

Fauquier, T., Romero, E., Picou, F., Chatonnet, F., Nguyen, X. N., Quignodon, L., et al. (2011). Severe impairment of cerebellum development in mice expressing a dominant-negative mutation inactivating thyroid hormone receptor alpha1 isoform. *Developmental Biology, 356*, 350–358.

Flamant, F., Poguet, A. L., Plateroti, M., Chassande, O., Gauthier, K., Streichenberger, N., et al. (2002). Congenital hypothyroid Pax8(-/-) mutant mice can be rescued by inactivating the TRalpha gene. *Molecular Endocrinology, 16*, 24–32.

Flamant, F., & Samarut, J. (2003). Thyroid hormone receptors: Lessons from knockout and knock-in mutant mice. *Trends in Endocrinology and Metabolism, 14*, 85–90.

Forrest, D., Hanebuth, E., Smeyne, R. J., Everds, N., Stewart, C. L., Wehner, J. M., et al. (1996). Recessive resistance to thyroid hormone in mice lacking thyroid hormone receptor beta: Evidence for tissue-specific modulation of receptor function. *The EMBO Journal, 15*, 3006–3015.

Fozzatti, L., Kim, D. W., Park, J. W., Willingham, M. C., Hollenberg, A. N., & Cheng, S. Y. (2013). Nuclear receptor corepressor (NCOR1) regulates in vivo actions of a mutated thyroid hormone receptor alpha. *Proceedings of the National Academy of Sciences of the United States of America, 110*, 7850–7855.

Fozzatti, L., Lu, C., Kim, D. W., Park, J. W., Astapova, I., Gavrilova, O., et al. (2011). Resistance to thyroid hormone is modulated in vivo by the nuclear receptor corepressor (NCOR1). *Proceedings of the National Academy of Sciences of the United States of America, 108*, 17462–17467.

Fraichard, A., Chassande, O., Plateroti, M., Roux, J. P., Trouillas, J., Dehay, C., et al. (1997). The T3R alpha gene encoding a thyroid hormone receptor is essential for post-natal development and thyroid hormone production. *The EMBO Journal, 16*, 4412–4420.

Francelle, L., Galvan, L., Gaillard, M. C., Guillermier, M., Houitte, D., Bonvento, G., et al. (2015). Loss of the thyroid hormone-binding protein Crym renders striatal neurons more

vulnerable to mutant huntingtin in Huntington's disease. *Human Molecular Genetics, 24*, 1563–1573.
Gauthier, K., Chassande, O., Plateroti, M., Roux, J. P., Legrand, C., Pain, B., et al. (1999). Different functions for the thyroid hormone receptors TRalpha and TRbeta in the control of thyroid hormone production and post-natal development. *The EMBO Journal, 18*, 623–631.
Gauthier, K., Plateroti, M., Harvey, C. B., Williams, G. R., Weiss, R. E., Refetoff, S., et al. (2001). Genetic analysis reveals different functions for the products of the thyroid hormone receptor alpha locus. *Molecular and Cellular Biology, 21*, 4748–4760.
Gautier, N. M., & Glasscock, E. (2015). Spontaneous seizures in Kcna1-null mice lacking voltage-gated Kv1.1 channels activate Fos expression in select limbic circuits. *Journal of Neurochemistry, 135*, 157–164.
Gentile, S., Darden, T., Erxleben, C., Romeo, C., Russo, A., Martin, N., et al. (2006). Rac GTPase signaling through the PP5 protein phosphatase. *Proceedings of the National Academy of Sciences of the United States of America, 103*, 5202–5206.
Gentile, S., Martin, N., Scappini, E., Williams, J., Erxleben, C., & Armstrong, D. L. (2008). The human ERG1 channel polymorphism, K897T, creates a phosphorylation site that inhibits channel activity. *Proceedings of the National Academy of Sciences of the United States of America, 105*, 14704–14708.
Gil-Ibanez, P., Garcia-Garcia, F., Dopazo, J., Bernal, J., & Morte, B. (2015). Global transcriptome analysis of primary cerebrocortical cells: Identification of genes regulated by triiodothyronine in specific cell types. *Cerebral Cortex*, 1–12. [Epub ahead of print].
Gil-Ibanez, P., Morte, B., & Bernal, J. (2013). Role of thyroid hormone receptor subtypes alpha and beta on gene expression in the cerebral cortex and striatum of postnatal mice. *Endocrinology, 154*, 1940–1947.
Goglia, F. (2005). Biological effects of 3,5-diiodothyronine (T(2)). *Biochemistry (Mosc), 70*, 164–172.
Gothe, S., Wang, Z., Ng, L., Kindblom, J. M., Barros, A. C., Ohlsson, C., et al. (1999). Mice devoid of all known thyroid hormone receptors are viable but exhibit disorders of the pituitary-thyroid axis, growth, and bone maturation. *Genes & Development, 13*, 1329–1341.
Grijota-Martinez, C., Samarut, E., Scanlan, T. S., Morte, B., & Bernal, J. (2011). In vivo activity of the thyroid hormone receptor beta- and alpha-selective agonists GC-24 and CO23 on rat liver, heart, and brain. *Endocrinology, 152*, 1136–1142.
Grontved, L., Waterfall, J. J., Kim, D. W., Baek, S., Sung, M. H., Zhao, L., et al. (2015). Transcriptional activation by the thyroid hormone receptor through ligand-dependent receptor recruitment and chromatin remodelling. *Nature Communications, 6*, 7048.
Guadano-Ferraz, A., Benavides-Piccione, R., Venero, C., Lancha, C., Vennstrom, B., Sandi, C., et al. (2003). Lack of thyroid hormone receptor alpha1 is associated with selective alterations in behavior and hippocampal circuits. *Molecular Psychiatry, 8*, 30–38.
Hallen, A., Cooper, A. J., Jamie, J. F., & Karuso, P. (2015). Insights into enzyme catalysis and thyroid hormone regulation of cerebral ketimine reductase/mu-crystallin under physiological conditions. *Neurochemical Research, 40*, 1252–1266.
Hashimoto, K., Curty, F. H., Borges, P. P., Lee, C. E., Abel, E. D., Elmquist, J. K., et al. (2001). An unliganded thyroid hormone receptor causes severe neurological dysfunction. *Proceedings of the National Academy of Sciences of the United States of America, 98*, 3998–4003.
Heuer, H., & Mason, C. A. (2003). Thyroid hormone induces cerebellar Purkinje cell dendritic development via the thyroid hormone receptor alpha1. *The Journal of Neuroscience, 23*, 10604–10612.

Hiroi, Y., Kim, H. H., Ying, H., Furuya, F., Huang, Z., Simoncini, T., et al. (2006). Rapid nongenomic actions of thyroid hormone. *Proceedings of the National Academy of Sciences of the United States of America, 103*, 14104–14109.

Hoopfer, E. D., Huang, L., & Denver, R. J. (2002). Basic transcription element binding protein is a thyroid hormone-regulated transcription factor expressed during metamorphosis in Xenopus laevis. *Development, Growth & Differentiation, 44*, 365–381.

Jones, I., Srinivas, M., Ng, L., & Forrest, D. (2003). The thyroid hormone receptor beta gene: Structure and functions in the brain and sensory systems. *Thyroid, 13*, 1057–1068.

Kalyanaraman, H., Schwappacher, R., Joshua, J., Zhuang, S., Scott, B. T., Klos, M., et al. (2014). Nongenomic thyroid hormone signaling occurs through a plasma membrane-localized receptor. *Science Signaling, 7*, ra48.

Kaneshige, M., Kaneshige, K., Zhu, X., Dace, A., Garrett, L., Carter, T. A., et al. (2000). Mice with a targeted mutation in the thyroid hormone beta receptor gene exhibit impaired growth and resistance to thyroid hormone. *Proceedings of the National Academy of Sciences of the United States of America, 97*, 13209–13214.

Kaneshige, M., Suzuki, H., Kaneshige, K., Cheng, J., Wimbrow, H., Barlow, C., et al. (2001). A targeted dominant negative mutation of the thyroid hormone alpha 1 receptor causes increased mortality, infertility, and dwarfism in mice. *Proceedings of the National Academy of Sciences of the United States of America, 98*, 15095–15100.

Kapoor, R., Ghosh, H., Nordstrom, K., Vennstrom, B., & Vaidya, V. A. (2011). Loss of thyroid hormone receptor beta is associated with increased progenitor proliferation and NeuroD positive cell number in the adult hippocampus. *Neuroscience Letters, 487*, 199–203.

Kim, D. W., Park, J. W., Willingham, M. C., & Cheng, S. Y. (2014). A histone deacetylase inhibitor improves hypothyroidism caused by a TRalpha1 mutant. *Human Molecular Genetics, 23*, 2651–2664.

Kyono, Y., Subramani, A., Ramadoss, P., Hollenberg, A. N., Bonett, R. M., & Denver, R. J. (2016). Liganded thyroid hormone receptors transactivate the DNA methyltransferase 3a gene in mouse neuronal cells. *Endocrinology, 157*, 3647–3657.

Lazar, M. A., Hodin, R. A., & Chin, W. W. (1989). Human carboxyl-terminal variant of alpha-type c-erbA inhibits trans- activation by thyroid hormone receptors without binding thyroid hormone. *Proceedings of the National Academy of Sciences of the United States of America, 86*, 7771–7774.

Lebrun, C., Avci, H. X., Wehrle, R., Doulazmi, M., Jaudon, F., Morel, M. P., et al. (2013). Klf9 is necessary and sufficient for Purkinje cell survival in organotypic culture. *Molecular and Cellular Neurosciences, 54*, 9–21.

Lindholm, D., Castren, E., Tsoulfas, P., Kolbeck, R., Berzaghi Mda, P., Leingartner, A., et al. (1993). Neurotrophin-3 induced by tri-iodothyronine in cerebellar granule cells promotes Purkinje cell differentiation. *The Journal of Cell Biology, 122*, 443–450.

Liu, Y. Y., Schultz, J. J., & Brent, G. A. (2003). A thyroid hormone receptor alpha gene mutation (P398H) is associated with visceral adiposity and impaired catecholamine-stimulated lipolysis in mice. *The Journal of Biological Chemistry, 278*, 38913–38920.

Machado, D. S., Sabet, A., Santiago, L. A., Sidhaye, A. R., Chiamolera, M. I., Ortiga-Carvalho, T. M., et al. (2009). A thyroid hormone receptor mutation that dissociates thyroid hormone regulation of gene expression in vivo. *Proceedings of the National Academy of Sciences of the United States of America, 106*, 9441–9446.

Mansouri, A., Chowdhury, K., & Gruss, P. (1998). Follicular cells of the thyroid gland require Pax8 gene function. *Nature Genetics, 19*, 87–90.

Manzano, J., Morte, B., Scanlan, T. S., & Bernal, J. (2003). Differential effects of triiodothyronine and the thyroid hormone receptor beta-specific agonist GC-1 on thyroid hormone target genes in the b ain. *Endocrinology, 144*, 5480–5487.

Marelli, F., Carra, S., Agostini, M., Cotelli, F., Peeters, R., Chatterjee, K., et al. (2016). Patterns of thyroid hormone receptor expression in zebrafish and generation of a novel model of resistance to thyroid hormone action. *Molecular and Cellular Endocrinology*, *424*, 102–117.

Martin, N. P., Fernandez de Velasco, E. M., Mizuno, F., Scappini, E. L., Gloss, B., Erxleben, C., et al. (2014). A rapid cytoplasmic mechanism for PI3 kinase regulation by the nuclear thyroid hormone receptor, TRbeta, and genetic evidence for its role in the maturation of mouse hippocampal synapses in vivo. *Endocrinology*, *155*, 3713–3724.

Mellstrom, B., Naranjo, J. R., Santos, A., Gonzalez, A. M., & Bernal, J. (1991). Independent expression of the alpha and beta c-erbA genes in developing rat brain. *Molecular Endocrinology*, *5*, 1339–1350.

Mitsuhashi, T., Tennyson, G. E., & Nikodem, V. M. (1988). Alternative splicing generates messages encoding rat c-erbA proteins that do not bind thyroid hormone. *Proceedings of the National Academy of Sciences of the United States of America*, *85*, 5804–5808.

Mittag, J., Wallis, K., & Vennstrom, B. (2010). Physiological consequences of the TRalpha1 aporeceptor state. *Heart Failure Reviews*, *15*, 111–115.

Moran, C., Schoenmakers, N., Agostini, M., Schoenmakers, E., Offiah, A., Kydd, A., et al. (2013). An adult female with resistance to thyroid hormone mediated by defective thyroid hormone receptor alpha. *The Journal of Clinical Endocrinology and Metabolism*, *98*, 4254–4261.

Mori, J., Suzuki, S., Kobayashi, M., Inagaki, T., Komatsu, A., Takeda, T., et al. (2002). Nicotinamide adenine dinucleotide phosphate-dependent cytosolic T(3) binding protein as a regulator for T(3)-mediated transactivation. *Endocrinology*, *143*, 1538–1544.

Moriyama, K., Yamamoto, H., Futawaka, K., Atake, A., Kasahara, M., & Tagami, T. (2016). Molecular characterization of human thyroid hormone receptor beta isoform 4. *Endocrine Research*, *41*, 34–42.

Morte, B., Manzano, J., Scanlan, T., Vennstrom, B., & Bernal, J. (2002). Deletion of the thyroid hormone receptor alpha 1 prevents the structural alterations of the cerebellum induced by hypothyroidism. *Proceedings of the National Academy of Sciences of the United States of America*, *99*, 3985–3989.

Neveu, I., & Arenas, E. (1996). Neurotrophins promote the survival and development of neurons in the cerebellum of hypothyroid rats in vivo. *The Journal of Cell Biology*, *133*, 631–646.

Ng, L., Cordas, E., Wu, X., Vella, K. R., Hollenberg, A. N., & Forrest, D. (2015). Age-related hearing loss and degeneration of cochlear hair cells in mice lacking thyroid hormone receptor beta1. *Endocrinology*, *156*, 3853–3865.

Ng, L., Hurley, J. B., Dierks, B., Srinivas, M., Salto, C., Vennstrom, B., et al. (2001). A thyroid hormone receptor that is required for the development of green cone photoreceptors. *Nature Genetics*, *27*, 94–98.

Nguyen, N. H., Apriletti, J. W., Cunha Lima, S. T., Webb, P., Baxter, J. D., & Scanlan, T. S. (2002). Rational design and synthesis of a novel thyroid hormone antagonist that blocks coactivator recruitment. *Journal of Medicinal Chemistry*, *45*, 3310–3320.

Nunez, J., Celi, F. S., Ng, L., & Forrest, D. (2008). Multigenic control of thyroid hormone functions in the nervous system. *Molecular and Cellular Endocrinology*, *287*, 1–12.

Ortiga-Carvalho, T. M., Shibusawa, N., Nikrodhanond, A., Oliveira, K. J., Machado, D. S., Liao, X. H., et al. (2005). Negative regulation by thyroid hormone receptor requires an intact coactivator-binding surface. *The Journal of Clinical Investigation*, *115*, 2517–2523.

Picou, F., Fauquier, T., Chatonnet, F., & Flamant, F. (2012). A bimodal influence of thyroid hormone on cerebellum oligodendrocyte differentiation. *Molecular Endocrinology*, *26*, 608–618.

Piehl, S., Hoefig, C. S., Scanlan, T. S., & Kohrle, J. (2011). Thyronamines—Past, present, and future. *Endocrine Reviews*, *32*, 64–80.

Pilhatsch, M., Winter, C., Nordstrom, K., Vennstrom, B., Bauer, M., & Juckel, G. (2010). Increased depressive behaviour in mice harboring the mutant thyroid hormone receptor alpha 1. *Behavioural Brain Research*, *214*, 187–192.

Plateroti, M., Gauthier, K., Domon-Dell, C., Freund, J. N., Samarut, J., & Chassande, O. (2001). Functional interference between thyroid hormone receptor alpha (TRalpha) and natural truncated TRDeltaalpha isoforms in the control of intestine development. *Molecular and Cellular Biology*, *21*, 4761–4772.

Portella, A. C., Carvalho, F., Faustino, L., Wondisford, F. E., Ortiga-Carvalho, T. M., & Gomes, F. C. (2010). Thyroid hormone receptor beta mutation causes severe impairment of cerebellar development. *Molecular and Cellular Neurosciences*, *44*, 68–77.

Quignodon, L., Grijota-Martinez, C., Compe, E., Guyot, R., Allioli, N., Laperriere, D., et al. (2007). A combined approach identifies a limited number of new thyroid hormone target genes in post-natal mouse cerebellum. *Journal of Molecular Endocrinology*, *39*, 17–28.

Quignodon, L., Vincent, S., Winter, H., Samarut, J., & Flamant, F. (2007). A point mutation in the activation function 2 domain of thyroid hormone receptor alpha1 expressed after CRE-mediated recombination partially recapitulates hypothyroidism. *Molecular Endocrinology*, *21*, 2350–2360.

Ramadoss, P., Abraham, B. J., Tsai, L., Zhou, Y., Costa-e-Sousa, R. H., Ye, F., et al. (2014). Novel mechanism of positive versus negative regulation by thyroid hormone receptor beta1 (TRbeta1) identified by genome-wide profiling of binding sites in mouse liver. *The Journal of Biological Chemistry*, *289*, 1313–1328.

Refetoff, S., Bassett, J. H., Beck-Peccoz, P., Bernal, J., Brent, G., Chatterjee, K., et al. (2014). Classification and proposed nomenclature for inherited defects of thyroid hormone action, cell transport, and metabolism. *Thyroid*, *24*, 407–409.

Remaud, S., Gothie, J. D., Morvan-Dubois, G., & Demeneix, B. A. (2014). Thyroid hormone signaling and adult neurogenesis in mammals. *Frontiers in Endocrinology*, *5*, 62.

Rovet, J. F. (2014). The role of thyroid hormones for brain development and cognitive function. *Endocrine Development*, *26*, 26–43.

Rusch, A., Ng, L., Goodyear, R., Oliver, D., Lisoukov, I., Vennstrom, B., et al. (2001). Retardation of cochlear maturation and impaired hair cell function caused by deletion of all known thyroid hormone receptors. *The Journal of Neuroscience*, *21*, 9792–9800.

Salto, C., Kindblom, J. M., Johansson, C., Wang, Z., Gullberg, H., Nordstrom, K., et al. (2001). Ablation of TRalpha2 and a concomitant overexpression of alpha1 yields a mixed hypo- and hyperthyroid phenotype in mice. *Molecular Endocrinology*, *15*, 2115–2128.

Shibusawa, N., Hashimoto, K., Nikrodhanond, A. A., Liberman, M. C., Applebury, M. L., Liao, X. H., et al. (2003). Thyroid hormone action in the absence of thyroid hormone receptor DNA-binding in vivo. *The Journal of Clinical Investigation*, *112*, 588–597.

Shrestha, S., Bloom, M. S., Yucel, R., Seegal, R. F., Rej, R., McCaffrey, R. J., et al. (2016). Thyroid function and neuropsychological status in older adults. *Physiology & Behavior*, *164*, 34–39.

Sotelo, C., & Dusart, I. (2009). Intrinsic versus extrinsic determinants during the development of Purkinje cell dendrites. *Neuroscience*, *162*, 589–600.

Storey, N. M., Gentile, S., Ullah, H., Russo, A., Muessel, M., Erxleben, C., et al. (2006). Rapid signaling at the plasma membrane by a nuclear receptor for thyroid hormone. *Proceedings of the National Academy of Sciences of the United States of America*, *103*, 5197–5201.

Suzuki, S. C., Bleckert, A., Williams, P. R., Takechi, M., Kawamura, S., & Wong, R. O. (2013). Cone photoreceptor types in zebrafish are generated by symmetric terminal divisions of dedicated precursors. *Proceedings of the National Academy of Sciences of the United States of America*, *110*, 15109–15114.

Suzuki, S., Suzuki, N., Mori, J., Oshima, A., Usami, S., & Hashizume, K. (2007). micro-Crystallin as an intracellular 3,5,3′-triiodothyronine holder in vivo. *Molecular Endocrinology*, *21*, 885–894.

Tagami, T., Yamamoto, H., Moriyama, K., Sawai, K., Usui, T., Shimatsu, A., et al. (2010). Identification of a novel human thyroid hormone receptor beta isoform as a transcriptional modulator. *Biochemical and Biophysical Research Communications*, *396*, 983–988.

Taneja, R., Roy, B., Plassat, J. L., Zusi, C. F., Ostrowski, J., Reczek, P. R., et al. (1996). Cell-type and promoter-context dependent retinoic acid receptor (RAR) redundancies for RAR beta 2 and Hoxa-1 activation in F9 and P19 cells can be artefactually generated by gene knockouts. *Proceedings of the National Academy of Sciences of the United States of America*, *93*, 6197–6202.

Tinnikov, A., Nordstrom, K., Thoren, P., Kindblom, J. M., Malin, S., Rozell, B., et al. (2002). Retardation of post-natal development caused by a negatively acting thyroid hormone receptor alpha1. *The EMBO Journal*, *21*, 5079–5087.

Venero, C., Guadano-Ferraz, A., Herrero, A. I., Nordstrom, K., Manzano, J., de Escobar, G. M., et al. (2005). Anxiety, memory impairment, and locomotor dysfunction caused by a mutant thyroid hormone receptor {alpha}1 can be ameliorated by T3 treatment. *Genes & Development*, *19*, 2152–2163.

Wallis, K., Dudazy, S., van Hogerlinden, M., Nordstrom, K., Mittag, J., & Vennstrom, B. (2010). The thyroid hormone receptor alpha1 protein is expressed in embryonic postmitotic neurons and persists in most adult neurons. *Molecular Endocrinology*, *24*, 1904–1916.

Wallis, K., Sjogren, M., van Hogerlinden, M., Silberberg, G., Fisahn, A., Nordstrom, K., et al. (2008). Locomotor deficiencies and aberrant development of subtype-specific GABAergic interneurons caused by an unliganded thyroid hormone receptor alpha1. *The Journal of Neuroscience*, *28*, 1904–1915.

Wang, Y., Fisahn, A., Sinha, I., Nguyen, D. P., Sterzenbach, U., Lallemend, F., et al. (2016). Hippocampal transcriptome profile of persistent memory rescue in a mouse model of THRA1 mutation-mediated resistance to thyroid hormone. *Scientific Reports*, *6*, 18617.

Wikstrom, L., Johansson, C., Salto, C., Barlow, C., Campos Barros, A., Baas, F., et al. (1998). Abnormal heart rate and body temperature in mice lacking thyroid hormone receptor alpha 1. *The EMBO Journal*, *17*, 455–461.

Wilcoxon, J. S., Nadolski, G. J., Samarut, J., Chassande, O., & Redei, E. E. (2007). Behavioral inhibition and impaired spatial learning and memory in hypothyroid mice lacking thyroid hormone receptor alpha. *Behavioural Brain Research*, *177*, 109–116.

Williams, G. R. (2000). Cloning and characterization of two novel thyroid hormone receptor beta isoforms. *Molecular and Cellular Biology*, *20*, 8329–8342.

Winter, H., Braig, C., Zimmermann, U., Geisler, H. S., Franzer, J. T., Weber, T., et al. (2006). Thyroid hormone receptors TRalpha1 and TRbeta differentially regulate gene expression of Kcnq4 and prestin during final differentiation of outer hair cells. *Journal of Cell Science*, *119*, 2975–2984.

Winter, H., Ruttiger, L., Muller, M., Kuhn, S., Brandt, N., Zimmermann, U., et al. (2009). Deafness in TRbeta mutants is caused by malformation of the tectorial membrane. *The Journal of Neuroscience*, *29*, 2581–2587.

Ying, M., Tilghman, J., Wei, Y., Guerrero-Cazares, H., Quinones-Hinojosa, A., Ji, H., et al. (2014). Kruppel-like factor-9 (KLF9) inhibits glioblastoma stemness through global transcription repression and integrin alpha6 inhibition. *The Journal of Biological Chemistry*, *289*, 32742–32756.

Yu, L., Iwasaki, T., Xu, M., Lesmana, R., Xiong, Y., Shimokawa, N., et al. (2015). Aberrant cerebellar development of transgenic mice expressing dominant-negative thyroid hormone receptor in cerebellar Purkinje cells. *Endocrinology*, *156*, 1565–1576.

Yuen, R. K., Thiruvahindrapuram, B., Merico, D., Walker, S., Tammimies, K., Hoang, N., et al. (2015). Whole-genome sequencing of quartet families with autism spectrum disorder. *Nature Medicine*, *21*, 185–191.

Zhu, X., Hill, R. A., Dietrich, D., Komitova, M., Suzuki, R., & Nishiyama, A. (2011). Age-dependent fate and lineage restriction of single NG2 cells. *Development*, *138*, 745–753.

Zucker, S. N., Fink, E. E., Bagati, A., Mannava, S., Bianchi-Smiraglia, A., Bogner, P. N., et al. (2014). Nrf2 amplifies oxidative stress via induction of Klf9. *Molecular Cell*, *53*, 916–928.

Zwart, W., Theodorou, V., Kok, M., Canisius, S., Linn, S., & Carroll, J. S. (2011). Oestrogen receptor-co-factor-chromatin specificity in the transcriptional regulation of breast cancer. *The EMBO Journal*, *30*, 4764–4776.

CHAPTER TWELVE

Resistance to Thyroid Hormone due to Heterozygous Mutations in Thyroid Hormone Receptor Alpha

Anja L.M. van Gucht*, Carla Moran†, Marcel E. Meima*, W. Edward Visser*, Krishna Chatterjee†, Theo J. Visser*,[1], Robin P. Peeters*

*Erasmus University Medical Center, Rotterdam, The Netherlands
†Wellcome-MRC Institute of Metabolic Science, University of Cambridge, Cambridge, United Kingdom
[1]Corresponding author: e-mail address: t.j.visser@erasmusmc.nl

Contents

Abstract

Background: Thyroid hormone (TH) acts via nuclear thyroid hormone receptors (TRs). TR isoforms (TRα1, TRα2, TRβ1, TRβ2) are encoded by distinct genes (*THRA* and *THRB*) and show differing tissue distributions. Patients with mutations in *THRB*, exhibiting resistance within the hypothalamic–pituitary–thyroid axis with elevated TH and nonsuppressed thyroid-stimulating hormone (TSH) levels, were first described decades ago. In 2012, the first patients with mutations in *THRA* were identified.

Scope of this review: This review describes the clinical and biochemical characteristics of patients with resistance to thyroid hormone alpha (RTHα) due to heterozygous mutations in *THRA*. The genetic basis and molecular pathogenesis of the disorder together with effects of levothyroxine treatment are discussed.

Conclusions: The severity of the clinical phenotype of RTHα patients seems to be associated with the location and type of mutation in *THRA*. The most frequent

Current Topics in Developmental Biology, Volume 125
ISSN 0070-2153
http://dx.doi.org/10.1016/bs.ctdb.2017.02.001

abnormalities observed include anemia, constipation, and growth and developmental delay. In addition, serum (F)T3 levels can be high-normal to high, (F)T4 and rT3 levels normal to low, while TSH is normal or mildly raised. Despite heterogeneous consequences of mutations in *THRA*, RTHα should be suspected in subjects with even mild clinical features of hypothyroidism together with high/high-normal (F)T3, low/low-normal (F)T4, and normal TSH.

1. INTRODUCTION

Thyroid hormone (TH) is essential for normal development, growth, and cellular metabolism. The importance of TH for the developing brain is exemplified by the severe defects in motor and mental development seen in patients with untreated congenital hypothyroidism (Bernal, 2005; Zhang & Lazar, 2000).

Serum TH levels are regulated by the hypothalamic–pituitary–thyroid (HPT) axis, where hypothalamic thyrotropin-releasing hormone (TRH) stimulates the production of pituitary thyroid-stimulating hormone (TSH), which in turn stimulates the thyroid gland to produce TH. The thyroid predominantly synthesizes the prohormone T4, with only a small fraction of T3, the bioactive hormone, being produced (Chiamolera & Wondisford, 2009; Fekete & Lechan, 2014). TRH and TSH synthesis is regulated by T3 and T4 as part of a negative-feedback mechanism.

The intracellular availability of TH is dependent on transmembrane transport, mediated by transporters such as monocarboxylate transporter 8. Intracellular TH concentrations are tightly regulated by three deiodinase enzymes (DIOs), mediating the conversion of T4 to T3 (DIO1 and DIO2) and of T3 and T4 into inactive metabolites (DIO3). In the nucleus, T3 binds to T3 receptors (TRs), which belong to the nuclear receptor superfamily of ligand-inducible transcription factors. TRs preferentially form heterodimers with retinoid X receptors (RXRs) and regulate target gene expression by binding to T3 response elements (TREs), usually located in their promoter region. In the absence of TH, unliganded TRs repress basal transcription of positively regulated genes by recruiting corepressors (e.g., nuclear receptor corepressor [NCoR], silencing mediator for retinoic acid and thyroid receptors [SMRT]) and histone deacetylase (HDAC). Binding of T3 to TRs results in dissociation of the corepressor complex and recruitment of coactivator proteins (e.g., steroid receptor coactivator 1 [SRC-1], and the histone acetyl transferase, CREB-binding protein [CBP] and CBP-associated factor [pCAF]), which induce transcriptional activation

(Brent, 2012; Dumitrescu & Refetoff, 2013; Refetoff & Dumitrescu, 2007; Schoenmakers et al., 2013). TH also exerts nongenomic effects (Bassett, Harvey, & Williams, 2003; Kalyanaraman et al., 2014), but these are beyond the scope of this review.

In humans, TRs are encoded by two genes (*THRA* and *THRB*) on chromosomes 17 and 3, respectively. Two receptor isoforms are generated from the *THRA* locus by alternative splicing: TRα1 is predominantly expressed in the central nervous system, bone, heart, skeletal muscle, and gastrointestinal tract, whereas the non-T3-binding isoform TRα2 is expressed in various tissues (e.g., brain and testis). Two different isoforms TRβ1 and TRβ2, which diverge in their amino-terminal regions, are generated from the *THRB* locus. TRβ1 is considered the major isoform in liver, kidney, and thyroid. TRβ1 is also predominantly expressed in the brain, pituitary, and inner ear, where it mediates overlapping functions with TRβ2 (Ng et al., 2015). TRβ2 has a more restricted expression pattern regulating neurosensory (hearing, vision) development as well as the HPT axis (Abel et al., 1999; Cheng, 2005; Cheng, Leonard, & Davis, 2010).

Since 1986 it has been known that heterozygous mutations within three, CpG-rich, hotspots of the ligand-binding domain (LBD) of TRβ result in resistance to thyroid hormone β (RTHβ), a clinical syndrome of refractoriness to TH that was recognized two decades earlier (Refetoff, DeWind, & DeGroot, 1967). The incidence of RTHβ is ~1 in 40,000 and several hundred mutations have been identified so far (Lafranchi et al., 2003; Refetoff & Dumitrescu, 2007). Patients with RTHβ are characterized by elevated serum TH levels with nonsuppressed TSH, and a phenotype ranging from being asymptomatic to having clinical features of thyrotoxicosis (e.g., failure to thrive in infancy; tachycardia and cardiac arrhythmia, raised metabolic rate, anxiety in adulthood) (Refetoff & Dumitrescu, 2007; Refetoff, Weiss, & Usala, 1993).

Human TRα and TRβ are highly homologous, with the proteins showing 80% amino acid identity. Taking into account the large number of different TRβ mutations associated with RTHβ, the identification of analogous mutations in human TRα had been anticipated. In an attempt to predict the clinical consequences of defective TRα, several knockin and knockout mouse models have been generated. Mice harboring different, heterozygous TRα mutations exhibit diverse phenotypes, depending on the severity and location of the mutations; however, all have near-normal thyroid function tests (Kaneshige et al., 2001; Liu, Schultz, & Brent, 2003; Quignodon, Vincent, Winter, Samarut, & Flamant, 2007;

Tinnikov et al., 2002). Perhaps due to this lack of a clear biochemical thyroid phenotype, no mutation in human *THRA* was reported until 2012.

In the last 4 years, 28 cases from 15 different families with mutations in *THRA* have been identified. In this review, we present an overview of the clinical features together with the underlying molecular mechanisms and effects of treatment in patients with resistance to TH due to heterozygous mutations in TRα (RTHα).

2. MOLECULAR MECHANISMS UNDERLYING RTHα

Table 1 shows all mutations in *THRA* identified in patients so far. Affected individuals are heterozygous for the mutations, which occurred de novo in eight cases or were familial in seven cases. All mutations identified so far are localized in the LBD of *THRA*, as illustrated in Fig. 1. Consistent with this, T3 binding by the mutant receptors is impaired, whereas DNA binding is unaffected.

The mutations can be categorized into three classes of receptor defects: (1) frame-shift/premature stop mutations, in which a deletion or insertion of one or several nucleotides results in a truncated receptor; (2) nonsense mutations, in which a nucleotide substitution results in a premature stop codon; and (3) missense mutations, in which a nucleotide substitution leads to an amino acid change without affecting protein length.

All patients are heterozygous, suggesting a dominant-negative effect of the mutant receptors on wild-type TRα. Indeed, when coexpressed the mutant receptors inhibit the function of their wild-type counterparts in a dominant-negative manner (Bochukova et al., 2012; Demir et al., 2016; Espiard et al., 2015; Moran et al., 2014, 2013; Tylki-Szymanska et al., 2015; van Gucht et al., 2016; van Mullem et al., 2013, 2012), similar to what has been reported for TRβ mutations in RTHβ. T3-induced dissociation of mutant TRα receptors from corepressors is retarded and ligand-dependent recruitment of coactivators impaired, likely contributing to the dominant-negative inhibition. Ex vivo analysis of mutation-containing, patient-derived, cells showed reduced T3-mediated induction of target gene expression, consistent with dominant-negative inhibition occurring in vivo (Bochukova et al., 2012; Moran et al., 2014, 2013).

In five families, affected individuals harbored missense mutations, localized further upstream in *THRA*, thereby involving both TRα1 and TRα2 proteins. At low T3 concentrations such missense mutant receptors showed reduced transcriptional activity together with moderate dominant-negative

Table 1 Mutations in *THRA* Identified in Patients

Missense	Nonsense	Frame Shift
G207E (unpublished) ♂ 15 months + mother (26 years)	C392X (Tylki-Szymanska et al., 2015) ♂ 18 years	C380fs387X (Demir et al., 2016) ♀ 12 years
D211G (van Gucht et al., 2016) ♀ 18 months + father (30 years)	E403X (Bochukova et al., 2012) ♀ 6 years	A382fs388X (Moran et al., 2013) ♀ 45 years
A263V (Moran et al., 2014) ♀ 60 years + 2 sons (26/30 years)	E403X (Tylki-Szymanska et al., 2015) ♀ 15 years	F397fs406X (van Mullem et al., 2012, 2013) ♀ 11 years + father (42 years)
A263S (Demir et al., 2016) ♂ 2.6 years + sister (7.4 years), mother (31 years), grandfather (55 years), aunt (35 years), 2 cousins (17/8.8 years)		
N359Y (Espiard et al., 2015) ♀ 27 years		
R384C (Yuen et al., 2015)		
R384H (Demir et al., 2016) ♂ 15 months + mother (30 years)		
E403K (Tylki-Szymanska et al., 2015) ♀ 6 years + father (39 years)		
P398R (Tylki-Szymanska et al., 2015) ♀ 5 years		

Mutations are organized by the type of mutation: frame shift, nonsense, and missense.

inhibition of wild-type TRα1. Interestingly, at higher T3 concentrations, mutant receptors exhibited transcriptional activity similar to wild-type TRα1 with reversal of their dominant-negative inhibitory activity (Demir et al., 2016; Moran et al., 2014; van Gucht et al., 2016). These observations suggest that patients harboring such milder, missense mutations may particularly benefit from TH treatment. Similar observations were made in mice with a missense mutation (R384C) in TRα1, with locomotor defects and behavioral abnormalities being alleviated by TH treatment (Tinnikov et al., 2002; Venero et al., 2005; Vennstrom, Mittag, & Wallis, 2008).

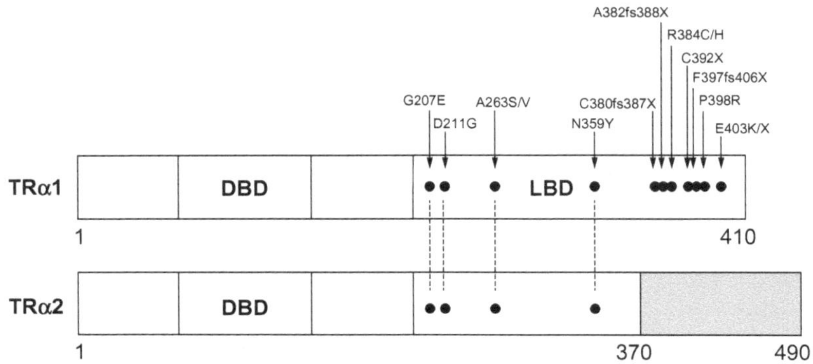

Fig. 1 Localization of identified mutations in *THRA*. Thyroid hormone receptor α1 (TRα1), the T3 receptor encoded by the *THRA* gene, and the splice variant TRα2, which does not bind T3 and does not act as a receptor, have identical amino-terminal and DNA-binding domains. The nonhomologous part of the ligand-binding domains (LBDs) is colored in *gray*. The location of all known TRα mutations is indicated. G207E, D211G, A263S, A263V, and N359Y affect both TRα1 and α2 transcripts.

The phenotype of patients with a mutation in both TRα1/2 resembles the clinical characteristics of RTHα patients with a mutation in TRα1 alone (Demir et al., 2016; Moran et al., 2014; van Gucht et al., 2016). One exception to this is a patient with a mutation (N359Y) in TRα1/2 who showed a distinct clinical phenotype comprising skeletal malformations (clavicular agenesis, metacarpal fusion, syndactyly of digits), chronic diarrhea, macrocytic anemia, and hypercalcemia (Espiard et al., 2015). It remains uncertain whether these additional phenotypic characteristics are solely due to the mutation in *THRA*. In all other RTHα cases described to date, the mutation in TRα2 does not seem to contribute to the phenotype, which is concordant with observations in TRα2KO mice. In these mice, phenotypic abnormalities are most probably not due to loss of TRα2, but are a consequence of compensatory overexpression of TRα1 (Salto et al., 2001). In addition, previous studies have suggested that TRα2 is unable to heterodimerize with RXR, bind TREs or exert dominant-negative activity via corepressor recruitment (Rentoumis et al., 1990; Tagami, Kopp, Johnson, Arseven, & Jameson, 1998; Yang, Burgos-Trinidad, Wu, & Koenig, 1996).

Noteworthy, the severity of the clinical RTHα phenotype appears to be associated with the type and localization of the mutation in TRα. In this, RTHα patients with frame-shift and nonsense mutations show a more affected phenotype in comparison to RTHα patients with milder missense mutations.

3. CLINICAL PHENOTYPE

3.1 Appearance

Most patients were born at term after an eventless pregnancy and displayed no abnormal characteristics, although recognized features of hypothyroidism (e.g., macroglossia, poor feeding, hoarse cry) were present in some cases at birth (Moran et al., 2014, 2013). Physical characteristics associated with hypothyroidism (e.g., macrocephaly, hypertelorism, a broad face, and flattened nasal bridge) became more evident with time, together with poor linear growth, resulting in a dysmorphic appearance (see Table 2) (Bochukova et al., 2012; Demir et al., 2016; Espiard et al., 2015; Moran et al., 2014, 2013; van Gucht et al., 2016; van Mullem et al., 2013, 2012). Several patients have numerous skin tags and moles, which seem to increase in number with age (Bochukova et al., 2012; Moran et al., 2014, 2013).

3.2 Neurological and Cognitive

In almost all index patients, a delay in reaching developmental milestones was the prime reason for consultation. Childhood cases show a slow initiation of motor movement together with impaired fine and gross motor coordination, resulting in dyspraxia and a broad-based ataxic gait. One young girl with a missense (D211G) mutation (van Gucht et al., 2016) initially presented with severe, predominantly axial, hypotonia. Two cases displayed childhood seizures (Moran et al., 2013; van Gucht et al., 2016). Several patients show mild to moderate/severe intellectual disability with a lower global IQ and reduced verbal and performance scores. A delay in speech development and dysarthric speech are consistent features in the majority of patients (see Table 2) (Bochukova et al., 2012; Demir et al., 2016; Moran et al., 2014, 2013; Tylki-Szymanska et al., 2015; van Gucht et al., 2016; van Mullem et al., 2013, 2012).

3.3 Skeletal

Most, but not all, RTHα patients have short stature; some have a disproportionate reduction in subischial leg length.

Skull radiographs of RTHα patients in childhood show delayed fontanelle fusion together with excessively serpiginious skull sutures (wormian bone appearance). Retarded ossification, resulting in delayed bone age or femoral epiphyseal dysgenesis, can be present. In addition, a thickened

Table 2 Clinical and Biochemical Characteristics of Patients With Resistance to Thyroid Hormone Alpha

System	Clinical/Biochemical Characteristics
Appearance: Dysmorphic syndrome	• Flat nasal bridge • Broad face, thickened lips • Macroglossia • Coarse facies • Skin tags
Neurological and cognitive	• Delayed developmental milestones • Delayed speech development, dysarthric speech • Slow initiation of movement • Fine and gross motor dyspraxia, broad-based ataxic gait • Dysdiadochokinesis • Reduced global IQ, reduced verbal, performance scores
Skeletal	• Disproportionate short stature: reduced total height, normal sitting height, reduced subischial leg length • Macrocephaly: delayed fontanelle fusion, thickened calvarium, wormian bones • Delayed tooth eruption • Delayed ossification: delayed bone age, femoral epiphyseal dysgenesis • Cortical hyperostosis in long bones, increased bone mineral density
Cardiac	• Bradycardia • Low/low-normal blood pressure
Gastrointestinal	• Constipation: decreased peristalsis, delayed intestinal transit, bowel dilatation
Biochemical and metabolic	• Thyroid function tests: normal or slightly raised TSH, low/low-normal (F)T4, high/high-normal (F)T3, low/low-normal rT3, decreased (F)T4/(F)T3 ratio, increased (F)T3/rT3 ratio • Hematological: low red cell mass or hematocrit with normal/raised MCV, B12, folate, reticulocyte count • SHBG: high/normal • IGF-1: low/normal • Low metabolic rate

calvarium and a delayed tooth eruption have been observed in most cases (Bochukova et al., 2012; Moran et al., 2014, 2013; Tylki-Szymanska et al., 2015; van Gucht et al., 2016; van Mullem et al., 2013, 2012). Especially in adults, cortical hyperostosis in long bones together with increased bone mineral density has been documented (see Table 2).

3.4 Cardiac and Gastrointestinal

Blood pressure and resting heart rate are normal or low/low-normal in most RTHα patients, which is remarkable in view of the high serum T3 levels (see later).

In the majority of both childhood and adult patients, constipation is a persistent problem due to decreased frequency of bowel movements (Bochukova et al., 2012; Demir et al., 2016; Moran et al., 2014, 2013; Tylki-Szymanska et al., 2015; van Gucht et al., 2016; van Mullem et al., 2013, 2012). When evaluated in patients, abdominal radiography shows a delayed intestinal transit together with bowel dilatation. In addition, reduced peristalsis is seen on colonic manometry (see Table 2) (Bochukova et al., 2012; Moran et al., 2013). Only one adult case has chronic diarrhea (Espiard et al., 2015).

3.5 Biochemical and Metabolic

Thyroid function tests show a consistent pattern, albeit with significant variation between different cases. The thyroid biochemical phenotype not only varies between patients with different mutations but also between patients carrying the same mutation. TSH levels are in the normal range in all patients, except for one adult female patient (Moran et al., 2013) who showed a slightly raised TSH while off LT4 treatment. In almost all index cases, (F)T4 levels are initially low or low-normal. A rise in (F)T4 levels into the normal range is observed in some of the cases during follow-up, possibly indicating attenuation of the biochemical phenotype with time. (F)T3 levels are high or high-normal in most patients, although normal (F)T3 levels have also been reported in several cases (Demir et al., 2016). Nevertheless, an abnormally high T3/T4 ratio seems a consistent finding among all patients. Reverse T3 (rT3) levels are normal or subnormal, resulting in a higher serum T3/rT3 ratio (see Table 2) (Bochukova et al., 2012; Moran et al., 2014, 2013; Tylki-Szymanska et al., 2015; van Gucht et al., 2016).

Other markers, reflecting TH action in peripheral tissues, can also be altered. A mild, usually normocytic normochromic, anemia is an almost universal characteristic in RTHα patients. Serum muscle creatine kinase levels and sex hormone-binding globulin (SHBG) levels may be raised, while circulating IGF-1 levels can be decreased (Bochukova et al., 2012; Demir et al., 2016; Espiard et al., 2015; Moran et al., 2014, 2013; Tylki-Szymanska et al., 2015; van Gucht et al., 2016; van Mullem et al., 2013, 2012). Total and LDL cholesterol levels may be increased, in both childhood and adult cases (van Mullem et al., 2013, 2012). In several patients, a low basal metabolic rate,

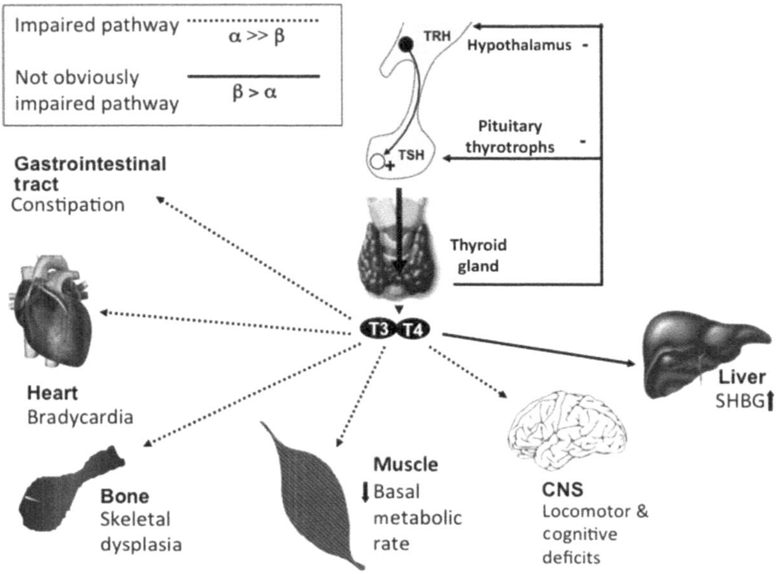

Fig. 2 Cartoon showing the major tissue actions of thyroid hormone together with the predominant receptor subtypes mediating these effects. In RTHα, tissues expressing mainly TRα are resistant to thyroid hormone action (impaired), while TRβ-expressing tissues are sensitive (not obviously impaired).

measured by indirect calorimetry, was recorded (see Table 2) (Bochukova et al., 2012; Espiard et al., 2015; Moran et al., 2014, 2013). Overall, both clinical and biochemical abnormalities seen in RTHα are consonant with TH resistance in tissues expressing predominantly TRα1 (e.g., heart, bone, muscle, intestine, brain), while TRβ-expressing tissues (e.g., hypothalamus, pituitary, liver) remain sensitive to TH (Fig. 2).

4. PATHOGENESIS

Many clinical features in RTHα patients, such as delayed motor development, delayed cranial suture closure resulting in macrocephaly, delayed tooth eruption, developmental skeletal abnormalities (e.g., femoral epiphyseal dysgenesis, wormian bones), delayed bone age, and lower segment growth retardation, are characteristic of untreated hypothyroidism (Beierwaltes, 1954; Huffmeier, Tietze, & Rauch, 2007). In addition, constipation due to decreased colonic motility and colonic dilatation is also clearly associated with hypothyroidism (Ebert, 2010).

Other characteristics of hypothyroidism, including slow speech, bradykinesia, dry skin, and slow relaxing reflexes are also reported in RTHα cases. Interestingly, many of these features (delayed growth, tooth eruption, fontanelle closure, and endochondral and intramembranous ossification) are also present in different mouse models with frame-shift mutations (TRα1-PV) or missense mutations in TRα1 (TRα1-R384C, TRα1-P398H, TRα1-L400R) (Bassett et al., 2014, 2007; Fraichard et al., 1997; Kaneshige et al., 2001; Liu et al., 2003; O'Shea et al., 2005; Quignodon et al., 2007; Tinnikov et al., 2002; Wojcicka, Bassett, & Williams, 2013). Abnormal motor development in patients resembles the psychomotor phenotype of TRα1-R384C mice, which show reduced grip strength, poor limb coordination, and an abnormal gait (Tinnikov et al., 2002; Vennstrom et al., 2008).

Several patients exhibit delayed mental development with cognitive deficits and two patients suffer from seizures, which have also been described in TRα1-R384C mice (Venero et al., 2005; Vennstrom et al., 2008). Constipation and bowel dilatation with reduced peristalsis that has been described in RTHα patients is very similar to the intestinal phenotype of $TR\alpha^{-/-}$ mice (Fraichard et al., 1997; Plateroti et al., 1999).

Despite the variation between patients, the biochemical abnormalities (increased T3/T4 and T3/rT3 ratios) found in most RTHα cases may reflect altered peripheral TH metabolism by deiodinases. As indicated by marked suppression of serum TSH by relatively small increases in serum (F)T4 in cases, central-feedback regulation of the HPT axis is preserved, favoring a minor contribution of TRα1 but predominant role for TRβ isoforms in regulation of the HPT axis (Langlois et al., 1997; Ng et al., 2015; van Mullem et al., 2013). Increased T3 and decreased rT3 levels could be due to upregulation of hepatic DIO1 or reduced DIO3 activity as has been shown in studies with TRα mutant mice (Barca-Mayo et al., 2011; Kaneshige et al., 2001). DIO1 is expressed in liver, kidney, and thyroid and plays an important role in the production of the bioactive T3 from T4 and the clearance of rT3. DIO3 is present in brain, skin, placenta, and pregnant uterus where it inactivates T3 and T4 and hereby protects these tissues from excess active TH. T3/T4 and T3/rT3 ratios are considered to be sensitive indicators of peripheral TH metabolism and are relatively independent of variations in serum-binding proteins or thyroidal T4 production.

The abnormalities in thyroid function seen in RTHα patients do not completely correlate with observations made in different TRα mutant mouse models, since overt thyroid dysfunction was only documented in

TRα1-PV and TRα1-P398H mice and not in TRα1-R384C and TRα1-L400R animals (Kaneshige et al., 2001; Liu et al., 2003; Quignodon et al., 2007; Tinnikov et al., 2002; Vennstrom et al., 2008). It has been shown that TRα1-PV mutant mice have markedly increased levels of kidney and liver DIO1, while T3-induced DIO3 expression in cortex is decreased, resulting in a decreased clearance of T3 (Zavacki et al., 2005).

Anemia, with normal hematinics (iron, vitamin B12, folate) and hemolytic indices (reticulocyte count, circulating haptoglobin, and lactate dehydrogenase), is a near-universal characteristic in RTHα. Usually a normocytic normochromic anemia, although the mean corpuscular volume was raised in three cases (Bochukova et al., 2012; Espiard et al., 2015; Moran et al., 2013). Several human studies have documented an association between hypothyroidism and anemia (Fein & Rivlin, 1975; Krause & Sokoloff, 1967). In addition, data from animal models support an important role for TRα in erythropoiesis (Angelin-Duclos et al., 2005; Gandrillon et al., 1989; Kendrick et al., 2008; Weinberger et al., 1986). The molecular mechanisms underlying anemia in RTHα patients remain to be elucidated.

Spontaneous conception and uneventful pregnancy has been observed in untreated females with RTHα, despite varying features of hypothyroidism (Demir et al., 2016). Maternal and paternal inheritance of TRα mutations suggests that transmission of TRα mutations from parents to offspring is less impaired in humans than in mice (Kaneshige et al., 2001).

The observation that several affected relatives of RTHα index cases reached adulthood without identification of medical problems supports the mild character of these mutations (Demir et al., 2016; van Gucht et al., 2016). Similar to RTHβ, individuals harboring the same mutation can exhibit significant variation in severity of clinical phenotypes and it is conceivable that additional variation (e.g., in genes encoding cofactors of TH action) could mediate this (Refetoff & Dumitrescu, 2007).

5. TREATMENT

Several RTHα patients have been treated with LT4 and as yet, no other therapeutic approaches have been explored (Bochukova et al., 2012; Demir et al., 2016; Espiard et al., 2015; Moran et al., 2014, 2013; van Gucht et al., 2016; van Mullem et al., 2013). Overall, 11 patients have been LT4-treated: in eight cases LT4 treatment was initiated from (early) childhood and in three cases LT4 was started in adult life. The duration of LT4 treatment in childhood cases ranges from 1 year in a girl with a

D211G mutation to 5 and 6 years in girls with F397fs406X and E403X mutations in TRα, respectively, and 12 years in a girl with a C380fs388X mutation (Bochukova et al., 2012; Demir et al., 2016; van Gucht et al., 2016; van Mullem et al., 2012). Adult cases have received LT4 supplementation for 1 month to over 40 years (Demir et al., 2016; Espiard et al., 2015; Moran et al., 2014, 2013). LT4 treatment had beneficial effects on several RTHα characteristics, as described later. In addition, initiation of LT4 therapy at a young age also had beneficial developmental effects in childhood cases.

LT4 treatment, even in typical physiological doses, resulted in suppressed TSH levels in all RTHα patients, implying an intact-feedback mechanism within the HPT axis (Bochukova et al., 2012; Demir et al., 2016; Espiard et al., 2015; Moran et al., 2014, 2013; van Gucht et al., 2016). In addition, serum (F)T4 and rT3 levels normalize, while serum T3 remains elevated. Some peripheral markers of TH action also responded to LT4 therapy, with a rise in serum SHBG and IGF1 levels together with a decrease in serum creatine kinase and LDL cholesterol levels (Bochukova et al., 2012; Demir et al., 2016; Moran et al., 2014, 2013; van Gucht et al., 2016; van Mullem et al., 2013). In most cases, LT4 therapy has a beneficial effect on constipation but has no effect on anemia and a blunted effect on cardiac function.

LT4 treatment initiated at young age resulted in a catch-up growth and improved overall height in several patients, and the addition of growth hormone did not further improve growth (Bochukova et al., 2012; van Gucht et al., 2016; van Mullem et al., 2013). Moreover, in one childhood case, harboring a TRα mutation whose dysfunction is reversible at higher TH levels (D211G), during LT4 treatment muscle hypotonia improved and motor development accelerated (van Gucht et al., 2016). LT4 treatment from early childhood in three related patients, with a relatively mild mutation (A263V) in TRα, may have ameliorated their phenotype and alleviated some symptoms (e.g., dyspraxia) in adulthood (Moran et al., 2014).

In contrast, initiation of LT4 at the age of 2.5 years in a girl with the most upstream C380fs387X frame-shift mutation did not ameliorate her growth retardation, and cardiac and renal problems, which may reflect ongoing tissue hypothyroidism, despite LT4 treatment. Concordant with this LT4 treatment of TRα1-PV mutant mice, harboring a deleterious frame-shift mutation, did not improve skeletal abnormalities or other abnormal phenotypes (Bassett et al., 2014; Itoh et al., 2001; Ying, Araki, Furuya, Kato, & Cheng, 2007). In patients with frame-shift mutations involving the carboxy-terminal region of TRα1, the severity of the clinical phenotype and also the effect of LT4 therapy may vary depending on the exact location of these mutations and the extent to which this domain is disrupted (Demir et al., 2016; Moran et al., 2013; van Mullem et al., 2013).

Several affected relatives of RTHα index patients with missense mutations, who had only been diagnosed after screening of family members of the index case, showed an amelioration of clinical features with time. Similar observations have been made in TRα1-R384C mice, which showed severe but transient impairment of postnatal development and growth. Alterations in TH levels were also only evident in these mice during the juvenile period (Tinnikov et al., 2002; Venero et al., 2005; Vennstrom et al., 2008). The mechanisms underlying such age-dependent amelioration of phenotypes associated with some TRα1 mutations are unknown.

Overall, the effects observed with LT4 treatment of RTHα patients correspond with the expected pattern, with TH resistance in organs predominantly expressing TRα1 (bone, skeletal muscle, gastrointestinal tract, myocardium, brain) and retention of TH sensitivity in TRβ-expressing tissues (hypothalamus, pituitary, liver). Indeed, chronic exposure to excess TH in LT4-treated RTHα patients could result in unwanted toxicities in TRβ-expressing tissues. The observation that LT4 treatment in RTHα further raises SHBG and (F)T3 levels with suppression of TSH, supports this possibility.

In this regard, future therapies that could selectively target TRα1, e.g., high-affinity TRα1-selective agonists, are of great interest. Such TRα1-specific thyromimetics could activate normal TRα1 and potentially, in higher dosages, overcome the impaired function of mutant TRα1. Alternative therapeutic strategies could involve targeting the dominant-negative effect of mutant TRα1 on its wild-type counterpart, either by inhibiting HDAC activity or by abrogating the interaction of mutant TRα1 with the corepressor complex. Studies in mice harboring the TRα1-PV mutation showed that aberrant recruitment of NCoR by mutant TRα1 contributes to the phenotype and the severity of this can be partially reversed by introducing a mutation in NCoR that abrogates its interaction with TRα1 (Fozzatti et al., 2013). In an alternative approach, TRα1PV-mediated repression could be relieved by administration of suberoylanilide hydroxyamic acid (SAHA), a HDAC inhibitor, also ameliorating the phenotypic abnormalities in these mice (Kim, Park, Willingham, & Cheng, 2014; Ocasio & Scanlan, 2006; Tan, Cang, Ma, Petrillo, & Liu, 2010).

6. CONCLUSIONS

The clinical consequences of mutations in *THRA* are variable in extent and severity. This applies both to patients with different TRα mutations and to individuals in a family harboring the same mutation. Similar to RTHβ, the observation that affected relatives with a relatively mild phenotype in some RTHα families were only diagnosed by genetic screening after initial diagnosis of the index cases supports the hypothesis that the incidence of RTHα could be much higher than currently estimated. Given the strong homology between the TRα1 and TRβ receptors and similar localization pattern of both TRα and TRβ mutations, the incidence of RTHα could correspond to that of RTHβ (1:40,000).

RTHα should be suspected in subjects when even mild clinical features of hypothyroidism are present along with a normal TSH, a high serum T3/T4 or T3/rT3 ratio. RTHα patients, particularly those with mild mutations or diagnosed in early life, may benefit particularly from LT4 treatment. Further research to develop alternative treatment options and clinical biomarkers which better reflect the effects of LT4 treatment in both TRα- and TRβ-expressing tissues will be of great value.

ACKNOWLEDGMENTS

A.L.M.v.G., M.E.M., and R.P.P. are supported by a Zon-MWTOP Grant (number 91212044) and an Erasmus MC MRACE Grant. C.M. and K.C. are supported by the NIHR Cambridge Biomedical Research Centre.

REFERENCES

Abel, E. D., Boers, M. E., Pazos-Moura, C., Moura, E., Kaulbach, H., Zakaria, M., … Wondisford, F. (1999). Divergent roles for thyroid hormone receptor beta isoforms in the endocrine axis and auditory system. *The Journal of Clinical Investigation*, *104*(3), 291–300. http://dx.doi.org/10.1172/jci6397.

Angelin-Duclos, C., Domenget, C., Kolbus, A., Beug, H., Jurdic, P., & Samarut, J. (2005). Thyroid hormone T3 acting through the thyroid hormone alpha receptor is necessary for implementation of erythropoiesis in the neonatal spleen environment in the mouse. *Development*, *132*(5), 925–934. http://dx.doi.org/10.1242/dev.01648.

Barca-Mayo, O., Liao, X. H., Alonso, M., Di Cosmo, C., Hernandez, A., Refetoff, S., & Weiss, R. E. (2011). Thyroid hormone receptor alpha and regulation of type 3 deiodinase. *Molecular Endocrinology*, *25*(4), 575–583. http://dx.doi.org/10.1210/me.2010-0213.

Bassett, J. H., Boyde, A., Zikmund, T., Evans, H., Croucher, P. I., Zhu, X., … Williams, G. R. (2014). Thyroid hormone receptor alpha mutation causes a severe and thyroxine-resistant skeletal dysplasia in female mice. *Endocrinology*, *155*(9), 3699–3712. http://dx.doi.org/10.1210/en.2013-2156.

Bassett, J. H., Harvey, C. B., & Williams, G. R. (2003). Mechanisms of thyroid hormone receptor-specific nuclear and extra nuclear actions. *Molecular and Cellular Endocrinology, 213*(1), 1–11. http://dx.doi.org/10.1016/j.mce.2003.10.033. S0303720703004131 [pii].
Bassett, J. H., Nordstrom, K., Boyde, A., Howell, P. G., Kelly, S., Vennstrom, B., & Williams, G. R. (2007). Thyroid status during skeletal development determines adult bone structure and mineralization. *Molecular Endocrinology, 21*(8), 1893–1904. http://dx.doi.org/10.1210/me.2007-0157.
Beierwaltes, W. H. (1954). Incomplete growth associated with hypothyroidism. *The Journal of Clinical Endocrinology and Metabolism, 14*(12), 1551–1559. http://dx.doi.org/10.1210/jcem-14-12-1551.
Bernal, J. (2005). Thyroid hormones and brain development. *Vitamins and Hormones, 71*, 95–122. http://dx.doi.org/10.1016/S0083-6729(05)71004-9. S0083-6729(05)71004-9 [pii].
Bochukova, E., Schoenmakers, N., Agostini, M., Schoenmakers, E., Rajanayagam, O., Keogh, J. M., … Chatterjee, K. (2012). A mutation in the thyroid hormone receptor alpha gene. *The New England Journal of Medicine, 366*(3), 243–249. http://dx.doi.org/10.1056/NEJMoa1110296.
Brent, G. A. (2012). Mechanisms of thyroid hormone action. *The Journal of Clinical Investigation, 122*(9), 3035–3043. http://dx.doi.org/10.1172/JCI60047.
Cheng, S. Y. (2005). Thyroid hormone receptor mutations and disease: Beyond thyroid hormone resistance. *Trends in Endocrinology and Metabolism, 16*(4), 176–182. http://dx.doi.org/10.1016/j.tem.2005.03.008.
Cheng, S. Y., Leonard, J. L., & Davis, P. J. (2010). Molecular aspects of thyroid hormone actions. *Endocrine Reviews, 31*(2), 139–170. http://dx.doi.org/10.1210/er.2009-0007.
Chiamolera, M. I., & Wondisford, F. E. (2009). Minireview: Thyrotropin-releasing hormone and the thyroid hormone feedback mechanism. *Endocrinology, 150*(3), 1091–1096. http://dx.doi.org/10.1210/en.2008-1795. en.2008-1795 [pii].
Demir, K., van Gucht, A. L., Buyukinan, M., Catli, G., Ayhan, Y., Bas, V. N., … Visser, T. J. (2016). Diverse genotypes and phenotypes of three novel thyroid hormone receptor-alpha mutations. *The Journal of Clinical Endocrinology and Metabolism, 101*(8), 2945–2954. http://dx.doi.org/10.1210/jc.2016-1404.
Dumitrescu, A. M., & Refetoff, S. (2013). The syndromes of reduced sensitivity to thyroid hormone. *Biochimica et Biophysica Acta, 1830*(7), 3987–4003. http://dx.doi.org/10.1016/j.bbagen.2012.08.005. S0304-4165(12)00225-5 [pii].
Ebert, E. C. (2010). The thyroid and the gut. *Journal of Clinical Gastroenterology, 44*(6), 402–406. http://dx.doi.org/10.1097/MCG.0b013e3181d6bc3e.
Espiard, S., Savagner, F., Flamant, F., Vlaeminck-Guillem, V., Guyot, R., Munier, M., … Wemeau, J. L. (2015). A novel mutation in THRA gene associated with an atypical phenotype of resistance to thyroid hormone. *The Journal of Clinical Endocrinology and Metabolism, 100*(8), 2841–2848. http://dx.doi.org/10.1210/jc.2015-1120.
Fein, H. G., & Rivlin, R. S. (1975). Anemia in thyroid diseases. *The Medical Clinics of North America, 59*(5), 1133–1145.
Fekete, C., & Lechan, R. M. (2014). Central regulation of hypothalamic-pituitary-thyroid axis under physiological and pathophysiological conditions. *Endocrine Reviews, 35*(2), 159–194. http://dx.doi.org/10.1210/er.2013-1087.
Fozzatti, L., Kim, D. W., Park, J. W., Willingham, M. C., Hollenberg, A. N., & Cheng, S. Y. (2013). Nuclear receptor corepressor (NCOR1) regulates in vivo actions of a mutated thyroid hormone receptor alpha. *Proceedings of the National Academy of Sciences of the United States of America, 110*(19), 7850–7855. http://dx.doi.org/10.1073/pnas.1222334110. 1222334110 [pii].

Fraichard, A., Chassande, O., Plateroti, M., Roux, J. P., Trouillas, J., Dehay, C., … Samarut, J. (1997). The T3R alpha gene encoding a thyroid hormone receptor is essential for post-natal development and thyroid hormone production. *The EMBO Journal, 16*(14), 4412–4420. http://dx.doi.org/10.1093/emboj/16.14.4412.

Gandrillon, O., Jurdic, P., Pain, B., Desbois, C., Madjar, J. J., Moscovici, M. G., … Samarut, J. (1989). Expression of the v-erbA product, an altered nuclear hormone receptor, is sufficient to transform erythrocytic cells in vitro. *Cell, 58*(1), 115–121.

Huffmeier, U., Tietze, H. U., & Rauch, A. (2007). Severe skeletal dysplasia caused by undiagnosed hypothyroidism. *European Journal of Medical Genetics, 50*(3), 209–215. http://dx.doi.org/10.1016/j.ejmg.2007.02.002. S1769-7212(07)00029-8 [pii].

Itoh, Y., Esaki, T., Kaneshige, M., Suzuki, H., Cook, M., Sokoloff, L., … Nunez, J. (2001). Brain glucose utilization in mice with a targeted mutation in the thyroid hormone alpha or beta receptor gene. *Proceedings of the National Academy of Sciences of the United States of America, 98*(17), 9913–9918. http://dx.doi.org/10.1073/pnas.171319498. 171319498 [pii].

Kalyanaraman, H., Schwappacher, R., Joshua, J., Zhuang, S., Scott, B. T., Klos, M., … Pilz, R. B. (2014). Nongenomic thyroid hormone signaling occurs through a plasma membrane-localized receptor. *Science Signaling*, 7(326), ra48. http://dx.doi.org/10.1126/scisignal.2004911.

Kaneshige, M., Suzuki, H., Kaneshige, K., Cheng, J., Wimbrow, H., Barlow, C., … Cheng, S. (2001). A targeted dominant negative mutation of the thyroid hormone alpha 1 receptor causes increased mortality, infertility, and dwarfism in mice. *Proceedings of the National Academy of Sciences of the United States of America, 98*(26), 15095–15100. http://dx.doi.org/10.1073/pnas.261565798.

Kendrick, T. S., Payne, C. J., Epis, M. R., Schneider, J. R., Leedman, P. J., Klinken, S. P., & Ingley, E. (2008). Erythroid defects in TRalpha−/− mice. *Blood, 111*(6), 3245–3248. http://dx.doi.org/10.1182/blood-2007-07-101105.

Kim, D. W., Park, J. W., Willingham, M. C., & Cheng, S. Y. (2014). A histone deacetylase inhibitor improves hypothyroidism caused by a TRalpha1 mutant. *Human Molecular Genetics, 23*(10), 2651–2664. http://dx.doi.org/10.1093/hmg/ddt660.

Krause, R. L., & Sokoloff, L. (1967). Effects of thyroxine on initiation and completion of protein chains of hemoglobin in vitro. *The Journal of Biological Chemistry, 242*(7), 1431–1438.

Lafranchi, S. H., Snyder, D. B., Sesser, D. E., Skeels, M. R., Singh, N., Brent, G. A., & Nelson, J. C. (2003). Follow-up of newborns with elevated screening T4 concentrations. *The Journal of Pediatrics, 143*(3), 296–301. http://dx.doi.org/10.1067/S0022-3476(03)00184-7. S0022-3476(03)00184-7 [pii].

Langlois, M. F., Zanger, K., Monden, T., Safer, J. D., Hollenberg, A. N., & Wondisford, F. E. (1997). A unique role of the beta-2 thyroid hormone receptor isoform in negative regulation by thyroid hormone. Mapping of a novel amino-terminal domain important for ligand-independent activation. *The Journal of Biological Chemistry, 272*(40), 24927–24933.

Liu, Y. Y., Schultz, J. J., & Brent, G. A. (2003). A thyroid hormone receptor alpha gene mutation (P398H) is associated with visceral adiposity and impaired catecholamine-stimulated lipolysis in mice. *The Journal of Biological Chemistry, 278*(40), 38913–38920. http://dx.doi.org/10.1074/jbc.M306120200.

Moran, C., Agostini, M., Visser, W. E., Schoenmakers, E., Schoenmakers, N., Offiah, A. C., … Chatterjee, K. K. (2014). Resistance to thyroid hormone caused by a mutation in thyroid hormone receptor (TR)α1 and TRα2: Clinical, biochemical, and genetic analyses of three related patients. *The Lancet Diabetes and Endocrinology, 2*(8), 619–626. http://dx.doi.org/10.1016/s2213-8587(14)70111-1.

Moran, C., Schoenmakers, N., Agostini, M., Schoenmakers, E., Offiah, A., Kydd, A., … Chatterjee, K. (2013). An adult female with resistance to thyroid hormone mediated by defective thyroid hormone receptor alpha. *The Journal of Clinical Endocrinology and Metabolism*, *98*(11), 4254–4261. http://dx.doi.org/10.1210/jc.2013-2215.

Ng, L., Cordas, E., Wu, X., Vella, K. R., Hollenberg, A. N., & Forrest, D. (2015). Age-related hearing loss and degeneration of cochlear hair cells in mice lacking thyroid hormone receptor beta1. *Endocrinology*, *156*(10), 3853–3865. http://dx.doi.org/10.1210/en.2015-1468.

Ocasio, C. A., & Scanlan, T. S. (2006). Design and characterization of a thyroid hormone receptor alpha (TRalpha)-specific agonist. *ACS Chemical Biology*, *1*(9), 585–593. http://dx.doi.org/10.1021/cb600311v.

O'Shea, P. J., Bassett, J. H., Sriskantharajah, S., Ying, H., Cheng, S. Y., & Williams, G. R. (2005). Contrasting skeletal phenotypes in mice with an identical mutation targeted to thyroid hormone receptor alpha1 or beta. *Molecular Endocrinology*, *19*(12), 3045–3059. http://dx.doi.org/10.1210/me.2005-0224.

Plateroti, M., Chassande, O., Fraichard, A., Gauthier, K., Freund, J. N., Samarut, J., & Kedinger, M. (1999). Involvement of T3Ralpha- and beta-receptor subtypes in mediation of T3 functions during postnatal murine intestinal development. *Gastroenterology*, *116*(6), 1367–1378.

Quignodon, L., Vincent, S., Winter, H., Samarut, J., & Flamant, F. (2007). A point mutation in the activation function 2 domain of thyroid hormone receptor alpha1 expressed after CRE-mediated recombination partially recapitulates hypothyroidism. *Molecular Endocrinology*, *21*(10), 2350–2360. http://dx.doi.org/10.1210/me.2007-0176.

Refetoff, S., DeWind, L. T., & DeGroot, L. J. (1967). Familial syndrome combining deaf-mutism, stuppled epiphyses, goiter and abnormally high PBI: Possible target organ refractoriness to thyroid hormone. *The Journal of Clinical Endocrinology and Metabolism*, *27*(2), 279–294. http://dx.doi.org/10.1210/jcem-27-2-279.

Refetoff, S., & Dumitrescu, A. M. (2007). Syndromes of reduced sensitivity to thyroid hormone: Genetic defects in hormone receptors, cell transporters and deiodination. *Best Practice & Research. Clinical Endocrinology & Metabolism*, *21*(2), 277–305. http://dx.doi.org/10.1016/j.beem.2007.03.005.

Refetoff, S., Weiss, R. E., & Usala, S. J. (1993). The syndromes of resistance to thyroid hormone. *Endocrine Reviews*, *14*(3), 348–399. http://dx.doi.org/10.1210/edrv-14-3-348.

Rentoumis, A., Chatterjee, V. K., Madison, L. D., Datta, S., Gallagher, G. D., Degroot, L. J., & Jameson, J. L. (1990). Negative and positive transcriptional regulation by thyroid hormone receptor isoforms. *Molecular Endocrinology*, *4*(10), 1522–1531. http://dx.doi.org/10.1210/mend-4-10-1522.

Salto, C., Kindblom, J. M., Johansson, C., Wang, Z., Gullberg, H., Nordstrom, K., … Vennstrom, B. (2001). Ablation of TRalpha2 and a concomitant overexpression of alpha1 yields a mixed hypo- and hyperthyroid phenotype in mice. *Molecular Endocrinology*, *15*(12), 2115–2128. http://dx.doi.org/10.1210/mend.15.12.0750.

Schoenmakers, N., Moran, C., Peeters, R. P., Visser, T., Gurnell, M., & Chatterjee, K. (2013). Resistance to thyroid hormone mediated by defective thyroid hormone receptor alpha. *Biochimica et Biophysica Acta*, *1830*(7), 4004–4008. http://dx.doi.org/10.1016/j.bbagen.2013.03.018.

Tagami, T., Kopp, P., Johnson, W., Arseven, O. K., & Jameson, J. L. (1998). The thyroid hormone receptor variant alpha2 is a weak antagonist because it is deficient in interactions with nuclear receptor corepressors. *Endocrinology*, *139*(5), 2535–2544. http://dx.doi.org/10.1210/endo.139.5.6011.

Tan, J., Cang, S., Ma, Y., Petrillo, R. L., & Liu, D. (2010). Novel histone deacetylase inhibitors in clinical trials as anti-cancer agents. *Journal of Hematology & Oncology*, *3*, 5. http://dx.doi.org/10.1186/1756-8722-3-5.

Tinnikov, A., Nordstrom, K., Thoren, P., Kindblom, J. M., Malin, S., Rozell, B., … Vennstrom, B. (2002). Retardation of post-natal development caused by a negatively acting thyroid hormone receptor alpha1. *The EMBO Journal*, *21*(19), 5079–5087.

Tylki-Szymanska, A., Acuna-Hidalgo, R., Krajewska-Walasek, M., Lecka-Ambroziak, A., Steehouwer, M., Gilissen, C., … Chrzanowska, K. H. (2015). Thyroid hormone resistance syndrome due to mutations in the thyroid hormone receptor alpha gene (THRA). *Journal of Medical Genetics*, *52*(5), 312–316. http://dx.doi.org/10.1136/jmedgenet-2014-102936.

van Gucht, A., Meima, M., Zwaveling-Soonawala, N., Visser, W. E., Fliers, E., Wennink, J., … van Trotsenburg, A. S. (2016). Resistance to thyroid hormone alpha in an 18 months old girl; clinical, therapeutic and molecular characteristics. *Thyroid*, *26*, 338–346. http://dx.doi.org/10.1089/thy.2015.0463.

van Mullem, A. A., Chrysis, D., Eythimiadou, A., Chroni, E., Tsatsoulis, A., de Rijke, Y. B., … Peeters, R. P. (2013). Clinical phenotype of a new type of thyroid hormone resistance caused by a mutation of the TRalpha1 receptor: Consequences of LT4 treatment. *The Journal of Clinical Endocrinology and Metabolism*, *98*(7), 3029–3038. http://dx.doi.org/10.1210/jc.2013-1050.

van Mullem, A., van Heerebeek, R., Chrysis, D., Visser, E., Medici, M., Andrikoula, M., … Visser, T. J. (2012). Clinical phenotype and mutant TRalpha1. *The New England Journal of Medicine*, *366*(15), 1451–1453. http://dx.doi.org/10.1056/NEJMc1113940.

Venero, C., Guadano-Ferraz, A., Herrero, A. I., Nordstrom, K., Manzano, J., de Escobar, G. M., … Vennstrom, B. (2005). Anxiety, memory impairment, and locomotor dysfunction caused by a mutant thyroid hormone receptor alpha1 can be ameliorated by T3 treatment. *Genes & Development*, *19*(18), 2152–2163. http://dx.doi.org/10.1101/gad.346105.

Vennstrom, B., Mittag, J., & Wallis, K. (2008). Severe psychomotor and metabolic damages caused by a mutant thyroid hormone receptor alpha 1 in mice: Can patients with a similar mutation be found and treated? *Acta Paediatrica*, *97*(12), 1605–1610. http://dx.doi.org/10.1111/j.1651-2227.2008.01031.x.

Weinberger, C., Thompson, C. C., Ong, E. S., Lebo, R., Gruol, D. J., & Evans, R. M. (1986). The c-erb-A gene encodes a thyroid hormone receptor. *Nature*, *324*(6098), 641–646. http://dx.doi.org/10.1038/324641a0.

Wojcicka, A., Bassett, J. H., & Williams, G. R. (2013). Mechanisms of action of thyroid hormones in the skeleton. *Biochimica et Biophysica Acta*, *1830*(7), 3979–3986. http://dx.doi.org/10.1016/j.bbagen.2012.05.005.

Yang, Y. Z., Burgos-Trinidad, M., Wu, Y., & Koenig, R. J. (1996). Thyroid hormone receptor variant alpha2. Role of the ninth heptad in dna binding, heterodimerization with retinoid X receptors, and dominant negative activity. *The Journal of Biological Chemistry*, *271*(45), 28235–28242.

Ying, H., Araki, O., Furuya, F., Kato, Y., & Cheng, S. Y. (2007). Impaired adipogenesis caused by a mutated thyroid hormone alpha1 receptor. *Molecular and Cellular Biology*, *27*(6), 2359–2371. http://dx.doi.org/10.1128/MCB.02189-06. MCB.02189-06 [pii].

Yuen, R. K., Thiruvahindrapuram, B., Merico, D., Walker, S., Tammimies, K., Hoang, N., … Scherer, S. W. (2015). Whole-genome sequencing of quartet families with autism spectrum disorder. *Nature Medicine*, *21*(2), 185–191. http://dx.doi.org/10.1038/nm.3792. nm.3792 [pii].

Zavacki, A. M., Ying, H., Christoffolete, M. A., Aerts, G., So, E., Harney, J. W., … Bianco, A. C. (2005). Type 1 iodothyronine deiodinase is a sensitive marker of peripheral thyroid status in the mouse. *Endocrinology*, *146*(3), 1568–1575. http://dx.doi.org/10.1210/en.2004-1392. en.2004-1392 [pii].

Zhang, J., & Lazar, M. A. (2000). The mechanism of action of thyroid hormones. *Annual Review of Physiology*, *62*, 439–466. http://dx.doi.org/10.1146/annurev.physiol.62.1.439.

CHAPTER THIRTEEN

TR2 and TR4 Orphan Nuclear Receptors: An Overview

Shin-Jen Lin*[,1], Dong-Rong Yang*[,†,1], Guosheng Yang*[,‡], Chang-Yi Lin*, Hong-Chiang Chang*, Gonghui Li*[,§,2], Chawnshang Chang*[,¶,2]

*George Whipple Lab for Cancer Research, The Wilmot Cancer Center, University of Rochester Medical Center, Rochester, NY, United States
[†]The Second Affiliated Hospital of Suzhou University, Suzhou, China
[‡]Guangdong 2nd Provincial People's Hospital, Guangzhou, China
[§]Sir Run Run Shaw Hospital, Zhejiang University, Hangzhou, China
[¶]Sex Hormone Research Center, China Medical University/Hospital, Taichung, Taiwan
[2]Corresponding authors: e-mail address: ligonghui1970@hotmail.com; chang@urmc.rochester.edu

Contents

Abstract

Testicular nuclear receptors 2 and 4 (TR2, TR4), also known as NR2C1 and NR2C2, belong to the nuclear receptor superfamily and were first cloned in 1989 and 1994, respectively. Although classified as orphan receptors, several natural molecules, their metabolites, and synthetic compounds including polyunsaturated fatty acids (PUFAs), PUFA

[1] Contribute equally.

Current Topics in Developmental Biology, Volume 125
ISSN 0070-2153
http://dx.doi.org/10.1016/bs.ctdb.2017.02.002

metabolites 13-hydroxyoctadecadienoic acid, 15-hydroxyeicosatetraenoic acid, and the antidiabetic drug thiazolidinediones can transactivate TR4. Importantly, many of these ligands/activators can also transactivate peroxisome proliferator-activated receptor gamma (PPARγ), also known as NR1C3 nuclear receptor. Both TR4 and PPARγ can bind to similar hormone response elements (HREs) located in the promoter of their common downstream target genes. However, these two nuclear receptors, even with shared ligands/activators and shared binding ability for similar HREs, have some distinct functions in many diseases they influence. In cancer, PPARγ inhibits thyroid, lung, colon, and prostate cancers but enhances bladder cancer. In contrast, TR4 inhibits liver and prostate cancer initiation but enhances pituitary corticotroph, liver, and prostate cancer progression. In type 2 diabetes, PPARγ increases insulin sensitivity but TR4 decreases insulin sensitivity. In cardiovascular disease, PPARγ inhibits atherosclerosis but TR4 enhances atherosclerosis through increasing foam cell formation. In bone physiology, PPARγ inhibits bone formation but TR4 increases bone formation. Together, the contrasting impact of TR4 and PPARγ on different diseases may raise a critical issue about drug used to target any one of these nuclear receptors.

ABBREVIATIONS

ACTH adrenocorticotropic hormone
AMPK adenosine monophosphate kinase
AR androgen receptor
ARA55 androgen receptor coactivator 55
CBP CREB-binding protein
CD36 cluster of differentiation 36
DBD DNA-binding domain
DR DNA response element
ER estrogen receptor
GR glucocorticoid receptor
HAT histone acetyltransferase
HREs hormone response elements
LBD ligand-binding domain
NR nuclear receptor
NR1C3 nuclear receptor superfamily 1 group C member 3
NR2C1 nuclear receptor superfamily 2 group C member 1
NR2C2 nuclear receptor superfamily 2 group C member 2
PCAF p300/CREB-binding protein-associated factor
PPARγ peroxisome proliferator-activated receptor gamma
PR progesterone receptor
PUFAs polyunsaturated fatty acids
RIP140 receptor-interacting protein 140
TR2 testicular nuclear receptor 2
TR4 testicular nuclear receptor 4
TRA TR4-associated protein
TZDs thiazolidinediones
VDR vitamin D receptor

1. INTRODUCTION

The testicular nuclear receptor 2 (*TR2*), also known as nuclear receptor superfamily 2 group C member 1 (*NR2C1*), is one of the first identified orphan receptors (Table 1). The cDNA was cloned in 1989 from human testis cDNA libraries as a new member of the nuclear receptor (NR) superfamily on the basis of homology to the DNA-binding domain (DBD) of

Table 1 Nuclear Receptors That Are Discussed in This Chapter

Gene ID	Name	Description	Alias
Nuclear Receptor			
367	NR3C4	Androgen receptor	AR
2099/2100	NR3A2	Estrogen receptor 1/2	ER
7181	NR2C1	Nuclear receptor subfamily 2 group C member 1	TR2
7182	NR2C2	Nuclear receptor subfamily 2 group C member 2	TR4
2908	NR3C1	Nuclear receptor subfamily 3 group C member 1, glucocorticoid receptor	GR
4306	NR3C2	Nuclear receptor subfamily 3 group C member 2, mineralocorticoid receptor	MCR
7025	NR2F1	Nuclear receptor subfamily 2 group F member 1	COUP-TFI
7026	NR2F2	Nuclear receptor subfamily 2 group F member 2	COUP-TFII
5241	NR3C3	Progesterone receptor	PR
5915	NR1B2	Retinoic acid receptor beta	RARβ
5469	NR1C3	Peroxisome proliferator-activated receptor gamma	PPARγ
6256	NR2B1	Retinoid X receptor alpha	RARA
7068	NR1A2	Thyroid hormone receptor beta	THRB
7421	NR1I1	Vitamin D (1,25-dihydroxyvitamin D3) receptor	VDR

steroid receptors (Chang & Kokontis, 1988; Chang et al., 1989). Later, a closely related testicular nuclear receptor 4 (*TR4*), also known as nuclear receptor superfamily 2 group C member 2 (*NR2C2*), was then cloned from human and rat hypothalamus, prostate, and testes libraries in 1994 (Chang et al., 1994).

Sequence comparison between TR4 and TR2 demonstrated that they shared a 65% identity in overall structure, with 51% homology in the N-terminus, 82% homology in the DBD, and 65% homology in the ligand-binding domain (LBD) (Chang et al., 1994). The high homology between these two orphan receptors suggests that they may constitute a unique subfamily within the NR superfamily.

The TR2 DBD shows a certain degree (50%–54%) of homology with several known NRs, including androgen receptor (AR), estrogen receptor (ER), glucocorticoid receptor (GR), progesterone receptor (PR), and mineralocorticoid receptor (Lee, Lee, & Chang, 2002). However, the TR2 LBD shows very low homology (<10%) with these same NRs (Lee et al., 2002).

Examination of structural relationships of the DBD of TR4 with other NRs shows TR2 is most closely related to TR4, followed by RXR and COUP-TF/ARP-1. All of them share the same type of direct repeat DNA response element (DR), AGGTCA, with 1 nucleotide spacing (DR1) (Chang et al., 1994).

TR2 has four isoforms with calculated molecular weights of 53 (TR2–5), 21 (TR2–7), 51 (TR2–9), and 67 (TR2–11) kDa, while TR4's molecular weight is 67 kDa. TR2 is mainly expressed in the prostate, seminal vesicle, and testis, while TR4 is expressed widely in the whole body including testis, prostate, ovary, cerebellum, and hippocampus (Bookout et al., 2006; Lee et al., 2002) and is regulated in a circadian rhythmic manner (Yang et al., 2006). TR4 locates on the chromosome 3p24, which is close to the other NRs including thyroid receptor beta (3p22–24), retinoic acid receptor beta (3p24), and peroxisome proliferator-activated receptor gamma (*PPARγ*, 3p25) (Yoshikawa, DuPont, Leach, & Detera-Wadleigh, 1996).

2. LIGANDS/ACTIVATORS THAT TRANSACTIVATE TR4

2.1 TR4 Natural and Synthetic Ligands/Activators

TR4 was initially classified as an orphan NR that regulates many physiological functions in fertility (Chen et al., 2008; Collins et al., 2004; Mu et al., 2004;

Mu, Yang, & Chang, 2006), neuron development (Chen, Collins, Uno, & Chang, 2005; Chen et al., 2007), metabolism (Kim et al., 2011; Liu et al., 2007, 2009), bone development (Lin et al., 2012), and skeletal muscle function (Liu et al., 2011). Later, during the study of the role of TR4 in controlling lipid uptake in macrophages, Xie et al. found that polyunsaturated fatty acid (PUFA), PUFA metabolites, and thiazolidinedione (TZD) might function as TR4 ligands/activators to transactivate the TR4. Peptide mapping assays revealed that PUFA metabolites, 13-hydroxyoctadecadienoic acid and 15-hydroxyeicosatetraenoic acid, could alter the TR4 conformation via direct binding to TR4 (Xie et al., 2009).

The TZD-rosiglitazone was identified as a ligand of PPARγ with the capacity to suppress diabetes (Lebovitz, 2002). The finding that TZD could also transactivate TR4 suggests these two NRs might share some similar physiological functions.

Taken together, these results suggest TR4 can function through transactivation and/or induction by these ligands/activators in either PPARγ-independent or -dependent manners to impact many pathophysiological functions, including metabolic syndrome, diabetes, and cardiovascular diseases. These conclusions were also confirmed in different studies showing the PUFA γ-linoleic acid could induce the TR4 target genes, including ApoE and PEPCK, via induction of TR4 transactivation (Mahajan et al., 2012; Tsai et al., 2009).

2.2 Phosphorylation and Acetylation Signals That Transactivate/Induce TR4

Early studies concluded that kinases/phosphatases could induce the NRs' transactivation by altering their conformation through phosphorylation or dephosphorylation (Jiang & Hunter, 1999). Several studies indicated that some phosphorylation and/or dephosphorylation signals could also influence TR4 transactivation. Kim et al. found adenosine monophosphate kinase (AMPK) could phosphorylate TR4 at serine-351 (S351), which is conserved in human, mouse, rat, and chicken, and mutation of S351 could abolish the phosphorylation (Kim et al., 2011). The consequence of such phosphorylation was to suppress the TR4 transactivation and to reduce its downstream target gene expression. The natural activator-AMP, AMPK inhibitor (compound C), and the AMP analog 5-aminoimidazole-4-carboxamide ribonucleotide could all influence the TR4 transactivation. Importantly, metformin, an antidiabetic drug for type 2 diabetes that could activate AMPK (Santomauro Junior, Ugolini, Santomauro, & Souto, 2008), could also suppress TR4 transactivation activity through AMPK-mediated TR4 phosphorylation (Kim et al., 2011).

These results suggest that via alteration of TR4 phosphorylation, metformin or newly screened small compounds may be developed as a novel therapy to better battle the metabolic syndrome and/or cardiovascular diseases.

In addition to phosphorylation, Xie et al. found that the alteration of acetylation could be another strategy to influence the TR4 transactivation (Xie et al., 2011). They found histone acetyltransferase (HAT) could acetylate TR4 at lysine-175 (K175) and lysine-176 (K176) to suppress its transactivation (Xie et al., 2011). Interestingly, the acetylation on TR4 K175/K176 is unique and is distinct from the classic function of acetylation of histones to induce downstream target genes via unwinding the chromatin for transcription (Lee & Workman, 2007). TR4 acetylation at K175 and K176, which are located in the DBD, actually suppresses TR4 binding to its HREs (Lee & Workman, 2007; Xie et al., 2011).

Furthermore, instead of enhancing AR transactivation via AR coactivator 55 (ARA55) (Fujimoto et al., 1999), Xie et al. found ARA55 could suppress the TR4 transactivation via enhanced acetylation of TR4 that involved the recruitment of HAT enzyme CREB-binding protein (CBP) (Xie et al., 2011).

2.3 Other Upstream Signals That Can Induce TR4 Expression

Several modulators can also influence TR4 function via inducing its expression. Liu et al. found cAMP/PKA signaling could induce TR4 expression by triggering CCAAT enhancer-binding proteins α and β binding to the selective cAMP response element located on the TR4 promoter (Liu et al., 2009). Retinoic acid (RA), a vitamin A metabolite, could also increase TR4 expression (Inui, Lee, Haake, Goldsmith, & Chang, 1999). The forkhead transcriptional factor, FOXO3a, a key stress-responsive factor that is negatively controlled by the PI3K/AKT pathway, could also induce the TR4 expression. The addition of selective PI3K inhibitors, including LY294002 that enhanced FOXO3a translocation and activity, could also enhance TR4 expression (Li et al., 2008).

Together, one can go through several ligands/activators and/or inducers to influence the TR4 function that may help us to better battle those TR4-related diseases.

3. TR4 DOWNSTREAM TARGET GENES

3.1 Heterodimerization With Other NRs

TR4, like many other NRs, is capable of forming heterodimers with various NRs to enhance or suppress its own or its partner's transactivation ability. TR4 prefers to form a heterodimer with TR2. The three leucine residues on helix 10 of TR2 are critical for this dimerization (Lee, Chinpaisal, & Wei, 1998). Coexpression of TR4 and TR2 can actually enhance their downstream target gene expression (Lee, Chinpaisal, et al., 1998). TR4 can also interact with thyroid hormone receptor (TR) by heterodimerization and enhance the expression of target genes downstream of the TR (Huang, Liao, Chen, Liao, & Lin, 2010). In contrast, TR4 can interact with other NRs, but in this case, heterodimerization leads to the inhibition of the function of its partner protein. With AR, heterodimerization prevents TR4 from binding to its response element and thus suppresses TR4 target genes expression (Lee, Shyr, Thin, Lin, & Chang, 1999). Interestingly, AR target gene expression was also repressed by the dimerization with TR4 (Lee, Shyr, et al., 1999). TR4 can also suppress ER transactivation through protein dimerization (Shyr, Hu, Kim, & Chang, 2002). The heterodimerization of TR4 and ER can actually suppress the homodimerization of ER itself and prevent it from binding to the estrogen response element. The consequence of this interaction would be the suppression of ER target gene expression and ER-mediated cell proliferation in MCF-7 cells (Shyr et al., 2002).

3.2 Competition for DNA-Binding Sites With Other NRs

TR4 binds specifically to the direct repeat DNA sequences located on the promoter of target genes. These DNA sequences are also the response elements for other NRs including TR, vitamin D receptor (VDR), RAR, RXR, and PPAR (Table 2) (Lee et al., 1997; Lee, Young, Burbach, & Chang, 1998; Lee, Young, et al., 1999). Through competing with other NRs for response elements, TR4 can either enhance or suppress downstream target gene expression, and TR4 increases the transcriptional induction of a reporter gene through a thyroid HRE; TR4 binding to the DR4 sequence was verified by the electrophoretic mobility shift assay (EMSA) and Scatchard analysis (Lee et al., 1997). In contrast, TR4 suppresses vitamin D3 target genes by competing with a vitamin D3 response

Table 2 TR4 Interactions With Direct Repeat Response Elements

Type of Direct Repeat	Regulated by (Nuclear Receptor Types)	Regulated by TR4 Heterodimer (Partner NR)	Examples of Downstream Genes	References
0	TR4		SCD1, NDUFAF1	Kim et al. (2011) and Liu et al. (2011)
1	TR4, RAR, RXR, PPAR	RXR	PEPCK, CD36	Liu et al. (2007) and Xie et al. (2009)
2	TR4, RAR		NDUFAF1, SV40	Lee, Lee, Burbach, and Chang (1995) and Liu et al. (2011)
3	TR4, VDR	RXR	Rat P450cc24	Lee, Young, Lin, Shyr, and Chang (1999)
4	TR4, TR		S14	Lee, Pan, Burbach, Morkin, and Chang (1997)
5	TR4, RAR	TR2, RXR	NDUFAF1, RARβ	Lee, Chinpaisal, et al. (1998) and Liu et al. (2011)

element (VDRE, DR3). Expression of Vitamin D3 target genes can be suppressed by TR4 through high affinity binding to the VDRE (Lee, Young, et al., 1999). EMSA and proteolytic analysis showed that TR4 adopts different conformations once bound to DR4–TRE or DR3–VDRE, suggesting that different conformations may allow TR4 to recruit different coregulators, therefore, either enhancing or suppressing the downstream targets (Lee, Young, et al., 1999).

TR4 also suppresses retinoid acid-induced transactivation in a dose-dependent manner via competing with the RARE/RXRE (Lee, Young, et al., 1998). In the PPAR family, TR4 plays opposite roles in terms of regulating downstream genes, i.e., suppresses PPARα targets (Yan et al., 1998), but enhances PPARγ targets (Xie et al., 2009). TR4, PPARα, and PPARγ can all bind to the DR1–PPRE. While TR4 binds to the DR1–PPRE of the PPARα target genes rat enoyl-CoA hydratase and peroxisomal fatty acetyl-CoA oxidase, it represses their expression in a dose-dependent manner (Yan et al., 1998). In contrast, TR4 binds to the DR1-PPRE of the PPARγ target gene cluster of differentiation 36 (*CD36*) to enhance its expression in a dose-dependent manner (Table 2) (Xie et al., 2009).

3.3 Interaction With Other Coregulators

To date, only one coactivator, p300/CREB-binding protein-associated factor (PCAF), has been identified that promotes TR4 transactivation. The consequences of such TR4–PCAF interaction may be to turn on the target gene, apoE (Huq, Gupta, Tsai, & Wei, 2006). Deletion of the TR4 LBD does not affect TR4 binding to PCAF, suggesting that interactions between TR4 and PCAF do not go through its LBD (Huq et al., 2006).

In contrast, there are several corepressors identified that suppress TR4 function. TIP27 functions as a TR4 corepressor via interaction with TR4 between Asp39 and Lys79. The binding of this region does not affect TR4 homodimerization or DNA binding but may affect TR4 and coactivator interaction (Nakajima, Fujino, Nakanishi, Kim, & Jetten, 2004). TR4-associated protein (TRA16) is another identified corepressor that can suppress TR4 activity via inhibiting TR4 binding to the DNA. The suppression of TR4-mediated transactivation is selective because TRA16 shows only a slight influence on the transactivation of AR, GR, and PR. TRA16 expression in the H1299 human lung cancer cell line was found to be higher than in normal human lung tissue, suggesting TR4 function is suppressed in the lung cancer (Yang et al., 2003). Receptor-interacting protein 140 (RIP140) has been shown to be a common coregulator of many NRs (Cavailles et al., 1995; Treuter, Albrektsen, Johansson, Leers, & Gustafsson, 1998). Huq et al. demonstrated RIP140 could be recruited once TR4 is phosphorylated by MAPK. The recruitment of RIP140 then suppresses TR4 transactivation activity (Huq et al., 2006). Kim et al. also reported that TR4-induced Bcl-2 gene expression can be suppressed by cotransfection with RIP140 in a dose-dependent manner (Kim, Ma, et al., 2007). Interestingly, Xie et al. found that the ARA55 coregulator might play dual roles in regulating AR vs TR4. ARA55 induced AR transactivation through direct binding (Fujimoto et al., 1999). In contrast, ARA55 inhibits TR4 function via recruitment of HAT enzyme CBP/p300 to acetylate TR4 at K175 and K176 (Xie et al., 2011).

4. TR4 ROLES IN PPARγ-RELATED DISEASES AND THEIR IMPACTS ON DRUG DEVELOPMENT

TR4 and PPARγ are two members of the NR superfamily sharing similar ligands/activators yet having distinct pathophysiological functions

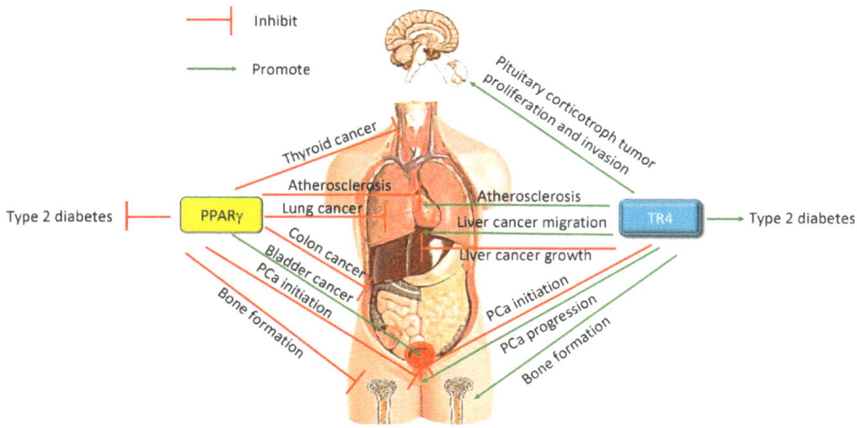

Fig. 1 The scheme of the TR4 and PPARγ functions in each organ.

in different diseases, including cancer development, metabolic syndromes, cardiovascular diseases, and bone physiology. These differing functions may complicate the future application of drug development to treat related diseases (Fig. 1).

4.1 TR4 and PPARγ in Cancer

The roles of TR4 and PPARγ in prostate cancer (PCa) are complicated. Jiang et al. reported that disruption of PPARγ signaling in prostate results in enhancing mouse prostatic intraepithelial neoplasia involving active autophagy (Jiang et al., 2010). In contrast, TR4 can be either a suppressor for PCa development or enhancer for PCa metastasis (Ding et al., 2014; Lin et al., 2014). Lin et al. found TR4 might act as a caretaker tumor suppressor to suppress PCa initiation through promoting DNA repair and maintaining genome integrity (Lin et al., 2014). In contrast, TR4 promotes various PCa cells migration/invasion via modulation of C-C motif chemokine ligand 2/C-C chemokine receptor type 2 axis, suggesting TR4 can enhance PCa metastasis (Ding et al., 2014).

Importantly, Lin et al. found that, in the absence of PPARγ, *TR4* might switch to an oncogene instead of a tumor suppressor gene to increase the stem cell population and epithelial–mesenchymal transition (Lin, Yang, et al., 2015). Reciprocally, in the absence of TR4, PPARγ may switch to an oncogene instead of a tumor suppressor gene through TZD signaling to increase HRAS expression (Lin, Lin, et al., 2015). Clinically, Lin et al. also found around 9% of PCa patients had either TR4 or PPARγ deletion

(Lin, Lin, et al., 2015; Lin, Yang, et al., 2015). Recently, TZD-pioglitazone (Actos) was suggested to increase bladder cancer risk. Yang et al. reported that about 38% of bladder cancer patients have PPARγ gene amplification and showed increased bladder cancer cell migration and invasion in vitro following TZD treatments (Yang et al., 2013). These data suggest that the imbalance of TR4 and PPARγ might contribute to the TZD-enhanced bladder cancer risk.

Besides PCa, PPARγ also acts as a tumor suppressor in tumors of the colon, breast, thyroid, and lung by inducing apoptosis (Kim, Hwang, et al., 2007; Koeffler, 2003; Philips, Petite, Willi, Buchegger, & Meier, 2004). In contrast, TR4 plays opposite roles in different cancers and different stages of cancer. In liver cancer, TR4 functions as a suppressor to inhibit liver cancer cell growth but functions as a stimulator to promote liver cancer cell migration (Huang et al., 2010). In breast cancer, TR4 can interact with ER and therefore suppress ER target genes cyclin D1 and pS2 that leads to the inhibition of ER-mediated cell proliferation in the MCF-7 cells (Shyr et al., 2002). In pituitary corticotroph tumors, TR4 is overexpressed which contributes to the transcription of the adrenocorticotropic hormone (ACTH) precursor polypeptide pro-opiomelanocortin, ACTH secretion, cellular proliferation, and tumor invasion (Du et al., 2013).

4.2 TR4 and PPARγ in Metabolic Syndromes

In the metabolic syndromes, TR4 and PPARγ function in an opposite manner, even though they share similar DNA-binding sites and activators. PPARγ increases insulin sensitivity, yet activation of TR4 decreases insulin sensitivity (Hevener et al., 2003; Kim et al., 2011; Olefsky & Saltiel, 2000). The development of insulin resistance is an early event of type 2 diabetes mellitus, and TZDs have been used to sensitize insulin through activating PPARγ (Olefsky & Saltiel, 2000). In contrast, reduced fat mass and smaller adipocyte size, low glucose levels at birth and in fasting conditions, and improved insulin sensitivity in $TR4^{-/-}$ mice suggest that loss of TR4 increases insulin sensitivity (Kim et al., 2011).

4.3 TR4 and PPARγ in Cardiovascular Diseases

In cardiovascular diseases, PPARγ, but not TR4, suppresses atherosclerosis, a lipid disorder disease in which TZD treatment can improve the lipid profiles of these patients (Subramanian, Golledge, Ijaz, Bruemmer, &

Daugherty, 2010; Xie et al., 2009). Xie et al. found that CD36 is a TR4 target gene, and thus TR4 might have direct and indirect influences on atherosclerosis via regulation of CD36 (Xie et al., 2009). They found TR4$^{-/-}$ mice have reduced foam cell formation with reduced CD36 expression and TR4 could function as a transcription factor to directly regulate CD36 gene expression in macrophage cells. In contrast with the TZD effect on atherosclerosis through PPARγ, Xie et al. showed that TZDs could activate the CD36 promoter through TR4 (Xie et al., 2009). These results suggested that TZDs might increase or decrease atherosclerosis depending on the activation of TR4 or PPARγ, which might lead to differential influence on atherosclerosis. Revealing details of the mechanisms by which these two NRs respond to TZDs in mediating lipid metabolism will be beneficial to our understanding of the complexity of TZDs activity and specificity.

4.4 TR4 and PPARγ in Bone Physiology

TR4 and PPARγ also function oppositely in bone formation. TR4 can promote bone formation, whereas PPARγ decreases bone formation. Lin et al. found that TR4$^{-/-}$ mice developed osteoporosis with reduced osteoblast activity. In vitro mechanistic dissection revealed that TR4 could transactivate osteocalcin to promote bone differentiation (Lin et al., 2012). In contrast, Marie et al. showed that PPARγ negatively regulates osteoblast differentiation, resulting in bone loss (Marie & Kaabeche, 2006). The opposite roles in bone formation between TR4 and PPARγ again strengthen the central dogma that TR4 and PPARγ need to be balanced in order to maintain the homeostasis, which may have potential implications in the treatment of bone diseases in clinical conditions.

5. SUMMARY AND FUTURE PERSPECTIVES

In early development, TR4 has been shown to play essential roles in skeleton, neuron, gonad, and skeletal muscle development (Chen et al., 2005, 2008; Collins et al., 2004; Liu et al., 2011; Mu et al., 2004). After birth, TR4 continues to play key roles in glucose and lipid metabolism (Kim et al., 2011; Liu et al., 2007). In the adult, TR4 controls the aging process and prevents the cancer initiation (Huang et al., 2010; Lee et al., 2011; Lin et al., 2014; Shyr et al., 2002). Together, these reports demonstrated that TR4 functions widely through the whole body in development and in aging. We have discussed that TR4 functions can be influenced by its cofactors and/or other NRs. The most critical example is in PCa that without PPARγ signaling TR4 switches its role to promote PCa initiation. On the other hand, without TR4 signaling, PPARγ switches its role to promote PCa by TZD stimulation. Differential TR4 and PPARγ effects were also reported in other cancers, metabolism, cardiovascular diseases, and bone physiology. With this knowledge in hand, the drug industry might be able to improve drug specificity to distinguish TR4 vs PPARγ and reduce the side effects of the old TZDs drugs.

ACKNOWLEDGMENT

This work was supported by NIH Grants (CA156700 and DK73414); Taiwan Ministry of Health and Welfare Clinical Trial and Research Center of Excellence (DOH102-TD-B-111-004) from China Medical University, Taichung, Taiwan.

Disclosure Statement: The authors have nothing to disclose.

REFERENCES

Bookout, A. L., Jeong, Y., Downes, M., Yu, R. T., Evans, R. M., & Mangelsdorf, D. J. (2006). Anatomical profiling of nuclear receptor expression reveals a hierarchical transcriptional network. *Cell, 126*(4), 789–799. http://dx.doi.org/10.1016/j.cell.2006.06.049.

Cavailles, V., Dauvois, S., L'Horset, F., Lopez, G., Hoare, S., Kushner, P. J., & Parker, M. G. (1995). Nuclear factor RIP140 modulates transcriptional activation by the estrogen receptor. *The EMBO Journal, 14*(15), 3741–3751.

Chang, C., Da Silva, S. L., Ideta, R., Lee, Y., Yeh, S., & Burbach, J. P. (1994). Human and rat TR4 orphan receptors specify a subclass of the steroid receptor superfamily. *Proceedings of the National Academy of Sciences of the United States of America, 91*(13), 6040–6044.

Chang, C., & Kokontis, J. (1988). Identification of a new member of the steroid receptor super-family by cloning and sequence analysis. *Biochemical and Biophysical Research Communications, 155*(2), 971–977.

Chang, C., Kokontis, J., Acakpo-Satchivi, L., Liao, S., Takeda, H., & Chang, Y. (1989). Molecular cloning of new human TR2 receptors: A class of steroid receptor with multiple ligand-binding domains. *Biochemical and Biophysical Research Communications, 165*(2), 735–741.

Chen, Y. T., Collins, L. L., Uno, H., & Chang, C. (2005). Deficits in motor coordination with aberrant cerebellar development in mice lacking testicular orphan nuclear receptor 4. *Molecular and Cellular Biology, 25*(7), 2722–2732.

Chen, Y. T., Collins, L. L., Uno, H., Chou, S. M., Meshul, C. K., Chang, S. S., & Chang, C. (2007). Abnormal cerebellar cytoarchitecture and impaired inhibitory signaling in adult mice lacking TR4 orphan nuclear receptor. *Brain Research, 1168*, 72–82.

Chen, L. M., Wang, R. S., Lee, Y. F., Liu, N. C., Chang, Y. J., Wu, C. C., … Chang, C. (2008). Subfertility with defective folliculogenesis in female mice lacking testicular orphan nuclear receptor 4. *Molecular Endocrinology, 22*(4), 858–867.

Collins, L. L., Lee, Y. F., Heinlein, C. A., Liu, N. C., Chen, Y. T., Shyr, C. R., … Chang, C. (2004). Growth retardation and abnormal maternal behavior in mice lacking testicular orphan nuclear receptor 4. *Proceedings of the National Academy of Sciences of the United States of America, 101*(42), 15058–15063.

Ding, X., Yang, D. R., Lee, S. O., Chen, Y. L., Xia, L. Q., Lin, S. J., … Chang, C. (2014). TR4 nuclear receptor promotes prostate cancer metastasis via up-regulation of CCL2/

CCR2 signaling. *International Journal of Cancer, 136*(4), 955–964. http://dx.doi.org/10.1002/ijc.29049.

Du, L., Bergsneider, M., Mirsadraei, L., Young, S. H., Jonker, J. W., Downes, M., … Heaney, A. P. (2013). Evidence for orphan nuclear receptor TR4 in the etiology of Cushing disease. *Proceedings of the National Academy of Sciences of the United States of America, 110*(21), 8555–8560. http://dx.doi.org/10.1073/pnas.1306182110.

Fujimoto, N., Yeh, S., Kang, H. Y., Inui, S., Chang, H. C., Mizokami, A., & Chang, C. (1999). Cloning and characterization of androgen receptor coactivator, ARA55, in human prostate. *The Journal of Biological Chemistry, 274*(12), 8316–8321.

Hevener, A. L., He, W., Barak, Y., Le, J., Bandyopadhyay, G., Olson, P., … Olefsky, J. (2003). Muscle-specific Pparg deletion causes insulin resistance. *Nature Medicine, 9*(12), 1491–1497. http://dx.doi.org/10.1038/nm956.

Huang, Y. H., Liao, C. H., Chen, R. N., Liao, C. J., & Lin, K. H. (2010). Human testicular orphan receptor 4 enhances thyroid hormone receptor signaling. *Journal of Cellular Physiology, 222*(2), 347–356. http://dx.doi.org/10.1002/jcp.21959.

Huq, M. D., Gupta, P., Tsai, N. P., & Wei, L. N. (2006). Modulation of testicular receptor 4 activity by mitogen-activated protein kinase-mediated phosphorylation. *Molecular & Cellular Proteomics, 5*(11), 2072–2082. http://dx.doi.org/10.1074/mcp.M600180-MCP200.

Inui, S., Lee, Y. F., Haake, A. R., Goldsmith, L. A., & Chang, C. (1999). Induction of TR4 orphan receptor by retinoic acid in human HaCaT keratinocytes. *The Journal of Investigative Dermatology, 112*(4), 426–431.

Jiang, M., Fernandez, S., Jerome, W. G., He, Y., Yu, X., Cai, H., … Hayward, S. W. (2010). Disruption of PPARgamma signaling results in mouse prostatic intraepithelial neoplasia involving active autophagy. *Cell Death and Differentiation, 17*(3), 469–481. http://dx.doi.org/10.1038/cdd.2009.148.

Jiang, G., & Hunter, T. (1999). Receptor signaling: When dimerization is not enough. *Current Biology, 9*(15), R568–R571.

Kim, H. J., Hwang, J. Y., Choi, W. S., Lee, J. H., Chang, K. C., Nishinaka, T., … Seo, H. G. (2007). Expression of a peroxisome proliferator-activated receptor gamma 1 splice variant that was identified in human lung cancers suppresses cell death induced by cisplatin and oxidative stress. *Clinical Cancer Research, 13*(9), 2577–2583. http://dx.doi.org/10.1158/1078-0432.CCR-06-2062. 13/9/2577 [pii].

Kim, E., Liu, N. C., Yu, I. C., Lin, H. Y., Lee, Y. F., Sparks, J. D., … Chang, C. (2011). Metformin inhibits nuclear receptor TR4-mediated hepatic stearoyl-CoA desaturase 1 gene expression with altered insulin sensitivity. *Diabetes, 60*(5), 1493–1503. http://dx.doi.org/10.2337/db10-0393.

Kim, E., Ma, W. L., Lin, D. L., Inui, S., Chen, Y. L., & Chang, C. (2007). TR4 orphan nuclear receptor functions as an apoptosis modulator via regulation of Bcl-2 gene expression. *Biochemical and Biophysical Research Communications, 361*(2), 323–328.

Koeffler, H. P. (2003). Peroxisome proliferator-activated receptor gamma and cancers. *Clinical Cancer Research, 9*(1), 1–9.

Lebovitz, H. E. (2002). Differentiating members of the thiazolidinedione class: A focus on safety. *Diabetes/Metabolism Research and Reviews, 18*(Suppl. 2), S23–S29.

Lee, C. H., Chinpaisal, C., & Wei, L. N. (1998). A novel nuclear receptor heterodimerization pathway mediated by orphan receptors TR2 and TR4. *The Journal of Biological Chemistry, 273*(39), 25209–25215. http://dx.doi.org/10.1074/jbc.273.39.25209.

Lee, H. J., Lee, Y., Burbach, J. P., & Chang, C. (1995). Suppression of gene expression on the simian virus 40 major late promoter by human TR4 orphan receptor. A member of the steroid receptor superfamily. *The Journal of Biological Chemistry, 270*(50), 30129–30133.

Lee, Y. F., Lee, H. J., & Chang, C. (2002). Recent advances in the TR2 and TR4 orphan receptors of the nuclear receptor superfamily. *The Journal of Steroid Biochemistry and Molecular Biology*, *81*(4–5), 291–308.

Lee, Y. F., Liu, S., Liu, N. C., Wang, R. S., Chen, L. M., Lin, W. J., ... Chang, C. (2011). Premature aging with impaired oxidative stress defense in mice lacking TR4. *American Journal of Physiology. Endocrinology and Metabolism*, *301*(1), E91–E98. http://dx.doi.org/10.1152/ajpendo.00701.2010.

Lee, Y. F., Pan, H. J., Burbach, J. P., Morkin, E., & Chang, C. (1997). Identification of direct repeat 4 as a positive regulatory element for the human TR4 orphan receptor. A modulator for the thyroid hormone target genes. *The Journal of Biological Chemistry*, *272*(18), 12215–12220.

Lee, Y. F., Shyr, C. R., Thin, T. H., Lin, W. J., & Chang, C. (1999). Convergence of two repressors through heterodimer formation of androgen receptor and testicular orphan receptor-4: A unique signaling pathway in the steroid receptor superfamily. *Proceedings of the National Academy of Sciences of the United States of America*, *96*(26), 14724–14729.

Lee, K. K., & Workman, J. L. (2007). Histone acetyltransferase complexes: One size doesn't fit all. *Nature Reviews. Molecular Cell Biology*, *8*(4), 284–295. http://dx.doi.org/10.1038/nrm2145.

Lee, Y. F., Young, W. J., Burbach, J. P., & Chang, C. (1998). Negative feedback control of the retinoid-retinoic acid/retinoid X receptor pathway by the human TR4 orphan receptor, a member of the steroid receptor superfamily. *The Journal of Biological Chemistry*, *273*(22), 13437–13443.

Lee, Y. F., Young, W. J., Lin, W. J., Shyr, C. R., & Chang, C. (1999). Differential regulation of direct repeat 3 vitamin D3 and direct repeat 4 thyroid hormone signaling pathways by the human TR4 orphan receptor. *The Journal of Biological Chemistry*, *274*(23), 16198–16205. http://dx.doi.org/10.1074/jbc.274.23.16198.

Li, G., Lee, Y. F., Liu, S., Cai, Y., Xie, S., Liu, N. C., ... Chang, C. (2008). Oxidative stress stimulates TR4 orphan receptor through forkhead transcription factor FOXO3a. *Endocrinology*, *149*(7), 3490–3499.

Lin, S. J., Ho, H. C., Lee, Y. F., Liu, N. C., Liu, S., Li, G., ... Chang, C. (2012). Reduced osteoblast activity in the mice lacking TR4 nuclear receptor leads to osteoporosis. *Reproductive Biology and Endocrinology*, *10*, 43. http://dx.doi.org/10.1186/1477-7827-10-43.

Lin, S. J., Lee, S. O., Lee, Y. F., Miyamoto, H., Yang, D. R., Li, G., & Chang, C. (2014). TR4 nuclear receptor functions as a tumor suppressor for prostate tumorigenesis via modulation of DNA damage/repair system. *Carcinogenesis*, *35*(6), 1399–1406. http://dx.doi.org/10.1093/carcin/bgu052.

Lin, S. J., Lin, C. Y., Yang, D. R., Izumi, K., Yan, E., Niu, X., ... Chang, C. (2015). The differential effects of anti-diabetic thiazolidinedione on prostate cancer progression are linked to the TR4 nuclear receptor expression status. *Neoplasia*, *17*(4), 339–347.

Lin, S. J., Yang, D. R., Wang, N., Jiang, M., Miyamoto, H., Li, G., & Chang, C. (2015). TR4 nuclear receptor enhances prostate cancer initiation via altering the stem cell population and EMT signals in the PPARG-deleted prostate cells. *Oncoscience*, *2*(2), 142–150.

Liu, S., Lee, Y. F., Chou, S., Uno, H., Li, G., Brookes, P., ... Chang, C. (2011). Mice lacking TR4 nuclear receptor develop mitochondrial myopathy with deficiency in complex I. *Molecular Endocrinology*, *25*(8), 1301–1310. http://dx.doi.org/10.1210/me.2010-0455.

Liu, N. C., Lin, W. J., Kim, E., Collins, L. L., Lin, H. Y., Yu, I. C., ... Chang, C. (2007). Loss of TR4 orphan nuclear receptor reduces phosphoenolpyruvate carboxykinase-mediated gluconeogenesis. *Diabetes*, *56*(12), 2901–2909.

Liu, N. C., Lin, W. J., Yu, I. C., Lin, H. Y., Liu, S., Lee, Y. F., & Chang, C. (2009). Activation of TR4 orphan nuclear receptor gene promoter by cAMP/PKA and C/EBP signaling. *Endocrine*, *36*(2), 211–217. http://dx.doi.org/10.1007/s12020-009-9220-0.

Mahajan, S., Dkhar, H. K., Chandra, V., Dave, S., Nanduri, R., Janmeja, A. K., ... Gupta, P. (2012). Mycobacterium tuberculosis modulates macrophage lipid-sensing nuclear receptors PPARgamma and TR4 for survival. *The Journal of Immunology*, *188*(11), 5593–5603. http://dx.doi.org/10.4049/jimmunol.1103038.

Marie, P. J., & Kaabeche, K. (2006). PPAR gamma activity and control of bone mass in skeletal unloading. *PPAR Research*, *2006*, 64807. http://dx.doi.org/10.1155/PPAR/2006/64807.

Mu, X., Lee, Y. F., Liu, N. C., Chen, Y. T., Kim, E., Shyr, C. R., & Chang, C. (2004). Targeted inactivation of testicular nuclear orphan receptor 4 delays and disrupts late meiotic prophase and subsequent meiotic divisions of spermatogenesis. *Molecular and Cellular Biology*, *24*(13), 5887–5899. http://dx.doi.org/10.1128/MCB.24.13.5887-5899.2004.

Mu, X., Yang, L., & Chang, C. (2006). Stage dependent and androgen inductive expression of orphan receptor TR4 in rat testis. *Biochemical and Biophysical Research Communications*, *341*(2), 464–469.

Nakajima, T., Fujino, S., Nakanishi, G., Kim, Y. S., & Jetten, A. M. (2004). TIP27: A novel repressor of the nuclear orphan receptor TAK1/TR4. *Nucleic Acids Research*, *32*(14), 4194–4204.

Olefsky, J. M., & Saltiel, A. R. (2000). PPAR gamma and the treatment of insulin resistance. *Trends in Endocrinology and Metabolism*, *11*(9), 362–368.

Philips, J. C., Petite, C., Willi, J. P., Buchegger, F., & Meier, C. A. (2004). Effect of peroxisome proliferator-activated receptor gamma agonist, rosiglitazone, on dedifferentiated thyroid cancers. *Nuclear Medicine Communications*, *25*(12), 1183–1186.

Santomauro Junior, A. C., Ugolini, M. R., Santomauro, A. T., & Souto, R. P. (2008). Metformin and AMPK: An old drug and a new enzyme in the context of metabolic syndrome. *Arquivos Brasileiros de Endocrinologia e Metabologia*, *52*(1), 120–125.

Shyr, C. R., Hu, Y. C., Kim, E., & Chang, C. (2002). Modulation of estrogen receptor-mediated transactivation by orphan receptor TR4 in MCF-7 cells. *The Journal of Biological Chemistry*, *277*(17), 14622–14628.

Subramanian, V., Golledge, J., Ijaz, T., Bruemmer, D., & Daugherty, A. (2010). Pioglitazone-induced reductions in atherosclerosis occur via smooth muscle cell-specific interaction with PPAR{gamma}. *Circulation Research*, *107*(8), 953–958. http://dx.doi.org/10.1161/CIRCRESAHA.110.219089.

Treuter, E., Albrektsen, T., Johansson, L., Leers, J., & Gustafsson, J. A. (1998). A regulatory role for RIP140 in nuclear receptor activation. *Molecular Endocrinology*, *12*(6), 864–881.

Tsai, N. P., Huq, M., Gupta, P., Yamamoto, K., Kagechika, H., & Wei, L. N. (2009). Activation of testicular orphan receptor 4 by fatty acids. *Biochimica et Biophysica Acta*, *1789*(11–12), 734–740. http://dx.doi.org/10.1016/j.bbagrm.2009.09.010.

Xie, S., Lee, Y. F., Kim, E., Chen, L. M., Ni, J., Fang, L. Y., ... Chang, C. (2009). TR4 nuclear receptor functions as a fatty acid sensor to modulate CD36 expression and foam cell formation. *Proceedings of the National Academy of Sciences of the United States of America*, *106*(32), 13353–13358. http://dx.doi.org/10.1073/pnas.0905724106.

Xie, S., Ni, J., Lee, Y. F., Liu, S., Li, G., Shyr, C. R., & Chang, C. (2011). Increased acetylation in the DNA-binding domain of TR4 nuclear receptor by the coregulator ARA55 leads to suppression of TR4 transactivation. *The Journal of Biological Chemistry*, *286*(24), 21129–21136. http://dx.doi.org/10.1074/jbc.M110.208181.

Yan, Z. H., Karam, W. G., Staudinger, J. L., Medvedev, A., Ghanayem, B. I., & Jetten, A. M. (1998). Regulation of peroxisome proliferator-activated receptor alpha-induced transactivation by the nuclear orphan receptor TAK1/TR4. *The Journal of Biological Chemistry*, *273*(18), 10948–10957.

Yang, X., Downes, M., Yu, R. T., Bookout, A. L., He, W., Straume, M., ... Evans, R. M. (2006). Nuclear receptor expression links the circadian clock to metabolism. *Cell*, *126*(4), 801–810. http://dx.doi.org/10.1016/j.cell.2006.06.050.

Yang, D. R., Lin, S. J., Ding, X. F., Miyamoto, H., Messing, E., Li, L. Q., … Chang, C. (2013). Higher expression of peroxisome proliferator-activated receptor gamma or its activation by agonist thiazolidinedione-rosiglitazone promotes bladder cancer cell migration and invasion. *Urology*, *81*(5). http://dx.doi.org/10.1016/j.urology.2012.12.027. 1109 e1101-1106.

Yang, Y., Wang, X., Dong, T., Kim, E., Lin, W. J., & Chang, C. (2003). Identification of a novel testicular orphan receptor-4 (TR4)-associated protein as repressor for the selective suppression of TR4-mediated transactivation. *The Journal of Biological Chemistry*, *278*(9), 7709–7717.

Yoshikawa, T., DuPont, B. R., Leach, R. J., & Detera-Wadleigh, S. D. (1996). New variants of the human and rat nuclear hormone receptor, TR4: Expression and chromosomal localization of the human gene. *Genomics*, *35*(2), 361–366.

CHAPTER FOURTEEN

The Role of COUP-TFII in Striated Muscle Development and Disease

Xin Xie*[,1], San-Pin Wu‡[,1], Ming-Jer Tsai*[,†,2], Sophia Tsai*[,†,2]
*Baylor College of Medicine, Houston, TX, United States
†Program in Developmental Biology, Baylor College of Medicine, Houston, TX, United States
‡Reproductive and Developmental Biology Laboratory, National Institute of Environmental Health Sciences, National Institute of Health, Research Triangle Park, NC, United States
[2]Corresponding authors: e-mail address: mtsai@bcm.tmc.edu; stsai@bcm.tmc.edu

Contents

Abstract

Skeletal and cardiac muscles are the only striated muscles in the body. Although sharing many structural and functional similarities, skeletal and cardiac muscles have intrinsic differences in terms of physiology and regenerative potential. While skeletal muscle possesses a robust regenerative response, the mammalian heart has limited repair capacity after birth. In this review, we provide an updated view regarding chicken ovalbumin upstream promoter-transcription factor II (COUP-TFII) function in vertebrate

[1] These authors contributed equally to this work.

Current Topics in Developmental Biology, Volume 125
ISSN 0070-2153
http://dx.doi.org/10.1016/bs.ctdb.2016.12.006

myogenesis, with particular emphasis on the skeletal and cardiac muscles. We also highlight the new insights of COUP-TFII hyperactivity underlying striated muscle dysfunction. Lastly, we discuss the challenges and strategies in translating COUP-TFII action for clinical intervention.

1. INTRODUCTION

The mammalian body contains two types of striated muscles, skeletal and cardiac muscles. The primary function of striated muscles is to generate force for locomotion, posture maintenance, and breathing (skeletal muscle) and for blood circulation (cardiac muscle). While skeletal muscles enable fast contraction and discontinuous voluntary action, cardiac muscles are capable of autonomous contraction. Despite having the similar "striated" pattern containing sarcomere, the microscopic structures of skeletal and cardiac muscles are different. Skeletal muscle fibers form by the fusion of individual myocytes, which generate multinucleated linear fibers, ranging from 0.1 cm to more than 30 cm in length. By contrast, cardiac muscles can be of mononucleated, binucleated, or multinucleated cells, and have relatively shorter length of about 0.1 mm long (Liu, Yue, Chen, Kubin, & Braun, 2010; Mollova et al., 2013). In the adult human, skeletal muscles comprise 30%–40% of body mass with an average weight ~27 kg, while cardiac muscles are much smaller, weighing approximately 270 g. With respect to energy metabolism, skeletal muscle is capable of powering contraction via aerobic and anaerobic metabolism. Conversely, cardiac muscle mainly relies on oxidative phosphorylation in the mitochondria to generate ATP (Stanley, Recchia, & Lopaschuk, 2005). In addition, healthy skeletal muscles hold a remarkable capacity for self-repair due to the resident muscle stem cell–satellite cell (SC) (Yin, Price, & Rudnicki, 2013). On the contrary, the adult mammals have an extremely limited or little ability for cardiac regeneration despite the discovery of endogenous cardiac stem/progenitor cells in the postnatal heart (Xiang & Kikuchi, 2016).

Chicken ovalbumin upstream promoter-transcription factor II (COUP-TFII), also known as NR2F2, is an orphan member of the nuclear receptor (NR) superfamily as specific ligands for COUP-TFII have not yet been identified (Tsai & Tsai, 1997). COUP-TFII protein is broadly and dynamically expressed in multiple tissues throughout fetal development and COUP-TFII-null mutants suffered early embryonic lethality due to deficient angiogenesis and heart development (Pereira, Qiu, Zhou,

Tsai, & Tsai, 1999). Consistently, COUP-TFII modulates a large variety of downstream targets involved in embryonic development and differentiation of the three germ layers (Lin, Qin, Tang, Tsai, & Tsai, 2011; Tsai & Tsai, 1997; Xie, Tang, Yu, Tsai, & Tsai, 2013). In adult animals, whole-body ablation of COUP-TFII does not generate discernible phenotypes under physiological conditions (unpublished results), consistent with the observation that COUP-TFII protein is preferentially expressed in undifferentiated stem/progenitor cells rather than mature epithelium of most structures (Xie et al., 2013). The effects of COUP-TFII in the development of the vasculature (Lin et al., 2011), of mesenchymal progenitors (Xie et al., 2013), and of eye (Tang, Tsai, & Tsai, 2015) have been reviewed previously and are beyond the scope of this report. Instead, we highlight some recent data on new and unexpected actions of COUP-TFII in striated muscle development and diseases. Given that many aspects of tissue regeneration recapitulate embryonic development, the role of COUP-TFII in the adult stages of muscle repair will be discussed as well.

2. COUP-TFII FUNCTIONS IN SKELETAL MUSCLE DEVELOPMENT

Skeletal myogenesis is initiated during early embryonic stages and continues throughout the neonatal growth period. In mammals, skeletal muscles of the body, excluding those of the head, arise from precursors residing in the somite, a transitory condensation of mesoderm cells that forms on either side of the neural tube and notochord (Tajbakhsh, 2003). As the somites mature, they generate three distinct cellular compartments, the dermomyotome, the myotome, and the sclerotome. Later on, the Pax3/Pax7 positive cells that originated from the dermomyotome and myotome migrate into limb buds, where they proliferate, commit to myogenic lineage, and subsequently differentiate into skeletal myocytes, which will further mature within the first 3–4 weeks after birth (Kassar-Duchossoy et al., 2005). Moreover, those somite-derived precursors also give rise to the adult muscle stem/progenitor cells, termed SCs (Bailey, Holowacz, & Lassar, 2001; Relaix, Rocancourt, Mansouri, & Buckingham, 2005). Several reports using genetic strategies to ablate the $Pax7^+$ population have demonstrated an absolute requirement of SCs for muscle repair and the renewal of myogenic lineage (Lepper, Partridge, & Fan, 2011; Murphy, Lawson, Mathew, Hutcheson, & Kardon, 2011; Sambasivan et al., 2011).

2.1 COUP-TFII Regulates Limb Myogenesis

Skeletal muscle in the mouse is established from embryonic day 8.5 (E8.5) to E18.5 (birth ~19 days). As the first step to dissect its biological significance in fetal myogenesis, we investigated the developmental profiles of the COUP-TFII protein in the murine embryo. Using *COUP-TFII-LacZ* knock-in mice in which a nuclear LacZ gene was placed under the control of the COUP-TFII regulatory region (Takamoto et al., 2005), COUP-TFII expression was first seen in the somites at E9.5, and persisted in myotome and dermomyotome at E10.5 (Lee et al., 2004). By E11.5, COUP-TFII protein was highly enriched in limb bud mesenchyme of myogenic lineage (Xie, Qin, Lin, Tsai, & Tsai, 2011). Chimera analyses using wild type and COUP-TFII-null embryonic stem cells revealed a cell-autonomous effect of COUP-TFII in limb bud outgrowth but not limb bud initiation (Lee et al., 2004), matching prior reports that the COUP-TFII$^{-/-}$ embryo, although it died before E10.5, still had an early limb bud (Pereira et al., 1999).

To circumvent the early embryonic lethality, Lee et al. employed *Prx1* Cre mice (Logan et al., 2002) to specifically ablate the *COUP-TFII* gene in the developing limb. Owing to the late onset of *Prx1* expression and its inadequate Cre recombinase activity, the COUP-TFII mutants had only mild phenotypes with hypoplastic musculature in the limb (Lee et al., 2004). Although it was apparent that the number of migrating progenitors was greatly reduced in the conditional mutants, targets of COUP-TFII, which presumably act within Pax3^{+}/Pax7^{+} population, remained undefined (Lee et al., 2004). Multiple factors, including hepatocyte growth factor/scatter factor and its receptor c-Met, the homeobox factor Lbx1 and Msx1, are essential to control the migratory behaviors of muscle progenitors (Birchmeier & Brohmann, 2000; Gross et al., 2000; Houzelstein et al., 1999; Schäfer & Braun, 1999), therefore, further studies are needed to clarify the molecular basis through which COUP-TFII coordinates the muscle cell migration.

Additional evidence pointing to a role of COUP-TFII in fetal muscle development emerged from gain-of-function studies. *Pax7*, a member of the paired/homeodomain family of transcriptional factors, has an early onset expression at E9.5 in the dermomyotome (Kassar-Duchossoy et al., 2005; Olguín & Pisconti, 2012). Enforced expression of COUP-TFII protein in Pax7 positive precursors caused profound deficits in fetal myogenesis and the mutant animals died at birth (unpublished results). A similar approach using *Myf5* Cre driving COUP-TFII activation also led to the loss of mature muscle fibers and postnatal death (unpublished results), confirming a detrimental effect on muscle development with COUP-TFII hyperactivity. In line with its inhibitory role on myogenesis, COUP-TFII heterozygous mice had small but significant increase in muscle mass compared with the wild-type controls (Li et al., 2009). Collectively, the above-described evidence suggests that a careful balance in COUP-TFII signaling is essential for appropriate embryonic myogenesis.

2.2 COUP-TFII Specifies Myogenic Fate of Mesenchymal Precursors

Prenatal muscle development consists of two phases, primary and secondary myogenesis, which are initiated from a pool of progenitor cells with mesodermal origin (Bailey et al., 2001). These multipotent mesenchymal cells can be committed to myogenic lineage as well as other cell types including adipocytes, cartilage, endothelial cell, and connective tissues (Almalki & Agrawal, 2016). Thus, genes involved in fate decision of mesenchymal precursors will have actual impacts on skeletal myogenesis.

The abundant expression of COUP-TFII protein in the centralized core of E11.5 and E12.5 limb mesenchyme provided the first clue to its function in mesenchymal cell development (Xie et al., 2011). Following this observation, we examined the differentiation potential of COUP-TFII-depleted cells, and found that inactivation of COUP-TFII promoted myogenic development of mesenchymal precursors (Xie et al., 2011). Work from other laboratories also documented the negative effects of COUP-TFII in myogenesis (Bailey, Sartorelli, Hamamori, & Muscat, 1998; Muscat, Rea, & Downes, 1995). In mesenchymal cells, the conversions of multipotent precursors to different mesodermal lineages are competitive or mutually exclusive (Almalki & Agrawal, 2016). COUP-TFII deficient cells failed to enter adipogenic and chondrogenic paths but adopted alternative cell fates such as myoblast and osteoblast (Lee et al., 2012; Xie et al., 2011), supporting a critical involvement of COUP-TFII in lineage specification of mesenchymal progenitors.

More evidences supporting COUP-TFII's action in mesenchymal cell destiny came from experiments searching for upstream regulators that modify COUP-TFII activity. For instance, microRNA-194, microRNA-302a, and FGF2 all act through COUP-TFII to define the osteogenic and adipogenic decisions of mesenchymal precursors (Jeong, Kang, Hwang,

Kim, & Koh, 2014; Kang et al., 2015; Lee et al., 2016). Of note, a recent work examining COUP-TFII function in cartilage also demonstrated a positive effect of COUP-TFII in driving chondrogenesis in human mesenchymal cells (Gao, Zhang, Hubbell, & Cui, 2017). More importantly, those in vitro studies were mirrored by the in vivo findings as COUP-TFII-deficient mice had higher bone density and increased muscle mass coupled with reduced fat adiposity and deficient cartilage formation (Li et al., 2009; Xie et al., 2011). Mechanistically, COUP-TFII specified the mesenchymal cell identity through direct control of tissue-specific transcription factors (Runx2, PPARγ, and Sox9) and canonical Wnt cascade (Li et al., 2009; Xie et al., 2011).

It is increasingly clear that SCs are multipotent precursors capable of developing into different mesodermal lineages including bone, fibroblast, and fat (Asakura, Komaki, & Rudnicki, 2001; Odelberg, Kolhoff, & Keating, 2000; Seale et al., 2008). Whether COUP-TFII plays a similar role in SC lineage specification is currently unknown. Nevertheless, our studies and others clearly ascertain a key role for COUP-TFII in defining cell fates of mesenchymal precursors.

3. COUP-TFII FUNCTIONS IN SKELETAL MUSCLE REGENERATION

Skeletal muscle, like other renewable tissue, relies on a stem/progenitor cell population to maintain tissue integrity. In vertebrates, this function is conferred by the SC, a descendant of myogenic progenitors with somitic origin (Scharner & Zammit, 2011). Therefore, morphogenetic pathways that dictate fetal myogenesis are largely redeployed for tissue reconstruction after trauma (Wang & Rudnicki, 2011). As discussed earlier, COUP-TFII actively participated in fetal muscle development, it is hence conceivable that COUP-TFII may function in regenerative myogenesis as well.

3.1 COUP-TFII Affects SC Activation and Proliferation

Skeletal muscle regeneration represents a highly defined sequence of events comprising SC activation, proliferation, differentiation, and fusion (Fig. 1). In response to damage signals, SCs exit their quiescent state and activate *MyoD*, a uniform marker of the activated stem cells and their descendant progenitors. *MyoD* is a master regulator of myogenesis, and its expression alone is sufficient to convert mesenchymal-like cells into muscle. The activation of SCs depends on the expression of MyoD protein and COUP-TFII

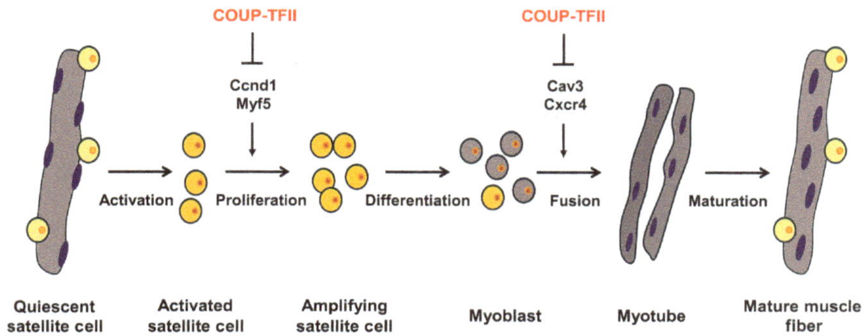

Fig. 1 Working model of COUP-TFII function in regenerative myogenesis. COUP-TFII regulates genes involved in cell proliferation and myoblast fusion to control satellite cell differentiation.

is known to inhibit MyoD activity transcriptionally and posttranscriptionally (Bailey et al., 1998; Muscat et al., 1995), raising an interesting possibility that COUP-TFII may affect stem cell activation. In keeping with this rationale, loss of COUP-TFII promoted SC activation and an elevated MyoD$^+$ population were seen in SC-specific COUP-TFII null muscles 2 days after injury (unpublished results). Consistently, COUP-TFII knockout mice displayed a hastened regenerative response with accelerated formation of bona fide myocytes (unpublished results).

Despite the induction of MyoD protein, activated SCs do not differentiate immediately, instead they undergo extensive division to provide the myonuclear input for regenerating fibers. One apparent defect in COUP-TFII-expressing SCs was the defective cell cycle control (Xie, Tsai, & Tsai, 2016). Two days after acute injury, an obvious proliferative failure was found in SCs carrying the COUP-TFII transgene (Xie et al., 2016). Accordingly, the Pax7$^+$ cell population was markedly reduced in COUP-TFII-overexpressing mice compared with control littermates (Xie et al., 2016). More definitive evidence was derived from in vitro cultures of primary SCs, whereby COUP-TFII-overexpressing myoblasts had a significantly slower rate of cell propagation compared to the control cells (Xie et al., 2016). The inhibition of stem cell expansion, at least in part, attributed to the dysregulation of *Ccnd1* and *Myf5* (Xie et al., 2016), key genes supporting myoblast proliferation during tissue reconstitution (Panda et al., 2016; Ustanina, Carvajal, Rigby, & Braun, 2007).

More recently, many NRs have been reported to be expressed in SCs and regulate stem cell behavior. Mutation of thyroid hormone receptor

(TR) α adversely affected myoblast proliferation and muscle regeneration (Milanesi et al., 2016). On the other hand, excessive intracellular T_3, demonstrated by SC-specific depletion of the *dio3* gene (the thyroid hormone-inactivating enzyme), led to extensive SC apoptosis and delayed muscle recovery (Dentice et al., 2014). Additionally, the peroxisome proliferator-activated receptor (PPAR) β/δ has been shown to promote myoblast expansion, thereby facilitating the SC-mediated reparative process (Angione, Jiang, Pan, Wang, & Kuang, 2011; Chandrashekar et al., 2015). Another NR, estrogen receptor β (ERβ), but not ERα, appeared to stimulate SC activation/proliferation during tissue repair (Velders, Schleipen, Fritzemeier, Zierau, & Diel, 2012). Furthermore, androgen signaling may also contribute to SC development as removal of endogenous androgen (gonadectomy) delayed stem cell activation in the early recovery phase (MacKrell et al., 2015).

3.2 COP-TFII Disrupts Myoblast Fusion In Vivo

The fusion of singly nucleated myoblasts into multinucleate myotubes is one of the most unusual aspects of skeletal myogenesis (Abmayr & Pavlath, 2012; Hindi, Tajrishi, & Kumar, 2013). Muscle cell fusion is fundamental in muscle growth during development and in fiber repair upon injury. The initial evidence for a role of COUP-TFII in myoblast fusion arose from the finding that conditional activation of COUP-TFII in the perinatal period impaired muscle development, causing a 25% reduction in muscle mass at advanced age (Xie et al., 2016). A general concept in skeletal myogenesis is that postnatal muscle growth is mainly due to the increase in myofiber size with little or no formation of new fibers (Tajbakhsh, 2009). In agreement, histological assays uncovered a 40% reduction in fiber cross-sectional area (CSA) in COUP-TFII-expressing myocytes at 20 months old (Xie et al., 2016), providing a connection between COUP-TFII hyperactivity and muscle cell fusion.

Myoblast fusion during regeneration is the best-studied model to examine the actual fusion events in vivo. Consistent with the above-mentioned developmental defects, forced expression of COUP-TFII attenuated muscle cell fusion, and the mean CSA of newborn fibers were drastically smaller in COUP-TFII transgenic animals than that in wild-type muscles (Xie et al., 2016). Lineage analysis tracing Myc-tagged COUP-TFII transgene confirmed a direct contribution of COUP-TFII hyperactivity to the fusion deficits (Xie et al., 2016). Most strikingly, high levels of COUP-TFII did not just temporarily delay, but permanently blocked the fusion of regenerating myocytes while maintaining the expression of early differentiation markers—eMyHC and myogenin (Xie et al., 2016). For the first time, we describe an animal model with complete destruction of myoblast fusion, which positions COUP-TFII being central in cell–cell fusion and muscle maturation. Despite the notable differences between forming muscle during embryo development, and rebuilding it after injury, COUP-TFII functions in myoblast proliferation and fusion are conserved in these two processes.

To unravel the molecular targets mediating COUP-TFII action, we screened fusion-related genes and uncovered *Cav3* and *Cxcr4* as direct targets of COUP-TFII (Xie et al., 2016). Cell migration is the first step prior to fusion and cultured myoblasts migrate extensively in response to chemoregulatory molecules (Abmayr & Pavlath, 2012). One chemokine–receptor pair, *Cxcr4* and its ligand *Sdf1*, regulated cell motility in vitro (Griffin, Apponi, Long, & Pavlath, 2010). Consistently, an in vivo function of *Cxcr4* in governing myogenic cell migration has been validated experimentally (Vasyutina et al., 2005). Caveolin-3 is a major mediator of β1D integrin-focal adhesion kinase pathway in mammalian myoblast fusion (Gazzerro, Sotgia, Bruno, Lisanti, & Minetti, 2010; Quach, Biressi, Reichardt, Keller, & Rando, 2009) and Cav3-deficient myoblasts did not fuse to myotubes in differentiating medium (Galbiati, Volonte, Engelman, Scherer, & Lisanti, 1999). Thus, by suppressing *Cav3* and *Cxcr4* transcription, COUP-TFII impeded the maturation of differentiating myoblasts. However, the repression of *Cav3* and *Cxcr4* can only contribute, but cannot alone account for COUP-TFII's action, because muscle fusion is far more defective in COUP-TFII transgenic mice than that of Cav3 or Cxcr4 mutants by itself. As such, large-scale genomic approaches such as microarray, RNA-seq, and ChIP-seq will be particularly informative in dissecting the molecular circuitry impacted by COUP-TFII and is a current research focus in the laboratory. Overall, our data paints a picture in which COUP-TFII lies upstream of *Ccnd1*, *Myf5*, *Cav3*, and *Cxcxr4* to control the cellular properties of SCs and regenerative myogenesis (Fig. 1).

In addition to COUP-TFII, other NRs are also implicated in muscle cell fusion. For example, skeletal muscles from TR mutants ($TR\alpha^{-/-}$ and $TR\beta^{-/-}$) displayed a significant reduction in their fiber diameters compared to that of wild-type animals (Milanesi et al., 2016; Yu et al., 2000). Accordingly, a fusion defect in regenerating myocytes was identified in TRα null mice suffering transient damage (Milanesi et al., 2016). Likewise, vitamin D receptor (VDR) may also function in myoblast fusion as

VDR-deficient mice exhibited structurally immature myotubes with aberrant expression of early differentiation markers, including myogenin, Myf5, as well as embryonic and neonatal myosin heavy chain (Bouillon, Gielen, & Vanderschueren, 2014; Endo et al., 2003).

Although myoblast fusion is a unique feature of muscle fibers, the biological processes and the molecular machinery driving cell fusion remain elusive, due to the difficulty to examine the fusion process in vivo. Recently, live-imaging systems that monitor the entire reparative program from injury to fiber replacement have been devised in zebrafish and mice (Gurevich et al., 2016; Webster, Manor, Lippincott-Schwartz, & Fan, 2016), it will be interesting to apply those technologies to directly visualize the behavior of COUP-TFII-expressing cells in vivo. Learning from such studies not only increases our knowledge of myogenesis, but also provides prospects on membrane and cell fusion for different organisms and cell types.

3.3 COUP-TFII Controls SC Commitment and Self-Renewal

SCs are heterogeneous in nature as they differ in gene expression, differentiation potential, and stemness (Yin et al., 2013). For instance, it has been proposed that approximately 10% of Pax7$^+$ cells, which have never expressed the myogenic commitment factor *Myf5*, represent true satellite stem cells (Kuang, Kuroda, Le Grand, & Rudnicki, 2007). SC proliferation not only assures a sufficient pool of progenitors necessary for muscle repair, but also reconstitutes the stem cell reservoir for later rounds of regeneration. As noted earlier, COUP-TFII clearly inhibits SC expansion during regeneration (Xie et al., 2016), it is thus possible that COUP-TFII may influence the replenishment of the stem cell niche. In support of this view, in a murine model of muscular dystrophy, there was an enhanced SC homeostasis in diaphragm muscles lacking COUP-TFII, whereas dystrophic muscles carrying the COUP-TFII minigene had the exact opposite behavior (Xie et al., 2016). Interestingly, an involvement of COUP-TFII in maintaining the self-renewing capacity and pluripotency was recently proven in human mesenchymal cells (Zhu et al., 2016).

SCs can divide and self-renew in two manners, asymmetric and symmetric cell division (Yin et al., 2013). Asymmetric division of SC is a fundamental property sustaining the stem cell pool and members of Wnt and Notch family proteins are specifically important for this process (Wang & Rudnicki, 2011). Given the multiple reports of COUP-TFII regulating Notch and Wnt signaling molecules (Li et al., 2009; Xie et al., 2011; You et al.,

2005), COUP-TFII is a potential candidate for the maintenance of SC stemness. In dividing precursors, asymmetric distribution of transcription factors, such as *MyoD* and *Numb* (Notch antagonist), directs the fate of SC progeny (Conboy, Karasov, & Rando, 2007; Conboy & Rando, 2002; Shinin, Gayraud-Morel, Gomès, & Tajbakhsh, 2006). In a preliminary analysis, COUP-TFII protein was asymmetrically located in some but not all daughter cells after cell division (unpublished results), raising a speculation that COUP-TFII may influence fate decision of daughter cells. In skeletal muscle, $Myf5^-$ SCs preferentially self-renew to replenish the stem cell niche, whereas $Myf5^+$ precursors more likely enter the differentiation program to support regenerative myogenesis (Kuang et al., 2007). Owing to the cell-intrinsic inhibition of Myf5 transcription, SCs bearing the COUP-TFII transgene were less committed to become bona fide myoblasts compared to its control counterparts (Xie et al., 2016). Taken together, the aforementioned observations point to a functional link between COUP-TFII dysregulation and SC self-renewal, which needs to be investigated further in the future.

3.4 Distinct Acute and Chronic Regenerative Responses in COUP-TFII Mutant SCs

Skeletal muscle regenerates in response to two types of injury, acute damage and chronic muscle diseases. A widely accepted view is that the same genetic hierarchy that controls the acute regenerative program, also happens during chronic muscle damage. Unexpectedly, COUP-TFII activation seems to provoke distinct regenerative responses under these two injury paradigms. In response to chronic muscle diseases, such as muscular dystrophy, COUP-TFII hyperactivity limited the muscle fate commitment as we barely detected $Myf5^+$ precursors as well as central-nucleated newborn fibers in dystrophic COUP-TFII overexpression mice (Xie et al., 2016). On the contrary, following a transient insult, COUP-TFII-expressing SCs were able to enter myogenic path but lineage progression was arrested at the fusion stage (Xie et al., 2016). At present, we do not know the precise cause for this discrepancy. A plausible interpretation invokes the variable degree of damage elicited by these two injury models. Genetic defect induces mild and progressive damage, and muscle tissue maintains a relatively intact environment. On the other hand, chemical injury triggers massive trauma and generates a hostile microenvironment in which the stem cell niche is largely destroyed. An alternative explanation might involve the loss of dystrophin protein in the dystrophic animals. Although both acute and chronic reparative processes are mediated by SCs, the fact that COUP-TFII-expressing SCs exhibit different behaviors strongly suggest distinct regenerative mechanisms may underlie these two injury paradigms.

4. COUP-TFII HYPERACTIVITY, SC DYSFUNCTION, AND MUSCULAR DYSTROPHY

Muscular dystrophies, a group of inheritable neuromuscular disorders, are some of the more difficult diseases for treatment because myonuclei are mitotically inactive and cannot replicate (Wallace & McNally, 2009). Duchenne muscular dystrophy (DMD), the most common and severe form of muscular dystrophy, is caused by mutations in the dystrophin gene (Hoffman, Brown, & Kunkel, 1987). Loss of dystrophin in patients renders myofibers highly susceptible to contraction-induced injury, and triggers repetitive degeneration/regeneration cycles, which gradually leads to muscle weakness, and eventually results in progressive paralysis and premature death (Bushby et al., 2010).

Although dystrophin loss in myofibers initiates and drives the pathological cascades, recent findings from Drs. Blau's and Rudnicki's group have suggested a direct contribution of SCs to DMD pathogenesis (Dumont et al., 2015; Sacco et al., 2010), raising an interesting concept that DMD is also a stem cell disease. Our studies strongly support this perception as activation of COUP-TFII in juvenile SCs prior to quiescence impaired SC development and provoked Duchenne-like phenotypes that progressively worsened with increasing age (Xie et al., 2016). During the perinatal growth period, $Pax7^+$ progenitors extensively proliferate, actively differentiate, and continuously fuse with existing fibers to support muscle growth (White, Biérinx, Gnocchi, & Zammit, 2010). SCs analyzed from COUP-TFII transgenic animals exhibited deficient cell proliferation and myoblast fusion, suggesting that the dystrophic phenotypes were primarily due to the developmental defects of $Pax7^+$ precursors (Xie et al., 2016). As such, SC dysfunction induced by COUP-TFII derepression is indeed a causative mechanism of muscular dystrophy. Reinforcing this, the dystrophic animals carrying the COUP-TFII transgene displayed exacerbated myopathies compared with the control counterparts (Xie et al., 2016), reflecting a synergistic effect when SC dysfunction cooperates with dystrophin loss. More importantly, blocking of COUP-TFII activity may confer a therapeutic benefit in muscular dystrophy as COUP-TFII loss stimulated SC function, attenuated dystrophic response, and counteracted dystrophin loss-induced muscle weakness in the murine model (Xie et al., 2016), suggesting that targeting COUP-TFII is a potential treatment strategy for degenerative muscle diseases. Taken together, our studies support a direct involvement of SCs in the DMD etiology, and uncover COUP-TFII as a novel contributing factor underlying muscular dystrophy.

5. COUP-TFII FUNCTIONS IN CARDIAC MUSCLE DEVELOPMENT

The process of forming a heart includes specification of cardiac progenitor cells from multiple sources, differentiation of precursor cells into nascent cardiomyocytes, determining the fate of myocyte subtypes, and maturation of cardiac muscles. Beginning at gastrulation, a subset of cells in the lateral mesoderm forms the early cardiogenic mesoderm that gives rise to the cardiac crescent, the primitive heart tube and then the left ventricle and portions of the atria (Evans, Yelon, Conlon, & Kirby, 2010). Apart from aforementioned cells from the first heart field, the developing heart is also joined by cardiac progenitors that reside in the second heart field (SHF), which contribute to the future atrial myocardium, the right ventricle, and the outflow tract (Meilhac, Lescroart, Blanpain, & Buckingham, 2014). A third cohort of precursors originated from the proepicardium, located ventrally to the sinus venosus, migrate to the surface of the looping heart, form epicardium that promotes myocardial growth, and provide cells that become cardiac fibroblasts and coronary vessels (Maya-Ramos, Cleland, Bressan, & Mikawa, 2013). During cardiac muscle development, progenitor cells lose the *Isl1* expression once they migrate into the heart tube, develop into nascent cardiomyocytes, and express early myocyte markers—*SMαA* and *SM22*. Subsequently, the nascent myocytes undergo a maturation process to acquire distinct morphological, metabolic, and electrical properties to form adult cardiomyocytes.

5.1 COUP-TFII Regulates Tubular Heart Development

Expression of COUP-TFII appeared in mouse embryos as early as E8.5 at the sinus venosus (Pereira et al., 1999). At E9.5, COUP-TFII was found in Isl1-positive cardiac progenitors of the posterior SHF as well as in SMαA-positive nascent cardiomyocytes of the sinus venosus and common atrium (Lin et al., 2012; Wu, Cheng, et al., 2013; Wu, Yu, Tsai, & Tsai, 2016). The importance of COUP-TFII in tubular heart development was first demonstrated by the COUP-TFII null mice, which failed to expand both the sinus venosus and the atrial compartment, leading to a dysmorphic heart (Pereira et al., 1999). Recent findings imply that the cause of this cardiac abnormality could, at least partially, result from defects in the SHF as described later. Lineage tracing experiments have shown that progenitor cells from the posterior SHF migrate into the heart tube and give rise to atrial myocytes (Dominguez, Meilhac, Bland, Buckingham, & Brown, 2012; Galli et al., 2008). Given the reported roles of COUP-TFII in controlling cell proliferation and survival in different cell types (Chen, Qin, Cheng, Tsai, & Tsai, 2012; Lin et al., 2010; Qin et al., 2013; Yu et al., 2012) as well as its expression in the Isl1-positive progenitors, it is reasonable to speculate that loss of COUP-TFII may have a negative impact on the cardiac progenitor cells in SHF, leading to a failure in supplying cells to the developing heart tube (Fig. 2). Furthermore, ablating COUP-TFII in the nascent atrial cardiomyocytes did not recapitulate the hypoplastic inflow tract phenotype found in COUP-TFII null mice (Wu, Cheng, et al., 2013), suggesting that the primary role of COUP-TFII in atrial cardiomyocytes is likely for fate determination (discussed later) instead of promoting cardiomyocyte proliferation.

The COUP-TFII-dependent inflow tract phenotype was also observed in the retinoic acid deficient animals. It has been shown that mice with pharmacological blockade or ablation of the retinaldehyde dehydrogenase, a key enzyme for retinoic acid production, resulted in hypoplastic atria around E9 (Niederreither et al., 2001; Xavier-Neto et al., 1999). Conversely, retinoic acid treatment expanded the expression domain of atrial-specific markers

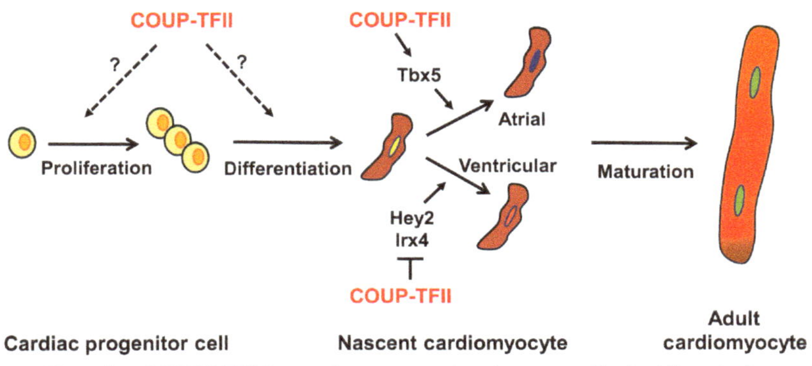

Fig. 2 The role of COUP-TFII in cardiomyocyte development. *Dashed lines* indicate the potential functions of COUP-TFII in directing proliferation and differentiation of cardiac progenitors in the posterior second heart field. COUP-TFII confers atrial identity by promoting Tbx5 and repressing Hey2 and Irx4 expression in nascent cardiomyocytes.

(Liberatore, Searcy-Schrick, & Yutzey, 2000; Yutzey, Rhee, & Bader, 1994). Moreover, prior studies have documented that retinoic acid may act through retinoid X receptor (RXR)/retinoic acid receptor (RAR) to induce COUP-TFII expression (Tsai & Tsai, 1997), directly bind to and modulate COUP-TFII activities (Kruse et al., 2008), or promote gene expression via COUP-TFII-dependent RXR–CREB binding protein interaction (Lin et al., 2000). Collectively, these observations implicate a possible link between COUP-TFII and retinoic acid signaling in the inflow tract development, which awaits further investigation in the future.

Recently, a sinus venosus hypoplasia was documented in *Dll4* heterozygote mutants, similar to that shown in *COUP-TFII* null mice (Duarte et al., 2004; Pereira et al., 1999). Considering COUP-TFII and Notch signaling were known to be mutually repressive during angiogenesis and adipogenesis (Chen et al., 2012; Diez et al., 2007; Scroyen, Bauters, Vranckx, & Lijnen, 2015; You et al., 2005), the interplay between these two signaling cascades on the expansion of sinus venosus myocardium constitutes an interesting topic for further investigation. Another gene found in the sinus venosus myocardium during tubular heart development is *PDGFα* (Bax et al., 2010). Deletion of *PDGFα* reduced the volume of the atrial and sinus venosus myocardium (Bax et al., 2010), despite that the phenotypes were less severe than that of the COUP-TFII$^{-/-}$ embryo (Pereira et al., 1999). Given that both *PDGFα* and *COUP-TFII* are expressed in the sinus venosus myocardium and share the similar inflow tract phenotype, it seems thus possible that a *PDGFα–COUP-TFII* interaction may underlie the development of inflow tract myocardium.

5.2 Fate Determination of Atrial Cardiomyocytes by COUP-TFII

Once the progenitor cells from SHF enter the myocardium, they differentiate into nascent cardiomyocytes at venous and arterial poles, adopting atrial and ventricular fates, respectively. While sharing the same function as contraction force providers, the atrial and ventricular cardiomyocytes differ in structural and operational aspects, reflecting the neurohormonally responsive nature of atria and the powerful pumping character of ventricles (Barth et al., 2005; de Bold, Borenstein, Veress, & Sonnenberg, 1981). Numerous reports have demonstrated that transcription factors *COUP-TFII* and *Tbx5* determined the atrial identity, whereas *Hey2*, *Irx4*, and *Nkx2.5/ Nkx2.7* maintained the ventricular fate (Bao, Bruneau, Seidman, Seidman, & Cepko, 1999; Bruneau, Bao, et al., 2001; Bruneau, Nemer, et al., 2001; Koibuchi & Chin, 2007; Lin et al., 2012; Targoff et al., 2013; Wu, Cheng, et al., 2013; Xin et al., 2007). In cardiomyocytes, COUP-TFII proteins were exclusively found in the atrial compartments and the COUP-TFII gene locus remains active there throughout the developmental course (Lin et al., 2012; Wu, Cheng, et al., 2013). Loss of COUP-TFII specifically in mouse cardiac muscles suppressed the transcripts of *Tbx5* as well as approximately 50% of atrial-specific genes, concomitant with gained expression of *Hey2*, *Irx4*, and 40% of ventricular markers (Wu, Cheng, et al., 2013) (Fig. 2). Amazingly, in addition to the key regulators *Tbx5*, *Hey2*, and *Irx4*, COUP-TFII binding was also detected in nearly one-third of its downstream targets (Wu, Cheng, et al., 2013), establishing a dominant role of COUP-TFII in regulating a broad spectrum of genes to shape the atrial cardiomyocytes. As a result, COUP-TFII loss transformed the atrial cardiomyocyte in cell dimensions, subcellular organelle organization, and electrical properties that resembled its ventricular counterpart. Furthermore, such an identity shift was cell autonomous and only permissible up to around E12.5 (Wu, Cheng, et al., 2013), suggesting that an unknown mechanism might kick in and lock the cell fate after an initial decision for atrial identity defined by COUP-TFII.

A recent study in zebrafish discovered the importance of Nkx2.5/Nkx2.7 in conferring the ventricular destiny (Targoff et al., 2013), raising a question on the relationship between COUP-TFII and Nkx genes in the decision of cardiomyocyte identity. Current findings support a context-dependent functional interaction between *COUP-TFII* and *Nkx* genes. In Drosophila, *seven-up* (a COUP-TF ortholog) suppresses *tinman* (an Nkx orthologue), and specifies cardiac precursor cells that give rise to ostia, while *tinman* determines the remaining part of a heart (Lo & Frasch, 2001; Reim, Mohler, & Frasch, 2005). In contrast, mouse *Nkx2.5* expression is not altered in the ventricularized atria upon COUP-TFII ablation (Wu, Cheng, et al., 2013). Interestingly, Tinman binding sites on the *seven-up* promoter were required for the activation of the promoter (Ryan, Hendren, Helander, & Cripps, 2007), suggesting that Tinman might also positively regulate *seven-up*. The interaction of such a complex genetic network that confers the fate of cardiomyocytes has yet to be fully elucidated.

It is still unknown how COUP-TFII participates in the diverging process that separate atrial myocardium and pacemaker lineages. Emerging data support a role of COUP-TFII in preventing nascent cardiomyocytes to enter the pacemaker path because COUP-TFII can bind to and repress the expression of *Id2*, key factor defining the cardiac conduction system

(Moskowitz et al., 2007; Wu, Cheng, et al., 2013). Moreover, in mouse embryonic atria, enriched COUP-TFII occupancies were seen in the genomic loci of *Tbx3* and *Tbx18*, two master regulators determining pacemaker specification (Kapoor, Liang, Marban, & Cho, 2012; Wiese et al., 2009; Wu, Dong, Regan, Su, & Majesky, 2013). Therefore, future studies are necessary to dissect the molecular interactions between COUP-TFII and aforementioned major regulators in cardiac cell fate determination.

5.3 COUP-TFII-Dependent Myocardial Growth

Epicardium contributes to myocardial growth via mitogenic paracrine signals and serves as a cell source for coronary vessels and cardiac fibroblasts (Braitsch & Yutzey, 2013). COUP-TFII protein was detected in the epicardium and its predecessor–proepicardium (Lin et al., 2012). Loss-of-function studies revealed a fundamental role for COUP-TFII in epicardium morphogenesis, epicardial-to-mesenchymal transformation, and subsequent coronary vessel formation (Lin et al., 2012). Consistently, epicardial-specific COUP-TFII mutants failed to develop compact myocardium, underscoring a requirement of epicardial COUP-TFII signaling for myocardium expansion (Lin et al., 2012). Several mechanisms may underlie the COUP-TFII action. First, loss of COUP-TFII in epicardium might affect cardiac fibroblast production, which is known to promote cardiac muscle proliferation via the β1 integrin signaling (Ali et al., 2014; Ieda et al., 2009; Moore-Morris et al., 2014). Given that the epicardial COUP-TFII depletion led to decreased numbers of WT1-positive cells in the ventricular myocardium (Lin et al., 2012), the defective compact myocardium formation might, partially if not all, result from the reduced mitogenic signal released by the cardiac fibroblasts. Second, COUP-TFII deficiency might impair retinoic acid mediated epicardial–myocardial crosstalk. Retinoic acid signaling on myocardial growth and embryo development has been extensively studied (Mark, Ghyselinck, & Chambon, 2009; Zaffran, Robrini, & Bertrand, 2014). Epicardial-derived retinoic acid signaling, presumably acting through NRs (RXRα, RARα, and RARγ) stimulated production of mitogenic FGF2 in both epicardial and myocardial compartments. Retinoic acid also promoted epicardial expression of FGF9, which induced cardiomyocyte proliferation via FGFR1 and FGFR2 (Lavine et al., 2005; Mendelsohn et al., 1994; Merki et al., 2005; Sucov et al., 1994). The phenotypic similarities between mutants of COUP-TFII and RARs strongly suggest that COUP-TFII and retinoic acid signaling may cooperate to control compact myocardium formation. A third aspect, the impaired coronary vessel development could hinder the myocardial compaction process. The COUP-TFII mutants displayed abnormal primitive vascular plexus and compromised vessel remodeling, which likely reduced blood supply and affected myocardial development (Lin et al., 2012). In summary, epicardial COUP-TFII may regulate myocardial growth through paracrine signaling that directs cardiomyocyte proliferation and coronary vessel formation.

Although interactions between COUP-TFII and major epicardial regulators are largely undefined, emerging data suggest potential links among COUP-TFII, *WT1*, *Tbx18*, and *Tcf21*. For instance, WT1 knockout mice exhibited deficient epicardial-to-mesenchymal transformation and myocardial noncompaction as observed in the COUP-TFII mutants (Lin et al., 2012; von Gise et al., 2011). Consistently, WT1 promotes epicardial retinoic acid signaling (Guadix et al., 2011). Although COUP-TFII has been shown to promote *WT1* transcription in the kidney (Yu et al., 2012), such regulation was not clear in COUP-TFII-depleted epicardial cells (Lin et al., 2012), indicating other factors aside from COUP-TFII may modulate *WT1* in the epicardium. In addition, *Tcf21* ablation also resulted in defective epicardial-to-mesenchymal transition and cardiac dysmorphogenesis (Acharya et al., 2012). Considering that epicardial *Tcf21* expression is regulated by retinoic acid (Braitsch, Combs, Quaggin, & Yutzey, 2012), and the interaction between COUP-TFII and retinoic acid signaling and the importance of COUP-TFII in epicardial-to-mesenchymal transformation (Lin et al., 2012), it is reasonable to speculate that *COUP-TFII* and *Tcf21* may act together in the epicardium. *Tbx18* is another epicardial transcription factor that marks progenitor cells for future cardiac fibroblasts and vascular smooth muscle (Cai et al., 2008). Interestingly, Tbx18 mutants displayed aberrant blebbing epicardium and impaired coronary vessel remodeling, phenotypes that resembled the COUP-TFII-deficient mice (Lin et al., 2012; Wu, Dong, et al., 2013). Furthermore, COUP-TFII occupancy was found in the *Tbx18* genomic locus in mouse embryonic hearts (Wu, Cheng, et al., 2013). Together, these observations point to a relevance of COUP-TFII and Tbx18 in epicardium homeostasis.

6. COUP-TFII OVEREXPRESSION, MITOCHONDRIA DYSFUNCTION, AND HEART FAILURE

While COUP-TFII is highly expressed and exerts profound effects in various aspects during embryogenesis, its expression often declines to basal levels in healthy adult tissues. Similarly, in the normal heart, COUP-TFII protein is predominantly expressed in the atria and its level is very low in the ventricular cardiomyocytes. However, increased COUP-TFII expression was seen in pathological states such as prostate cancer and heart failure in humans (Kittleson et al., 2005; Qin et al., 2013; Sack, Disch, Rockman, & Kelly, 1997). The causal effect of heightened COUP-TFII levels in adult ventricles was assessed in a mouse model via overexpressing COUP-TFII specifically in cardiac muscles after heart cell maturation (Wu et al., 2015). COUP-TFII overexpression resulted in dilated cardiomyopathy with decreased contractility, and eventually heart failure. From a mechanistic perspective, COUP-TFII repressed, directly or indirectly, the transcription of *PGC-1α*, *PGC-1β*, *ERRα*, *ERRγ*, and *PPARα*, well-studied regulators for fatty acid oxidation pathways (Eichner & Giguere, 2011; Riehle & Abel, 2012). In addition, COUP-TFII also adversely affected the expression of key glycolytic genes including *Glut4*, *Hk2*, and *Pfkm* (Wu et al., 2015). The concurrent suppression of both fatty acid and glucose utilization pathways resulted in loss of fuel usage flexibility and decreased energy production that were essential for normal myocardial function. On the other hand, mutant hearts showed increased production of reactive oxygen species (ROS), presumably due to reduced expression of electron transport chain and ROS scavenger genes, which subsequently led to mitochondrial dysfunction through excess production of and inefficient clearance of ROS (Wu et al., 2015). Normally, a Pink1-dependent mitochondria quality control program will remove damaged mitochondria, followed by Mfn1, Mfn2, and Opa1-dependent fission and fusion processes to generate new mitochondria, thereby sustaining mitochondrial homeostasis (Dorn & Kitsis, 2014). However, due to the reduced expression of these mitochondria dynamic genes, damaged mitochondria were accumulated in COUP-TFII overexpressing hearts, rendering it unable to maintain healthy mitochondrial population and oxidative phosphorylation ability (Wu et al., 2015). In conclusion, in diseased adult cardiomyocytes, COUP-TFII suppresses muscle contraction via limiting the fuel utilization and damaging mitochondria homeostasis.

7. CONCLUDING REMARKS AND FUTURE PERSPECTIVES

In this review, we present the most current aspects regarding COUP-TFII function in the context of striated muscle development, maintenance, and pathology. Consistent with its spatiotemporal expression pattern, COUP-TFII is indispensable for embryonic myogenesis, while its aberrant activities in adults are associated with muscle disorders including muscular dystrophy and cardiomyopathy (Wu, Cheng, et al., 2013; Wu et al., 2015; Xie et al., 2016).

The discovery that COUP-TFII regulates SC properties has provided new mechanistic insights into the etiology of muscular dystrophy. However, some questions concerning COUP-TFII's role in skeletal muscle remain to be clarified. For instance, cells derived from non-Pax7^{+} lineages such as bone marrow-derived mesenchymal cells and pericytes, where COUP-TFII is highly expressed, also possess myogenic potential and participate in muscle reconstitution (Yin et al., 2013), it is unclear whether COUP-TFII is also involved in Pax7^{-} cell-mediated reparative processes. Fibrosis is a specific hallmark of advanced muscular dystrophy, and muscle-resident mesenchymal progenitors are presumably the major source of fibroblasts (Pessina et al., 2015; Uezumi et al., 2011). It is, thus, interesting to examine whether and how COUP-TFII controls the fibrogenic commitment of mesenchymal cells in the setting of muscular dystrophy. As eluded to earlier, COUP-TFII may restrict the myogenic entry of SCs. If that is the case, what is the fate of stem cells that are unable to enter the myogenic path? Although our studies focus on the action of COUP-TFII in the stem cell population, emerging evidence suggests that aberrant COUP-TFII activity on myofibers may also influence muscle performance. In that scenario, is there a function of COUP-TFII beyond stem/progenitor cell stage?

Similar to its critical impact on skeletal muscles, COUP-TFII also serves as a versatile regulator controlling cardiac cell destiny, intercellular communication, cell transformation, and energy metabolism. As noted, COUP-TFII is expressed in the SHF. Then, what is the mechanism that restricts COUP-TFII expression in the SHF and what does COUP-TFII do there? A big hurdle for engineering stem cell-derived cardiomyocytes is the maturation of myocytes to the extent similar to real adult cardiomyocytes (Furtado, Nim, Boyd, & Rosenthal, 2016). Interestingly, the COUP-TFII-dependent switch of chamber identity is lost between E12.5 and E15.5, as only activation of COUP-TFII via the early expressing SM22-cre can reprogram the embryonic ventricular myocytes toward atrial fate, but not the Myh6-ERT2-driving COUP-TFII expression at 2 months of age (Wu, Cheng, et al., 2013 and unpublished results). Therefore, the failure in altering cardiomyocyte identity is more likely due to closure of window of plasticity, rather than issues regarding COUP-TFII overexpression levels, suggesting there are multiple steps between nascent and fully matured cardiomyocytes. A closer examination on the timely change of myocyte properties, perhaps by Drop-seq, in response to COUP-TFII expression should reveal the missing piece to achieve full maturation.

Despite the facts that we present compelling data establishing the critical roles of COUP-TFII in muscular dystrophy and dilated cardiomyopathy in animal models, is COUP-TFII really a major player in disease progression in the natural pathological context in humans? How is COUP-TFII expression induced in patients suffering muscular dystrophy and cardiomyopathy? Furthermore, does the aberrant COUP-TFII activity found in SCs extend to other tissues that contribute to muscular dystrophy pathophysiology? It has become clear that dystrophic patients develop dilated cardiomyopathy with aging (Kamdar & Garry, 2016), and increased COUP-TFII activity in cardiac myocytes elicits heart failure (Wu et al., 2015), raising a possibility that dystrophin loss in heart muscles may activate COUP-TFII expression which likely contributes to cardiac dilation in patients. Finally, the causal effects of COUP-TFII hyperactivity in muscular dystrophy and heart failure as well as the protective effects of COUP-TFII depletion against striated muscle disorders (Wu et al., 2015; Xie et al., 2016), raise the question of whether blunting of COUP-TFII signaling after onset of diseases would improve skeletal and cardiac function in patients. Given that COUP-TFII harbors a ligand-binding pocket whose activity can be modulated by small, drug-like molecules (Kruse et al., 2008), pharmacological blockade of COUP-TFII action using its inhibitors/antagonists may hold promising potential for the treatment of muscular dystrophy and its associated cardiomyopathy.

ACKNOWLEDGMENTS

We thank Ms. Jodie R. Hebert for critical review of the manuscript. This work was supported by grants from NIH DK59820 and HL114539 to S.Y.T. and M.J.T., DK45641 to M.J.T.

REFERENCES

Abmayr, S. M., & Pavlath, G. K. (2012). Myoblast fusion: Lessons from flies and mice. *Development*, *139*, 641–656.

Acharya, A., Baek, S. T., Huang, G., Eskiocak, B., Goetsch, S., Sung, C. Y., et al. (2012). The bHLH transcription factor Tcf21 is required for lineage-specific EMT of cardiac fibroblast progenitors. *Development*, *139*, 2139–2149.

Ali, S. R., Ranjbarvaziri, S., Talkhabi, M., Zhao, P., Subat, A., Hojjat, A., et al. (2014). Developmental heterogeneity of cardiac fibroblasts does not predict pathological proliferation and activation. *Circulation Research*, *115*, 625–635.

Almalki, S. G., & Agrawal, D. K. (2016). Key transcription factors in the differentiation of mesenchymal stem cells. *Differentiation, 21*, S0301–S4681.
Angione, A. R., Jiang, C., Pan, D., Wang, Y. X., & Kuang, S. (2011). PPARdelta regulates satellite cell proliferation and skeletal muscle regeneration. *Skeletal Muscle, 1*, 33.
Asakura, A., Komaki, M., & Rudnicki, M. (2001). Muscle satellite cells are multipotential stem cells that exhibit myogenic, osteogenic, and adipogenic differentiation. *Differentiation, 68*, 245–253.
Bailey, P., Holowacz, T., & Lassar, A. B. (2001). The origin of skeletal muscle stem cells in the embryo and the adult. *Current Opinion in Cell Biology, 13*, 679–689.
Bailey, P., Sartorelli, V., Hamamori, Y., & Muscat, G. E. (1998). The orphan nuclear receptor, COUP-TF II, inhibits myogenesis by post-transcriptional regulation of MyoD function: COUP-TF II directly interacts with p300 and myoD. *Nucleic Acids Research, 26*, 5501–5510.
Bao, Z. Z., Bruneau, B. G., Seidman, J. G., Seidman, C. E., & Cepko, C. L. (1999). Regulation of chamber-specific gene expression in the developing heart by Irx4. *Science, 283*, 1161–1164.
Barth, A. S., Merk, S., Arnoldi, E., Zwermann, L., Kloos, P., Gebauer, M., et al. (2005). Functional profiling of human atrial and ventricular gene expression. *Pflügers Archiv, 450*, 201–208.
Bax, N. A., Bleyl, S. B., Gallini, R., Wisse, L. J., Hunter, J., Van Oorschot, A. A., et al. (2010). Cardiac malformations in Pdgfralpha mutant embryos are associated with increased expression of WT1 and Nkx2.5 in the second heart field. *Developmental Dynamics, 239*, 2307–2317.
Birchmeier, C., & Brohmann, H. (2000). Genes that control the development of migrating muscle precursor cells. *Current Opinion in Cell Biology, 12*, 725–730.
Bouillon, R., Gielen, E., & Vanderschueren, D. (2014). Vitamin D receptor and vitamin D action in muscle. *Endocrinology, 155*, 3210–3213.
Braitsch, C. M., Combs, M. D., Quaggin, S. E., & Yutzey, K. E. (2012). Pod1/Tcf21 is regulated by retinoic acid signaling and inhibits differentiation of epicardium-derived cells into smooth muscle in the developing heart. *Developmental Biology, 368*, 345–357.
Braitsch, C., & Yutzey, K. (2013). Transcriptional control of cell lineage development in epicardium-derived cells. *Journal of Developmental Biology, 1*, 92.
Bruneau, B. G., Bao, Z. Z., Fatkin, D., Xavier-Neto, J., Georgakopoulos, D., Maguire, C. T., et al. (2001). Cardiomyopathy in Irx4-deficient mice is preceded by abnormal ventricular gene expression. *Molecular and Cellular Biology, 21*, 1730–1736.
Bruneau, B. G., Nemer, G., Schmitt, J. P., Charron, F., Robitaille, L., Caron, S., et al. (2001). A murine model of Holt-Oram syndrome defines roles of the T-box transcription factor Tbx5 in cardiogenesis and disease. *Cell, 106*, 709–721.
Bushby, K., Finkel, R., Birnkrant, D. J., Case, L. E., Clemens, P. R., Cripe, L., et al. (2010). Diagnosis and management of Duchenne muscular dystrophy, part 1: Diagnosis, and pharmacological and psychosocial management. *Lancet Neurology, 9*, 77–93.
Cai, C., Martin, J., Sun, Y., Cui, L., Wang, L., Ouyang, K., et al. (2008). A myocardial lineage derives from Tbx18 epicardial cells. *Nature, 454*, 104–108.
Chandrashekar, P., Manickam, R., Ge, X., Bonala, S., McFarlane, C., Sharma, M., et al. (2015). Inactivation of PPARβ/δ adversely affects satellite cells and reduces postnatal myogenesis. *American Journal of Physiology. Endocrinology and Metabolism, 309*, E122–E131.
Chen, X., Qin, J., Cheng, C. M., Tsai, M. J., & Tsai, S. Y. (2012). COUP-TFII is a major regulator of cell cycle and Notch signaling pathways. *Molecular Endocrinology, 26*, 1268–1277.
Conboy, M. J., Karasov, A. O., & Rando, T. A. (2007). High incidence of non-random template strand segregation and asymmetric fate determination in dividing stem cells and their progeny. *PLoS Biology, 5*, 1120–1126.

Conboy, I. M., & Rando, T. A. (2002). The regulation of Notch signaling controls satellite cell activation and cell fate determination in postnatal myogenesis. *Developmental Cell, 3*, 397–409.
de Bold, A. J., Borenstein, H. B., Veress, A. T., & Sonnenberg, H. (1981). A rapid and potent natriuretic response to intravenous injection of atrial myocardial extract in rats. *Life Sciences, 28*, 89–94.
Dentice, M., Ambrosio, R., Damiano, V., Sibilio, A., Luongo, C., Guardiola, O., et al. (2014). Intracellular inactivation of thyroid hormone is a survival mechanism for muscle stem cell proliferation and lineage progression. *Cell Metabolism, 20*, 1038–1048.
Diez, H., Fischer, A., Winkler, A., Hu, C. J., Hatzopoulos, A. K., Breier, G., et al. (2007). Hypoxia-mediated activation of Dll4-Notch-Hey2 signaling in endothelial progenitor cells and adoption of arterial cell fate. *Experimental Cell Research, 313*, 1–9.
Dominguez, J. N., Meilhac, S. M., Bland, Y. S., Buckingham, M. E., & Brown, N. A. (2012). Asymmetric fate of the posterior part of the second heart field results in unexpected left/right contributions to both poles of the heart. *Circulation Research, 111*, 1323–1335.
Dorn, G. W., & Kitsis, R. N. (2014). The mitochondrial dynamism-mitophagy-cell death interactome: Multiple roles performed by members of a mitochondrial molecular ensemble. *Circulation Research, 116*, 167–182.
Duarte, A., Hirashima, M., Benedito, R., Trindade, A., Diniz, P., Bekman, E., et al. (2004). Dosage-sensitive requirement for mouse Dll4 in artery development. *Genes and Development, 18*, 2474–2478.
Dumont, N. A., Wang, Y. X., von Maltzahn, J., Pasut, A., Bentzinger, C. F., Brun, C. E., et al. (2015). Dystrophin expression in muscle stem cells regulates their polarity and asymmetric division. *Nature Medicine, 21*, 1455–1463.
Eichner, L. J., & Giguere, V. (2011). Estrogen related receptors (ERRs): A new dawn in transcriptional control of mitochondrial gene networks. *Mitochondrion, 11*, 544–552.
Endo, I., Inoue, D., Mitsui, T., Umaki, Y., Akaike, M., Yoshizawa, T., et al. (2003). Deletion of vitamin D receptor gene in mice results in abnormal skeletal muscle development with deregulated expression of myoregulatory transcription factors. *Endocrinology, 144*, 5138–5144.
Evans, S. M., Yelon, D., Conlon, F. L., & Kirby, M. L. (2010). Myocardial lineage development. *Circulation Research, 107*, 1428–1444.
Furtado, M. B., Nim, H. T., Boyd, S. E., & Rosenthal, N. A. (2016). View from the heart: Cardiac fibroblasts in development, scarring and regeneration. *Development, 143*, 387–397.
Galbiati, F., Volonte, D., Engelman, J. A., Scherer, P. E., & Lisanti, M. P. (1999). Targeted down-regulation of caveolin-3 is sufficient to inhibit myotube formation in differentiating C2C12 myoblasts. Transient activation of p38 mitogen-activated protein kinase is required for induction of caveolin-3 expression and subsequent myotube formation. *Journal of Biological Chemistry, 274*, 30315–30321.
Galli, D., Dominguez, J. N., Zaffran, S., Munk, A., Brown, N. A., & Buckingham, M. E. (2008). Atrial myocardium derives from the posterior region of the second heart field, which acquires left-right identity as Pitx2c is expressed. *Development, 135*, 1157–1167.
Gao, G., Zhang, X. F., Hubbell, K., & Cui, X. (2017). NR2F2 regulates chondrogenesis of human mesenchymal stem cells in bioprinted cartilage. *Biotechnology and Bioengineering, 114*(1), 208–216.
Gazzerro, E., Sotgia, F., Bruno, C., Lisanti, M. P., & Minetti, C. (2010). Caveolinopathies: From the biology of caveolin-3 to human diseases. *European Journal of Human Genetics, 18*, 137–145.
Griffin, C. A., Apponi, L. H., Long, K. K., & Pavlath, G. K. (2010). Chemokine expression and control of muscle cell migration during myogenesis. *Journal of Cell Science, 123*, 3052–3060.

Gross, M. K., Moran-Rivard, L., Velasquez, T., Nakatsu, M. N., Jagla, K., & Goulding, M. (2000). Lbx1 is required for muscle precursor migration along a lateral pathway into the limb. *Development*, *127*, 413–424.

Guadix, J. A., Ruiz-Villalba, A., Lettice, L., Velecela, V., Munoz-Chapuli, R., Hastie, N. D., et al. (2011). Wt1 controls retinoic acid signaling in embryonic epicardium through transcriptional activation of Raldh2. *Development*, *138*, 1093–1097.

Gurevich, D. B., Nguyen, P. D., Siegel, A. L., Ehrlich, O. V., Sonntag, C., Phan, J. M., et al. (2016). Asymmetric division of clonal muscle stem cells coordinates muscle regeneration in vivo. *Science*, *353*(6295), aad9969-1–aad9969-11.

Hindi, S. M., Tajrishi, M. M., & Kumar, A. (2013). Signaling mechanisms in mammalian myoblast fusion. *Science Signaling*, *6*, 1–9.

Hoffman, E. P., Brown, R. H., Jr., & Kunkel, L. M. (1987). Dystrophin: The protein product of the Duchenne muscular dystrophy locus. *Cell*, *51*, 919–928.

Houzelstein, D., Auda-Boucher, G., Chéraud, Y., Rouaud, T., Blanc, I., Tajbakhsh, S., et al. (1999). The homeobox gene Msx1 is expressed in the lateral dermomyotome of a subset of somites, and in muscle progenitor cells migrating into the forelimb. *Development*, *126*, 2689–2701.

Ieda, M., Tsuchihashi, T., Ivey, K. N., Ross, R. S., Hong, T. T., Shaw, R. M., et al. (2009). Cardiac fibroblasts regulate myocardial proliferation through beta1 integrin signaling. *Developmental Cell*, *16*, 233–244.

Jeong, B. C., Kang, I. H., Hwang, Y. C., Kim, S. H., & Koh, J. T. (2014). MicroRNA-194 reciprocally stimulates osteogenesis and inhibits adipogenesis via regulating COUP-TFII expression. *Cell Death & Disease*, *5*, e1532.

Kamdar, F., & Garry, D. J. (2016). Dystrophin-deficient cardiomyopathy. *Journal of the American College of Cardiology*, *67*, 2533–2546.

Kang, I. H., Jeong, B. C., Hur, S. W., Choi, H., Choi, S. H., Ryu, J. H., et al. (2015). MicroRNA-302a stimulates osteoblastic differentiation by repressing COUP-TFII expression. *Journal of Cellular Physiology*, *230*, 911–921.

Kapoor, N., Liang, W., Marban, E., & Cho, H. C. (2012). Direct conversion of quiescent cardiomyocytes to pacemaker cells by expression of Tbx18. *Nature Biotechnology*, *31*, 54–62.

Kassar-Duchossoy, L., Giacone, E., Gayraud-Morel, B., Jory, A., Gomes, D., & Tajbakhsh, S. (2005). Pax3/Pax7 mark a novel population of primitive myogenic cells during development. *Genes and Development*, *19*, 1426–1431.

Kittleson, M. M., Minhas, K. M., Irizarry, R. A., Ye, S. Q., Edness, G., Breton, E., et al. (2005). Gene expression analysis of ischemic and nonischemic cardiomyopathy: Shared and distinct genes in the development of heart failure. *Physiological Genomics*, *21*, 299–307.

Koibuchi, N., & Chin, M. T. (2007). CHF1/Hey2 plays a pivotal role in left ventricular maturation through suppression of ectopic atrial gene expression. *Circulation Research*, *100*, 850–855.

Kruse, S. W., Suino-Powell, K., Zhou, X. E., Kretschman, J. E., Reynolds, R., Vonrhein, C., et al. (2008). Identification of COUP-TFII orphan nuclear receptor as a retinoic acid-activated receptor. *PLoS Biology*, *6*, e227.

Kuang, S., Kuroda, K., Le Grand, F., & Rudnicki, M. A. (2007). Asymmetric self-renewal and commitment of satellite stem cells in muscle. *Cell*, *129*, 999–1010.

Lavine, K. J., Yu, K., White, A. C., Zhang, X., Smith, C., Partanen, J., et al. (2005). Endocardial and epicardial derived FGF signals regulate myocardial proliferation and differentiation in vivo. *Developmental Cell*, *8*, 85–95.

Lee, K. N., Jang, W. G., Kim, E. J., Oh, S. H., Son, H. J., Kim, S. H., et al. (2012). Orphan nuclear receptor chicken ovalbumin upstream promoter-transcription factor II (COUP-TFII) protein negatively regulates bone morphogenetic protein 2-induced

osteoblast differentiation through suppressing runt-related gene 2 (Runx2) activity. *Journal of Biological Chemistry, 287*, 18888–18899.
Lee, M. N., Kim, J. W., Oh, S. H., Jeong, B. C., Hwang, Y. C., & Koh, J. T. (2016). FGF2 stimulates COUP-TFII expression via the MEK1/2 pathway to inhibit osteoblast differentiation in C3H10T1/2 cells. *PLoS One, 11*, 7.
Lee, C. T., Li, L., Takamoto, N., Martin, J. F., Demayo, F. J., Tsai, M. J., et al. (2004). The nuclear orphan receptor COUP-TFII is required for limb and skeletal muscle development. *Molecular and Cellular Biology, 24*, 10835–10843.
Lepper, C., Partridge, T. A., & Fan, C. M. (2011). An absolute requirement for Pax7-positive satellite cells in acute injury-induced skeletal muscle regeneration. *Development, 138*, 3639–3646.
Li, L., Xie, X., Qin, J., Jeha, G. S., Saha, P. K., Yan, J., et al. (2009). The nuclear receptor COUP-TFII plays an essential role in adipogenesis, glucose homeostasis, and energy metabolism. *Cell Metabolism, 9*, 77–87.
Liberatore, C. M., Searcy-Schrick, R. D., & Yutzey, K. E. (2000). Ventricular expression of tbx5 inhibits normal heart chamber development. *Developmental Biology, 223*, 169–180.
Lin, F. J., Chen, X., Qin, J., Hong, Y. K., Tsai, M. J., & Tsai, S. Y. (2010). Direct transcriptional regulation of neuropilin-2 by COUP-TFII modulates multiple steps in murine lymphatic vessel development. *Journal of Clinical Investigation, 120*, 1694–1707.
Lin, B., Chen, G. Q., Xiao, D., Kolluri, S. K., Cao, X., Su, H., et al. (2000). Orphan receptor COUP-TF is required for induction of retinoic acid receptor beta, growth inhibition, and apoptosis by retinoic acid in cancer cells. *Molecular and Cellular Biology, 20*, 957–970.
Lin, F. J., Qin, J., Tang, K., Tsai, S. Y., & Tsai, M. J. (2011). Coup d'Etat: An orphan takes control. *Endocrine Reviews, 32*, 404–421.
Lin, F. J., You, L. R., Yu, C. T., Hsu, W. H., Tsai, M. J., & Tsai, S. Y. (2012). Endocardial cushion morphogenesis and coronary vessel development require chicken ovalbumin upstream promoter-transcription factor II. *Arteriosclerosis, Thrombosis, and Vascular Biology, 32*, e135–e146.
Liu, Z., Yue, S., Chen, X., Kubin, T., & Braun, T. (2010). Regulation of cardiomyocyte polyploidy and multinucleation by CyclinG1. *Circulation Research, 106*, 1498–1506.
Lo, P. C., & Frasch, M. (2001). A role for the COUP-TF-related gene seven-up in the diversification of cardioblast identities in the dorsal vessel of Drosophila. *Mechanisms of Development, 104*, 49–60.
Logan, M., Martin, J. F., Nagy, A., Lobe, C., Olson, E. N., & Tabin, C. J. (2002). Expression of Cre recombinase in the developing mouse limb bud driven by a Prxl enhancer. *Genesis, 33*, 77–80.
MacKrell, J. G., Yaden, B. C., Bullock, H., Chen, K., Shetler, P., Bryant, H. U., et al. (2015). Molecular targets of androgen signaling that characterize skeletal muscle recovery and regeneration. *Nuclear Receptor Signaling, 13*, e005.
Mark, M., Ghyselinck, N. B., & Chambon, P. (2009). Function of retinoic acid receptors during embryonic development. *Nuclear Receptor Signaling, 7*, e002.
Maya-Ramos, L., Cleland, J., Bressan, M., & Mikawa, T. (2013). Induction of the proepicardium. *Journal of Developmental Biology, 1*, 82–91.
Meilhac, S. M., Lescroart, F., Blanpain, C., & Buckingham, M. E. (2014). Cardiac cell lineages that form the heart. *Cold Spring Harbor Perspectives in Medicine, 4*, a013888.
Mendelsohn, C., Lohnes, D., Decimo, D., Lufkin, T., LeMeur, M., Chambon, P., et al. (1994). Function of the retinoic acid receptors (RARs) during development (II). Multiple abnormalities at various stages of organogenesis in RAR double mutants. *Development, 120*, 2749–2771.
Merki, E., Zamora, M., Raya, A., Kawakami, Y., Wang, J., Zhang, X., et al. (2005). Epicardial retinoid X receptor alpha is required for myocardial growth and coronary

artery formation. *Proceedings of the National Academy of Sciences of the United States of America*, *102*, 18455–18460.
Milanesi, A., Lee, J. W., Kim, N. H., Liu, Y. Y., Yang, A., Sedrakyan, S., et al. (2016). Thyroid hormone receptor α plays an essential role in male skeletal muscle myoblast proliferation, differentiation, and response to injury. *Endocrinology*, *157*, 4–15.
Mollova, M., Bersell, K., Walsh, S., Savla, J., Das, L. T., Park, S. Y., et al. (2013). Cardiomyocyte proliferation contributes to heart growth in young humans. *Proceedings of the National Academy of Sciences of the United States of America*, *110*, 1446–1451.
Moore-Morris, T., Guimaraes-Camboa, N., Banerjee, I., Zambon, A. C., Kisseleva, T., Velayoudon, A., et al. (2014). Resident fibroblast lineages mediate pressure overload-induced cardiac fibrosis. *Journal of Clinical Investigation*, *124*, 2921–2934.
Moskowitz, I. P., Kim, J. B., Moore, M. L., Wolf, C. M., Peterson, M. A., Shendure, J., et al. (2007). A molecular pathway including Id2, Tbx5, and Nkx2-5 required for cardiac conduction system development. *Cell*, *129*, 1365–1376.
Murphy, M. M., Lawson, J. A., Mathew, S. J., Hutcheson, D. A., & Kardon, G. (2011). Satellite cells, connective tissue fibroblasts and their interactions are crucial for muscle regeneration. *Development*, *138*, 3625–3637.
Muscat, G. E., Rea, S., & Downes, M. (1995). Identification of a regulatory function for an orphan receptor in muscle: COUP-TF II affects the expression of the myoD gene family during myogenesis. *Nucleic Acids Research*, *23*, 1311–1318.
Niederreither, K., Vermot, J., Messaddeq, N., Schuhbaur, B., Chambon, P., & Dolle, P. (2001). Embryonic retinoic acid synthesis is essential for heart morphogenesis in the mouse. *Development*, *128*, 1019–1031.
Odelberg, S. J., Kolhoff, A., & Keating, M. T. (2000). Dedifferentiation of mammalian myotubes induced by msx1. *Cell*, *103*, 1099–1109.
Olguín, H. C., & Pisconti, A. (2012). Marking the tempo for myogenesis: Pax7 and the regulation of muscle stem cell fate decisions. *Journal of Cellular and Molecular Medicine*, *16*, 1013–1025.
Panda, A. C., Abdelmohsen, K., Martindale, J. L., Di Germanio, C., Yang, X., Grammatikakis, I., et al. (2016). Novel RNA-binding activity of MYF5 enhances Ccnd1/Cyclin D1 mRNA translation during myogenesis. *Nucleic Acids Research*, *44*, 2393–2408.
Pereira, F. A., Qiu, Y., Zhou, G., Tsai, M. J., & Tsai, S. Y. (1999). The orphan nuclear receptor COUP-TFII is required for angiogenesis and heart development. *Genes and Development*, *13*, 1037–1049.
Pessina, P., Kharraz, Y., Jardí, M., Fukada, S., Serrano, A. L., Perdiguero, E., et al. (2015). Cell plasticity blunts tissue regeneration and aggravates muscular dystrophy. *Stem Cell Reports*, *4*, 1046–1060.
Qin, J., Wu, S. P., Creighton, C. J., Dai, F., Xie, X., Cheng, C. M., et al. (2013). COUP-TFII inhibits TGF-beta-induced growth barrier to promote prostate tumorigenesis. *Nature*, *493*, 236–240.
Quach, N. L., Biressi, S., Reichardt, L. F., Keller, C., & Rando, T. A. (2009). Focal adhesion kinase signaling regulates the expression of caveolin 3 and beta1 integrin, genes essential for normal myoblast fusion. *Molecular Biology of the Cell*, *20*, 3422–3435.
Reim, I., Mohler, J. P., & Frasch, M. (2005). Tbx20-related genes, mid and H15, are required for tinman expression, proper patterning, and normal differentiation of cardioblasts in Drosophila. *Mechanisms of Development*, *122*, 1056–1069.
Relaix, F., Rocancourt, D., Mansouri, A., & Buckingham, M. A. (2005). Pax3/Pax7-dependent population of skeletal muscle progenitor cells. *Nature*, *435*, 948–953.
Riehle, C., & Abel, E. D. (2012). PGC-1 proteins and heart failure. *Trends in Cardiovascular Medicine*, *22*, 98–105.

Ryan, K. M., Hendren, J. D., Helander, L. A., & Cripps, R. M. (2007). The NK homeodomain transcription factor Tinman is a direct activator of seven-up in the Drosophila dorsal vessel. *Developmental Biology, 302*, 694–702.

Sacco, A., Mourkioti, F., Tran, R., Choi, J., Llewellyn, M., Kraft, P., et al. (2010). Short telomeres and stem cell exhaustion model Duchenne muscular dystrophy in mdx/mTR mice. *Cell, 143*, 1059–1071.

Sack, M. N., Disch, D. L., Rockman, H. A., & Kelly, D. P. (1997). A role for Sp and nuclear receptor transcription factors in a cardiac hypertrophic growth program. *Proceedings of the National Academy of Sciences of the United States of America, 94*, 6438–6443.

Sambasivan, R., Yao, R., Kissenpfennig, A., Van Wittenberghe, L., Paldi, A., Gayraud-Morel, B., et al. (2011). Pax7-expressing satellite cells are indispensable for adult skeletal muscle regeneration. *Development, 138*, 3647–3656.

Schäfer, K., & Braun, T. (1999). Early specification of limb muscle precursor cells by the homeobox gene Lbx1h. *Nature Genetics, 23*, 213–216.

Scharner, J., & Zammit, P. S. (2011). The muscle satellite cell at 50: The formative years. *Skeletal Muscle, 1*, 1–13.

Scroyen, I., Bauters, D., Vranckx, C., & Lijnen, H. R. (2015). The anti-adipogenic potential of COUP-TFII Is mediated by downregulation of the Notch target gene Hey1. *PLoS One, 10*, e0145608.

Seale, P., Bjork, B., Yang, W., Kajimura, S., Chin, S., Kuang, S., et al. (2008). PRDM16 controls a brown fat/skeletal muscle switch. *Nature, 454*, 961–967.

Shinin, V., Gayraud-Morel, B., Gomès, D., & Tajbakhsh, S. (2006). Asymmetric division and cosegregation of template DNA strands in adult muscle satellite cells. *Nature Cell Biology, 8*, 677–687.

Stanley, W. C., Recchia, F. A., & Lopaschuk, G. D. (2005). Myocardial substrate metabolism in the normal and failing heart. *Physiological Reviews, 85*, 1093–1129.

Sucov, H., Dyson, E., Gumeringer, C., Price, J., Chien, K., & Evans, R. (1994). RXRa mutant mice establish a genetic basis for vitamin A signaling in heart morphogenesis. *Genes and Development, 8*, 1007–1018.

Tajbakhsh, S. (2003). Stem cells to tissue: Molecular, cellular and anatomical heterogeneity in skeletal muscle. *Current Opinion in Genetics and Development, 13*, 413–422.

Tajbakhsh, S. (2009). Skeletal muscle stem cells in developmental versus regenerative myogenesis. *Journal of Internal Medicine, 266*, 372–389.

Takamoto, N., You, L. R., Moses, K., Chiang, C., Zimmer, W. E., Schwartz, R. J., et al. (2005). COUP-TFII is essential for radial and anteroposterior patterning of the stomach. *Development, 132*, 2179–2189.

Tang, K., Tsai, S. Y., & Tsai, M. J. (2015). COUP-TFs and eye development. *Biochimica et Biophysica Acta, 1849*, 201–209.

Targoff, K. L., Colombo, S., George, V., Schell, T., Kim, S. H., Solnica-Krezel, L., et al. (2013). Nkx genes are essential for maintenance of ventricular identity. *Development, 140*, 4203–4213.

Tsai, S. Y., & Tsai, M. J. (1997). Chick ovalbumin upstream promoter transcription factors (COUP-TFs): Coming age. *Endocrine Reviews, 18*, 229–240.

Uezumi, A., Ito, T., Morikawa, D., Shimizu, N., Yoneda, T., Segawa, M., et al. (2011). Fibrosis and adipogenesis originate from a common mesenchymal progenitor in skeletal muscle. *Journal of Cell Science, 124*, 3654–3664.

Ustanina, S., Carvajal, J., Rigby, P., & Braun, T. (2007). The myogenic factor Myf5 supports efficient skeletal muscle regeneration by enabling transient myoblast amplification. *Stem Cells, 25*, 2006–2016.

Vasyutina, E., Stebler, J., Brand-Saberi, B., Schulz, S., Raz, E., & Birchmeier, C. (2005). CXCR4 and Gab1 cooperate to control the development of migrating muscle progenitor cells. *Genes and Development, 19*, 2187–2198.

Velders, M., Schleipen, B., Fritzemeier, K. H., Zierau, O., & Diel, P. (2012). Selective estrogen receptor-β activation stimulates skeletal muscle growth and regeneration. *FASEB Journal, 26*, 1909–1920.

von Gise, A., Zhou, B., Honor, L. B., Ma, Q., Petryk, A., & Pu, W. T. (2011). WT1 regulates epicardial epithelial to mesenchymal transition through beta-catenin and retinoic acid signaling pathways. *Developmental Biology, 356*, 421–431.

Wallace, G. Q., & McNally, E. M. (2009). Mechanisms of muscle degeneration, regeneration, and repair in the muscular dystrophies. *Annual Review of Physiology, 71*, 37–57.

Wang, Y. X., & Rudnicki, M. A. (2011). Satellite cells, the engines of muscle repair. *Nature Reviews Molecular Cell Biology, 13*, 127–133.

Webster, M. T., Manor, U., Lippincott-Schwartz, J., & Fan, C. M. (2016). Intravital imaging reveals ghost fibers as architectural units guiding myogenic progenitors during regeneration. *Cell Stem Cell, 18*, 243–252.

White, R. B., Biérinx, A., Gnocchi, V. F., & Zammit, P. S. (2010). Dynamics of muscle fibre growth during postnatal mouse development. *BMC Developmental Biology, 10*, 21–32.

Wiese, C., Grieskamp, T., Airik, R., Mommersteeg, M. T., Gardiwal, A., de Gier-de Vries, C., et al. (2009). Formation of the sinus node head and differentiation of sinus node myocardium are independently regulated by Tbx18 and Tbx3. *Circulation Research, 104*, 388–397.

Wu, S. P., Cheng, C. M., Lanz, R. B., Wang, T., Respress, J. L., Ather, S., et al. (2013). Atrial identity is determined by a COUP-TFII regulatory network. *Developmental Cell, 25*, 417–426.

Wu, S. P., Dong, X. R., Regan, J. N., Su, C., & Majesky, M. W. (2013). Tbx18 regulates development of the epicardium and coronary vessels. *Developmental Biology, 383*, 307–320.

Wu, S. P., Kao, C. Y., Wang, L., Creighton, C. J., Yang, J., Donti, T. R., et al. (2015). Increased COUP-TFII expression in adult hearts induces mitochondrial dysfunction resulting in heart failure. *Nature Communications, 6*, 8245.

Wu, S. P., Yu, C. T., Tsai, S. Y., & Tsai, M. J. (2016). Choose your destiny: Make a cell fate decision with COUP-TFII. *Journal of Steroid Biochemistry and Molecular Biology, 157*, 7–12.

Xavier-Neto, J., Neville, C. M., Shapiro, M. D., Houghton, L., Wang, G. F., Nikovits, W., Jr., et al. (1999). A retinoic acid-inducible transgenic marker of sino-atrial development in the mouse heart. *Development, 126*, 2677–2687.

Xiang, M. S., & Kikuchi, K. (2016). Endogenous mechanisms of cardiac regeneration. *International Review of Cell and Molecular Biology, 326*, 67–131.

Xie, X., Qin, J., Lin, S. H., Tsai, S. Y., & Tsai, M. J. (2011). Nuclear receptor chicken ovalbumin upstream promoter-transcription factor II (COUP-TFII) modulates mesenchymal cell commitment and differentiation. *Proceedings of the National Academy of Sciences of the United States of America, 108*, 14843–14848.

Xie, X., Tang, K., Yu, C. T., Tsai, S. Y., & Tsai, M. J. (2013). Regulatory potential of COUP-TFs in development: Stem/progenitor cells. *Seminars in Cell and Developmental Biology, 24*, 687–693.

Xie, X., Tsai, S. Y., & Tsai, M. J. (2016). COUP-TFII regulates satellite cell function and muscular dystrophy. *The Journal of Clinical Investigation, 126*, 3929–3941.

Xin, M., Small, E. M., van Rooij, E., Qi, X., Richardson, J. A., Srivastava, D., et al. (2007). Essential roles of the bHLH transcription factor Hrt2 in repression of atrial gene expression and maintenance of postnatal cardiac function. *Proceedings of the National Academy of Sciences of the United States of America, 104*, 7975–7980.

Yin, H., Price, F., & Rudnicki, M. A. (2013). Satellite cells and the muscle stem cell niche. *Physiological Reviews, 93*, 23–67.

You, L. R., Lin, F. J., Lee, C. T., DeMayo, F. J., Tsai, M. J., & Tsai, S. Y. (2005). Suppression of Notch signalling by the COUP-TFII transcription factor regulates vein identity. *Nature*, *435*, 98–104.

Yu, F., Göthe, S., Wikström, L., Forrest, D., Vennström, B., & Larsson, L. (2000). Effects of thyroid hormone receptor gene disruption on myosin isoform expression in mouse skeletal muscles. *American Journal of Physiology. Regulatory, Integrative and Comparative Physiology*, *278*, R1545–R1554.

Yu, C. T., Tang, K., Suh, J. M., Jiang, R., Tsai, S. Y., & Tsai, M. J. (2012). COUP-TFII is essential for metanephric mesenchyme formation and kidney precursor cell survival. *Development*, *139*, 2330–2339.

Yutzey, K. E., Rhee, J. T., & Bader, D. (1994). Expression of the atrial-specific myosin heavy chain AMHC1 and the establishment of anteroposterior polarity in the developing chicken heart. *Development*, *120*, 871–883.

Zaffran, S., Robrini, N., & Bertrand, N. (2014). Retinoids and cardiac development. *Journal of Developmental Biology*, *2*, 50.

Zhu, N., Wang, H., Wang, B., Wei, J., Shan, W., Feng, J., et al. (2016). A member of the nuclear receptor superfamily, designated as NR2F2, supports the self-renewal capacity and pluripotency of human bone marrow-derived mesenchymal stem cells. *Stem Cells International*, *2016*, 5687589.

Kolibri 360 GmbH
Ferdinand-Jühlke-Straße 7
99095 Erfurt, Deutschland
produktsicherheit@kolibri360.de